CELL FUNCTION AND DISEASE

CELL FUNCTION AND DISEASE

Edited by

L. E. Cañedo
Department of Biochemistry
School of Medicine
University of Nuevo Leon
Monterrey, Nuevo Leon, Mexico

L. E. Todd
Magnetic Resonance Unit
University Hospital
Monterrey, Nuevo Leon, Mexico

L. Packer
Lawrence Berkeley Laboratory
University of California
Berkeley, California

and

J. Jaz
International Biomedical Institute
Bari, Italy

PLENUM PRESS • NEW YORK AND LONDON

Library of Congress Cataloging in Publication Data

International Symposium on Cell Function and Disease (1988: Monterrey, Mexico)
  Cell function and disease / edited by L. E. Cañedo . . . [et al.].
      p.   cm.
  "Proceedings of an International Symposium on Cell Function and Disease, held
April 18–22, 1988 in Monterrey, Nuevo Leon, Mexico" — T.p. verso.
  Includes bibliographies and index.
  ISBN-13:978-1-4612-8095-8   e-ISBN-13:978-1-4613-0813-3
  DOI: 10.1007/978-1-4613-0813-3

  1. Diseases — Causes and theories of causation — Congresses. 2. Pathology, Molecular
— Congresses. 3. Pathology, Cellular — Congresses. I. Cañedo, L. E. II. Title.
  [DNLM: 1. Brain — transplantation — congresses. 2. Cells — physiology — congresses.
3. Cytogenetics — congresses. 4. Disease — physiopathology — congresses. 5. Molecular
Biology — congresses. QZ 40 I61c 1988]
RB151.I57   1988
611'.01815 — dc20
DNLM/DLC   89-3965
for Library of Congress   CIP

Proceedings of an International Symposium on Cell Function and Disease,
held April 18–22, 1988, in Monterrey, Nuevo Leon, Mexico

© 1988 Plenum Press, New York
Softcover reprint of the hardcover 1st edition 1988

A Division of Plenum Publishing Corporation
233 Spring Street, New York, N.Y. 10013

PREFACE

    The new experimental tools and approaches of modern
biology have allowed us to better understand many fundamental
properties of the eukaryotic cells. These significant
discoveries have drastically changed the diagnostic and
therapeutic approaches of modern clinical practice. On April
18-22, 1988, an International Symposium on Cell Function and
Disease was held in Monterrey, Nuevo León, México, aimed at
reviewing some of the most recent advances made in the
following five areas: Genes and Human Diseases; Cellular and
Molecular Pathology; Infectious Diseases; Brain Transplants
and the New Approaches and Techniques with Potential
Application to Cell Function and Disease. This book is based
on the contributed papers of the symposium. To underline the
importance of the clinical approach to the study of cell
function and disease a section on this subject was added at
the end of the book. The chapters in this volume include
contributions by some of the leading scientists of the
international scientific community and México.

    During the course of this international conference,
numerous discussions were held by the local and international
representatives of the scientific community concerning the
creation of an International Center of Molecular Medicine
aimed at stimulating further interaction between molecular
biologists, biochemists, biophyscists and clinicians. Such
ideas received the endorsement and support of the Director
General of the United Nations Educational and Scientific
Organization (UNESCO), Federico Mayor, the Governor of the
State of Nuevo León, Jorge Treviño, and the Secretary of
Health of México, Guillermo Soberon. In this context, then,
this conference and these proceedings represent the first
tangible results of the International Center of Molecular
Medicine.

    The organizers of this conference wish to acknowledge
the President of México, Miguel de la Madrid, the Director
General of UNESCO, Federico Mayor, the Governor of Nuevo
León, Jorge Treviño and the Secretary of Health of México,
Guillermo Soberon for inaugurating the conference.

    We owe special thanks to the sponsors of the symposium:
UNESCO, the Government of the State of Nuevo León, the
National Council for Science and Technology (CONACYT), the
University of Nuevo León (UANL), the Monterrey Technological
Institute (ITESM), the International Biomedical Institute
(Bari) and the National Foundation for Cancer Research.

In addition, we express appreciation to the other members of the organizing committee comprised of A. Azzi, H. Barrera, R. Drucker-Colín, G. Elizondo, C. Hazlewood, A. Kotyk, J. Kumate, S. Papa, B. Pullman, T. Slater and P. Ts'o.

We were also fortunate to have the help of many persons of the scientific and local community in the Monterrey area. In particular we express appreciation to Mrs. Angelina Decanini de Viesca for organizing the social program, to Mr. And Mrs. Eugenio and Eva Garza Laguera, Mr. And Mrs. David and Yolanda Garza Laguera and Mr. And Mrs. Generoso and María Elena Villareal, for hosting delegations of the visiting scientists. Also we thank colleagues and students at the Autonomous University of Nuevo León and the group of the Monterrey Technological Institute lead by Mr. Carlos Jiménez and integrated by Raúl Morales, Adrían Herrera, Alberto Sada, Leonel Dignowity, Bernardo Robles, Cristina Riojas, Esther Riojas, Jorge Salinas, José Bosco and Mauricio Belden, who took care of the visiting scientists and who helped with the numerous local arrangements. We appreciate the skillful administrative capacity of J. Gonzalez Miller and his staff as well as the group led by Professor Pamanes and José Díaz Chacón in supporting the logistical arrangement. Finally, we especially thank Ms. Martha T. Riojas for her help with all of the details that are part of running a conference and for her assistance with the editorial work of this book.

L.E. Cañedo
L.E. Todd
L.Packer
J.Jaz

Monterrey, México
August 1988.

CONTENTS

GENES AND HUMAN DISEASES

CELLULAR AND MOLECULAR PATHOLOGY

CELL FUNCTION AND BRAIN TRANSPLANT

NEW APPROACHES AND TECHNIQUES WITH POTENTIAL APPLICATION TO
CELL FUNCTION AND DISEASE

TOWARDS ACTIVE CHROMATIN STRUCTURE

Henryk Eisenberg

Polymer Department
The Weizmann Institute of Science
Rehovot 76100, Israel

INTRODUCTION

Nucleic acids and proteins are major protagonists in the continuously ongoing drama of life[1]. According to the now classical dogma, almost always respected and followed, DNA makes RNA, RNA makes protein, and protein closes the circle by providing activator and control elements necessary for proper DNA and RNA operation. In addition, proteins (in the form of enzymes) provide the more mundane function of catalyzing a variety of metabolic reactions, which culminate in the synthesis of vital components for the molecules of life.

Though proteins and nucleic acids interact closely together, they are basically different. Globular proteins, synthesized as linear polypeptide chains on the ribosome, quickly fold into stable three-dimensional structures, sometimes associate in dimers or higher units composed of identical or nonidentical subunits, occasionally form multienzyme complexes, capable of performing series of consecutive reactions with preordained space and time-coordinated activity. The globular, somewhat flexible protein subunit, stabilized by a variety of noncovalent and covalent - disulphide - bonds, extends a few nanometers into space; on it the catalytic active site can be identified as well as other accessory binding sites and surfaces capable of associating with other protein or nucleic acid structures. Throughout its lifetime in the cell the protein molecule maintains its structure and pursues its function, until it is replaced by an identical macromolecule following natural turnover, in keeping with a preassigned lifespan.

Nucleic acids carry the genetic information and control instructions for the complete organism, from the lowest unicellular to the highest eukaryote including man, encoded in a linear nucleotide code. A single DNA chain of a chromosome of a higher eukaryote, for instance man, composed of about 2×10^8 bp (base pairs), extends about 7 cm in length, though its width is not more than 2 nm. To maintain its integrity and to fit into the narrow confines of the nucleus in the cell, the DNA molecule must thus be highly folded and packaged into chromosomal structures. This packaging, in eukaryotic systems, is achieved with histone and sometimes non-histone proteins, in a broad hierarchy of structures. In the packaged form DNA is not active and the message it carries cannot be read. To allow the processes of transcription and replication, it is thus necessary to unfold selected portions of the chromosomal structure which can eventually be refolded, in a resting phase of the cell cycle of chromosomal activity. The necessity to achieve a quick transition from a totally folded to a partially unfolded chromosomal structure prescribes the necessity of potential lability associated with these structures. Folding and unfolding of chromatin is believed to be due to postsynthetic modifications such as acetylation of non-structured histone tails, phosphorylation, ubiquitinization or the binding and release of non-histone high mobility group (HMG) proteins.[2] In chromatin, isolated as chromosomal fragments from nuclear structures by the use of nucleases or site specific restriction enzymes, the transition from more to less folded structures is mimicked *in vitro* by a decrease in the concentration of salt,[3] yet it is by no means established that the structure of chromatin at low concentrations of salt, far away from physiological conditions, is related to the physiologically unfolded form in a meaningful way.

In 1973 it was found that limited endonuclease action in rat liver chromatin led to distinct particles which were then defined as nucleosome core particles consisting of 145 bp complexed with eight core histone molecules (two each of the H2a, H2b, H3 and H4 core histones).[4,5] The structure of the nucleosome core particle has been characterized by X-ray diffraction[6] and controversies around its validity may be due to flexibility and structural rearrangements of the protein core in various solvent systems.[7] The chromatosome,[8] the complete chromatin repeating unit, consists of 160 bp of DNA, wound in two superhelical turns around the core histone octamer, locked at the common entry and exist point by an H1 (H5, or other linker histone variant) linker histone. The histones, in particular the H4 and H3 histones, are amongst the evolutionary most conserved proteins known[2]. The first stage in DNA compaction is the formation of the DNA core histone linker histone complex, representing a succession of nucleosomes separated by linker DNA stretches. The linker DNA stretches vary within a given organism around an average value which is different in different organisms.

A commonly accepted view, which has found its way securely into biochemistry textbooks is the concept of the folding of the linear chains of nucleosomes from a flexible random coil structure at low ionic strength (the 10 nm LOS lower order structure - 10-11 nm is the large diameter of the flat cylinder nucleosome), into the 30 nm diameter rigid HOS higher order solenoid.[9] In the Finch and Klug solenoid the pitch of the single-start helix is 11 nm (the height of the nucleosome stacked head-to-head), there are about six nucleosomes per turn of the helix, their faces roughly radially parallel to the helix axis (a conclusion strengthened by electric dichroism studies[10]). At best this structure is an idealization considering the variable and irregular linker length and conflicting opinion with respect to the dependence of the nominal 30 nm higher order structure diameter on linker length.[11,12] The controversies surrounding the basic chromatin structures were summarized in 1986[13,14,22] and further discussion here will deal with more recent developments. In a soccer game one side usually wins, but sometimes there is a draw. I would conclude that while we seem to be getting a better view of the lower order structure at low ionic strength, whose relevance to biological function is moot, our views on the higher order structure appear to be locked in a temporary draw.

THE STRUCTURE OF BULK CHROMATIN

Bulk, or total genomic, chromatin obtained by limited nuclease action on nuclei and subsequent lysis is prone to aggregation and precipitates with increase in concentration of monovalent and multivalent cations.[15-17] It is an extremely sensitive material and its properties often depend on ways in which it has been handled both before and after isolation. Chicken erythrocyte chromatin is more stable and less prone to attack by endogenous nucleases than chromatin from other sources and has therefore become a favorite system for study. At monovalent salt concentrations of about 0.15 M non-histone proteins are mostly removed and it is believed that the basic structure formed by the DNA and the core and linker histones is maintained. It is therefore essential to maintain the integrity of these components and to avoid migration of the linker histones which dissociate earlier than the core histones with increasing ionic strength. In general two philosophies have been developed to study the structure of the isolated bulk chromatin. In one approach, attempts were made to obtain a highly soluble subfraction of chromatin and then to apply a range of physical methods in solution, at variable conditions of electrolyte type and concentration under conditions in which aggregation was completely avoided.[15] Once clear, reproducible and interpretable results are obtained, the question relating to relevance cannot be avoided. For electrolyte concentrations approaching physiological conditions the question is less severe than, as already mentioned, for the very low electrolyte environment, sometimes investigated. In the other approach, deriving from attempts to study chromatin *in situ* in intact cells,[18] the guiding concept was not to isolate chromatin fragments but to recreate conditions of cellular aggregation in the capillary test-tube, by choosing conditions and concentrations leading to benign orientation of the aggregated chromatin strands, without application of an external orientating force.[19] Previously attempts of producing oriented chromatin fibers had failed because of the already mentioned inherent instability of these structures.

Aided by modern image analysis to improve the quality of their X-ray diffraction patterns, Widom and Klug[19] succeeded in identifying equatorial and meridional reflections which essentially confirmed the validity of the Finch and Klug[9] solenoid. Earlier,[10] support for the solenoid structure was derived from electric dichroism studies, undertaken at millimolar $MgCl_2$ concentrations, in the absence of other electrolytes. Conflicting claims as to the sign and the size of dichroism corresponding to full orientation following application of electric or hydrodynamic fields have plagued these activities and are discussed in a recent work.[20]

Solution studies at increasing electrolyte concentrations comprising sedimentation in the ultracentrifuge,[21] quasielastic and total intensity light scattering[15] pointed to compaction with increasing ionic strength or addition of $MgCl_2$, though the inherent low resolution of these methods precluded meaningful analysis in terms of a significant unique molecular model. The considerably lower wavelength of X-ray and of neutrons leads to increased resolution in solution scattering studies[22] and significant experimental advantages (reliable extrapolation to lower scattering vectors) accrued from the use of synchrotron radiation for X-rays[23] and sophisticated neutron scattering facilities.[24] A major conclusion from these latter studies is that the low electrolyte LOS is not the disorganized 10 nm coil as previously believed but has enough stiffness and structure to yield a considerably larger cross-sectional radius in a cross section scattering plot, not much different from that of the HOS into which it folds quickly and easily. A true transition from a 10 nM LOS into a 30 nm HOS thus does not appear to be a good description of the folding process. A major alternate model to the Finch and Klug solenoid is the crossed-linker double start helix[11], which has lost the central hole characterizing the solenoid, and claims linker-length dependent diameter. The latter conclusion, as well as the correct value of the mass-per-unit length have not satisfactorily been settled from either scattering or STEM measurements.[24] In an experimental X-ray scattering study in our laboratory, in which the lowest scattering angles were not available, we could show that the chromatin folding process could be simulated by the compaction of a wormlike coil or a collection of rigid cylinders.[22] Koch et al.[12] confirmed that convolution of the wormlike coil with the nucleosome scattering yielded the features observed in the low-angle scattering curve.

Presently it does not appear that methods in hand will allow satisfactory resolution of the points under discussion. Much of the uncertainty may be due to the irregularity of the structure and it is therefore of significant interest that a construct has been achieved[25] with nucleosomes spaced regularly along a defined-size self-repeating DNA. So far it has not been possible to attach a properly located linker histone to this construct and recreate a complete well-behaved chromatin chain. This leads to the inevitable conclusion that the DNA entrance and exit sections on these reconstituted nucleosomes may not be properly arranged, as it has been shown that H1 can be removed from native chromatin and reaffixed with complete restoration of the folding properties in the HOS.[26] When linker histone is carefully removed from native chromatin the LOS behaves like a 10nm coil and loses the ability to fold into the HOS, which is regained upon readdition of H1.

The structural study of total genomic bulk chromatin represented a necessity arising from the fact that relatively large amounts of material are required for physical studies. Difficulties encountered in the determination of the crystal structure of the nucleosome core particle, heterogeneous with respect to DNA composition, were overcome by genetically engineering a uniform DNA fragment, which folds precisely into a crystallisable nucleosome.[27] A similar result has not yet been achieved for a chromatin fragment. The concentration of single-copy genes is minute (two copies of the gene per nucleus) yet current molecular biology technology has made it possible to study chromatin structure near an expressed gene, in distinction to the study of essentially inactive bulk chromatin.[28]

DNA fragments comprising the adult β-globin gene from chicken erythrocytes have been isolated, cloned, sequenced and used as probes in the study of β-globin chromatin. A basic property of chromatin near an expressed gene is its sensitivity and hypersensitivity to nuclease action[29] though nucleosomes from the β-globin gene were found to be similar in structure to nucleosomes from bulk chromatin.[28] Hypersensitivity most probably arises from the fact that DNA stretches surrounding the gene are not covered by nucleosomes but interact with transcription factors relating to gene expression.[30] In consequence of this structural feature it could be shown by careful analysis of migration in sucrose gradients of random nuclease digests of chicken chromatin that at 100 mM NaCl β-globin chromatin from erythrocytes remains partially unfolded whereas ovalbumin gene containing chromatin from the same tissue folds well.[31] At lower electrolyte concentrations (25 mM NaCl) both chromatin samples migrate identically. The opposite result was obtained when the two genes were isolated from oviduct chromatin. Kimura et al.[32] had earlier shown that an EcoR1 6.2 kb (kilo base pair) chicken erythrocyte fragment, enclosing the β-globin gene, sediments more slowly than bulk chromatin fragments of similar size, whereas ovalbumin and α2-collagen gene fragments in erythrocyte chromatin and β-globin in spleen chromatin sediment with bulk chromatin fragments of the same DNA size. More recently,[33] in an elaboration of these observations it was shown that the specific retardation of the chicken β-globin chromatin fragments cannot be reversed by adding extra linker histones to native chromatin. Globin and bulk chromatin behave identically upon unfolding by removal of linker histones or lowering of the ionic strength. The original difference in sedimentation can be restored by readdition of the linker histones and elevation of the ionic strength. Cleavage of the EcoR1 chicken globin fragment to remove the hypersensitive ends, in the 5' and 3' flanking regions, leads to a fragment sedimenting normally.

Both active and less transcriptionally active genes are organized in a nuclear matrix in loops of varying sizes[34] and it not clear to date whether transcribing RNA polymerase moves along these loops in the transcription process, or remains bound close to the base of the loops.[35,36] For partial unfolding to allow transcription, temporary removal of linker histone is necessary, though the question whether an RNA polymerase transcribes through a nucleosome core without its transient release from the DNA has received opposing answers.[37,38] There have only been limited attempts to purify unique genes as chromatin[39,40] and additional efforts in this direction will help to answer some of the questions raised.

Although DNA binding proteins have been isolated from prokaryotic organisms and we would expect its chromatin organization to be much simpler than that of eukaryotes, very little is known with certainty about the structure and modulation of bacterial chromatin.[41]

This study is supported by grants from the Israel-U.S. Binational Science Foundation, Jerusalem, Israel and from the Minerva Foundation, Munich, Germany.

REFERENCES

1. J.D. Watson, N.H. Hopkins, J.W. Roberts, J.A. Steitz and A.M. Weiner, "Molecular Biology of the Gene", 4th Ed., Benjamin/Cummings, Menlo Park (1987).

2. E.M. Bradbury, N. Maclean and H.R. Matthews, "DNA, Chromatin and Chromosomes", Wiley, New York (1981).

3. J.D. McGhee and G. Felsenfeld, Nucleosome structure, *Ann. Rev. Biochem.* 49:1115-1156 (1980).

4. R.D. Kornberg, Structure of chromatin, *Ann. Rev. Biochem.* 46:931-954 (1977).

5. D.M.J. Lilley and J.F. Pardon, Structure and function of chromatin, *Ann. Rev. Genet.* 13:197-233 (1979).

6. T.J. Richmond, J.T. Finch, B. Rushton, D. Rhodes and A. Klug, Structure of the nucleosome core particle at 7Å resolution, *Nature* 311:532-537 (1984).

7. K. Park and G.D. Fasman, The histone octamer, a conformationaly flexible structure, *Biochemistry* 26:8042-8045 (1987).

8. R.T. Simpson, Structure of the chromatosome, a chromatin particle, containing 160 base pairs of DNA and all the histones, *Biochemistry* 17:5524-5531 (1978).

9. J.T. Finch and A. Klug, Solenoidal model for superstructure in chromatin, *Proc. Natl. Acad. Sci. USA* 73:1897-1901 (1976).

10. J.D. McGhee, J.M. Nickol, G. Felsenfeld and D.C. Rau, Higher order structure of chromatin: Orientation of nucleosomes within the 30 nm chromatin solenoid is independent of species and spacer length, *Cell* 33:831-841 (1983).

11. S.P. Williams, B.D. Athey, L.J Muglia, R.S. Schappe, A.H. Gough and J.P. Langmore, Chromatin fibers are left-handed double helices with diameter and mass-per-unit length that depend on linker length, *Biophys. J.* 49:233-248 (1986).

12. M.H.J. Koch, M.C. Vega, Z. Sayers and A.M. Michon, The superstructure of chromatin and its condensation mechanism III: Effect of monovalent and divalent cations, X-ray solution scattering and hydrodynamic studies, *Eur. Biophys. J.* J. 14:307-319 (1987).

13. D.S. Pederson, F. Thoma and R.T. Simpson, Core particle, fiber, and transcriptionally active chromatin structure, *Ann. Rev. Cell Biol.* 2:117-147 (1986).

14. G. Felsenfeld and J.D. McGhee, Structure of the 30nm chromatin fiber, *Cell* 44:375-377 (1986).

15. J. Ausio, N. Borochov, D. Seger and H. Eisenberg, Interaction of chromatin with NaCl and MgCl$_2$, *J. Mol. Biol.* 177:373-398 (1984).

16. N. Borochov, J. Ausio and H. Eisenberg, Interaction and conformational changes of chromatin with divalent ions, *Nucl. Acid Res.* 12:3089-3096 (1984).

17. J. Widom, Physicochemical studies of the folding of the 100Å nucleosome filament into the 300Å filament. Cation dependence, *J. Mol. Biol.* 190:411-424 (1986).

18. J.P. Langmore and J.R. Paulson, Low angle X-ray diffraction studies of chromatin structure *in vivo* and in isolated nuclei and metaphase chromosomes, *J. Cell Biol.* 96:1120-1131 (1983).

19. J. Widom and A. Klug, Structure of the 300Å chromatin filament: X-ray diffraction from oriented samples, *Cell* 43:207-213 (1985).

20. S.I. Dimitrov, I.V. Smirnov and V.L. Makarov, Optical anisotropy of chromatin. Flow linear dichroism and electric dichroism study. *J. Biomol. Struct. Dyn.* 5:1135-1148 (1988).

21. P.J.G. Butler, The folding of chromatin, *CRC Crit. Rev. Biochem.* 15:57-91 (1983).

22. K.O. Greulich, E. Wachtel, J. Ausio, D. Seger and H. Eisenberg, Transition of chromatin from the "10 nm" lower order structure to the "30 nm" higher order structure as followed by small angle X-ray scattering, *J. Mol. Biol.* 193:709-721 (1987).

23. J.Bordas, L. Perez-Grau, M.H. J. Koch, C. Nave and M.C. Vega, The superstructure of chromatin and its condensation mechanism I: Synchrotron radiation X-ray scattering results, *Eur. Bioph. J.* 13:157-174 (1986).

24. S.E.Gerchman and V. Ramakrishnan, Chromatin higher order structure studies by neutron scattering and scanning transmission electron microscopy, *Proc.Natl.Acad.Sci. USA* 84:7802-7806 (1987).

25. R.T. Simpson, F. Thomas and J.M. Brubaker, Chromatin reconstituted from tandemly repeated cloned DNA fragments and core histones: A model system for study of higher order structure. *Cell* 42:799-808 (1985).

26. J.Allan, D.Z.Staynov and H. Gould, Reversible dissociation of linker histone from chromatin with preservation of internucleosomal repear, *Proc.Natl.Acad.Sci. USA* 77:885-889 (1980).

27. T.J.Richmond, M.A. Searles, and R.T. Simpson, Crystals of nucleosome core particle containing defined sequence DNA, *J. Mol. Biol.* 199:161-170 (1988).

28. G.Felsenfeld, B.M. Emerson, P.D. Jackson, C.D. Lewis, J.E. Hesse, M.R. Lieber and J.M. Nickol, Chromatin structure near an expressed gene, *In:* "New Frontiers in the Study of Gene Functions", P. Poste and S.T. Crooke, Eds., Plenum, New York, 99-109 (1987).

29. W.I. Wood and G. Felsenfeld, Chromatin structure of the chicken β-globin region: Sensitivity to DNaseI, micrococcal nuclease, and DnaseII, *J. Biol. Chem.* 257:7730-7736 (1982).

30. B.M.Emerson and G. Felsenfeld, Specific factors conferring nuclease hypersensitivity at the 5' end of the chicken adult β-globin gene, USA, 81:95-99 (1984).

31. E.A. Fisher and G. Felsenfeld, A comparison of the folding of β-globin and ovalbumin gene-containing chromatin from chicken oviduct and erythrocytes, *Biochemistry* 25:8010-8016 (1986).

32. T.Kimura, F.C. Mills, J. Allan and H. Gould, Selective unfolding of erythroid chromatin in the region of the active β-globin gene, *Nature* 306:709-712 (1983).

33. A.Caplan, T. Kimura, H. Gould and J. Allan, Perturbation of chromatin structure in the region of the adult β-globin gene in the chicken erythrocyte chromatin, *J. Mol. Biol.* 193:57-70 (1987).

34. S.M. Gasser and U.K. Laemmli, A glimpse at chromosomal order, *TIG* 3:16-22 (1987).

35. D.A. Jackson and P.R. Cook, Transcription occurs at the nucleoskeleton, *EMBO J.* 4:919-925 (1985).

36. M. Roberge and E.M. Bradbury, Chromosomal loop/nuclear matrix organization of the transcriptionally active and inactive RNA polymerases in Hela nuclei, *J. Cell Biochem. Suppl.* 12D:147 (1988).

37. R.Losa and D.D. Brown, A bacteriophage RNA polymerase transcribes *in vitro* through a nucleosome core without displacing it, *Cell* 50:801-808 (1987).

38. Y.Lorch, J.W. LaPointe and R.D. Kornberg, Nucleosomes inhibit the initiation of transcription but allow chain elongation with the displacement of histones, *Cell* 49:203-210 (1987).

39. J.L. Workman and J.P. Langmore, Efficient solubilization and partial purification of sea urchin histone genes as chromatin, *Biochemistry* 24:4731-4738 (1985).

40. D.S. Pederson, M. Venkatesan, F. Thoma and R.T. Simpson, Isolation of an episomal yeast gene and replication origin as chromatin, *Proc.Natl.Acad.Sci. USA* 83:7206-7210 (1986).

41. M.B. Schmid, Structure and function of the bacterial chromosome, *TIBS* 13:131-135 (1988).

MOLECULAR GENETICS OF WILMS' TUMOR

Vicki Huff[1], Duane A. Compton[1], Michael M. Weil[1],
Louise C. Strong[2], and Grady F. Saunders[1]

[1]Department of Biochemistry and Molecular Biology
[2]Department of Pediatrics
University of Texas System Cancer Center
Houston, Texas 77030

Based on present rates, some type of cancer will affect three out of ten individuals in the United States in his or her lifetime (Amer. Cancer Soc., 1988). The impact of cancer on the health of the population has made an understanding of the etiology and biology of malignancy a primary research concern. The role of chromosome abnormalities in tumorigenesis was first proposed by Boveri in 1914. The subsequent observation of the clonal nature of most tumors supported the notion that genetic alterations were critical in carcinogenesis. Thus, the identification, isolation, and characterization of genes involved in carcinogenesis has become an important approach for understanding cancer etiology and biology. Advances in the past decade in molecular biology, cytogenetics, and somatic cell genetics have allowed the localization and cloning of many human disease loci. The isolation of genes involved in carcinogenesis is confounded by the heterogeneous nature of neoplasia and the absence of a single gene that is clearly involved. One approach to circumvent these complications is to study cancers in which only one or a few genes are hypothesized to have a major role in tumorigenesis. Wilms' tumor (WT) is one such cancer.

Wilms' tumor is an embryonal renal neoplasm which affects about 1 in 10,000 children (Matsunaga, 1981). One or both kidneys can be affected with 5-10% of Wilms' tumors being bilateral. In ~8% of cases, Wilms' tumor is associated with aniridia or genitourinary anomalies (Breslow and

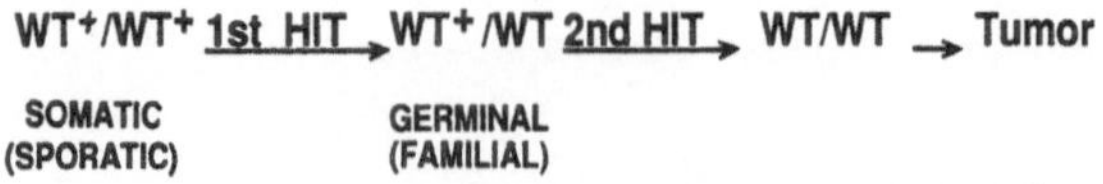

$$WT^+/WT^+ \xrightarrow{\underline{\text{1st HIT}}} WT^+/WT \xrightarrow{\underline{\text{2nd HIT}}} WT/WT \rightarrow \text{Tumor}$$

SOMATIC   GERMINAL
(SPORATIC)   (FAMILIAL)

Fig. 1. Two-hit model for tumorigenesis as proposed by Knudson and Strong (1972). Two
mutations are required for the development
of a tumor. In somatic (sporadic) Wilms'
tumor, both mutations are somatic. In
germinal (familial) Wilms' tumor, an
individual inherits the first mutation; only
one subsequent somatic mutation is required
for tumorigenesis.

Beckwith, 1982). Epidemiological studies have revealed that a vast
majority of tumors occur sporadically with no family history of Wilms'
tumor. About 1% of the cases, however, are familial, most often
affecting siblings or cousins. Familial cases are more frequently
bilateral than sporadic cases (Strong, 1984). In general, familial and
bilateral cases are diagnosed at an earlier age (median of 25 months for
bilateral tumors vs. 36 months for unilateral cases; Breslow and
Beckwith, 1982). From these observations Knudson and Strong (1972)
proposed that, like retinoblastoma, another pediatric tumor, the development of Wilms' tumor requires two mutations (Fig. 1). In nonheritable
(somatic) WT, the initial mutation is a somatic event, whereas in
heritable (germinal) WT the mutation is germinal. In both nonheritable
and heritable WT, the second mutation is somatic. Because individuals
have inherited the first mutation, only a single additional mutation may
be sufficient for tumor development, leading to an earlier age of onset
and the more frequent development of bilateral tumors. From the observed
data on age of onset and laterality, it was estimated that in heritable
WT, ~63% of carriers of the predisposing gene would be affected (Knudson
and Strong, 1972).

GENETIC ALTERATIONS IN WILMS' TUMOR

Cytogenetic studies first indicated that the genomic location of at
least one of the hypothesized mutations was at chromosomal band 11p13.
Karyotypic analyses of patients with Wilms' tumor and aniridia (which is
associated with Wilms' tumor in 1-2% of cases; Breslow and Beckwith,

8

1982) revealed chromosome deletions in the short arm of chromosome 11 (Riccardi et al., 1978). Although the size of the deletions varied among the three patients studied, a common region of 11p13 was always deleted. Patients with aniridia and an 11p13 deletion have a high frequency of

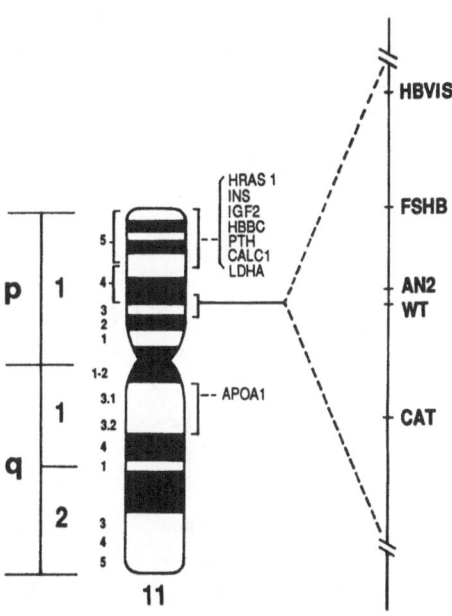

Fig. 2. Position of chromosome 11 loci in relation to the proposed location of the Wilms' tumor loci. HRAS1: oncogene Harvey-ras-1; INS: insulin; IGF2: insulin-like growth factor II; HBBC: hemoglobin beta gene cluster; PTH: parathyroid hormone; CALC1: calcitonin; LDHA: lactate dehydrogenase A; HBVIS: hepatitis B virus insertion site; FSHB: beta subunit of follicle stimulating hormone; AN2: aniridia-2; WT: Wilms' tumor; CAT: catalase; APOAI: apolipoprotein AI.

Wilms' tumor (~50%) (Narahara et al., 1984). Furthermore, the age of onset and frequency of bilateral tumors in aniridia/11p13 deletion patients is similar to that of familial Wilms' tumor patients (Knudson and Strong, 1972). These data suggested that a locus at 11p13 was involved in Wilms' tumor. Observations of 11p13 deletions in tumors from patients with no constitutional karyotypic abnormality (Kaneko et al., 1981; Slater and deKraker, 1982) supported this hypothesis.

Studies and observations of other familial cancers (Festa et al., 1979; Benedict et al., 1983; Cavenee et al., 1983) suggested that, by analogy, the second mutation in Wilms' tumor development could be the loss of the normal 11p13 allele by chromosome loss or somatic recombination. This hypothesis was investigated by molecular analyses using chromosome 11-specific DNA probes that detect restriction fragment length polymorphisms (RFLPs). The location of some of these probes on chromosome 11 is shown in Figure 2. When tumor and normal tissue from WT patients were compared in several different studies, 55% of the tumors

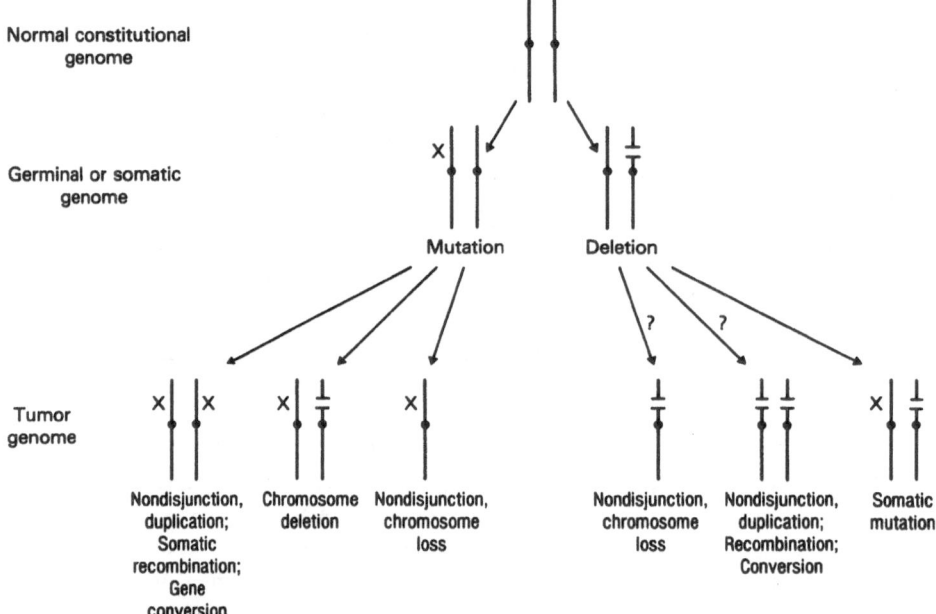

Fig. 3. Mechanisms for loss of normal alleles in Wilms' tumor. The normal genome sustains a point mutation or a deletion. This first mutation can be either somatic or germinal. The remaining normal allele is either lost by chromosome nondis-recombination, gene conversion, chromosome deletion, chromosome loss, or somatic mutation, resulting in the various indicated genomes. The question marks indicate possible mechanisms that have not been demonstrated experimentally.

were monomorphic at loci at which the normal tissue was polymorphic (Reeve et al., 1984; Fearon et al., 1984; Orkin et al., 1984; Raizis et al., 1985; Dao et al., 1987). This loss of heterozygosity in the tumors suggested that the second mutation hypothesized by Knudson and Strong (1972) was the physical loss of the normal chromosome 11 allele in at least half of the cases. Thus, a refinement of the two-hit hypothesis states that tumorigenesis involves 1) a mutation (germinal or somatic) at 11p13 and 2) subsequent somatic loss or alteration of the normal allele.

The loss of normal alleles can occur by several mechanisms including chromosome loss, chromsome deletion, chromosome loss and reduplication of the remaining chromosome, somatic recombination, and gene conversion

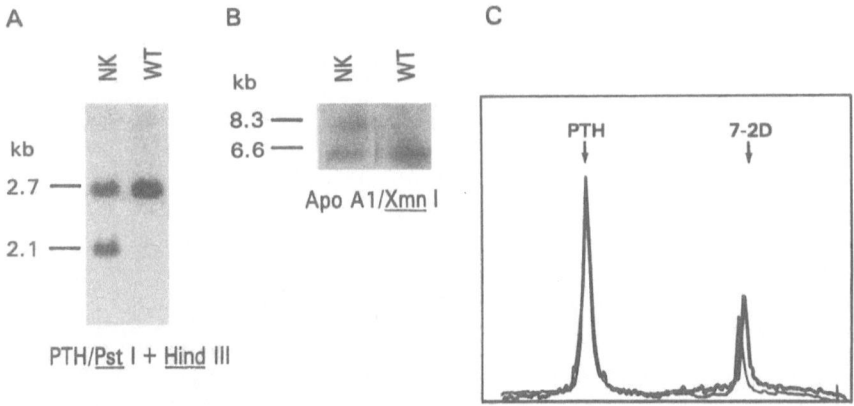

Fig. 4. Loss of heterozygosity by chromosome loss and duplication. (A) Autoradiogram of DNA from normal kidney (NK) and tumor (WT) restricted with Pst I and Hind III and hybridized with a probe for the parathyroid hormone (PTH) gene. The normal kidney genome is heterozygous, containing both the 2.7 kb and 2.1 kb alleles. Only the 2.1 kb allele is observed in the tumor DNA. Similarly (B) tumor DNA has lost heterozygosity at the locus for apolipoprotein AI (apo AI). (C) Densitometric scan of autoradiograms of normal (thin line) and tumor (thick line) DNA restricted with Eco RI and hybridized with the probes PTH and 7-2D, a probe for the HLA class II-associated invariant chain gene which is located on chromosome 5. Densitometric scanning of the two blots was standardized using the 7-2D band. The intensity of the PTH bands in the tumor and the normal kidney DNA are the same, indicating that there are two copies of the PTH gene in the tumor genome. Patient is reference number 197830-000. (From Dao et al., 1987).

(Fig. 3). Chromosome loss and chromosome deletion have been detected cytogenetically. Data from Dao et al. (1987) provide confirmation for some of the proposed mechanisms for chromosome 11 allelic loss. Loss of heterozygosity at loci both at 11p (PTH) and 11q (apoAI) in conjunction with the presence of two copies of the 11p probe, PTH, suggests that in this tumor the entire chromosome 11 has been lost and the remaining chromosome has been duplicated (Fig. 4). In a tumor from another patient, loss of heterozygosity at 11p loci (c-Ha-ras1, PTH) and retention of heterozygosity at the apoAI locus on 11q indicate that somatic recombination involving the short arm of chromosome 11 is responsible for the loss of the normal 11p13 allele (Fig. 5). Roughly half of the Wilms' tumors examined to date do not have any detectable chromosome 11 alterations. Presumably these tumors have sustained subtle

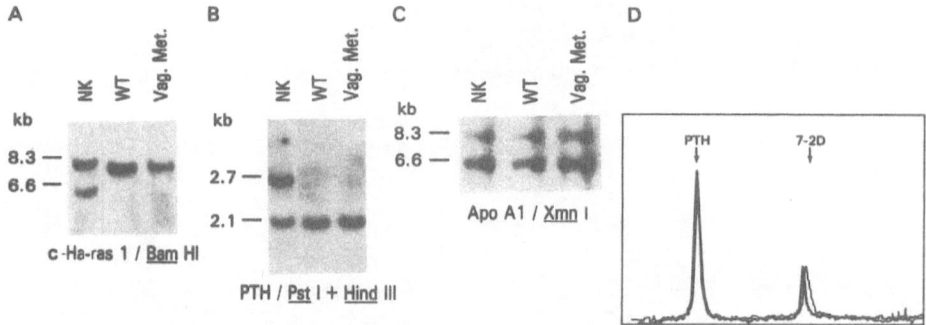

Fig. 5. Loss of heterozygosity by somatic recombination. (A-D) Auto-radiograms of DNA from Wilms' tumor (WT), vaginal metastasis (Vag. Met.), and normal kidney (NK) from a patient. DNA from the tissues was restricted and hybridized with the indicated restriction enzyme/probe combinations. Normal kidney DNA is heterozygous at two 11p loci: oncogene Harvey-ras-1 (HRAS1), and parathyroid hormone (PTH); and one 11q locus: apolipoprotein AI (apoAI). Hetero-zygosity is lost at both 11p loci in both the tumor and the me-tastasis. Heterozygosity is retained at the 11q locus, apoA1, in the tumor and metastasis. (E) Densitometric scan of autoradiograms of normal (thin line) and tumor (thick line) DNA restricted with Eco RI and hybridized to the probes PTH and 7-2D, a probe for the HLA class II-associated invariant-chain gene located on chromosome 5. The two scans were standardized using the 7-2D band. The intensity of the PTH bands from tumor and normal kidney DNA are the same, indicating two copies of the PTH gene in the tumor genome. Patient is reference number 198871-000. (From Dao et al., 1987).

mutations at 11p13 which are not detectable by present means. These
mutations may be small deletions, point mutations, or very localized
somatic recombination or gene conversion events.

Of interest is the observation that in the tumors in which parental
origin of the polymorphic alleles can be determined, all the tumors that
lost heterozygosity lost the maternally derived alleles (Schroeder et
al., 1987). This observation extends to Wilms' tumors studied in other
laboratories (Reeve et al., 1984), to retinoblastoma (Dryja, 1984), and
to one soft tissue sarcoma (V. Huff, unpublished data). This nonrandom
loss of maternal alleles suggests a possible higher paternal mutation
rate, a possible selection for paternally-derived alleles, or a possible
requirement of maternally-derived alleles for normal development. The
first explanation implies that the initial mutation in the tumors is
germinal. However, only three out of the eight patients display a
phenotype of multiple tumors or associated abnormalities which would
suggest a germinal origin. The other five tumors are unilateral and
presumed to be the result of two somatic mutations. The observation that
familial WT can be either maternally or paternally inherited (Matsunaga,
1981) argues against the latter two possibilities. Further study of
sporadic and familial tumors and a better estimate of the contribution of
germinal and somatic mutations to Wilms' tumor may help to understand
this phenomenon.

LOCALIZATION OF THE WILMS' TUMOR GENE

Many approaches have been taken to identify and isolate genes
involved in disease. When little or nothing is known about the defective
protein or mRNA, identification and cloning of the disease gene can be
approached by making use of a knowledge of its location in the genome.
In an effort to identify and isolate the putative WT gene, our laboratory
has developed a panel of 11p13-specific probes. These probes are being
used to identify 11p13 alterations in tumors and to characterize the
extent of chromosome deletions in patients with both microscopic and
submicroscopic constitutional deletions. This information will be
important for sublocalizing the WT gene in 11p13. In conjunction, a
physical map of 11p13 is being constructed using the 11p13 probes and
infrequently cutting restriction enzymes. These data are important in
our efforts to identify and isolate the Wilms' tumor gene.

Isolation of 11p13-specific DNA Probes

Probes from the short arm of chromosome 11 were obtained from a
genomic library derived from a Chinese Hamster Ovary (CHO)-human somatic

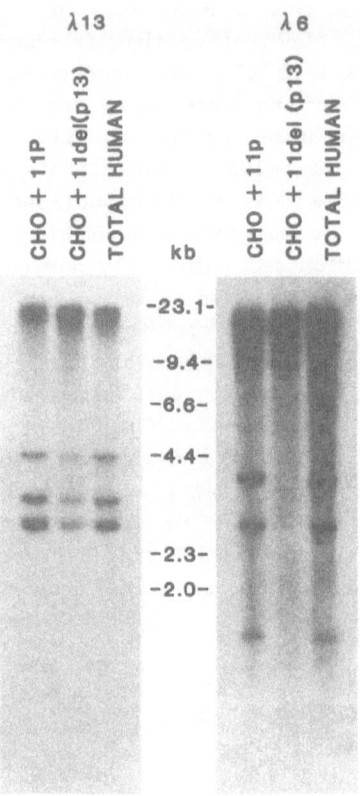

Fig. 6. Selection of 11p13-specific probes. DNA from human
chromosome 11p-specific clones was radiolabeled and
prehybridized to an excess of total human DNA to
competively bind repetitive sequences in the phage. The
radiolabeled phage DNA was then used as a hybridization
probe for filters containing Eco RI-restricted DNA from
(left to right): cell line 901-110 containing human
chromosome 11p; cell line C2-1, a CHO-human somatic cell
hybrid containing a human del(11)(p12-p14) chromosome;
or normal human lymphocytes. λ13 DNA hybridizes to DNA
from all three cell lines, indicating that it is homologous to sequences on human chromosome 11, but outside
of chromosomal band 11p13. λ6 DNA hybridizes to DNA from
the somatic cell hybrid containing 11p and to total
human DNA. It does not hybridize to DNA from the somatic
cell hybrid which contains the human del(11)(p13)
chromosome. The pattern of hybridization for λ6
indicates that this clone is derived from the short arm
of human chromosome 11 and is localized to band 11p13.

cell hybrid, cell line 901-110. The only human component in this hybrid cell line is the short arm of chromosome 11 (Jones et al., 1978). Human clones were identified by hybridization to the human Alu repeat clone, BLUR8. The isolated human clones were used as probes for hybridization to normal human DNA, DNA from the original hybrid cell line (901-110), and DNA from a CHO-human somatic cell hybrid (C2-1) which contains a human chromosome 11 deleted of band p13 (Michalopoulos et al., 1985). 11p13-specific human probes were identified by their ability to hybridize to normal human DNA but not to the DNA from the cell hybrid carrying the 11p13 deletion chromosome (Fig. 6). In this manner, sixteen 11p13-specific DNA probes were isolated (Compton et al., 1987).

Localization of the probes within band 11p13 was accomplished by use of DNA from a series of human cell lines carrying various 11p13 deletions or translocations (Fig. 7). Because each cell line contained one normal chromosome 11 and one altered chromosome 11, gene dosage studies were necessary to determine if a probe hybridized to DNA present in one or two copies in the cell (Fig. 7). If only one copy was detected (hemizygosity), the probe was assigned to the interval spanned by the deletion. In this manner 11p13-specific probes were divided into three major groups (Fig. 7 and Table 1). Of particular interest are the probes p5 and p60 which are hemizygous in LCS036, a lymphoblastoid cell line from a patient with Wilms' tumor and aniridia but no cytogenetically detectable deletion (Riccardi et al., 1982). The loss of p5 and p60 sequences in this cell line demonstrates that a small deletion is present. The small size of the LCS036 deletion relative to the cytogenetically detectable deletions in the other cell lines make this cell line very useful for sublocalizing the WT locus. In addition, the sequences deleted in patient LCS036 DNA are important starting points for closing in on the Wilms' locus by isolating other 11p13 probes via chromosome walking and chromosome hopping.

Construction of an 11p13 Physical Map

As part of our effort to identify and isolate the WT locus, we are using pulsed field gel electrophoresis of large DNA restriction fragments along with our 11p13 probes to construct a physical map of 11p13. The large scale of a physical map generated by this method allows us to detect alterations in WT patient DNA which would not be detected by other means. The ability to localize 11p13 deletions and deletion breakpoints in various WT DNA samples has enabled us to sublocalize the Wilms' tumor locus within 11p13.

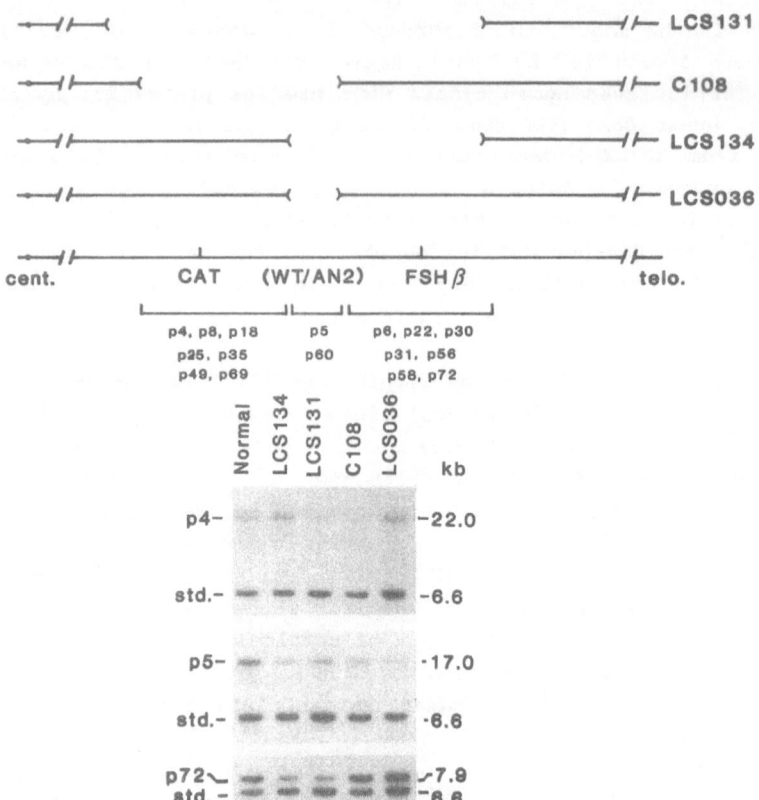

Fig. 7. Sublocalization of 11p13 probes. Single copy
subclones from 11p13-specific lambda clones were
used in gene dosage experiments using DNA from
lymphoblastoid cell lines derived from patients
with visible or suspected chromosome 11 deletions.
A chromosome 5-specific probe was used as a hybridization standard (std.) for the dosage blots. TOP:
Diagram of 11p13 deletions in cell lines: LCS131,
del(11)(p11.2p14.1), Wilms' tumor and ani-ridia;
C108, del(11)(p13)t(1;11)(p11.5;p12), Wilms' tumor
and aniridia; LCS134, del(11)(p13.2p14.1), aniridia
and genitourinary anomalies; LCS036, no visible
chromosome 11 deletion, Wilms' tumor and anirida.
BOTTOM: autoradiograms of cell line DNA restricted
with Bam HI and hybridized to probes p4, p5, and
p72. Dosage analysis indicates that p4 is present
in only one copy in LCS131 and C108 DNA, p5 is
present in only one copy in all but the normal cell
line DNA, and p72 is present in one copy in LCS131
and LCS134 DNA. By this type of analysis sixteen
11p13 probes were sublocalized to three regions of
11p13.

Table 1. Summary of 11p13 DNA probe localization through gene dosage analysis.

WT/AN Lymphoblastoid Cell Line		CAT	p4	p8	p18	p25	p49	p69	p5	p60	p6	p22	p30	p31	p56	p58	p72	FSHβ
																		11p13 DNA Probes
LCS134 46,XY, del 11(p13.2p14.1) Inv7q		2	2	2	2	2	2	2	–	–	–	–	–	–	–	–	–	–
LCS036 46,XX		2	2	2	2	2	2	2	–	–	2	2	2	2	2	2	2	2
C108 46, XY, del 11(p13)t(1:11)(p11.5;p12)		–	–	–	–	–	–	–	–	–	2	2	2	2	2	2	2	2
FAMILIAL ANIRIDA	der (11)	+	+	+	+	+	+	+	+	+	–	–	–	–	–	–	–	–
46, XY, t(4:11)(q22:p13)	der(4)	–	–	–	–	–	–	–	–	–	+	+	+	+	+	+	+	+

17

To construct the physical map, high molecular weight DNA from various cell lines was restricted with one of several restriction endonucleases whose recognition sites occur infrequently in the genome. Restriction enzymes we routinely use are Not I, Sfi I, Mlu I, and Nru I. The resulting large (150-7000 kb) restriction fragments were separated by pulsed field gel electrophoresis. The electrophoresed DNA was tranferred to nylon membranes which were hybridized to the 11p13 probes.

The autoradiograms shown in Figure 8 are of filters of cell line DNA restricted with Not I and hybridized with several 11p13 probes: a probe for the hepatitis B virus insertion site (HBVIS), the gene for the beta subunit of follicle stimulating hormone (FSHB), catalase (CAT), or an 11p13 random probe, p72. The cell lines are from patients with Wilms' tumor, aniridia, and/or genitourinary anomalies. The Not I fragments detected with these probes range in size from >7000 kb to 1000 kb. Variant fragments are observed in the DNA from two of the cell lines probed with p72 (Fig. 8). DNA from patient C108 gives a Not I fragment of ~1200 kb in addition to the normal Not I fragment of 1400 kb. Cytogenetically, patient C108 has a deletion of a part of 11p13 in conjunction with an (1;11) translocation. The presence of a variant fragment suggests that one of the 11p13 breakpoints occurs within the normal 1400 kb Not I fragment detected by p72. Similarly, a variant Not I fragment of ~1000 kb is observed in patient LCS036. DNA from LCS036, a patient with Wilms' tumor and aniridia, carries a submicroscopic deletion which was detected by virtue of being hemizygous for probes p5 and p60 (Fig. 7 and Table 1). The abnormal 1000 kb fragment seen in patient LCS036 DNA suggests that one of the breakpoints also occurred within the normal 1400 kb Not I/p72 fragment. In addition to localizing the breakpoints in DNA from patients C108 and LCS036 to a particular Not I restriction fragment, these data also help to order the 11p13 probes and their respective restriction fragments. Since the p72 probe appears to detect restriction fragment alterations that the FSHB probe does not, p72 must be proximal (centromeric) to FSHB. Thus, by coordinately using probe dosage information (Fig. 7 and Table 1) and restriction fragment data from both normal DNA and 11p13 deletion DNA, a physical map of 11p13 can be constructed.

This approach has been expanded by use of additional rare-cutting restriction enzymes. Figure 9 is a diagram of the relative order of Not I and Mlu I restriction fragments in 11p13. Because a fragment will not be detected without the appropriate probe, the diagrammed Not I and Mlu I fragments may or may not be contiguous. Use of a second enzyme does, however, help to estimate the distances between restriction fragments. For example, Not I restriction analyses indicated that probes p5 and p60 reside on a 325 kb fragment. However, the distance between them and the next distal (telomeric) probe (p72) and the next proximal probe could not be estimated based only on the Not I data. Mlu I restriction analysis

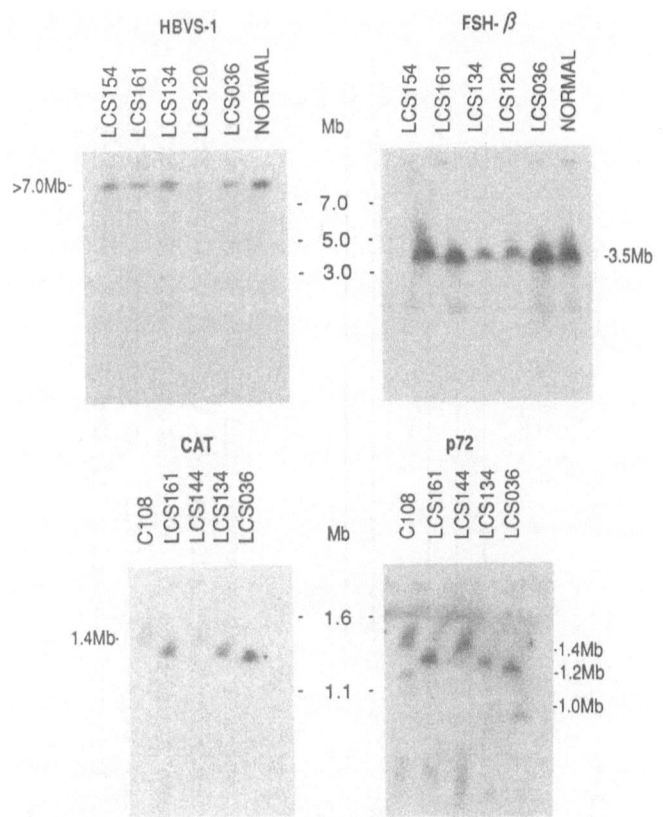

Lymphoblastoid Cell Line	Karyotype	Phenotype
C108	46, XY, del11 (p13) t(1;11) (p11.5;p12)	WT/AN
LCS 036	46,XX	WT/AN
LCS 120	46,XY,del11 (p12.8p15.1)	WT/AN/GU
LCS 134	46,XY, del11(p13.2 p14.1)	AN/GU
LCS 144	46, XY, del11(p13.4 p14.1)t(8p;11p;11q)	WT/AN
LCS 154	46,XY	WT/GU
LCS 161	46,XX	WT

Fig. 8. Autoradiograms of cell line DNA restricted with Not I, pulse field gel electrophoresed, and probed with the indicated chromosome 11 probes. HBVS-1, probe for the hepatitis B virus insertion site; FSH-beta, probe for the gene for the beta subunit of follicle stimulating hormone; CAT, probe for the catalase gene; p72, 11p13-specific probe. Migration of yeast chromosome size markers are indicated between autoradiograms. The karyotypes and phenotypes of the patients from whom the lymphoblastoid cell lines were derived are indicated in the table at the bottom.

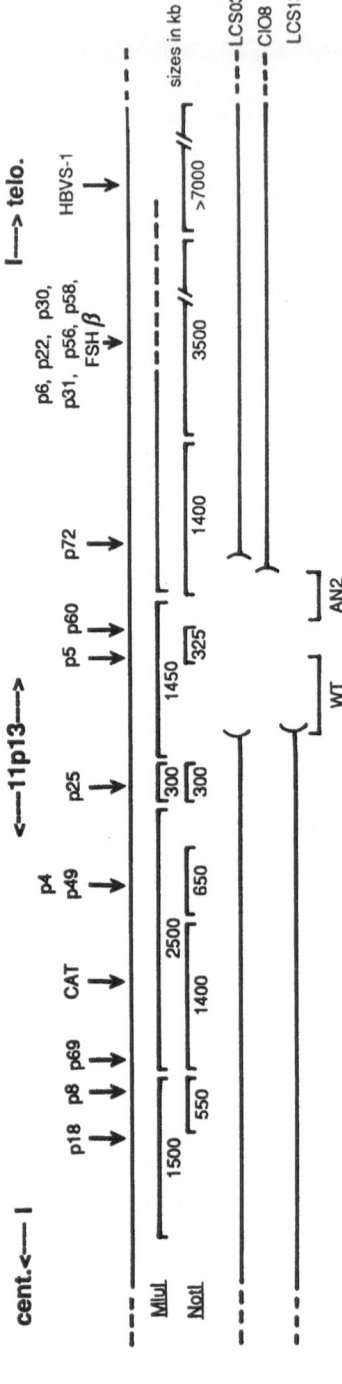

Fig. 9. Not I and Mlu I restriction map for chromosomal band 11p13. Restriction fragments shown are those detected with the 11p13 probes indicated above the restriction map. 11p13 probes were localized by hybridization with DNA from lymphoblastoid cell lines LCS036, C108, and LCS134 as diagrammed at the bottom and described in Fig. 7. Inferred location of the Wilms' tumor and aniridia genes are shown.

revealed that p5 and p60 reside on the same 1450 kb fragment which was not detected by any other probes. This information serves to expand the region around p5 and p60. The use of enzymes that produce overlapping fragments serves to localize probes that could not be ordered relative to each other using the 11p13 deletion cell lines. For example, probes p4 and p49 detect a 650 kb Not I fragment and probes p69 and CAT detect a 1400 kb Not I fragment. All four hybridize to a 2500 kb Mlu I fragment. Thus, the two Not I fragments can be localized to the same Mlu I fragment. The identification of other restriction fragments which overlap the known fragments will be useful for linking up the known fragments as will information on partial restriction products.

Strategies for Sublocalization and Isolation of the Wilms' Tumor Locus

Chromosome 11 deletions in DNA from Wilms' tumor patients enabled us to orient the 11p13 probes and restriction fragments relative to each other. Of more interest is the ability to map disease loci within 11p13. From the data shown in Figure 7 and Table 1, it is apparent that in the deletion cell lines from Wilms' tumor patients, the region commonly deleted is that surrounding probes p5 and p60. This area includes a 325 kb Not I restriction fragment which is contained within a 1450 kb Mlu I restriction fragment (Fig. 9). The commomly deleted region does not affect Not I or Mlu I restriction fragments detected by the catalase probe or by other probes on the catalase region. However, the deletion in cell line LCS036 does extend into the 1400 kb Not I fragment detected by probe p72. This information identifies probe p72 as the closest marker on the distal side of the deletion. To date, with 11p13-specific probes we have detected restriction fragments the sizes of which total almost 10,000 kb. Approximately 8,000 kb of this can be excluded from the critical WT region as defined by the cell line deletions.

These data can also be used to sublocalize the aniridia-2 gene. Probes p5 and p60 have been placed proximal to the translocation seen in a familial aniridia patient without Wilms' tumor (Simola et al., 1983), whereas probe p72 is distal to the translocation (Table 1). This information, in addition to the deletions observed in other individuals with aniridia, places the gene for chromosome 11-associated aniridia to the region distal to the Wilms' tumor locus as indicated in Figure 7.

Knowing the orientation of restriction fragments and 11p13 probes relative to the Wilms' tumor locus enables us to take a more selective approach in isolating DNA clones from critical regions of p13. Not I restriction fragments that, by our deletion analyses, are hypothesized to contain the Wilms' tumor gene can be selectively cloned. One approach for doing this is to restrict the CHO-plus human 11p (somatic cell hybrid, 901-110) DNA with Not I, electrophoretically separate the restriction fragments, and isolate appropriately sized restriction fragments from the gel. A library can then be constructed from the isolated DNA and the resulting clones screened for human sequences that are located in the targeted 11p13 Not I fragment.

A knowledge of the physical map of 11p13 also allows us to identify which of the currently available 11p13 probes are appropriate starting points for chromosome walking and chromosome hopping. Because of the relatively close proximity of probes p5 and p60 to the Wilms' tumor locus, we are currently using them for screening libraries. Although the probe p72 is not as close to the Wilms' tumor gene, it is close to the deletion breakpoints in cell lines LCS036 and C108 (Fig. 9). Chromosome walking and hopping from this probe can identify DNA sequences that are located at the breakpoints. Analysis of these clones will yield information about whether chromosome deletions occur at particular DNA sequences. More importantly for the goal of isolating the Wilms' tumor gene, chromosome walking or hopping from p72 will allow us to cross the breakpoint in the DNA from patient LCS036, leading to the isolation of DNA sequences on the opposite side of the breakpoint. Beside being useful for refining the physical map, these sequences will define the closest proximal markers for the Wilms' tumor locus and will allow directed movement towards the Wilms' tumor gene.

CONCLUSION

Sublocalization and isolation of the WT gene will therefore be facilitated by a knowledge of the physical map of chromosomal band 11p13. The isolated gene will be an invaluable tool for investigating the biology of Wilms' tumor. The role of the normal 11p13 gene in kidney development, the incomplete penetrance of the inherited predisposing gene, and the association of Wilms' tumor with genitourinary abnormalities, aniridia, mental retardation and other embryonal tumors are all research areas that will be expanded once the gene is obtained and its protein product is identified. In addition, understanding the way in which a single gene plays a key role in the development of Wilms' tumor will be important for elucidating the etiology and biology of more complex and heterogeneous cancers.

REFERENCES

American Cancer Society, 1988, Cancer Facts & Figures-1988.

Benedict, W. F., Murphree, A. L., Banerjee, A., Spina, C. A., Sparkes, M. C., and Sparkes, R. S., 1983, Patient with 13 chromosome deletion: evidence that the retinoblastoma gene is a recessive cancer gene, Science, 219:973.

Boveri, T., "Zur Trage der Entstehung malignen Tumoren." G. Fischer, Jena (1914).

Breslow, N. E. and Beckwith, J. B., 1982, Epidemiological features of Wilms' tumor: results of the National Wilms' Tumor Study, J. Natl. Cancer Inst., 68:429.

Cavenee, W. K., Dryja, T. P., Phillips, R. A., Benedict, W. F., Godbout, R., Gallie, B. L., Murphree, A. L., Strong, L. C., and White, R. L., 1983, Expression of recessive alleles by chromosomal mechanisms in retinoblastoma, Nature, 305:779.

Compton, D. A., Weil, M. M., Chao, L.-Y., Lewis, W. H., Jones, C., Riccardi, V. M., Strong, L. C., and Saunders, G. F., 1987, Physical analysis of chromosomal band 11p13 using random probes and field inversion gel electrophoresis. Am. J. Hum. Genet., 41:A162.

Dao, D. D., Schroeder, W. T., Chao, L.-Y., Kikuchi, H., Strong, L. C., Riccardi, V. M., Pathak, S., Nichols, W. W., Lewis, W. H., and Saunders, G. F., 1987, Genetic mechanisms of tumor-specific loss of 11p13 DNA sequences in Wilms' tumor, Am. J. Hum. Genet., 41:202.

Dryja, T. P., Cavenee, W. K., White, R., Rapaport, J. M., Petersen, D., Albert, D. M., and Bruns, G. A., 1984, Homozygosity of chromosome 13 in retinoblastoma, N. Engl. J. Med., 310:550.

Fearon, E. R., Vogelstein, B., and Feinberg, A. P., 1984, Somatic deletion and duplication of genes on chromosome 11 in Wilms' tumours, Nature, 309:176.

Festa, R. S., Meadows, A. T., and Boshes R. A., 1979, Leukemia in a black child with Bloom's syndrome, Cancer, 44:1507.

Jones C., and Kao, F. T., 1978, Regional mapping of the gene for human lysosomal acid phosphatase (ACP2) using a hybrid clone panel containing segments of human chromosome 11, Hum. Genet., 45:1.

Kaneko, Y., Egues, M. C, and Rowley, J. D., 1981, Interstitial deletion of short arm of chromosome 11 limited to Wilms' tumor cells in a patient without aniridia, Cancer Res. 41:4577.

Knudson, A. G., and Strong, L. C., 1972, Mutation and cancer: a model for Wilms' tumor of the kidney, J. Natl. Cancer Inst., 48:313.

Matsunaga, E., 1981, Genetics of Wilms' tumor, Hum. Genet. 57:231.

Michalopoulos, E. E., Bevilacqua, P. J., Stokoe, N., Powers, V.E., Willard, H.F., and Lewis, W. H., 1985, Molecular analysis of gene deletion in aniridia-Wilms tumor association, Hum. Genet. 70:157.

Narahara, K., Kikkawa, K., Kimira, S., Kimoto, H., Ogata, M., Kasai, R., Hamawaki, M., and Marsuoka, K., 1984, Regional mapping of catalase and Wilms tumor -- aniridia, genitourinary abnormalities, and mental retardation triad loci to the chromosome segment 11p1305 p1306, Hum. Genet. 66:181.

Orkin, S. H., Goldman, D.S., and Sallan, S. E., 1984, Development of homozygosity for chromosome 11p markers in Wilms' tumour, Nature, 309:172.

Raizis, A. M., Becroft, D. M., Shaw, R. L., and Reeve, A. E., 1985, A mitotic recombination in Wilms tumor occurs between the parathyroid hormone locus and 11p13, Hum. Genet., 70:344.

Reeve, A. E., Housizux, P. J., Gardner, R. J. M., Chewings, W. E., Grindley, R. M., and Millow, L. J., 1984, Loss of a Harvey ras allele in sporadic Wilms' tumour, Nature, 309:174.

Riccardi, V. M., Sujansky, E., Smith, A. C., and Francke, U., 1978, Chromosomal imbalance in the aniridia-Wilms' tumor association: 11p interstitial deletion, Pediatrics, 61:604.

Riccardi, V. M., Hittner, H. M., Strong, L. C., Fernbach, D. J., Lebo, R., and Ferrell, R. E., 1982, Wilms tumor with aniridia/iris dysplasia and apparently normal chromosomes, J. Pediatr., 100:574.

Simola, K. O. J., Knuutila. S., Kaitila, I., Pirkola, A., and Pohja, P., 1983, Familial aniridia and translocation t(4;11)(q22;p13) without Wilms' tumor, Hum. Genet. 63:158.

Slater, R. M., and de Kraker, J., 1982, Chromosome 11 and Wilms' tumor. Cancer Genet. Cytogenet., 5:237.

Strong, L. C., 1984, Genetics, etiology, and epidemiology of childhood cancer, in: "Clinical Pediatric Oncology, 3rd ed." W. W. Sutow, T. J. Vietti, and D. J. Fernback, eds, Mosby, St. Louis.

MECHANISMS OF ALTERED GROWTH REGULATION IN HUMAN LEUKEMIA[1]

Bruno Calabretta and Donatella Venturelli

Department of Pathology, Temple University Medical School
3400 N. Broad Street
Philadelphia, PA 19140

The process of normal growth requires the orderly occurrence of many
events such as interaction of (a) stimulatory growth factor(s) with its
(their) own receptors(s), signals transmission throughout the cytoplasm and
into the nucleus and finally DNA synthesis and cell division. In the
hemopoietic system cell growth is tightly linked to terminal cell differentiation; as soon as the choice between self renewal and lineage commitment
is made, terminally differentiated cells are generated throughout a
limited number of cells divisions.

Leukemic cells are characterized by impairment of the capability to
terminally differentiate and the fact that they belong to a pool of potentially self-renewing cells; growth advantage of leukemic cells over normal
marrow cells in vivo is the resulting clinical manifestation.

It is likely that alteration in the regulated expression of any gene
or genes controlling each step of cell growth in the hemopoietic system will
lead to uncoupling of proliferation and differentiation which in the case of
leukemia is manifested as growth advantage in vivo.

In a few cases of leukemia this growth advantage may be the result of
autocrine secretion of GM-CSF (1,2) or dependence on GM-CSG (3) but for the
majority of cases the mechanisms is not clear. In other cases the growth
advantage of leukemic cells may result from abnormal expression of genes
controlling pivotal aspects of cell proliferation such as DNA synthesis and
cell division. The genes controlling cell proliferation are broadly defined
here as growth regulated; the study of these genes is providing insights on
the altered mechanisms of growth in leukemic cells.

IDENTIFICATION OF GROWTH-REGULATED GENES

In the past few years, a considerable number of animal genes have been
identified whose expression is cell-cycle dependent; that is, genes whose
RNA levels are markedly increased in a specific phase of the cell cycle or
when quiescent cells are stimulated to proliferate. In a broad sense
growth-regulated genes are defined as genes whose expression is higher in
proliferating cells than in quiescent cells. The most scrutinized members
of the growing list of growth-regulated genes include the cellular homologues of at least six oncogenes: c-myc, c-fos, c-myb, c-ras, c-fgr, and
p53 (4,11). Other well known growth-regulated genes are calmodulin (12,

---

[1]Supported by a grant from W.W. Smith Charitable Trust to B.C.

β-actin (8), thymidine kinase (13), dihydrofolate reductase (13), thymidy-late synthase (14), ornithine decarboxylase (15,16), PCNA (17,18) and histones (19,20). Furthermore, other cell-cycle dependent genes have been identified as cDNA clones by differential screening of cDNA libraries (21-24). Some of these genes correspond to previously characterized proteins (25,26,27), but the function of other genes belonging to this group is still unknown (28,29).

Growth-regulated expression does not necessarily mean that genes behaving in this fashion regulate cell cycle progression. However, the fact that the expression of some oncogenes is growth-regulated (4-11) suggests that some of these may have a regulatory role in the control of cell prolifera-tion. Growing experimental evidence suggests such a role for c-myc, c-ras, p53 and c-fos (30-37). The example of p53 is particularly striking (10, 38-39,40,41); this gene was first found to play a role in the control of cell proliferation (42,43) and subsequently to cooperate with an activated human ras sequence to transform primary embryonic cells (44-46).

EXPRESSION OF GROWTH-REGULATED GENES IN HUMAN LEUKEMIAS

A feature common to most oncogenes is their overexpression in a variety of human malignancies including acute leukemias (47-50). The altered expression of the oncogenes, however it might be achieved, is currently thought to be a major molecular mechanism associated with the origin and progression of human malignancies (51). The close relationship between oncogenes and growth regulated genes suggests that studying the mRNA levels of growth regulated genes in human leukemia is necessary in order to esta-blish a relationship between altered growth regulation in leukemic cells and clinical manifestations of the disease.

In normal tissues, only a small percentage of cells are cycling, while the majority are in a quiescent state which is called G_0 (for a review 52). The fraction of cycling cells is called the growth fraction (53). The growth fraction is often increased in tumors (54,55). If, in a given cell population, more cells are cycling, the expression of growth regulated genes would be increased with respect to cell populations in which fewer cells are cycling. The overexpression of growth regulated genes would then be simply a measure of the increased proliferating activity of tumors rather than a demonstration that their expression is actually deregulated. Since the expression of the core histone genes is restricted to the S phase of the cell cycle, (21-22) the level of histone H3 mRNA can be taken as a measure of growth fraction (54-58).

In the past three years, this laboratory has analyzed the expression of growth regulated genes in normal and malignant cells in order to develop a systematic approach study of the control of proliferation in human leukemias.

The work has consisted of two steps:

a) Identification of growth-regulated pattern of expression of putative growth-regulated genes in two human systems: peripheral blood mononuclear cells (PBMC) stimulated by PHA and/or IL-2 and serum-stimulated WI-38 diploid fibroblasts (56,57,58,59).

b) Measurement of the mRNA level of growth-regulated genes in fresh human leukemic cells in relation to the growth fraction of each leukemic population. (4).

We reasoned that the expression of growth-regulated genes and S-phase specific genes may vary from one tissue to another, reflecting the fraction of cycling cells, but their ratio should be constant. However, if their ratio were to increase it would be consistent with a true deregulation of growth-regulated genes.

In most human leukemias, the large majority of growth-regulated genes are expressed at a level which reflects the number of cycling cells (5). Cytofluorimetric studies of leukemic cells revealed a very good correlation between the number of cells in S phase and the mRNA levels of the histone H3 gene (5). This observation rules out an overexpression of growth-regulated genes as common feature of the leukemic phenotype per se. On the other hand, c-myc and calcyclin are truly overexpressed in a sizeable number of leukemic patients and other growth-regulated genes including p53 and vimentin are occasionally overexpressed with no apparent correlation to the number of cycling cells (4,5). To determine the existence of altered mRNA levels of these genes each leukemic RNA was simultaneously hybridized with a growth-regulated gene (c-myc, calcyclin or p53) and histone H3. The ratio of expression c-myc/H3, calcyclin/H3 and p53/H3 distinguishes the over-expression of oncogenes or growth regulated genes due to a high proportion of cycling cells, from that resulting from a true temporal deregulation of these genes due to an inappropriate expression of growth-regulated genes by functionally noncycling leukemic cells. The altered expression of growth-regulated genes by functionally noncycling leukemic cells may provide these cells with some kind of "competence" to proliferate. From this putative stage they could be susceptible to proliferative stimuli which may determine the reentry into the pool cycling cells.

Therefore a fraction of functionally noncycling leukemic cells may undergo self renewal rather than enter into the pathway leading to terminal differen-tiation and doing so may sustain the leukemic phenotype.

EARLY EFFECTS OF CHEMOTHERAPY ON THE mRNA LEVELS OF GROWTH-REGULATED GENES: HUMAN LEUKEMIAS

Since most chemotherapy kills cycling cells, while sparing noncycling cells, it is expected that chemotherapy interfering with DNA transcription will have early effects on the mRNA levels of growth regulated genes. It is likely that an early decrease in the expression of growth regulated genes, post-chemotherapy, will occur in patients in whom the expression of these genes is restricted to cycling cells.

However, there are patients in whom a significant number of function-ally noncycling leukemic cells express growth regulated genes; in these patients post-chemotherapy changes in the mRNA levels of growth-regulated genes might be less significant or may not occur at all. Nevertheless it is likely that in some of these patients, chemotherapy will render more evident the extent of the alteration in the ratio growth-regulated gene/ histone H3, since the expression ratio of growth-regulated genes to histone H3 will be increased when noncycling leukemic cells express the former at high levels.

In the past two years, we have studied early effects of chemotherapy on the mRNA levels of growth-regulated genes in patient leukemic cells and we have tried to correlate the observed changes with therapeutic outcome.

Our studies can be summarized as follows:

1) Chemotherapy alters mRNA levels of growth-regulated genes in leukemic blast cells;

2) the early decrement (24 hours post chemotherapy) of c-myc and histone H3 mRNA levels may predict which patients will achieve complete remission, whereas lack of reduction in these mRNA's correlates with failure to achieve remission.

If the relationships observed in the small group studied until now (20 cases) are confirmed in a larger study, our analysis could form the basis of a sensitive test for predicting post chemotherapy outcome. By extension, in those patients who fail to respond to a first dose of chemotherapy with a decrement in c-myc and H3 mRNA levels, a more aggressive treatment of their disease might be indicated.

CONCLUSIONS

We studied the altered growth regulated of leukemic cells by analyzing in these cells mRNA levels of growth-regulated genes.

We have also investigated the effects of chemotherapy on the mRNA levels of growth-regulated genes and attempted a correlation with the attainment of hematological remission. Our studies revealed that functionally non-cycling leukemic cells express at high level growth regulated and this might be a mechanism which sustains the leukemic phenotype.

We have also found that absence of changes in the mRNA levels of growth regulated genes post-chemotherapy correlates with failure to obtain hematology remission, while an early decrement may predict hematological remission.

Our investigation suggests a mechanism of altered growth regulation of leukemic cells dependent on the inappropriate expression of growth regulated genes by noncycling leukemic cells; of more practical interest, growth regulated genes, independently of the role they play in the control of cell proliferation provide an important tool to predict the effectiveness of chemotherapy in the induction of hematological remission in leukemic patients.

REFERENCES

1. D. C. Young, J. Griffin, Autocrine secretion of GM-CSF in acute myelo-blastic leukemia, Blood 68:1178-1181 (1986).
2. D. C. Young, K. Wagner, J.D. Griffin, Constitutive expression of the granucyte-macrophage colony stimulating factor genes in acute myelobla-stic leukemia, J. Clin, Invest. 79:100-106 (1987).
3. B. Lang, Growth factor requirements of childhood acute leukemia: establishment of FM-CSF dependent cell lines, Blood 70:192-199, (1987).
4. K. Kelly, B. Cochran, C.D. Stiles, P. Leder, Cell-specific regulation of the c-myc gene by lymphocyte mitogens and platelet-derived growth factor, Cell 35:603-610, (1983).
5. M.E. Greenberg, E.B. Ziff, Stimulation of 3T3 cells induces transcription of c-fos proto-oncogene, Nature 311:433-438, (1984).
6. G. Torelli, L. Selleri, A. Donelli, S. Ferrari, G. Emilia, D. Venturelli, L. Moretti, U. Torelli, Activation of c-myc expression by phytohemagglu-tinin stimulation in normal human T lymphocytes, Mol Cell Biol. 5:2874-2878, (1985).
7. C.B. Thompson, P.B. Challoner, P.E. Neiman, and M. Goudine, Expression of the c-myb proto-oncogene during cellular proliferation, Nature 319: 374-380, (1986).
8. J. B. Stern, and K.A. Smith, Interleukin-2 induction of T-cell G_1 progression and c-myb expression, Science 233:203-206, (1986).
9. J. Campisi, H.E. Gray, A.B. Pardee, M. Dean, G.E. Sonenshein, Cell cycle control of c-myc but not c-ras expression is lost following chemical transformation, Cell 36:241-248, (1984).
10. M. Goyette, C.J. Petropoulos, P.R. Shank, N. Fausto, Regulation transcription of c-Ki-ras and c-myc during compensatory growth of rat liver, Mol. Cell. Biol. 4:1498-1502, (1984).

11. N.E. Reich, A.J. Levine, Growth regulation of a cellular tumor antigen p53 in nontransformed cells, Nature 308:199-201, (1984).
12. J.G. Chafouleas, L. Lagace, W.E. Bolton, A.E., III Boyd, A.R. Means, Changes in calmodulin and its mRNA accompany re-entry of quiescent (G_0) cells into the cell cycle, Cell 36:73-84, (1985).
13. H-T., Liu, C.W. Gibson, R.R. Hirschhorn, S. Rittling, R. Baserga, W.E. Mercer, Expression of thymidine kinase and dihydrofolate reductase genes in mammlian ts mutants of the cell cycle, J. Biol. Chem. 260:3269-3275, (1985).
14. C-H., Jehn, P.K. Geyer and L.F. Johnson, Control of thymidylate synthesis and mRNA content and gene transcription in an overlapping mouse cell line. Mol. Cell. Biol. 5:5375-5384, (1985).
15. C. Kahana, D. Nathans, Nucleotide sequence of murine ornithine decarboxylase mRNA, Proc. Natl. Acad. Sci. USA 82:1673-1678, (1985).
16. L. Kaczmarek, B. Calabretta, S. Ferrari, and J.K. deRiel, Cell cycle dependent expression of human ornithine decarboxylase, J. Cell Physiol. 132:545-551, (1985).
17. J.M. Almendral, D. Huebsch, P.A. Blundell, H. MacDonald-Bravo and R. Bravo, Cloning and sequence of the human nuclear protein cyclin: Homology with DNA binding proteins, Proc. Natl. Acad. Sci. USA 84:1575-1579, (1987).
18. D. Jaskulski, C. Gatti, S. Travali, B. Calabretta and R. Baserga, Regulation of the PCNA cyclin and thymidine kinase mRNA levels by growth factors, J. Biol. Chen, in press.
19. M. Plumb, J. Stein, G. Stein, Coordinate regulation of multiple histone mRNAs during the cell cycle of HeLa cells. Nucl. Acid. Res. 11:2391-2410, (1983).
20. R.R. Hirschhorn, F. Marashi, R. Baserga, J. Stein and G. Stein, Expression of histone genes on G_1 specific temperature sensitive mutant of the cell cycle. Biochemistry 23:3731-3735, (1984).
21. D.I.H. Linzer, D. Nathans, Growth-regulated changes in specific mRNAs of cultured mouse cells. Proc. Natl. Acad. Sci. USA 83:4271-4275, (1983).
22. B.H. Cochran, A.C. Reffel, C.D, Molecular cloning of gene sequences regulated by platelet-derived growth factor, Cell. 33:939-951, (1983).
23. R.R. Hirschhorn, P. Aller, Z-A., Yuan, C.W. Gibson, R. Baserga, Cell-cycle-specific cDNAs from mammalian cells temperature sensitive for growth, Proc. Natl. Acad. Sci. USA 81:6004-6008, (1984).
24. S.K. Arya, F, Wong-Staal, R.C. Gallo, Transcriptional regulation of a tumor promoter and mitogen-inducible gene in human lymphocytes, Mol. Cell Biol. 4:2540-2542, (1984).
25. S. Ferrari, R, Battini, L. Kaczmarek, S. Rittling, B. Calabretta, J.K. DeRiel, V. Philiponis, J.F. Wei, and R. Baserga, Coding sequence and growth regulation of the human vimentin gene, Mol. Cell. Biol. 61:3614-3620, (1986).
26. L.M. Matrisian, G. Rautmann, B.E. Magun, and R. Breathnach, Epidermal growth factor or serum stimulation of rat fibroblasts induces an elevation in mRNA levels for lactate dehydrogenase and other glycolytic enzymes, Nucl. Acids Res. 13:711-726, (1985).
27. B. Calabretta, R. Battini, L. Kaczmarek, J.K. deRiel, and R. Baserga, Molecular cloning of the cDNA for a growth factor-inducible gene with strong homology to S-100, a calcium binding protein. J. Biol. Chem 261:12628-12632, (1986).
28. D.R. Edwards, P. Waterhouse, M.L. Holman, and D.T. Dehardt, A growth responsive gene (16B) in normal mouse fibroblasts homologous to a human collagenase inhibitor with erythroid potentiating activity: evidence for inducible and constitute transcripts, Nucl Acids Res. 14:8863-8878, (1986).
29. H.A. Armelin, M.C.S. Armelin, K. Kelly, T. Stewart, P. Leder, B.K. Cochran, C.D. Stiles, Functional role for c-myc in mitogenic response to platelet-derived growth factor, Nature, 310:655-660, (1984).

30. L. Kaczmarek, J.K. Hyland, R. Watt, M. Rosenberg, R. Baserga, Microinjected c-myc as a competence factor, Science 228:1313-1315, (1985).

31. G.P. Studzinski, Z.S. Brelvi, S.C. Feldman, and R.A. Watt, Participation of c-myc protein in DNA synthesis of human cells. Science 234:467-470,

32. J.R. Feramisco, M, Gross, R. Kamata, M. Rosenberg, R.W. Sweet, Microinjection of the oncogene form of the human H-ras (T24) protein results in rapid proliferation of quiescent cells. Cell 38:109-121, (1984).

33. J.K. Hyland, C.M. Rogers, E.M. Scolnick, R.B. Stein, R. Ellis, R. Baserga, Microinjection ras family oncogenes stimulate DNA synthesis in quiescent mammlian cells. Virology, 141:333-338.

34. L. Kaczmarek, M. Oren, R. Baserga, Cooperation between the p53 protein tumor antigen and platelet-poor plasma in the induction of cellular DNA synthesis, Exp. Cell Res. 162:268-270, (1986).

35. T.J. Holt, T. Venkatgopal, A.D. Moulton, and A.W. Nienhaus, Inducible production of c-fos antisense RNA inhibits 3T3 cells proliferation. Proc. Natl. Acad. Sci. 83:4797-4798, (1986).

36. K. Nishikura and J.M. Murray, Antisense RNA of proto-oncogene c-fos blocks renewed growth of quiescent 3T3 cells, Mol. Cell. Biol 7:639-649.

37. W.E. Mercer, R. Baserga, Expression of the p53 protein during the cell cycle of human peripheral blood lymphocytes, Exp. Cell Res. 160:31-42, (1985).

38. J. Minler, F. McCormick, Lymphocyte Stimulation: Concanavalin A. induces the expression of a 53K protein. Cell Biol. 7:639-649, (1980).

39. J.C. Reed, D.J. Alpers, P.C. Nowell and R.J. Hoover, Sequential expression of proto-oncogenes during normal human lymphocytes mitogenesis, Proc. Natl. Acad. Sci. 83:3982-3896, (1986).

40. B. Calabretta, L. Kaczmarek, L. Selleri, G. Torelli, PM.L. Ming, S.C. Ming, and W.E. Mercer, Growth dependent expression of human p53 tumor antigen mRNA in normal and neoplastic cells, Cancer Res., 46:5738-5742, (1986).

41. W.E. Mercer, D. Nelson, A.B. DeLeo, L.J. Old, R. Baserga, Microinjection of monoclonal antibody to protein p53 inhibits serum-induced DNA synthesis in 3T3 cells. Proc. Natl. Acad. Sci. USA 79:6309-6313, (1982).

42. W.E. Mercer, C. Avignolo, R. Baserga, Role of the p53 protein in cell proliferation as studied by microinjection of monoclonal antibodies, Mol. Cell. Biol. 4:276-280.

43. D. Eliyahu, A. Ras, P. Gruss, D. Givol, M. Oren, Participation of p53 tumor antigen in transformation of normal embryonic cells. Nature 312: 646-649, (1984).

44. L. Parada, H. Land, R.A. Weinberg, D. Wolf, V. Rotter, Cooperation between gene encoding p53 tumor antigen and ras in cellular transformation. Nature, 312:649-651, (1984).

45. J.R. Jenkins K. Rudge, G.A. Currie, Cellular immortalization by a cDNA clone encoding the transformation association phosphorprotein p53. Nature 312:651-653, (1984).

46. E.H. Westin, F. Wong-Staal, E.P. Gelman, R. Dalla Favera, T.S. Papas, J.A. Lautenberg, A. Eva, E.P. Reddy, S.R. Tronick, S.A. Aaronson, and R.C. Gallo, Expression of cellular homologues of retroviral oncogenes in human hematopoietic cells. Proc. Natl. Acad. Sci. USA 79:2490-2494, (1982).

47. R. Eva, K.C. Robbins, P.R. Anderson, A. Srinivasan, S.R. Tronick, E.P. Reddy, N.W. Ellmore, A.T. Galen, J.A. Lautenberg, T.S. Papas, E.H. Westin, F. Wong-Staal, R.C, Gallo and Aaronson, S.A. Cellular genes analogous to retroviral onc genes are transribed in human tumor cells, Nature 295:116-119, (1982).

48. G.P. Rothberg, M.D. Erisman, R.E. Diehl, V.G. Rovigatti and S.M. Astriun, Structure and expression of the oncogene c-myc in fresh tumor material from patients with hematopoietic malignancies, Mol. Cell. Biol. 4:1036-1103.

49. D.S. Slamon, J.B. deKernion, I.M. Verma, and M.S. Cline, Expression of cellular oncogenes in human malignancies, Science 224:256-261, (1984).

50. R.A. Weinberg, The action oncogenes in the cytoplasm and nucleus, Science 230:770-776, (1985).

51. R. Baserga, The biology of Cell Reproduction (Harvard Univ. Press Cambridge, MA) (1985).

52. M.L. Mendelsonn, Autoradiography analysis of cell proliferation in spontaneous breast cancers of C3H mouse. III The Growth Faction, J. Natl. Cancer Int. 28:1015-1024, (1962).

53. F. Bresciani, R. Pauluzzi, M. Benassi, C. Nerve, C. Casale, E. Ziparno, Cell kinetics and growth of squamous cells carcinomas in man. Cancer Res. 34:2405-2415, (1974).

54. G.G. Steel, Growth kinetics of tumors. Clarendon Press, (1977).

55. R. Baserga, The cell-cycle. N. Engl. J. Med. 304:453-459, (1981).

56. B. Calabretta, L. Kaczmarek, W. Mars, D. Ochoa, C.W. Gibson, R.R. Hirschhorn and R. Baserga, Cell cycle specific genes differentially expressed in human leukemias, Proc. Natl. Acad. Sci. USA 82:4463-4467, (1985).

57. L. Kaczmarek, B. Calabretta, R. Baserga, Expression of cell-cycle-dependent genes in phytohemagglutinin-stimulated human lymphocytes. Proc. Natl. Acad. Sci. USA 82:5375-5379.

58. L. Kaczmarek, B. Calabretta, and R. Baserga, Effect of interleukin-2 on the expression of cell cycle genes in human T lymphocytes, Biochem. Biophys. Res. Comm. 133:410-416, (1985).

59. L. Kaczmarek, L. B. Calabretta, I.B. Elfenbein, and W.E. Mercer, Cell cycle analysis of human peripheral blood T lymphocytes in long-term culture. Exp. Cell Res. 173:70-79, (1987).

60. B. Calabretta, D. Venturelli, L. Kaczmarek, F. Narni, M. Talpaz, B. Anderson, M. Beran, R. Baserga, Altered Expression of G_1 specific genes in human malignant myeloid cells. Proc. Natl. Acad. Sci. USA 83:1495-1498, (1986).

61. S. Ferrari, F. Narni, W.L. Kaczmarek, D. Venturelli, B. Anderson, and B. Calabretta, Expression of growth-regulated genes in human acute leukemias, Cancer Res. 46:5162-5166, (1986).

THE HUMAN PLACENTAL LACTOGEN AND GROWTH HORMONE MULTI-GENE FAMILY

Hugo A. Barrera-Saldaña[1,2], Ramiro Ramírez-Solís[1], William H. Walker[3], Susan L. Fitzpatrick[3], Diana Reséndez-Pérez[1] and Grady F. Saunders[3]

[1]Unidad de Laboratorios de Ingeniería y Expresión Genéticas (ULIEG) del Departamento de Bioquímica. Facultad de Medicina de la U.A.N.L., Monterrey, N.L. México. And [2]The International Center for Molecular Medicine, Monterrey, N. L. México

[3]Department of Biochemistry and Molecular Biology M.D. Anderson Hospital and Tumor Institute. The University of Texas System Cancer Center, Houston, Texas, 77030 USA

INTRODUCTION

The central problem of molecular biology is the understanding of how the genetic information coded in the nucleic acid is expressed and, what are the mechanisms that regulate such expression. Much progess has been made in understanding gene regulation in prokaryotic systems, however, in eukaryotic organisms, the advances have been much slower and much more recent. This is closely related in part to the degree of evolutionary complexity of the eukaryotic cell.

The regulation of genetic expression is an essential characteristic of living cells. Of the total amount of genetic informacion possessed in all cells, only a small fraction is differentially transcribed within a certain time and space.

Studies on the expression of procaryotic genes showed that besides the region of DNA to be transcribed into RNA, sequences located both to the 5' side (operators, promoters, etc.) and 3' side (terminators) of this structural gene, are involved in regulating gene expression. In addition, according to the operon theory (1), for the regulation of the expression of bacterial genes, various elements are required such as: DNA-dependent RNA polymerase (the enzyme responsible for transcribing the gene); several types of protein factors associated with the enzyme (σ rho, etc.) as well as metabolites (inducers, repressors, etc.) which interact with DNA to regulate bacterial gene expression (1).

Eukaryotic cells have approximately 10^3 times more DNA than a bacterium, multiple forms of DNA-dependent RNA polymerases (at least three types, each transcribing a different subset of genes) and compartmentalization of the processes of transcription (in the nucleus) and translation (in the cytoplasm). In addition to these differences, the discovery in eucaryotes of split genes (2), and RNA processing, made it clear that regulation of eukaryotic gene expression is complex and can be exerted at a variety of different levels (3). The picture becomes even more complicated considering the organization of the eukaryotic genome into chromatin (4). In this regard, genes being expressed are said to be in an "active" chromatin configuration, although what determines this configuration is unclear.

THE PLACENTA AS AN EXPERIMENTAL MODEL

In evolutionary terms the placenta is recently acquired and a very efficient organ that functions to aid in the survival of the offspring and thus in the perpetuation of the species. Development of placental mammals allowed the mother to carry the unborn young with her while searching for food, thus protecting the fetus from predators. It also increased the area that could be covered during the searching process and facilitated migration. Thus, the placenta played a crucial role in the success of mammals colonizing the earth (5).

The placenta (6) is a remarkable organ in that is created from the same fertilized ovum that gives rise to the fetus but functions independently. The placenta exhibits unique characteristics that make it an excellent model for biological research. These properties include: a) its rapid growth and invasion of the maternal uterine tissue, b) the sudden stop of this invasion (by an unknown mechanism), c) the immunological processes that protect the placenta and the fetus from rejection, d) its hormonal regulation of pregnancy, etc.

In addition, the placenta is the organ with the highest rate of protein synthesis (7); and, since it develops and matures in less than 40 weeks, it constitutes an excellent system to study changes in gene expression during development and cell differentiation.

The placenta synthesizes a large variety of hormones (8). Probably the best characterized are chorionic gonadotropin (hCG) and placental lactogen (hPL, also know as chorionic somatomammotropin; HCS). While first trimester placental tissue is highly active in the synthesis of hCG with only low levels of hPL, in term placenta the situation is reversed (7). High levels of hPL (up to one gram per day) but low levels of hCG are produced by placenta at term.

In 1962, Josimovich and MacLaren (9) defined and characterized human placental lactogen as a polypeptide hormone present in extracts of human term placental and retroplacental blood that exhibited both potent lactogenic activity and an inmunochemical reaction of partial identity with human growth hormone. hPL maternal blood levels are used to reflect the functional integrity of the placenta during pregnancy (10). hPL influences mammogenesis and lactogenesis as well as many aspects of the maternal intermediary metabolism directly related to the supply of nutrients for the metabolism of the fetus (11). However, the primary action of this hormone has not been defined.

The hPL molecule is a single-chain polypeptide of 191 aminoacids, produced in the syncytiotrophoblast layer (12). It contains two intramolecular disulfide bonds and no carbohydrate or lipid.

The hPL production is coupled to the development of the placenta, reaching its maximum towards the end of pregnancy. The great quantities in which this hormone is produced, makes this hormone ideal for research and biochemical manipulations. For these reasons, we choose the placenta as the ideal organ to carry out the studies described in this paper.

RECOMBINANT DNA: A NEW AND POWERFUL TECHNOLOGY TO STUDY GENE STRUCTURE AND EXPRESSION

The birth of recombinant DNA technology in the early 1970's, marked the beginning of a new era in Molecular Biology. Recombinant DNA simply means the recombination in the test tube of different DNA molecules. This technology has provided us with very powerful tools and methods for the isolation, characterization and manipulation of gene sequences.

With the aid of this technology, it has been possible to begin the analysis of highly complex genomes of eukaryotic cells at the molecular level. Studies are being carried out to analyze the molecular structure and organization of genes in order to understand their function, regulation and origin.

The essential elements that constitute the group of recombinant DNA thechniques include:

I. Enzymes to modify DNA and RNA

Such as restriction enzymes, ligase, phosphatase, reverse transcriptase, DNA polymerase, polynucleotide kinase, etc. These proteins are employed to carry out the manipulation process of the DNA to be cloned. These genetic manipulations consist of specific cleavages along the DNA molecule or, modifications, covalent unions, radioactive labeling, etc.

II. Molecular Hybridizations

They can be performed using a liquid or solid support. This technique (13) consists of the detection, through the use of radioactive probes containing complementary sequences, of desired molecular species that are present in complex mixtures of DNA or RNA. These radioactive probes when denatured and later renatured in the presence of the mixture, form molecular hybrids with the desired single DNA chain or RNA. This coupling is stable due to the establishment of hydrogen bonds between the complementary nucleoside bases of the hybrid molecule.

III. Molecular vehicles

They include: plasmids (14), cosmids (15), lambda bacteriophage derivatives (16) and M13 bacteriophage derivatives (17). These are used to clone foreign DNA fragments (such as: human genes) and permit its propagation in bacteria; thus exploiting the following three qualities:

a. DNA fragments can autonomously replicate in host cells as they are inserted into vectors containing replication origins.

b. They can be separated from the bacterial nucleic acids and easily purified.

c. They contain DNA regions that are not essential for its propagation in bacteria. Foreign DNA inserted in these regions is replicated and propagated as if they were a normal component of the vector.

IV. Determination of the sequence and synthesis of DNA

This can be carried out either by enzymatic (18) or chemical (19) techniques. The sequencing methods for DNA (or including RNA) generate a great quantity of valuable information concerning the primary structure, organization, regulation, and evolution of the genes and proteins which they code. Thanks to these techniques, it is now much easier to determine the amino acid sequence of a protein through the sequence of its cloned messenger RNA (transformed to DNA through reverse transcription). On the other hand, it is possible to sequence a part of a protein whose gene we wish to isolate and characterize. With the information of the amino acid sequence and genetic code, it is possible to synthesize an oligonucleotide capable of serving as a probe to carry out molecular hybridizations. In this manner, the desired gene is isolated and characterized from a gene bank.

V. Gene library or bank

It is created (20) starting from either plasmids, bacteriophages, or cosmids; all containing either natural genes (gene bank) or DNA complementary (cDNA) to the messenger RNA population of a particular tissue or cell (cDNA bank). Basically, what is done with these banks is to take advantage of the classical bacterial models that have been so useful to elucidate the molecular basis of the regulation and expression in prokaryotics, to study the molecular genetics of higher organisms.

Briefly, to clone and isolate a gene, the following steps are carried out:

1. The genome under study is isolated

2. It is then fragmented by the use of restriction enzymes

3. The resulting DNA fragments are introduced in molecular vehicles, thus constructing gene banks.

4. A radioactive probe containing a complementary sequence to the gene portion that we propose to isolate, is used to identify the clone that contains the gene under study.

5. Sufficient quantities of the desired gene is purified.

6. The desired gene is characterized.

After carrying out these steps, the gene is used to perform the pertinent studies which help us understand the evolutionary history and gene functions in the living organism.

From the numerous studies carried out concerning the molecular structure and organization of the genes in eukaryotics, the term, split gene, has emerged (2). The majority of the genes in higher organisms are discontinuous. This means that the DNA which codes for the protein is interrupted by non-coding regions known as introns. These introns form part, as well as the code regions (or exons), of the primary transcript of the gene; but, are eventually eliminated to produce the mature mRNA which later is translated into a specific protein. In fig. 1, the molecular anatomy of the split gene model is outlined. Besides the introns and exons, some DNA regions or sequences are described. These DNA regions are also important to achieve a precise and efficient expression of these type of genes.

The experiments described in this section, were possible due to the development of Recombinant DNA techniques. Starting from the idea of exploiting the human placenta as the experimental model to study the mechanisms which regulate the specific genetic expression of the tissue, and using recombinant DNA techniques; our effort was focused towards the biogenesis of the most abundant protein in this organ, the Placental Lactogen hormone (hPL).

The particular objectives of our experiments were to identify the components and to elucidate the different steps involved in the genetic flow of information responsible for the synthesis and regulation of this polypeptide hormone.

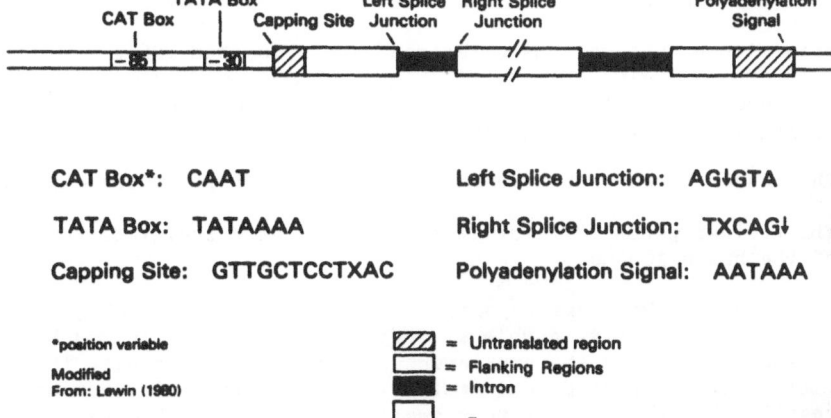

CAT Box*: CAAT   Left Splice Junction: AG↓GTA

TATA Box: TATAAAA   Right Splice Junction: TXCAG↓

Capping Site: GTTGCTCCTXAC   Polyadenylation Signal: AATAAA

*position variable

Modified
From: Lewin (1980)

▨ = Untranslated region
☐ = Flanking Regions
■ = Intron
☐ = Exon

Fig. 1.   Canonical protein-encoding mammalian gene. The figure shows the structural and regulatory sequences that characterize this type of genes.

In 1978, we started to investigate the genetic expression of hPL by trying to isolate and characterize the DNA dependent RNA polymerase type II of human placenta. This enzyme is responsible in carrying out the transcription of the genes that code for proteins. Because of this, it plays a key role in the functioning of the cell. However, this approach was somewhat premature since we first had to establish how to isolate and then biochemically and structurally characterize placental nuclei, which were to be our source of the enzyme. Once we accomplished this requisite (21) the isolation of the enzyme turned out to be a difficult project. Besides the inherent difficulties related with the technical procedures to be used, problems related with endogenous proteases and a lack of knowledge of the quaternary structure, were added on to our already existing problems. In spite of this, we were able to partially purify the enzyme and at least four of its subunits were detected (22). This was achieved through the use of electrophoresis in polyacrylamide gels with sodium dodecil sulphate (SDS).

With the coming of the DNA recombinant technology, the manner to attack the problem acquired a new focus. The molecular cloning of genes of higher organisms and the use of genetic engineering caused a revolution in the laboratories dedicated to the study of the gene expression and regulation. With a new plan in mind, adding these new techniques to the existing ones in our laboratory, we began to identify the elements that were involved in the different regulation levels of the biosynthesis of hPL (23). Briefly, we will mention our experimental strategy which can be used as a useful guide to present similar studies.

1) The in vitro synthesis of hPL and of its precursor form (pre-hPL) was studied (24).

2) The mRNA and pre-mRNA's for hPL were characterized (24).

3) The construction of a cDNA bank (DNAs which were copied from mRNA using the reverse transcriptase enzyme) was performed using terminal human placenta mRNAs (24).

4) The cDNA derived from the hPL mRNA was isolated (24).

5) This cloned cDNA was characterized, and sequenced (25).

6) The chromosomal localization of the hPL and hGH genes was carried out by in situ hybridization (26).

7) The number of genes for these two hormones was estimated (26).

8) The nuclear genes (27) for these hormones were isolated from a library of the human genome.

9) The cloned cDNA was hybridized against cloned hPL genes to form heteroduplexes (i.e. double stranded DNA molecules formed between one strand of cDNA and the complementary strand of the gene). The examination of these hybrid DNA molecules with the electron microscope, revealed the presence of four small introns in the hPL genes (24).

10) Two genes for hPL, two for hGH and one hPL-like gene were identified (23).

11) The expression of the genes for hPL were analyzed (25).

12) The hPL - like gene was sequenced showing similarities with the genes for hPL.

13) Elements involved in regulation of hPL transcription were identified.

The experimental projects and their results are described as follows:

MOLECULAR STRUCTURE

Molecular cloning of the mRNA for hPL

Using the same experimental model (the placenta) but with more powerful methods at hand, we proposed new and more ambitious questions:

Why is it that the placenta at term contains four to five times more translatable mRNA for hPL than first trimester placenta?. What were the possible mechanisms involved in the specific regulation of the expression of the gene (or genes) for hPL?.

In an effort to respond to these questions, we decided to study the structure, abundance and origin of the mRNA for hPL.

Projects carried out by other investigators (28) indicated that the mRNA for hPL should contain approximately 900 nucleotides. In addition, a fragment of 550 base pairs corresponding to a portion of DNA complementary to the hPL mRNA (synthesized using reverse transcriptase and the DNA polymerase I of Escherichia coli) was already cloned in plasmid (29).

We isolated the total nucleic acids from term placenta. The high molecular weight RNAs were purified by the use of selective precipitation with 3M sodium acetate, pH 5.2. From these RNAs, the mRNA were selected through affinity chromatography in columns of oligo-dT-cellulose. This was achieved by exploiting the characteristic property of mRNA's possessing poly A "tails" in the 3' end.

The RNA messengers were translated in a cell-free system prepared from rabbit reticulocytes and mouse cell cultures. The synthesized proteins (which were radioactively labeled) were analyzed by the use of electrophoresis in polyacrylamide gels with SDS. By carrying out immuno-precipitation using anti-hPL serum two bands were observed: one that co-migrated with purified hPL and another more prominent, that most likely represented a pre-hPL (i.e. the immature form of hPL containing the signal peptide). The sum of these two bands represented approximately 15% of the total radioactively labeled protein.

When the RNA messengers were analyzed through electrophoresis in urea-acid-agarose gels, a prominent band of approximately 860 nucleotides was observed (24). A band of the same magnitude was observed when a recombinant plasmid, which contained the cDNA fragment of 550 nucleotides of hPL (30), was labeled with ^{32}P and hybridized against total RNAs fixed within filters. When the nuclear RNA was analyzed using this same method, four additional bands were observed of 990, 1200, 1460 and 1760 nucleotides. These most likely are the hPL mRNA precursors. The RNA messengers were also used to construct a cDNA bank.

Approximately 5% of the recombinant clones, which constitute the bank, hybridized with sequences of hPL DNA. This indicated that the RNA messenger for hPL is certainly abundant in terminal placental tissue. One of the clones which scored positive in the hybridization, contained a cDNA of approximately 815 base pairs. This clone was isolated and characterized by the use of restriction enzymes forming the map seen in fig. 2.

cDNA molecules were hybridized with molecules of the hPL gene to construct what is known as a heteroduplex. The electronic microscope analysis of the heteroduplex reveled the presence, in the hPL gene, of four small introns or intervenient sequences (fig. 3) which explains the presence and sizes of the four precursors for the hPL mRNA present in the nuclear RNA.

The importance of these results is that they established for the first time the following:

1. The complete map of the recognition sites of the restriction enzymes present in the complementary DNA of the mRNA for hPL.

2. The existence of precursors (pre-mRNA) of the mRNA for hPL.

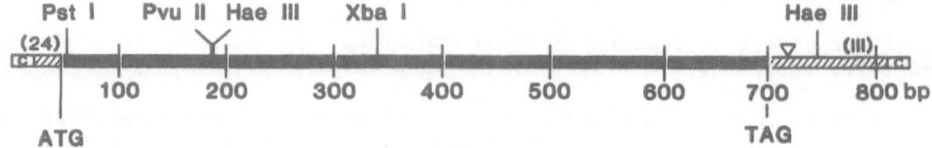

coding region

untranslated regions

dC-tail

▽ tetramer insertion

Fig. 2.   Map and organization of the hPL cDNA insert of phPL815. It contains the 651 nucleotides (nuc.) coding for the 26 aminoacids of the signal peptide and 191 aminoacids of the mature hPL hormone. In addition, the insert also includes 24 nuc. of the 5' unstranslated region and 111 nuc. of the 3' untranslated region.

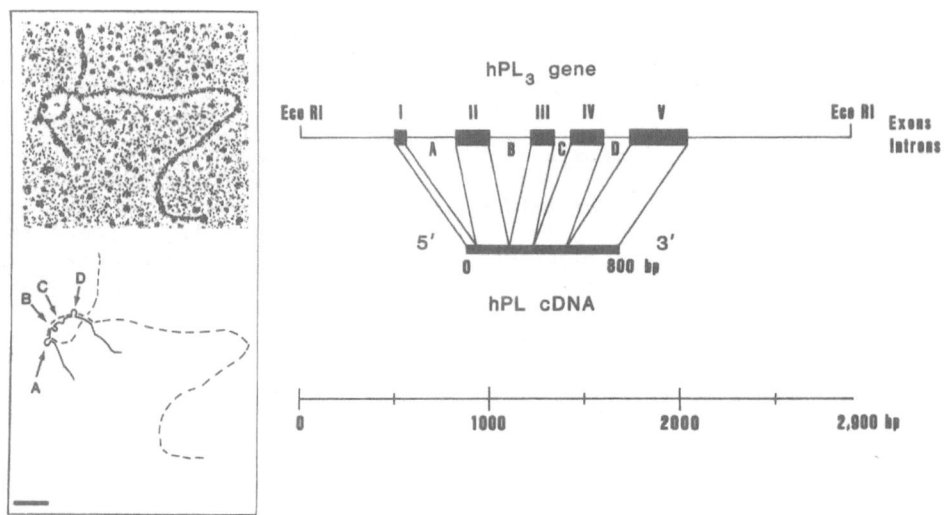

Fig. 3.   Molecular structure of the human placental lactogen genes. The information obtained from the analysis of 16 heteroduplexes, as the one of the electron micrograph at the left, was used to determine the molecular structure of the human placental lactogen genes (represented by hPL₃). The heteroduplexes were formed by hybridization of plasmid DNA (phPL815, dashed line) containing the cDNA to hPL mRNA, with the DNA containing the hPL₃ gene (solid line). Four small intervening sequences are detected in the heteroduplex region at positions indicated by arrows labeled A-D. Magnification is indicated by bar length of 0.1 μm in lower left corner.

3.   The presence of four introns in the gene for hPL.

Chromosomal localization of the hPL-hGH gene complex

On which chromosoma are the genes for hPL located on and how many are coding for this hormone?.

To directly visualize the genes in their chromosomal location(s) an experiment involving hybridization of DNA in metaphase chromosomes with a radioactive labeled probe (^{3}H) was carried out. The probe consisted of a portion of complementary DNA to the messenger RNA for the (hPL) which was cloned in the molecular vehicle known as pBR322 (14).

The molecular hybridization was carried out in the presence of dextran sulphate 10%, which accelerates by ten- fold the velocity of the process of hybridization.   Another important factor which decisively contributed to our success in the chromosomal localization of genes with very few copies per genome, was the presence of the DNA chains of the vector covalently linked to the insert. These single chain DNAs hanging from the site of hybridization where the insert is hybridized   served as an anchor for multiple hybridizations between complementary chains of the denatured plasmid, resulting in an increased accumulation of specific radioactivity.

The number of copies of genes for hPL and for human growth hormone (hGH) was determined through hybridization experiment on nitrocelullose filter. In this experiment, samples of chicken cell DNA were tested in parallel mixed with quantities of the recombinant plasmid (containing the cDNA for the messenger RNA of hPL) calculated in such a way that they represented 1,2,4 and 6 gene copies per human haploid genome.

Even though the experiments described above involve DNA sequences for hPL, the results could also be applied to the hGH genes. These two hormones are very closely related with respect to their evolution and their respective DNA demonstrate a high degree of sequence similary; due to this, they easily hybridize with one another. It was discovered that there existed approximately three genes for hPL and three genes for hGH per human haploid genome and all of them were located in the chromosomic segment known as 17q22-24 (long arm of chromosome 17)

These experiments demonstrated the great utility of the hybridization in situ for the mapping of genes found in very few copies in the human genome.   Besides, they demonstrate for the first time, the subchromosomal localization of the genes for hPL and hGH.

Isolation and characterization of the members of the multi-gene family

Having verified the number, the chromosome localization, the mole-cular anatomy of the hPL and hGH genes, and counting with the complete messenger RNA for hPL made from cDNA and finally cloned; we proceeded to isolate clones from the human gene bank (27). These clones contained complementary sequences to a portion of the cDNA for the mRNA of hPL cloned (30) into pBR322. This recombinant plasmid, proportioned by Dr. Peter Seeburg, is known as pBR322-HCS, (HCS stands for Human Somatomammo-tropin, name also given for hPL).

Kidd took DNA from the recombinant plasmid, labeled it with ^{32}P, denaturalized it, and immediately hybridized it against phage recombinant DNAs originating from the gene bank. From a total of 900,000 plaques of phages analyzed, the seven that resulted positive, were grown and their DNAs purified. The DNA obtained from each recombinant phage, was characterized using restriction enzymes and hybridized against the hPL cDNA to locate regions containing hPL or hGH genes. Seven different genes were identified: hPL_1, hPL_2, hPL_3, hPL_4, hGH_1, hGH_2 and hGH_3.

In at least two phages, we were able to verify the connection between a hPL and a hGH gene. In addition, in another phage, we confirmed the linkage between two hPL genes. This indicated that probably all the genes were related with each other; thus suggesting, that all of these genes evolved from a common ancestor and that they originated by mechanisms of gene duplication and diversion. In more recent studies we established that hPL_2 and hPL_4 were the same gene, reducing to three the number of genes for hPL (hPL_1, hPL_4 and hPL_3). Also, the existance of the hGH_3 gene was not confirmed.

The Gene Structure

The molecular structures of the genes were obtained by four different methods. These were: 1) hybridizations on nitrocellulose filters (13) of the labeled cDNA (probe), against DNA from the fragmented genes obtained by the use of restriction enzymes. 2) comparing maps that contained various restriction enzyme cutting positions carried out for every gene, 3) confirming information obtained by determining the nucleotide sequence of the regions that flank the initiation and termination points of the genes, as well as, the borders between the exons and introns. 4) and finally, obtaining formation from the literature concerning the sequences of the cDNA and the gene for growth hormone, described by other groups of investigators (31,32). Four of the genes showed a very high degree of nucleotide sequence similarity with each other, as well as quite similar restriction enzyme maps. However, by: 1) the presence of repeated sequences adjacently to the 3' end of the gene, 2) the length of the fragments flanked by EcoRI sites and 3) the presence of characteristic restriction sites, it was possible to distinguish and identify each of these genes fig. 4). For example, while the two hGH genes are contained within Eco RI fragments of 2.6 kilobases (kb), both contain repetitive sequences near their 3' end region. Furthermore, both have unique sites for BglII. Even though these genes possess these similarities, one of them (hGH_1) has only one BamHI site in its fourth intron. As another example, we can mention the following. Two of the genes for hPL are contained within fragments Eco RI of 2.9 kb and possess unique sites for both XbaI and BamHi. However, they can be distinguished from each other by the presence of one (hPL_3) or two (hPL_4) PvuII sites. As a final example, the fifth characterized gene (hPL_1) is contained in a EcoRI fragment of 8.5 kb, which can be cut with XbaI, liberating a fragment of 3.5 kb containing the gene. In addition, this gene is characterized by the absence of XbaI and BamHi sites.

Molecular Anatomy of the Multi-gene Complex

The results obtained through the in situ hybridization experiments indicated that the genes were grouped within the region between the bands q22 and q24 of chromosome 17. Due to the quantity of DNA contained within this chromosomal region (several millions of base pairs), little could we deduce concerning the organization of these genes. Thus various questions arise: how close are these genes with respect to each other? What is the spatial relationship that the hGH genes have with respect to one another and with the hPL genes? Is each gene transcribed from the same DNA chain thus maintaining the same sense of transcript direction; or, do members exist that are transcribed from the opposite chain?

42

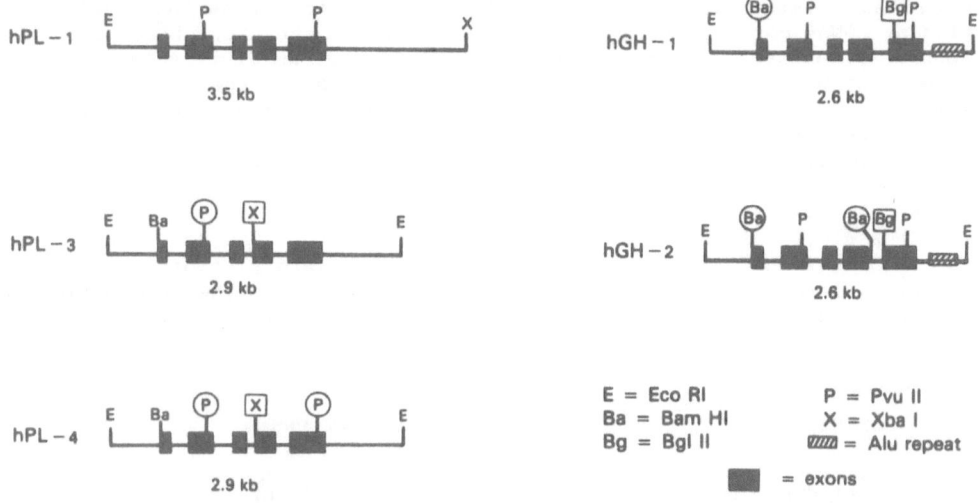

Fig. 4. The members of the hPL-hGH multi-gene family. Those
enzymes useful to destinguish the different genes
are in circles or boxes. hGH-1 is the same as hGH_N,
while hGH-2 is also known as hGHv.

To solve these inquiries not only the data obtained from the isolated
recombinant phages were analyzed; but, also we used information from other
laboratories who were also working on the same theme. The data obtained
from the molecular genetic analysis of patients who presented a congenital absence of hGH or hPL was also very useful to us. For example, a
patient who presented with an absence of hPL (see Fig. 5), was shown to
possess a deletion of approximately 35 kb with the loss of the genes hPL₄,
hGH₂ and hPL₃. It was concluded that at least these three genes were quite
close to each other. Having found recombinant phages containing two genes
which were always one of the hPL type and the other of the hGH type, was
particulary valuable in the establishment of the molecular anatomy of the
genetic complex for hPL and hGH.   This indicated that not only are the
genes close to each other but, that they are also intermixed with respect
to their spatial arrangement along the chromosome.   In fig. 5, the map
proposed for the genetic complex is illustrated. This map was later
confirmed by isolating cosmid clones containing all the gene members of
the hGH-hPL gene complex (33).

GENETIC EXPRESSION

hPL gene expression

What is the reason for the multiplicity of the hPL genes? What might
be the function of these genes? Might it be that all are transcriptionally
active in the term placenta or, are some of them pseudogenes? To answer
these questions the following experiments were designed.

Fifteen recombinant plasmids were selected from a cDNA bank of term
placenta RNA, that tested positive when they were hybridized to detect
complementary sequences to hPL cDNA. Having knowledge of the map indicating the characteristic cleavage sites for the restriction enzymes of each
hPL gene (see fig. 4); we proceeded to cut the DNAs of these 15 plasmids

with the restriction enzymes. This was performed with the objetive of
finding cDNAs corresponding to hPL transcripts of the hPL₁, hPL₃ and hPL₄
genes. By using this method, these genes could be distinguished by the
absence or presence of certain sites of the diagnostic restriction enzymes
for each gene. We were able to detect plasmids with insert characteristic
of hPL₄ and hPL₃ but not however for the characteristics expected for a
transcript of hPL₁. In this manner, we were able to establish for the
first time, that hPL₄ and hPL₃ were transcriptionally active in term
placenta; and that possibly, hPL₁ was a pseudogene or that it is expressed
in other stages of the placental development. Both cDNAs were characterized and sequenced. Their representation in the term placenta, determined
by three different experiments, was estimated to be in the proportion of
two to three for the RNA messengers of hPL₃ and hPL₄ respectively.

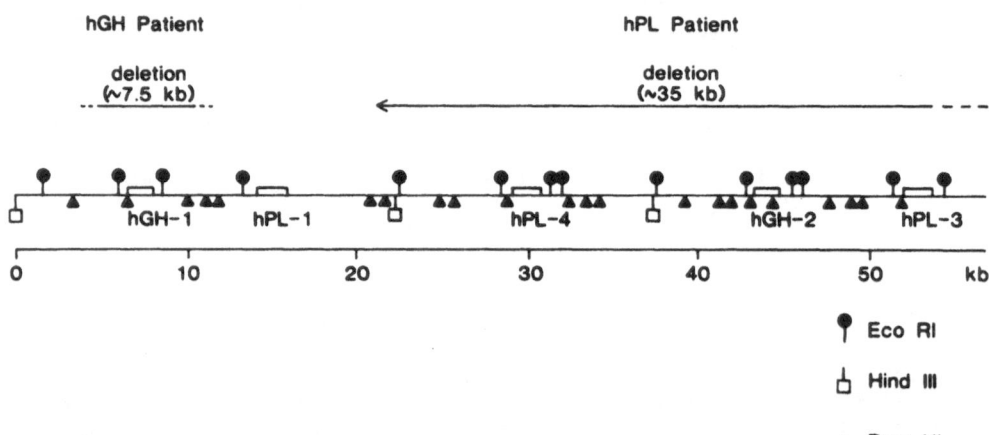

Fig. 5.   Linkage map of the hPL-hGH multi-gene cluster.
          This map was constructed using information from
          restriction enzyme analysis of recombinant phages
          containing hPL and hGH genes, and the analysis of
          information from a molecular study of DNA from a
          patient with antenatal deficiency of hPL (see
          ref.46). The areas believed to be deleted in this
          patient and in a patient with familial isolated
          growth hormone deficiency are indicated above the
          map.

     Finally, comparing the sequences of both cDNAs, 10 differences in the
nucleotide positions were detected; although only one of them, cuased a
change in the amino acid sequence. This change is located in the signal
peptide, which means that when this protein is processed before secretion,
the mature protein coded by each of the two different genes is identical.

DNA sequences involved in regulation of the hPL gene.

     Most cellular and viral genes have been shown to contain cis-acting
sequences which regulate transcription of the gene. Some of these sequence
include the TATA box and CAAT box located in the upstream promoter
regions. Also hormone receptor sites, enhancers, and sequences which bind
general or tissue specific factors can alter gene expression. These
regulatory regions may be found upstream, downstream or within a gene.

44

We surveyed the entire hPL-hGH gene cluster for the presence of transcriptional enhancers (34). The gene cluster was digested with EcoRI and each restriction fragment was tested, using a transformation assay, for enhancer activity in hPL producing choriocarcinoma cells. The only enhancer detected was located 2 kb 3' to the hPL₃ gene (fig. 6).

A vector construct (fig. 7) containing the hPL enhancer linked to the reporter gene chloramphenicol acetyl transferase (CAT) under the control of the SV40 promoter was used in transfection studies to further characterize the hPL enhancer. These studies showed the enhancer was tissue specific as it was active only in cells which produce hPL. Transfection studies using choriocarcinoma cell lines have sublocalized the hPL enhancer to a 730 bp AccI-AvaI restriction fragment.

Deletion analysis of the enhancer suggests that multiple sequences throughout the 730 bp enhancer are necessary for enhancer activity. One region may be more important than others as over 60% of enhancer activity was found within the 5' most 210 base pares (bp). Protein binding studies suggest another 265 bp region (region II, fig. 6) may also be important. This region specifically binds protein found only in placental nuclear extracts. Regions I and III bind nuclear proteins common to both cells which produce hPL and cells which do not produce hPL.

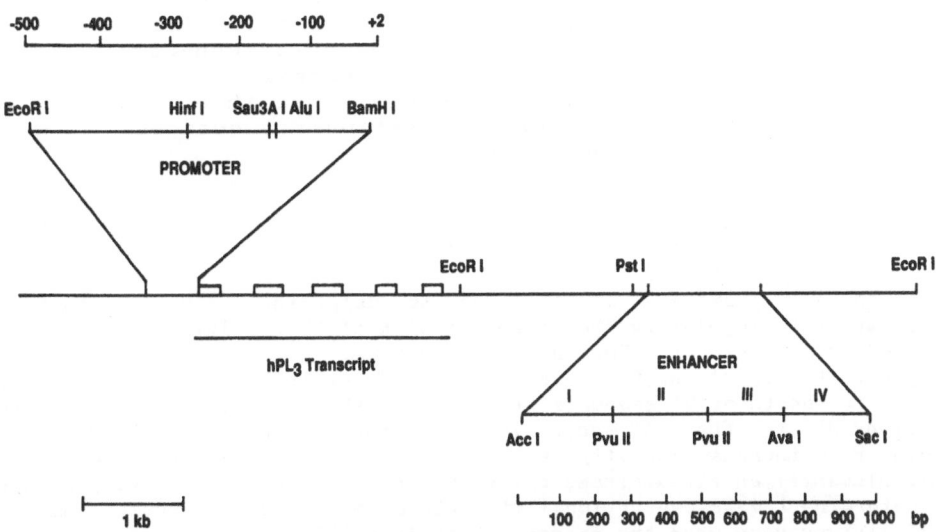

Fig. 6.   hPL gene and regulatory sequences. Restriction enzyme digestion maps of both the promoter and enhancer regions are shown. Boxes correspond to hPL₃ exons. Enhancer restriction fragments tested for nuclear protein - DNA interactions are denoted as I, II, III and IV.

DNA sequences extending 500 bp upstream of the hPL and hGH genes are 95% homologous. When these 500 bp regions were inserted 5' to the CAT gene they were each shown to have low promoter activity. However, when the hPL enhancer was linked to either the hPL or hGH promoter, a marked difference in transcription activity was noted. Transcription activity of the hPL-promoter hPL-enhancer pair was 10-fold higher compared to that of the promoter alone, and 5 fold higher than the hGH-promoter hPL-enhancer combination. This suggests that the hPL enhancer may act preferentially on the hPL promoter.

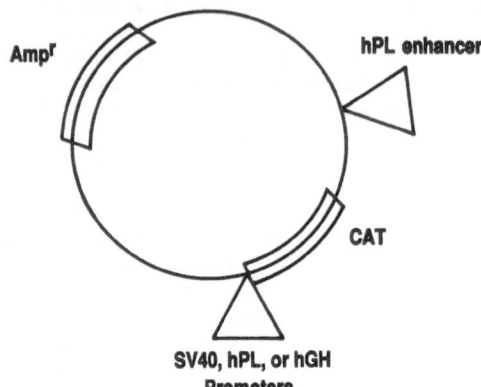

Fig. 7. CAT vector for hPL promoter and enhancer studies. Plasmid vector containing the promoter less bacterial gene for chloramphenicol acetyl transferase (CAT). Promoter sequences were inserted 5' to CAT and enhancer sequences added 3' to CAT.

The hPL promoter was analyzed to determine sequences that are important for regulating the transcription of hPL. Deletion mutants were created by digesting the 500 bp hPL promoter with various restriction enzymes. The resulting DNA fragments all contained the same 3' end but varied in length of 5' sequence (fig. 8). The deletion fragments were ligated 5' to the CAT gene in a vector that also contained the hPL enhancer to increase activity (see fig. 7). The plasmids were transfected into placental choriocarcinoma cells and the level of CAT activity assayed to determine the transcriptional strength of each promoter mutant. Maximal activity was seen when the promoter contained 390 bp of DNA and decreasing the size of the promoter to 152 bp did not lower this high activity. Initial experiments showed lower activity when the promoter fragments contained 2300 bp or 500 bp of upstream DNA sequence. A pronounced decrease (7-10 fold) in CAT activity was seen when the promoter was reduced to 129 bp. This low activity was also seen when the sequences between -152 bp and -129 bp were removed from the 390 bp fragment (390 SA). A new 142 bp clone was prepared and it had activity similar to the 152 bp clone indicating that an important region for transcription is contained between the DNA sequences -142 and -129 bp. This sequence may be important for binding trans-acting factors which stimulate the transcription of the hPL gene.

The hPL genes contain sequences both 5' and 3' to the genes that are important for its transcriptional control. Further study of these regions may elucidate their mechanism(s) of action in the regulation of hPL expression.

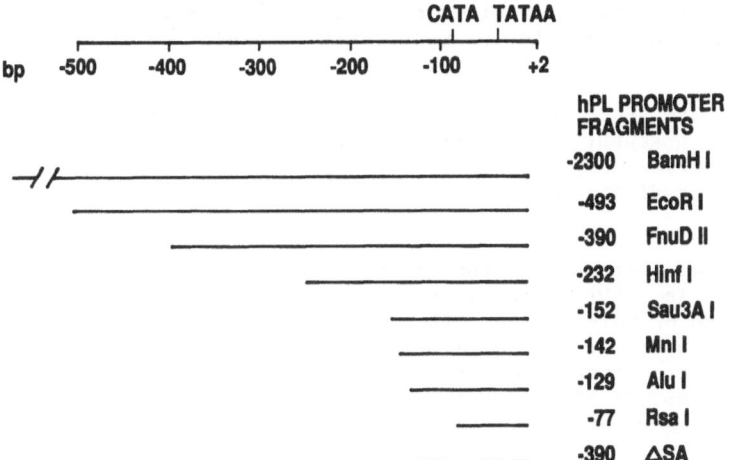

Fig. 8. Deletion mutants of the hPL promoter. Deletion mutants of the hPL promoter were created by digesting the DNA with various restriction enzymes. The fragments (solid line) all had identical 3' ends but decreasing amounts of 5' DNA.

Is hPL₁ a pseudogene?

What was the nature of the hPL₁ gene? What type of protein was coded in its nucleotide sequence? Was this gene, whose transcriptional products we could not find in term placenta, potentially functional? or did it contain mutations that rendered it as a pseudogene? To obtain an insight into these uncertainties, we decided to dissect the hPL₁ gene.

A 3.5 kb EcoRI - XbaI DNA fragment containing the hPL₁ gene was digested with various restriction enzymes. Aliquots of each of these digestions were ligated to M13 vectors that had already been cut with the appropriate restriction enzymes to generate compatible ends for cloning. Recombinants for each of the different fragments spanning the gene were identified and their single stranded DNAs were purified. The method of Sanger (18) was used to determine the sequence of the inserts in these M13 recombinants. The analysis of the hPL₁ sequence indicated that the characteristics of this gene are very similar to those of the other members of hPL-hGH gene complex. However, in spite of these similarities this gene possesses unique features. Both, the common and unique characteristics of the hPL₁ gene with regard to the other hPL genes are listed below.

1. It contains five exons of identical length to those in the other hPL and hGH genes.

2. It contains normal TATA (-30) and CAT (-84) boxes.

3. It contains a normal polyadenylation signal which is followed by a characteristic truncated Alu element located 100 nucleotides further downstream.

4. The protein that it encodes differs in 16 amino acid positions, as compared to the mature hPL protein coded by the hPL_4 and hPL_3.

5. The hPL_1 gene contains a G to A transition at the 5' splice consensus site of the second intron; thus, preventing pre-mRNA processing at this site and classifying hPL_1 as a pseudogene candidate.

When we searched for the presence of hPL_1 gene cDNA clones in a human term placenta cDNA library, we did not detect them (25). The search for transcripts of this gene by other group of investigators (35), who used hPL_1 gene specific oligonucleotides, was also unsuccessful. This evidence along with the presence of the point mutation at 5' splice site of the second intron of hPL_1, that in other genes (36) has been proven to be a cause of gene inactivity; strongly suggested that this gene was a pseudogene.

More direct experimental evidence was required to reach a definitive conclusion regarding hPL_1 gene expression. We decided to test in a transient expression experiment, if the splice site point mutation in hPL_1 rendered it unable to generate a mature mRNA. We rationalized that adequate controls for this experiment should be to introduce the mutation present in hPL_1 into a normal hPL gene and see if it now becomes defective in the production of mature mRNA, and to revert to wild type the mutation in hPL_1 to see if it now becomes active. The assumption behind these experiments was that the point mutation was the only cause of hPL-1 apparent gene inactivity. Next, we went on testing this assumption.

Comparisons of the nucleotide sequences around the region of the 5' splice donor site, contained in the active hPL-3 gene and the putative hPL-1 pseudogene, made it clear that we could easily exchange the mutated area between these two genes. We found PvuII and SacI sites located 30 base pairs upstream and 86 base pairs downstream of the mutation site, respectively. The sequences in this restriction fragment, of approximately 120 bp, differs among the two genes in only four nucleotide positions. In addition to the single point mutation in the 5' donor splice site, there were three other nucleotides sites within this region where these genes differ. The additional changes, however, are located inside the intron at 6,27 and 64 bp downstream from the exon-intron border, positions that could be taken as of little importance for the pre-mRNA processing.

We carried out the shuffling of this restriction fragment among hPL_1 and hPL_3 genes. To our advantage, the differences between these two genes inside the second intron sequences, were associated with an AluI site. Thus, the successful exchange of the PvuII-SacI fragment could be monitored by AluI digestions of the recombinant hybrid genes.

Furthermore, to leave no doubt of the identity of the recombined genes, we determined their nucleotide sequence in the area around the exchange site. In this manner we were sure we had our recombined and wild type genes to test our hypothesis.

We next constructed a new plasmid vector derived from pCMVCat (a plasmid containing cytomegalovirus enhancer-promoter control sequences fused to chloramphenicol acetyl transferase structural gene). This construction was shown previously to be a powerful expression system (37). Our new vector called pAVE-1 contains: pBR322 (14) sequences from the EcoRI site to the AccI site; the enhancer-promoter sequences of cytomegalovirus (37); and the polylinker region of pUC18 (38). We then

subcloned into pAVE-1 the structural regions of our test genes and were
ready to assay their expression (fig. 9).

To determine the expression of our hybrid genes, we choose to introduce them into cultured cells (transient expression-assay) by the technique of calcium-phosphate-DNA coprecipitation (39). The efficiency of
transfection was evaluated by using the CAT assay carried out on a
fraction of the cultured cells cotransfected with both the test gene and
pCMVCat.

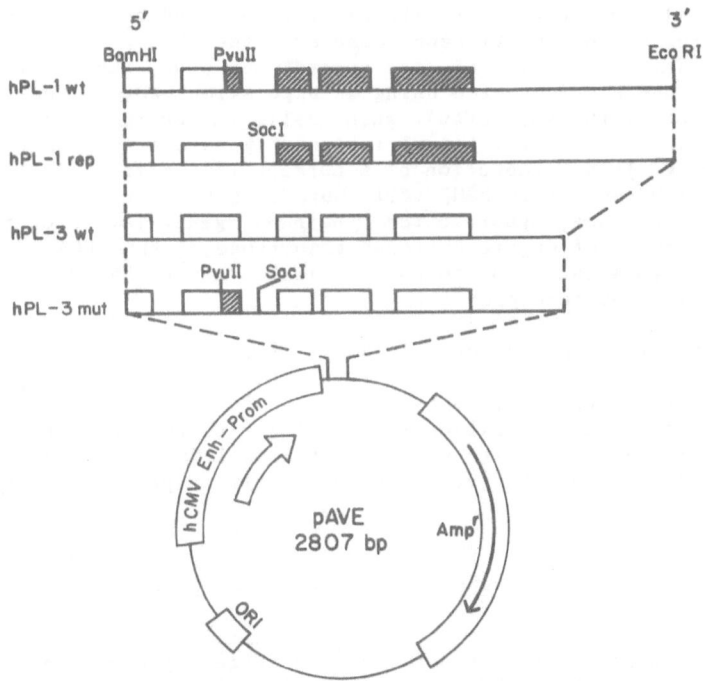

Fig. 9. Subcloning of hPL genes into the
expression vector. The hPL$_3$ genes
were directly cloned in pAVE-1 from
their Bam HI site at beginning of
the first exon to the Eco RI site
at the 3' end of the gene. A
different strategy was used for the
hPL-1 genes since they lack the
Bam HI site. To clone them, we
replaced the 5' end of these genes
with that of the hPL-3 gene, from
the Bam HI site to the Pvu II in
the second exon.

We isolated total RNA from the remaining transfected cells by the
method of guanidinium isothyocianate-lithium chloride (40). Then we
glyoxalated the RNA and separated it in an agarose gel. We transferred
the RNA to a nitrocellulose membrane and hybridized it with an oligolabe-
led hPL cDNA probe. The preliminary results from this type of analysis
indicated that the wild-type hPL-1 gene is not capable of producing a
mature mRNA and that when its mutated splice donor site is substituted
with a normal one this gene deffect is reversed. On the other hand, the
wild-type hPL-3 gene produces good levels of hPL mRNA and which are

drastically decresed when this gene harbors the mutation present in hPL-1 gene.

Therefore, we concluded that hPL-1 is a pseudogene because, even when it has a high similarity to the hPL active genes, does not produce a mature mRNA.

hGH gene expression

Even though the 22 kilodaltons form of human growth hormone, the one which normally circulates in the blood, is coded by the gene located in the 5' region of the multi-gene complex (the hGH_1, or hGH_N, N being "normal"). The expression of the second hGH gene, hGH_2 or hGHv (v being "variant"), has been detected using an expression vector derived from the monkey simian virus 40 (SV40). When cells in culture are infected with this recombinant virus the transcription of gene can be observed; and as result, there is a production of a hormone that differs in 13 amino acid positions with respect to hGH_1 (41). More recently, by the use of synthetic oligonucleotides specific for the hGH_2 gene, levels of hGH_2 mRNA at least four orders of magnitude lower than those of hPL mRNA, were detected in term placenta (42). Up to now, it is not known what possible function may have this hGHv gene product.

The complications do not end here since the primary transcription product of the hGH_1 gene, is regulated in its processing. Due to the presence of two alternative splicing sites in the elimination of one of its intron (43); the additional mRNA that is produced, gives rise to a variant of the normal growth hormone (of 20 kilodaltons), which represents approximately 10% of the growth hormone activity present in the pituitary gland.

Expression and functional analysis of transfected hGH and hPL genes

hGH and hPL are the two well characterized products of the hGH and hPL multigene complex. They share 85% of their aminoacids. hGH is unique among the animal growth hormones in that it possesses prolactin-like activity. hPL also shows prolactin-like activity and in spite of its sequence similarity to hGH, it is virtually inactive as GH. The structural similarities and functional differences of these hormones offers a good opportunity to study the evolution of functional domains in proteins.

We are interested in studying the possible functions of the less characterized members of this multigene complex. While the major product of the hGH_N gene (the 22 kd form of hGH) and the identical mature protein encoded by both hPL_4 and hPL_3 genes are easily obtained from pituitary gland and placenta, respectively, this is not true for the putative protein products of the remaining gene members. The proteins encoded by the hGHv and the hPL_1 genes, and any other minor products generated through differential splicing from any of the gene members can only be studied by recombinant DNA approaches.

As a neccesary step to pursue our objective we decided to stablish an efficient expression system based in the transfection of genes into cultured cells. Starting with the hGH_N gene, we have constructed new hybrid genes (Fig. 10) by joining the structural region of this gene with various types of transcriptional control elements. The novel joints plasmids harbor the following promoter and/or enhancer elements: the mouse metallothionein 1 promoter (pMThGH), the mouse metalothionein promoter together with the SV40 enhancer (pNUThGH), the promoter-enhancer of the immediate early gene of human cytomegalovirus (pCMVhGH), as well as the natural promoter of the gene in conjunction with the SV40 enhancer (pSVgpthGH). As a result, we have achieved a good level of expression of this structural gene by transfection into COS-7 cells. Both Northern blotting and radioimmunoassay were performed to evaluate the strenght of these transcriptional control elements and hormone secretion, respectively. The best secretion of hGH was achieved using the pNUThGH plasmid. The potency of the remaining novel joints can be ordered, going from the strongest to the weakest as follows: enhancer-promoter of CMV, natural promoter in combination with the SV40 enhancer, and finally, we obtained the lowest secretion of hGH into the medium when using alone the promoter of MT-I.

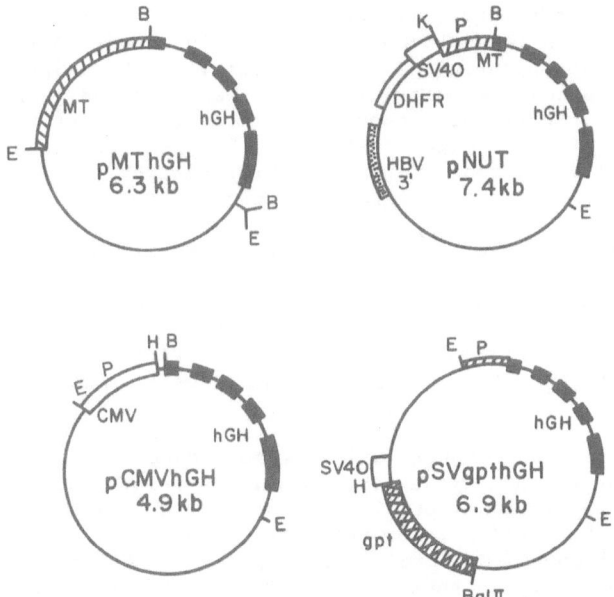

Fig. 10. Structure of hGH expression vectors. With the exception of pSVgpthGH, which contains the entire hGH gene (including its natural promoter), all the other vectors shown here were constructed by fusing the hGH gene structural region (BamHI site at the beginning of the first exon to EcoRI site at the 3' end) to various transcriptional control elements. In pMThGH expression of hGH promoter-less gene is directed by the mouse metalothionein promoter. In pNUT act both SV40 enhancer sequences and mouse metalothionein promoter. Finally, in pCMVhGH the transcriptional control element is the enhancer-promoter of the inmediate early gene of human cytomegalovirus.

Our results confirm those obtained by Pavlakis et al (41). Using SV40 vectors carrying the structural region of the hGH$_M$ and hGHv genes, they showed that monkey kidney cells were able to process hGH prehormone and secret the mature hormone into the culture medium. The amount of hGH secreted in our transient expression experiments was comparable to that secreted by Vero cells lines permanently transfected with a plasmid carrying an SV40-hGH hybrid gene (44). Therefore, we are now ready to produce enough quantities of the other protein products encoded by the hPL-hGH gene family members, to start an analysis of their possible functions.

MOLECULAR BASIS OF THE DEFICIENCIES OF hGH AND hPL

In 1982, a case of a pregnant woman was described who presented a total absence of hPL (determined immunologically). Even so, her pregnancy, birth and child were completely normal (45). A molecular study (46) of the hPL-hGH complex in this placenta, revealed the cause of this lack of hPL. A DNA deletion of approximately 35 kb (see fig. 5) which eliminated the hPL_4 and hPL_3 genes, as well as hGHv, was detected. However, the fifth gene (hPL_1) which contains sequences related to hPL, was not eliminated in the deletion. It was thought that maybe this gene produced a protein with activities similar to that of hPL; although chemically slightly different since it did not react to anti-hPL antiserum.

As described above, information obtained at present from the molecular dissection of the hPL_1 gene strongly suggests hPL_1 is a pseudogene. This is because it contains a mutation in regulatory regions (donor site for the processing of the second intron) that in other examples would inactivate the gene (36). This immediately suggested that probably hPL may not be necessary for the pregnancy and normal fetal and extrauterine growth.

Contrary to what happens with the absence of the genes hPL_4 and hPL_3 responsible for the hPL synthesis, in a 1981 report (47), it was demonstrated that in a case of type A of human growth hormone deficiency (characterized by dwarfism, the complete absence of immunoreactive hGH and by the generation of antibodies in response to hGH treatment), the cause was a deletion of 7.5 Kb that eliminates only the gene for hGH (see fig. 5).

MOLECULAR EVOLUTION

The sequence analysis of the genes has given us valuable information to allow us to try to reconstruct the evolutionary process of the structure of the hGH-hPL complex. In 1971, Nial et al. (48) noticed the presence of similar aminoacid regions in the growth, placental lactogen and prolactin hormones. This observation indicated that these hormones were probably related to each other with respect to their evolution. This was also suggested by their physiological activities, since the three show overlapping activities in different degrees. These observations lead to the hypothesis the genes for these hormones constitute a family, from an evolutionary point of view, having originated from a common ancestor. This common ancestor, which is believed to be similar to prolactin (since it exists in all vertebrates while the other two hormones evolved later in higher vertebrates and placental mammals), underwent a gene duplication and subsequent divergence of the products. This implies that there existed at least two events of genetic duplication during the evolution of these hormones (fig. 11). The first established the branches of prolactin and of growth hormone. The second, most likely occurred in the growth hormone branch, giving rise to the placental lactogen hormones, the youngest member of this family.

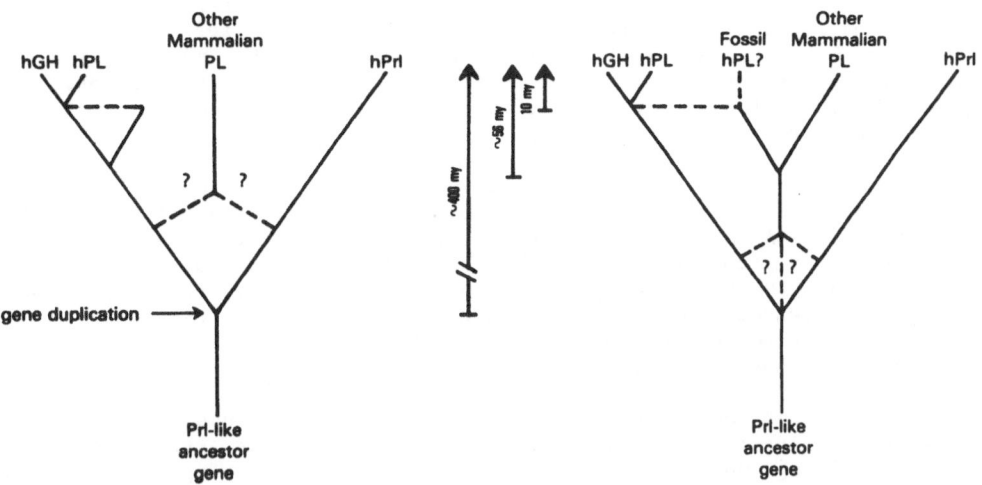

Fig. 11. Possible pathway for evolution of hPL in mammalians. Two altenative hypothesis are presented. In
both of them a gene conversion mechanisms has
rendered the aminoacid sequences in hPL and hGH
very similar (85%).

Using taxonomic, amino acid, and nucleotide sequence comparisons, it
has been estimated that the first genetic duplication occurred approximately 400 million years ago. This estimation goes back to the times when
the evolutionary divergence of fish and tetrapods occurred; since the
pituitary glands of amphibians, reptiles and birds contain different
molecules similar to prolactin and growth hormone.

Estimations of the evolutionary origin of the hPL gene, using two
different criteria, give conflicting results. The necessity of placental
lactogen accompanied the origin of the principal orders of placental
mammals, approximately 75 million years ago. On the other hand, the
branching of the primates occurred about 60 million years ago, considered
to be the approximate time when the hPL appeared. However, the analysis of
the amino acid and nucleotide sequence suggest that the appearance of an
hPL gene by duplication from its ancestor hGH gene occurred approximately
10 million years ago. To resolve this paradox, it has been postulated (49)
that gene conversion must have occurred (a non-reciprocal, recombinant
mechanism where a gene serves as a mold to correct mutations in another
gene). The fact that the genes for these two hormones are contiguous on
the same chromosome, could have facilitated this gene conversion; furthermore this event must have happened about 10 million years ago (see fig.
11).

Finally, the observations of the internal homologies of these three
hormones, suggest that the common ancestral gene could have resulted from
a repetitive duplication process of a gene that coded for a primordial
peptide of approximately 20 amino acids (48).

PERSPECTIVES

The results presented here allow us to elucidate, at a molecular level, the route of the genetic flow of information for the Placental Lactogen hormone. Since its starting point in the nucleus of human placental cells and through the different levels of expression, this genetic information is a product of the structure, function, and evolution of this hormone.

Now we know the structure of the two active genes responsible for the production of the hPL. They are relatively small genes (the sequences present in mRNA are approximately 800 bp distributed in five exons interrupted by four small introns). These two genes are within a cluster that includes two genes for the growth hormone and a fifth gene that contains sequences that identify it as similar to the hPL genes. These genes are distributed in approximately 50,000 bp of DNA in the q22-24 region of the human chromosome 17.

In spite of the advances achieved, there is still much to be investigated concerning the expression and regulation of this genetic complex. We know nothing about the expression of the hPL genes during pregnancy nor in trophoblast pathologies. We are just begining to understand the mechanisms (in spite of the proximity and very high similarity of its genes) by which the cell manages to control the specific expression of the genes such that hPL is produced in tha placenta, while hGH be produced in the pituitary gland. The causes of sudden increase of hPL synthesis during the second trimester, as well as the regulation of the synthesis of hPL are still unknown. We do not know if hPL is involved in the mechanisms of initiation of labor during birth, nor if it has any role in the immunological concealment that the placenta and fetus possess to avoid being rejected by the maternal immune system.

It is necessary to perform more definitive studies in relation to the expression of the hGHv gene, to better understand physiological role that its gene products might have.

It is clear that the studies described in this assay, allow a better understanding of the molecular basis that control biological processes so important such as cellular differentiation and development. A better understanding of these normal processes would help clarify the causes of genetic and other type of ailments, the aging process, and the origins of cancer.

Certainly, this is a good model to carry out studies related to genetic regulation and we hope to continue learning from it.

ACKNOWLEDGMENTS

We would like to express our gratitude to all those investigators that in one way or another, collaborated with us in carrying out these experiments. Our gratitude also extends to all those members of both U.L.I.E.G. and those pertaining to Dr. Saunders' laboratory.

H.A.B.S. and R.R.S. thank the National Council of Science and Technology of Mexico for their support through post-graduate scholarships.

This project was supported by grants given by: the Subsecretaría de Educación Superior e Investigación Científica and the Subsecretaría de Educación e Investigación Tecnológicas of the Mexican Ministry of

Education, the National Council of Science and Technology of the Mexican Government, and finally, the Fondo de Estudios e Investigación Ricardo J. Zevada.

H.A.B.S. would also like to thank the Faculty of Medicine of the Universidad Autónoma de Nuevo León for their continuous support.

The authors wish to thank Julie B. Silva for her valuable help in translating and proofreading for this manuscript.

REFERENCES

1.- F. Jacob, and J. Monod, Genetic regulatory mechanisms in the synthesis of proteins, J.Mol. Biol. 3:318 (1961).

2.- P. Chambon, Split genes, Sci. Am. 244(5):60 (1981).

3.- J.E. Darnell, Variety in the level of gene control in eukaryotic cells, Nature 297:365 (1982).

4.- T. Igo-Kemenes, W. Horz, and H.G. Zachau, Chromatin, Ann. Rev. Biochem. 51:89 (1982).

5.- D. Attenborough, Life on earth, Little Brown Co., Boston, (1979).

6.- P. Beaconsfield, G. Birdwood, and R. Beaconsfield, The placenta, Sci. Am. 243(2):80 (1980).

7.- M. Chatterjee, and H.N. Munro, Structure and biosynthesis of placental peptides, Vitamines and Hormones, 35:149 (1977).

8.- E.R. Simpson, and P.C. MacDonald, Endocrine physiology of the placenta, Ann. Rev. Physiol. 43:163 (1981).

9.- J.B. Josimovich, and J.A. MacLaren, Presence in human placenta and term serum of highly lactogenic substance immunologically related to pituitary growth hormone, Endocrinology, 71:209 (1962).

10.- B.N. Saxena, K. Emerson, and H.A. Selenkow, Serum placental lactogen (hPL) levels as an index of placental function, New Eng. J. Med. 281:225 (1969).

11.- H.N. Munro, Placental protein and peptide hormone synthesis: impact of maternal nutrition, Federation Proc. 39:255 (1980).

12.- J.J. Sciarra, S.L. Kaplan, and M.M. Grumbach, Localization of anti-human growth hormone serum within the placenta: evidence for a human chorionic "Growth hormone-prolactin", Nature 199:1005 (1963).

13.- E.M. Southern, Detection of sepecific sequences among DNA fragments separated by gel electrophoresis, J. Mol. Biol. 98:503 (1975).

14.- F. Bolivar, R.L. Rodríguez, P.J. Greene, M.C. Betlach, H.L. Heyneker, H.W. Boyer, J.H. Crosa, and S. Falkows, Construction and Characterization of new cloning vehicles, II A multipurpose cloning system, Gene 2:95 (1977).

15.- J. Collins, and B. Hohn, Cosmids: A type of plasmid gene-cloning vector that is packagable in vitro in bacteriophage lambda heads, Proc. Natl. Acad. Sci. USA. 75:4242 (1978).

16.- F.R. Blattner, B.G. Williams, A.E. Blechl, K. Deniston-Thompson, H.E. Faber, L.A. Furlong, D.J. Grunwald, D.O. Keifer, D.D. Moore, E.L. Sheldon, and O. Smithies, Charon phages: Safer derivatives of bacteriophage lambda for DNA cloning, Science 196:161 (1977).

17.- J. Messing, R. Crea, and P.H. Seeburg, A system for shotgun DNA sequencing, Nucl. Acids Res. 9:309 (1981).

18.- F. Sanger, S. Nicklen, and A.R. Coulson, DNA sequencing with chain terminating inhibitors, Proc. Natl. Acad. Sci. USA. 74:5463 (1977).

19.- A.M. Maxam, and W. Gilbert, Sequencing end-labeled DNA with base-specific chemical cleavages, In "Methods in enzymology" Grossman and K. Moldave, eds., 65:499 Academic Press, New York (1980).

20.- T. Maniatis, R.C. Hardison, E. Lacy, J. Laver, C. O'Connell, D. Qoun, D.K. Sim, and A. Efstratiadis, The isolation of structural genes from libraries of eukaryotic DNA, Cell 15:687 (1978).

21.- D. Resendez-Perez, H.A. Barrera-Saldaña, M.R. Morales-Vallarta, E. Ramírez-Bon, C.H. Leal-Garza, A. Feria-Velazco, and F.J. Sánchez-Anzaldo, Low speed purification of placental nuclei, Placenta 5:523 (1984).

22.- H.A. Barrera-Saldaña, Isolation and characterization of human placental nuclei, BS Thesis, Universidad Autónoma de Nuevo León, Monterrey, N.L., Mexico (1979).

23.- H.A. Barrera-Saldaña, Expression of the human placental lactogen genes, Ph.D. Thesis, The University of Texas Health Science Center, Houston, Texas (1982).

24.- H.A. Barrera-Saldaña, D.L. Robberson, and G.F. Saunders, Transcriptional Products of the Human Placental Lactogen Gene, J. Biol. Chem. 257:12399 (1982).

25.- H.A. Barrera-Saldaña, P.H. Seeburg, and G.F. Saunders, Two structurally different genes produce the same secreted human placental lactogen hormone, J. Biol. Chem. 258:3787 (1983).

26.- M.E. Harper, H.A. Barrera-Saldaña, and G.F. Saunders, Chromosomal Localization of the Human Placental Lactogen - Growth Hormone Gene Cluster to 17q22-24, Am. J. Hum. Genet. 34: 229 (1982).

27.- V.J. Kidd, H.A. Barrera-Saldaña, and G.F. Saunders, The human growth hormone and Placental Lactogen gene complex, In "Perspectives on genes and the Molecular Biology of Cancer," D.L. Robberson and G.F. Saunders, eds., Raven Press, New York (1983).

28.- I. Boime, and S. Boguslawski, The synthesis of human placental lactogen by ribosomes derived from human placenta, Proc. Natl. Acad. Sci. USA. 71:1322 (1974).

29.- J. Shine, P.H. Seeburg, J.A. Martial, J.D. Baxter, and H.M. Goodman, Construction and analysis of recombinant DNA for human chorionic somatomammotropin, Nature 270:494 (1977).

30.- P.H. Seeburg, J. Shine, J.A. Martial, A. Ullrich, J.D. Baxter, and H.M. Goodman, Nucleotide sequence of part of the gene for human chorionic somatomammotropin: purification of DNA complementary to predominant mRNA species, Cell 12:157 (1977).

31.- J.A. Martial, R.A. Hallawell, J.D. Baxter, and H.M. Goodman, Human growth hormone: Complementary DNA cloning and expression in bacteria, Science 205: 602 (1979).

32.- F.M. DeNoto, D.D. Moore, and H.M. Goodman, Human growth hormone DNA sequence and mRNA structure: possible alternative splicing, Nucl. Acids Res. 9:3719 (1981).

33.- G.S. Barsh, P.H. Seeburg, and R.E. Gelinas, The human growth hormone gene family: Structure and evolution of the chromosomal locus, Nucl. Acids Res. 11:3839 (1983).

34.- B.L. Rodgers, M.G. Sobnosky, and G.F. Saunders, Transcriptional enhancer within the human placental lactogen and growth hormone multigene cluster, Nucl. Acids Res. 14: 7647 (1986).

35.- P.H. Seeburg, Structural and functional features of the human growth hormone locus, Am. Soc. Endoc. (Abstract) (1984).

36.- M. Baird, C. Priscoll, H. Schreiner, G.V. Sciarratta, G. Sansone, G. Niazi, F. Ramírez, and A. Bank, A nucleotide change at a splice junction in the human β-globin gene is associated with β° thalassemia, Proc. Natl. Acad. Sci. USA. 78:4218 (1981).

37.- M.K. Foecking, and H. Hofstetter, Powerful and versatile enhancer promoter unit for mammalian expression vector, Gene 45:101 (1986).

38.- C. Yanish-Perron, J. Vieira, and J. Messing, Improved M13 phage vectors and host strains: Nucleotide sequences of the M13mp18 and pUCl9 vectors, Gene 33:103 (1985).

39.- F.L. Graham, and A.J. VanderEb, A new technique for the assay of infectivity of human adenovirus 5 DNA, Virology 52:456 (1973).

40.- G. Cathala, J.S. Savooret, B. Mendez, I. West, M. Karim, J. Martial and J.D. Baxter, A method for isolation of intact traslationally active ribonucleic acid, DNA, 2:329 (1983).

41.- G.N. Pavlakis, N. Hizuka, P. Gorden, P. Seeburg, and D.H. Hamer, Expression of two human growth hormone genes in monkey cells infected by simian virus 40 recombinants, Proc. Natl. Acad. Sci. USA 78:7398 (1981).

42.- F. Frankene, F. Rentier-Delrue, M-L Scippo, J. Martial, and G. Hennen, Expression of the Growth hormone variant gene in human placenta, J. Clin. Endoc. Metab. 64:635 (1987).

43.- N. Masuda, W. Masanori, M. Tanaka, M. Yamakawa, K. Shimizu, N. Nagai, and K. Nakashima, Molecular cloning of cDNA encoding 20 kd variant human growth hormone and the alternative splicing mechanism. Biochem. Biophys. Acta, 949:125 (1988).

44.- J. H. Lupker, W. G. Roskam, B. Miloux, P. Liauzun, M. Yaniv, and J. Jouannau, Abundant excretion of human growth hormone by recombinant-plasmid-transformed monkey kidney cells, Gene 24:281 (1983).

45.- P.V. Nielsen, H. Pedersen, and E.M. Kampmann, Absence of human placental lactogen in an otherwise uneventful pregnancy, Am. J. Obstet. Gynecol. 135:322 (1979).

46.- J.M. Wurzel, J.S. Parks, J.E. Herd, and P.V. Nielsen, A gene deletion is responsible for absence of human chorionic somatomammotropin, DNA, 1:251 (1982).

47.- J.A. PhillipsIII, B.L. Hjelle, P.H. Seeburg, and M. Zachmann, Molecular basis for familial isolated growth hormone deficiency, Proc. Natl. Acad. Sci. USA. 78: 6372 (1981).

48.- H.D. Niall, M.L. Hogan, R. Sauer, I.Y. Rosenblum, and F.C. Greenwood, Sequence of pituitary and placental lactogenic and growth hormones: Evolution from a primordial peptide by gene reduplication, Proc. Natl. Acad. Sci. USA. 68:866 (1971).

49.- N.E. Cooke, D. Coit, J. Shine, J.D. Baxter, and J.A. Martial, Human prolactin: cDNA structural analysis and evolutionary comparisons, J. Biol. Chem. 256:4007 (1981).

OVERPRODUCTION OF PROTEINS BY RECOMBINANT DNA: HUMAN INSULIN

Paulina Balbas, Ramon de Anda, Noemi Flores, Xochitl Alvarado, Norberto Cruz, Fernando Valle and Francisco Bolivar

Centro de Investigacion sobre Ingenieria Genetica y Biotecnologia, Universidad Nacional Autonoma de Mexico, Apdo Postal 510-3 Cuernavaca Mor., C.P. 62270, MEXICO

INTRODUCTION

An important goal of recombinant DNA research is to develop methods for the overproduction of useful gene products. Cloning a gene of interest for its expression in a particular host system is not an easy task, for the manipulations require extensive knowledge of the regulatory mechanisms of gene expression, basic understanding of the vector-host relationships, skills in the recombinant DNA techniques, adequate technical support for cell growth, and finally, suitable purification procedures for the protein products of interest.

Today, it is possible to clone and express a gene in a variety of different host systems, such as bacteria, yeast, fungi, insect cells, mammalian cells, animals and plants (Inouye, 1983; Rodriguez and Denhardt, 1987; Luckow and Summers, 1988). With so many alternatives, the selection of the best system for producing a particular protein is crucial. One concept is now clear: there is no "perfect host". Not only does every protein have its own unique properties, but so does every host-vector system.

Nevertheless, E. coli has been and continues to be the most utilized host in the gene expression field. No other system has been developed which can express such a large number of known gene products, as well as express undefined coding sequences (open reading frames) in amounts sufficient to determine the identity of the gene product, to raise specific antibodies and to characterize its function (Weinstock et al., 1983; Shatzman and Rosenberg, 1987). Furthermore, E. coli exhibits a rapid cell growth rate, and low cost culture conditions are available, when compared to eukaryotic expression systems.

The most important drawbacks in the use of Escherichia coli as host for the overproduction of proteins are its restricted capacity to secrete proteins and its inability to exert certain post-transcriptional modification of proteins (such as disulfide bond formation and glycosilation), which usually lead to wrong protein folding. Thus, proteins may be obtained in high quantities, but their purification and renaturalization, in order to obtain biological activity, may be difficult and expensive.

59

The general strategy used in the design of expression systems involves the insertion of the gene or DNA segment of interest into a vector system in such a way that the DNA is efficiently transcribed and translated (for a review of expression plasmid vectors see Balbas et al., 1986; Rodriguez and Denhardt, 1987). An expression vector should supply a strong promoter (usually regulable, not constitutive), a ribosome binding site, and in some cases, the amino terminal coding region of a bacterial gene, to insure high rates of transcription initiation and efficient translation. Sophisticated expression vectors, however, contain DNA fragments that help reduce problems such as premature termination of transcription (antiterminators), mRNA stability (stabilizing sequences), protein instability (genes for fusion proteins) and plasmid instability (transcriptional terminators) (Itakura et al., 1977; Gentz et al., 1981; Reznikoff and Gold, 1986; Schoner et al., 1986; Wong and Chang, 1986; Shatzman and Rosenberg, 1987). Table I shows some of the most important factors that influence the high-rate synthesis of heterologous proteins in <u>Escherichia coli</u>.

Once the gene is inserted and enzymatically tailored into the expression vector, the gene product can be obtained in a suitable host strain, containing the genotypic features needed for promoter regulation and growth. Expression may be monitored and quantitated by gel electrophoresis, immunological detectors, HPLC, or functional assays (Balbas and Bolivar, 1988). Finally, the expressed protein can be purified and characterized in terms of its physical and biological properties.

Table I. Factors affecting the overall product yield of an expression system in <u>E. coli</u>.

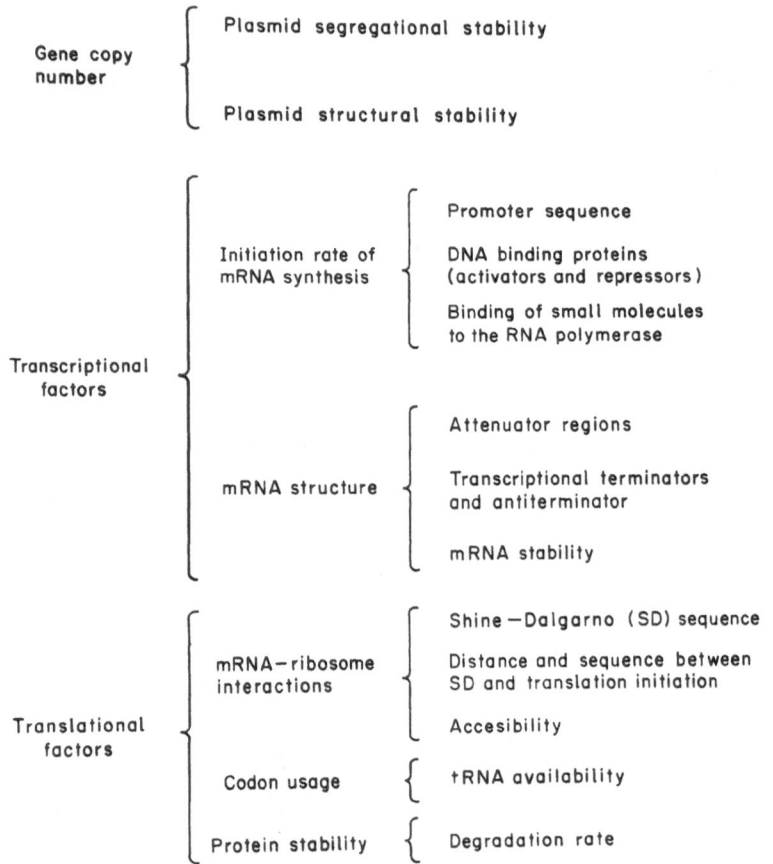

In this paper, we describe a process for the production of the human hormone insulin. The two polypeptides that conform the protein, chain A (21 aa) and chain B (30 aa), are synthesized independently from two synthetic genes, as the carboxy end of hybrid proteins, where the amino terminus is the first third part of the bacteriophage lambda repressor, called cI. These hybrid polypeptides were purified from the whole cells as inclusion bodies and cleaved with cyanogen bromide (CNBr) in the presence of a reducing agent, in order to liberate the insulin chains. After their purification, the chains were reassociated, in controlled oxidizing conditions, and the correct isomeric form of insulin was then purified by HPLC. Finally, this product was characterized by physicochemical and immunological methods, and assayed for biological activity in rats.

DESIGN AND CONSTRUCTION OF THE EXPRESSION VECTORS

The expression vectors for the synthesis of insulin chains A and B were ensambled with the following DNA fragments: plasmid pBR712, a 674 bp fragment of bacteriophage lambda DNA containing the coding region for the amino terminus of the cI monomer, and synthetic genes coding for chains A or B, respectively. The construction strategy is shown in figure 1.

Plasmid vector pBR712 is a pBR327 derivative (Soberon et al., 1980), that carries the gene for ampicillin resistance, in which the tetracycline promoter (P2, according to Balbas et al., 1986), was destroyed and replaced by a 300 bp DNA fragment containing the promoter-operator region of the Escherichia coli tryptophan operon (trp). In this operon, transcription initiation is controlled by two mechanisms: repression and attenuation. As bacterial cells become deficient in the amino acid tryptophan, both repression and attenuation are relieved, resulting in a several hundred-fold elevation of the overall expression of the genes downstream the control region. In order to achieve even higher levels of expression, the trp attenuator region can be deleted to avoid premature transcription termination. Repression can therefore be achieved by high levels of tryptophan in the culture medium, whereas induction may be accomplished by tryptophan depletion or chemical induction with either of two chemical analogs: the 3-B-indoleacrilic acid or the indole-3-propionic acid. Complete repression of the trp promoter is difficult to achieve, which in some instances, may cause plasmid or cell instability. Despite this fact, the trp promoter-operator region has been widely used for the expression of several cloned genes (Nichols and Yanofsky, 1983; Harris, 1983).

Bacteriophage lambda repressor cI monomer is a 236 aa protein consisting of an amino terminal DNA binding domain and a carboxy terminal oligomerization domain. These two domains are structurally and dinamically independent, and proteolytic fragments corresponding to each domain remain stably folded and functionally active. The "de novo" folding of the amino terminal domain, is also independent of the carboxy terminal domain since cloned amino terminus fragments are active in vivo (Pabo et al., 1979). Furthermore, it has been demonstrated that in vivo, cI repressor appeared stable for at least two generations (Roberts and Roberts, 1975). All these data supported the hypothesis that the amino terminus of the cI monomer could act as a good carrier protein.

The design of the DNA oligonucleotides that comprise the synthetic genes included certain important considerations. Both genes include flanking heterologous restriction sites at the ends, as well as an ATG codon immediately before the first chain codon. In this way, a methionine residue connects the carrier protein to the insulin chain. A cyanogenolysis reaction with CNBr is enough to liberate the hormone chain from the hybrid. Similar synthetic genes have been used successfully for the biosynthesis of the

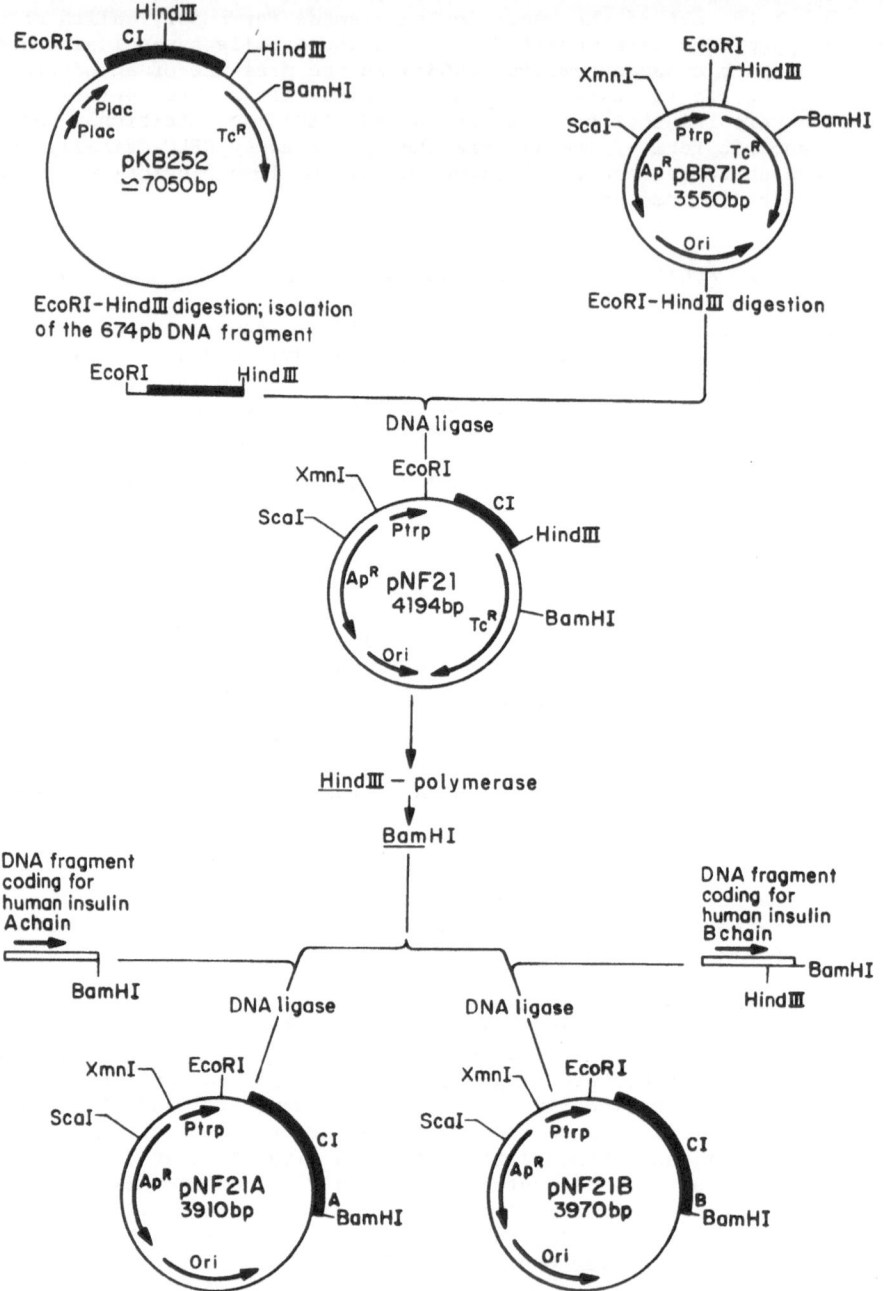

Fig. 1. Construction of recombinant plasmids pNF21A and pNF21B that code for the hybrid proteins CI-A and cI-B, respectively (adapted from Flores et al., 1986).

human insulin chains in a variety of expression systems (Goeddel et al., 1979; Rosteck Jr., et al., 1983).

STRAIN SELECTION

The properties of any microbial cell are ultimately determined by its genome, which carries the information that enables the organism to mantain its functional and structural integrity. It also contains the potential to respond to changes in its environment, hence, adapting to different conditions (Meyer, 1985). The host genetic background can play an important role in the final yield of any accumulated gene product. Up to date, the reasons for the rather dramatic differences seen in product yields in different host strains are not well understood, although plasmid stability and product degradation often appear to be determining factors.

In many cases, strains carrying recombinant high-level expression systems have been found to be unstable (Kim, 1984; Nichols and Yanofsky, 1983). The host which carries a plasmid inflicting a high metabolic load, tends to be overgrown by cells without the extra energy requirement. Two general types of plasmid instability have been characterized. The first type, segregational instability, arises due to defective cell partitioning of plasmid DNA between daughter cells during cell division, resulting in the loss of the plasmid from one of the cells. The second type, structural instability, is the result of physical changes in the plasmid DNA, such as deletions, insertions and rearrangements. Structural instabilities are more insidious because the changes usually result in the loss or alteration of the DNA of interest, while the bulk of plasmid DNA, including the selectable markers, is retained (Ensley, 1985).

Besides the genetic background of the host, there are two other important issues in the selection of a strain as a host: growth requirements and the mode of regulation of gene expression. Growth requirements include nutrients (vitamins, aa, metals, etc.), temperature and oxygen, all of which incide directly on the costs of the fermentation process. Therefore, a strain with fewer specific requirements is preferred over others with several auxotrophies or specific growth needs. In addition to the genotype of the host strain, the mode of induction can also affect the overall accumulation of a gene product. Such modes of induction of an expression system will undoubtedly bias the selection of a host strain, since a particular genotype is usually needed to maximise induction. The most widely used induction methods include heat (lambda pL), chemical inducers (trp, lac, lambda pL) and amino acid starvation (trp) (Balbas et al., 1986). The host response to any of these induction methods will lead to different cellular states, that range from mild alterations (modulation of metabolic pathways) to dramatic alterations (SOS or heat shock response mechanisms) (Gottesman, 1984; Shatzman and Rosenberg, 1987). Moreover, it has been demonstrated that the same expression vector exhibits different product yields in a variety of host strains (Kaytes et al., 1986).

Based in the above considerations, it is always desireable to have a series of strains with the necessary genetic traits for the particular expression system, in order to scan for the highest productivity yields.

For plasmids pNF21A and pNF21B the host strains that were used are W3110 (trp-), C600 (trp+), CS520 (trp-) and 294 (trp+). As shown in Table II, both plasmids exhibited dramatically different stabilities in these strains when grown in rich medium at high rates, eventhough the trp promoter was not induced. Both plasmids were very unstable when they were transformed into strains W3110 and 294. The other two strains did not show significant segregation of the recombinant plasmids under the same conditions.

Table II. Segregation rates of plasmids pNF21A and pNF21B in different host strains growing in Luria Broth at 37°C.

Escherichia coli strain	Plasmid	% of cells containing plasmid DNA after 25 generations*
W3110	pNF21A pNF21B	11.70 14.09
294	pNF21A pNF21B	0.0 75.66
C600	pNF21A pNF21B	89.86 94.70
CS520	pNF21A pNF21B	100.00 99.50

* The numerical values are the average of at least ten different determinations. The variability of the values did not exceed +/- 10% in every case except in the case of 294/pNF21A

FERMENTATION CONDITIONS

Once a hypothetically good expression system has been constructed and an adequate host strain has been selected, the establishment of the growth and induction conditions are necessary to obtain the maximal efficiency of the system (Zabriskie and Arcuri, 1986).

The environmental parameters that commonly influence the growth of microbial cells can be divided into two categories, namely physical (temperature, oxygen flow rate) and chemical (media composition, pH). Usually, the alteration of one of these parameters will lead to a different behaviour of the culture, and therefore to a change in the final productivity of the system.

A good productivity of a specific expression system is obtained when the result of a fermentation is a high biomass containing a large amount of the desired protein. Plasmid stability is then an essential issue in the fermentation of strains carrying recombinant DNA. The instability of a recombinant plasmid in a culture may reduce the overall levels of the desired product in the fermentation, increasing production costs, since growth substrates are consumed by non productive cells.

In general, the segregational instability of a plasmid may be affected not only by the genetic features of the host cell, as demonstrated with pNF21A and pNF21B in the strains tested, but also by the copy number of the plasmid, the culture conditions and the genes contained in the plasmid.

Although there are theoretical copy numbers for almost every plasmid used for DNA cloning and expression, it has become clear that the actual copy number of a recombinant plasmid in a growing culture is influenced at least by the growth rates (Seo and Bailey, 1985). In fact, this relationship is inversely proportional, so at high growth rates the copy number of plasmids per cell is lower, thus, inducing a higher segregational instability (Klotsky and Shwartz, 1987).

64

Based upon this observation, the culture conditions must be settled as to restrain the growth rates to avoid plasmid loss. Moreover, fast growth rates are usually achieved at the expense of product yield. Therefore, optimal culture conditions for product formation are usually incompatible with optimal growth conditions (Meyer et al., 1985). The obvious consecuence is that product formation needs to be evaluated at different growth rates.

The growth rate may be modulated by nutrient input, aereation rate and temperature. Nutrient limitation, however, may hamper the product yield, so low temperature seems to be a better growth modulator. The strains carrying plasmids pNF21A and pNF21B were grown at 30°C instead of the optimal 37°C in order to improve plasmid stability, and therefore product yield (data not shown).

The genes contained in the plasmid will also determine plasmid stability and the overall product yield as well. Cloning and expression vectors usually contain at least one gene for selection. In general terms, selection refers to the use of growth environments in which only cells possessing certain genetic traits are able to grow. If a single gene is involved in the selection scheme, its presence accounts for the survival of the organism. However, when considering genes in multiple copies, the conceptual basis of selection becomes more complicated due to the probability of substantial variability in single cell gene content. When the selective agent is decomposed, deactivated or tightly bound to the product of the selection gene, cells with high content in the selection genes reduce the concentration of the selective agent, diminishing its effects on the more sensitive cells. This is a general limitation and practical difficulty for all selection strategies of this type (Dennis et al., 1985). The ampicillin resistance gene present in the pBR322 derivatives has this drawback. An efficient way to insure selective pressure in these systems, is the addition of ampicillin in various stages of the culture. However, this may result expensive in a large scale fermentation process, so the use of another selection scheme is usually necessary.

The DNA region comprised between the XmnI and BamHI in the plasmids pNF21A and pNF21B has been recently cloned into the ScaI site of the ampicillin resistance gene of pBR327, in order to use the tetracycline resistance gene as a selective marker. These plasmids have shown better stability than their predecesors, after only one addition of the antibiotic at the begining of the fermentaton (data not shown).

A final comment on plasmid structure is that high levels of transcription of the cloned genes, may interfere with plasmid replication (Shatzman and Rosenberg, 1987). It is then desireable to either isolate the origin of replication region with a transcription terminator, or, in the case of pBR322 derivatives, to direct transcription in the same orientation of the RNAII replication primer (Bolivar et al, 1977; Balbas et al., 1986). In any case, the expression should be as tightly shut off as possible during the fermentation process.

Once the culture conditions have been settled to obtain a fully grown culture, where almost the totality of the cells contain the expression plasmid of interest, the following step is to accomplish total induction of the genetic system in order to obtain high quantities of the protein of interest.

As it was previously mentioned, the trp promoter-operator region may be induced by either tryptophan starvation or chemical induction, both methods being very effective in relieving repression. However, when the system is induced by both methods symultaneously, the expression of the protein of interest is achieved faster.

An induction profile of plasmids pNF21A and pNF21B is shown in figure 2. As it can be seen, the hybrid protein cI-B chain (gel B) is obtained at higher levels than the hybrid cI-A chain (gel A). This difference may be due to differential stability of either the mRNA or the hybrid protein itself (Nilsson et al., 1981). Rosteck et al. (1983) reported the same phenomenon for chains A and B, fused to a fragment of the protein product coded by trpE, which suggests that hybrids with chain A may be inherently unstable.

ISOLATION OF INCLUSION BODIES AND PURIFICATION OF INSULIN CHAINS A AND B

The Escherichia coli protein-degradation system can interfere with the attempts to obtain accumulation of heterologous proteins, especially if they have low molecular weight (Itakura et al., 1977). As previously discussed, the strategy of fusing two genes to obtain a stable hybrid protein, has been successfully utilized in a variety of opportunities, although this approach requires a proteolytic cleavage procedure to liberate the desired peptide from the hybrid.

However, it has been demonstrated that in many cases, high levels of expression of hybrid proteins, lead to the formation of inclusion bodies, which in turn may be easily purified by differential centrifugation of disrupted cells (Schoner et al., 1985, Flores et al., 1986; Primrose, 1986).

The term inclusion bodies refer to a morphologically characteristic intracellular product, whose accumulation in form of tight aggregates, corresponds to the formation of the chimeric protein. Moreover, the scanning electron micrographs of cells carrying fully induced, high-level expression plasmids, indicate that inclusion aggregates are sufficiently rigid and cohesive to cause a prominent distention of the bacterial cell wall (William et al., 1982).

For the cultures producing the cI-A and cI-B hybrid proteins, the inclusion bodies were released intact from cells by sonication without using lysozime. After a single washing step, it was possible to obtain the hybrid products highly purified, as shown in figure 2 lane b.

In order to purify the insulin A and B chains from the cI amino terminal region, the inclusion bodies obtained as described, were solubilized and treated with CNBr. This compound is capable of cleaving thioethers, and its action upon proteins is unique in its selective attack on the amino acid methionine at acidic pH (Gross, 1967). The CNBr cleavage products were purified by HPLC using previously reported conditions. Also, the identities of A and B chains obtained by this method, were confirmed using physicochemical methods, amino acid composition analysis and radioimmunoanalysis as previously described (Ladron de Guevara et al., 1985).

REASSOCIATION OF BACTERIAL A AND B CHAINS AND CRYSTALIZATION OF HUMAN INSULIN

It has been established that synthesis of insulin can be accomplished by reassociation or combination of its individual chains, natural, chemically synthesized or obtained via recombinant DNA, in a variety of hosts (Katsoyannis et al., 1967; Goeddel et al., 1979; Chance et al., 1981).

However, an extraordinary variety of products may result from the reassociation of the two insulin polypeptide chains, particularly if the disulfide bonds between the cysteine residues are formed at random.

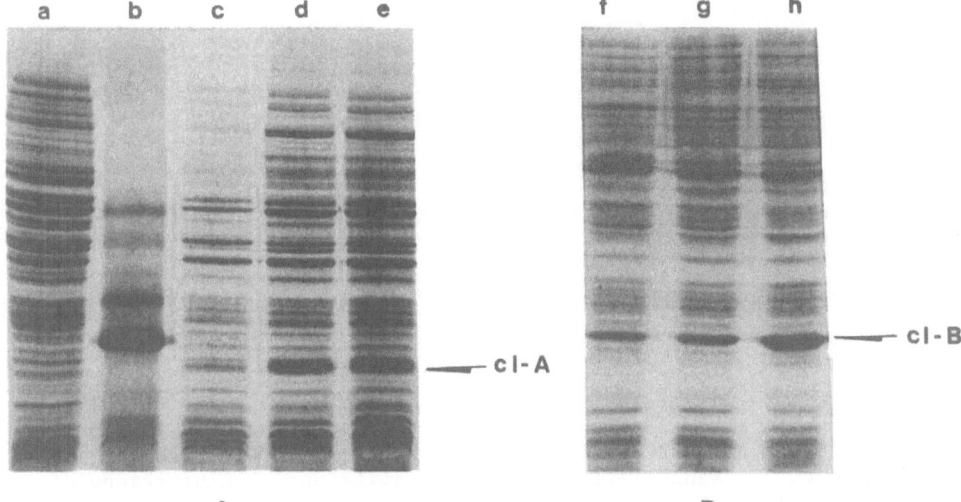

a   b   c   d   e   f   g   h

cI-A

cI-B

A   B

Fig. 2. SDS–PAGE of total proteins from Escherichia coli strain W3110
carrying recombinant plasmids. The gel shows the induction
profiles for the cI-A and cI-B hybrid proteins. An actively
growing colony was picked from a plate containing the proper
selective media (100 µg/ml of ampicillin), and inoculated in
a 2 l fermentor containing M-9 media supplemented with 100 µg/
ml of tryptophan. Fermentations were carried out at 30°C, with
an aeration rate of 0.4 vvm. A growth rate of 0.23/h was
obtained. When the optical density of the culture (A=590)
reached 0.8, the culture was centrifuged, washed in M-9 salts
and resuspended in minimal medium without tryptophan. One hour
later, the inducer 3-B-indoleacrylic acid was added to a final
concentration of 20 µg/ml. Induction was monitored at
different time spans.

The lanes in gel A are: (a) wild type strain W3110; (b) one
step purification of inclusion bodies containing cI-B;
(c) W3110/pNF21A before induction; (d) W3110/pNF21A, 60 min
after induction; (e) W3110/pNF21A, 120 min after induction.
The lanes in gel B are: (f) W3110/pNF21B, 30 min after
induction; (g) W3110/pNF21B, 60 min after induction; (h)
W3110/pNF21B, 120 min after induction. The arrows indicate
the migration of the cI-A and cI-B hybrids, respectively.

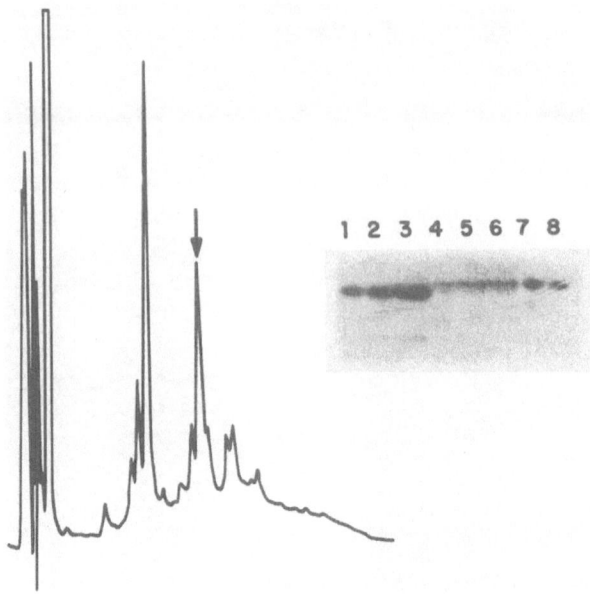

Fig. 3. HPLC profile of the purification of the correct isomeric form of
insulin after reassociation of the A and B chains. The insert
shows a polyacrylamide gel electrophoresis of the reassociation
mixture. Lanes 1 to 3 show three different concentrations of
the mixture prior to the thiolysis oxidation procedure. The
prominent band corresponds to chain B. Lanes 4 to 6 show three
different concentrations of the reassociation products. Lanes
7 and 8 are porcine insulin standards from Sigma.

The reassociation procedure used in this work was previously reported
(Chance et al., 1981), and it consists of the thiolysis of the S-sulfonate
cysteine derivatives (SSO_3) and their subsequent oxidation for the formation
of the intrachain disulfide bonds. The analysis by HPLC of a reassociation
mixture is shown in figure 3 and it usually displays considerable
heterogeneity. However, the high yield of the correct isomeric form of
insulin clearly shows that the combination reaction does not proceed in a
random fashion. Some of the factors that may lead to a high yield of insulin
include the specific chain interactions, disulfide interchanges prior to
bonding and the apparent conformational stability of the insulin molecule.
A gel electrophoresis showing the results of such reassociation is presented
in the figure insert. The prominent band corresponding to human insulin in
lanes 4 to 6, was used to perform a radioimmunoanalysis with specific
antibodies raised in goat against porcine insulin.

Radioimmunoassay combines the extreme sensibility of detection of
radioactively labeled substances with the high specificity of immunological
reactions. Particularly, radioimmunoassay of insulin using a two-antibody
system offers to be a simple, rapid and versatile method permitting large
numbers of determinations to be carried out with high precision and great
sensitivity (Morgan and Lazarow, 1962). The procedure used to detect
radioimmune activity from the insulin obtained by the reconstitution
procedure of the S-sulfonated derivatives, was from Radioassay System
Laboratories. The results are shown in figure 4. As it can be seen, the
standard curve of porcine insulin radioimmunoanalysis against the results of
recombinant human insulin radioimmunoanalysis, showed parallelism with the
dose-response curve.

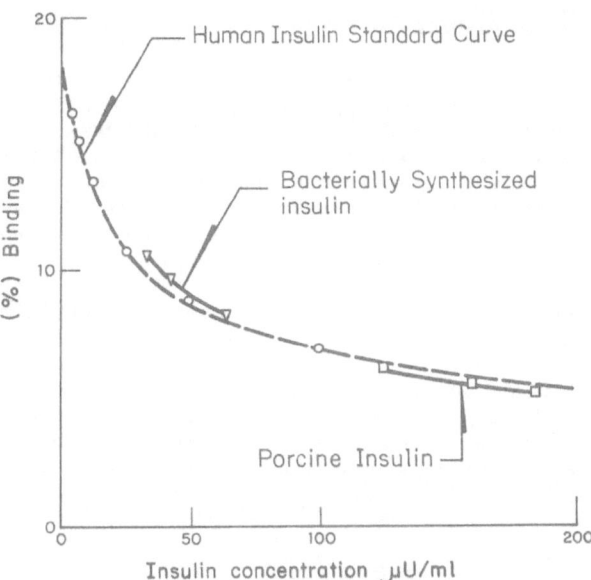

Fig. 4. Radioimmunoassay results of the reassociated recombinant human
insulin. The procedure was performed as indicated by Radioassay
System Laboratories (RSL). Radiolabelled insulin was purchased
from Amersham, and the second antibody, goat anti-guinea pig
gamma globulin, was diluted 1:10.

Once the insulin is purified by HPLC, the hormone molecules are
crystalized, in the presence of zinc, to eliminate impurities. Figure 5
shows an electron micrograph of the crystals obtained according to the
methods reported (Rommans et al., 1940; Randall, 1964).

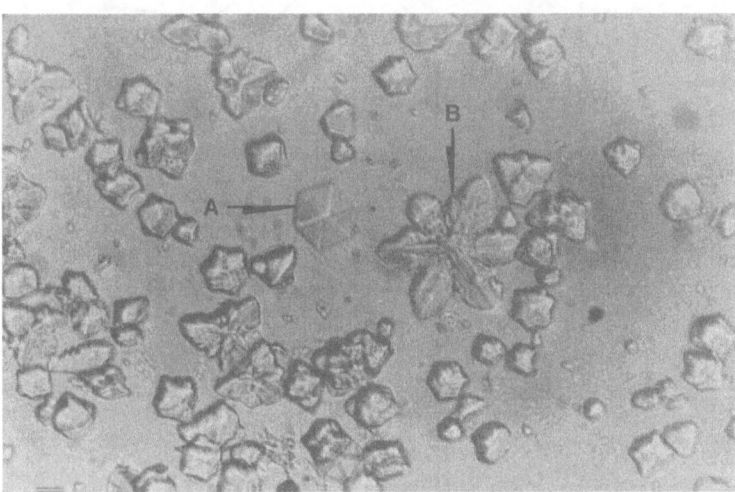

Fig. 5. Electron micrograph of the recombinant human insulin crystals in
the presence of zinc. The crystal labelled as A corresponds to
the rhombic form, whereas the crystal labelled as B corresponds
to the star-like form. These patterns are obtained according to
the crystallization rates, being the rhombic form a result of
slow rates, while the star-like form is dependent on fast rates
(Schlichtkrull, 1956).

The biological trials of pharmaceutical products in animal models, is the first step in the evaluation of new products as potential medical agents. The animal model system utilized in this work, used a particular rat strain named Wistar (male animals). The method used was originally reported by Halban et al. (1981), although the anesthesis step was omitted in order to obtain blood samples free of contaminants. Also, the assay was performed under non-stressing conditions to provide accurate measurements of the plasma glucose depression over a period of 60 min.

Figure 6 shows the results of the bioassay of the recombinant insulin. As it can be seen, this type of insulin displayed similar biologic potency (plasma glucose depression) on a weight basis when compared to porcine insulin. These data demonstrates that the human insulin obtained after this process, has full biologic activity and it is cleared from the blood circulation in a similar fashion to porcine insulin. These results indicate that this recombinant product consists of authentic insulin.

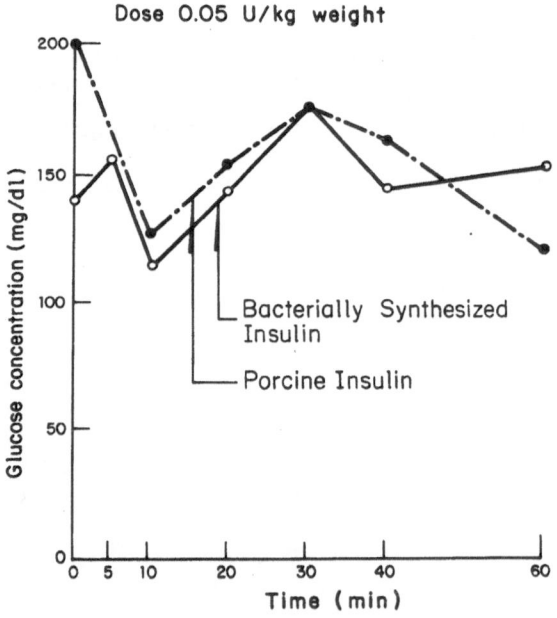

Fig. 6. Biological activity of recombinant human insulin by cumulative plasma glucose depression over a 60-min test period. Male Wistar rats (180-230 g in weight) are surgically catheterized three days prior to the test and trained to stay still in a specially designed container. Blood samples are taken prior to the insulin intravenous injection, and then every 10 minutes after insulin addition to monitor glucose depression. Groups of five rats were sampled for the determination of the standard with human insulin (pancreatic), as well as porcine insulin, and recombinant human insulin, respectively.

DISCUSSION

Since the advent of recombinant DNA technology it has become possible to produce for the first time, substantial quantities of human proteins normally present in the body in trace amounts, highly specific diagnostic kits, and safer vaccines for a variety of human and animal diseases.

Despite the fact that over 200 biotechnology companies throughout the world are developing human proteins as therapeutic agents, only four proteins have reached the marketplace in a decade: human insulin, human somatotropin, interferon-alpha and human tissue plasminogen activator factor. The delay in developing other proteins as human medicines is due to a mixture of technical problems, cost considerations and the need to administer the proteins parenterally (Primrose, 1986).

A summary of the technical problems usually encountered in the development of a strategy for overproducing proteins in Escherichia coli has been presented in this work, along with the particular aspects of the production of insulin chains A and B and their association to generate the biologically active hormone. It is clear that the process is quite complex, and while some of its components are well understood, others remain obscure.

The DNA manipulation techniques have been so efficiently developed in the past few years, that they do not represent a particularly difficult problem. The vast array of enzymes, plasmids and methods available make it possible to design and construct almost any recombinant DNA molecule for the expression of a particular gene. Furthermore, the generation of basic information about genetic regulation and host-vector interactions, have permitted the construction of more versatile, improved expression systems.

The two particularly difficult problems that may be encountered when overproducing a protein in Escherichia coli are the instability of the plasmid DNA, the mRNA transcript and the protein, and the renaturalization of the protein to obtain biological activity. Although some solutions are available for the stability problems (special strains, growth conditions, stabilizing sequences, etc.), there is no way in which an accurate prediction can be made on wether a particular gene, DNA construction or protein will be stable. The design of alternative strategies is, up to date, the best way to cope with a potential stability problem.

Protein renaturalization and enzymatic modification in Escherichia coli is yet, a quite unsolvable problem. Prokaryotic organisms are naturally unable to mimic the folding and the post-translational modification patterns of eukaryotic organisms. When a protein exhibits a high degree of complexity, such as several disulfide bonds or extensive glycosilation, the strategy should be selection of an eukaryotic expression system. This, however, is not usually a simple step in terms of costs and efficiency.

If a process is to be scaled-up to the industrial level, fermentation and purification procedures must also be scaled-up and optimized. Computarized equipment is available for the performance of controlled cell growth, as well as several choices of purification procedures (charge, size, hydrophobicity and affinity for ligands or antibodies) (Datar, 1986, Bonnereja et al., 1986; Naveh, 1986). Finally, the strict safety requirements of a new drug also accounts for a large period of time to elapse before a new pharmaceutical product reaches the market (Primrose, 1986). The above considerations lead us to the fact that the costs of developing a new biological entity as a drug for human therapy are enormous. Therefore, before choosing to develop a recombinant DNA-derived protein, attention must be paid to its marketability and its potential to produce sales far in excess of the development costs.

REFERENCES

Balbas, P., Soberon, X., Merino, E., Zurita, M., Lomeli, H., Valle, F., Flores, N. and Bolivar, F. ,1986, Plasmid vector pBR322 and its special-purpose derivatives- a review, Gene, 50:3-40.

Balbas, P. and Bolivar, F., 1988, Molecular cloning by plasmid vectors, in, Principles of recombinant DNA methodology, J.J. Greene, ed., Marcel Dekker, Inc, New York, NY, in press.

Bolivar, F. Rodriguez, R.L., Greene, P.J., Betlach, M.C., Heyneker, H.L., Boyer, H.W., Crosa, J.H. and Falkow, S., 1977, Construction and characterization of new cloning vehicles. II. A multipurpose cloning system, Gene, 2:95-113.

Bonnereja, J., Oh, S., Hoare, M. and Dunnhill, P., 1986, Protein purification: the right step at the right time, Bio/technology, 4:954-957.

Chance, R.E., Hoffmann, J.A., Kroeff, E.P., Johnson, M.G., Schirmer, E.W., Bromer, W.W., Ross, M.J. and Wetzel, R., 1981, The production of human insulin using recombinant DNA technology and a new chain combination procedure, Proc. of the Seventh American Peptide Symposium, 7:721-728.

Datar, R., 1986, Economics of primary separation steps in relation to fermentation and genetic engineering, Process Biochem., Feb:19-26

Dennis, K., Srienc, F. and Bailey, J.E., 1985, Ampicillin effects on five recombinant Escherichia coli strains: implications for selection pressure design, Biotechnol. Bioeng., 27:1490-1494.

Ensley, B.D., 1985, Stability of recombinant plasmids in industrial microorganisms, CRC Biotechnol., 4:263-277.

Flores, N., de Anda R., Guereca, L., Cruz, N., Antonio, S., Balbas, P., Bolivar, F. and Valle, F., 1986, A new expression vector for the production of fused proteins in Escherichia coli, Appl. Microbiol. Biotechnol., 25:267-271.

Gentz, R., Langer, A., Chang, A.C., Cohen, S.N. and Bujard, H., 1981, Cloning and analysis of strong promoters is made possible by the downstream placement of a RNA termination signal, Proc. Natl. Acad. Sci. USA, 78:4936-4940.

Goeddel, D.V., Kleid, D.G., Bolivar, F., Heyneker, H.L., Yansura, D.G., Crea, R., Hirose, T., Krazewski, A., Itakura, K. and Riggs, A.D., 1979, Expression in Escherichia coli of chemically synthesized genes for human insulin, Proc. Natl. Acad. Sci. USA, 76:106-110.

Gottesman, S., 1984, Bacterial regulation: global regulatory networks, Ann. Rev. Genet., 18:415-441.

Gross, E., 1967, The cyanogen bromide reaction, Meth. Enzymol., 11:238-255.

Harban, P.A., Berger, M., Gjinovci, A. and Renold, A.E., 1981, Biological activity and pharmacokynetics of biosynthetic human insulin in the rat, Diabetes Care, 4:238-243.

Harris, T.J.R., 1983, Expression of eukaryotic genes in Escherichia coli, in, Genetic engineering, R. Williamson, ed., Academic Press, New York, NY, pp. 127-183.

Inouye, I., 1983, Experimental manipulation of gene expression, Academic Press, New York, NY.

Itakura, K., Hirose, T. Crea, R., Riggs, A.D., Heyneker, H.L., Bolivar, F. and Boyer, H.W., 1977, Expression in Escherichia coli of a chemically synthesized gene for the human hormone somatostatin, Science, 198:1056-1063.

Katsoyannis, P.G., Trakatellis, A.C., Johnson, S., Zalut, C. and Schwartz, G., 1967, Studies on the synthesis of insulin from natural and synthetic A and B chains, Biochemistry, 6:2642-2655.

Kaytes, P.S., Theriault, N.Y., Poorman, R.A., Murakami, K. and Tomich, C.C., 1986, High-level expression of human rennin in Escherichia coli, J. Biotechnol., 4:205-218.

Kim, S.H. and Ryu, D.D.Y., 1984, Instability kinatics of trp operon plasmid ColE1-trp in recombinant Escherichia coli MV12(pVH5) and MV12trpR(pVH5), Biotechnol. Bioeng., 26:497-502.

Klotsky, R.A. and Schwartz, I. , 1987, Measturement of cat expression from growth-rate-regulated promoters employing B-lactamase activity as an indicator of plasmid copy number, Gene, 55:141-146.

Ladron de Guevara, O., Estrada, G., Antonio, S., Alvarado, X., Guereca, L., Zamudio, F. and Bolivar, F., 1985, Identification and isolation of human insulin A and B chains by high-performance liquid chromatography, J. Chromatogr., 349:91-98.

Luckow, V.A. and Summers M.D., 1988, Trends in the development of Baculovirus expression vectors, Bio/technology, 6:47-55.

Meyer, H.P., Kappeli, O. and Fiechter, A., 1985, Growth control in microbial cultures, Ann. Rev. Microbiol., 39:299-319.

Morgan, C,R. and Lazarow, A., 1962, Immunoassay of insulin using a two-antibody system, Proc. Soc. Exp. Biol. Med., 3:29-32.

Naveh, D. , 1986, Scale-up of fermentation for recombinant DNA products, Food Technol., 11: 102-109.

Nichols, B.P. and Yanofsky, C., 1983, Plasmids containing the trp promoters of Escherichia coli and Serratia marcescens and their use in expressing cloned genes, Meth. Enzymol., 101:155-164.

Nilsson, G., Belasco, J.G., Cohen, S.N. and von Gabain, A., 1984, Growth-rate dependent regulation of mRNA stability in Escherichia coli, Nature, 312:75-76.

Pabo, C.O., Sauer, R.T., Sturtevant, J.M. and Ptashne, M., 1979, Lambda repressor contains two domains, Proc. Natl. Acad. Sci. USA, 76:1608-1611.

Primrose, S.B., 1986, The application of genetically engineered microorganisms in the production of drugs, J. Appl. Bacteriol., 61: 99-116.

Randall, S.S., 1964, The small-scale preparation of crystalline insulin, Biochim. Biophys. Acta, 90:472-476.

Reznikoff, W. and Gold, L., 1986, Maximizing gene expression, Butterworths Pub., Stoneham, Mass.

Roberts, J.W. and Roberts, C.W., 1975, Proteolytic cleavage of bacteriophage lambda repressor in induction, Proc. Natl. Acad. Sci. USA, 72:147-151.

Rodriguez, R.L. and Denhardt, D.T., 1987, Vectors: a survey of molecular cloning vectors and their uses, Butterworths Pub., Stoneham, Mass.

Romans, R.G., Scott, D.A. and Fisher, A.M., 1940, Preparation of crystalline insulin, Indust. Engin. Chem., 32:908-910.

Rosteck Jr., P.R. and Hershberger, C.L., 1983, Selective retention of recombinant plasmids coding for human insulin, Gene, 25:29-38.

Schlichtkrull, J., 1956, Insulin crystals, Acta Chemica Scandinavia, 10:1455-1458.

Schoner, B.E., Belagaje, R.M. and Schoner, R.G., 1986, Translation of a synthetic two-cystron mRNA in Escherichia coli, Proc. Natl. Acad. Sci. USA, 83:8506-8510.

Schoner, R.G., Ellis, L.F. and Schoner, B.E., 1985, Isolation and purification of protein granules from Escherichia coli from cells overproducing bovine growth hormone, Bio/technology, 3:151-154.

Seo, J. and Bailey, J.E., 1985, Effects of recombinant plasmid content on growth properties and cloned gene product formation in Escherichia coli, Biotechnol. Bioeng., 27:1668-1674.

Shatzman, A.R. and Rosenberg, M., 1987, Expression, identification and characterization of recombinant gene products in Escherichia coli, Meth. Enzymol., 152:661-673.

Soberon, X., Covarrubias, L. and Bolivar, F., 1980, Construction and characterization of new cloning vehicles. IV. Deletion derivatives of pBR322 and pBR325, Gene, 9:287-305.

Swami, K.H.S. and Goldberg, A.L., 1981, Escherichia coli contains eight soluble proteooytic activities, one being ATP dependent, Nature, 292-652-654.

Weinstock, G.M., Rhys, C., Berman, M.L., Hamper, B., Jackson, D., Silhavy, T.J., Weisman, J. and Zweig, M., 1983, Open reading frame expression vectors: a general method for antigen production in Escherichia coli using protein fusions to B-galactosidase, Proc. Natl. Acad. Sci., USA 80:4432-4436.

Williams, D.C., Van Frank, R.M. Muth, W. and Burnett, J.P., 1982, Cytoplasmic inclusion bodies in Escherichia coli producing biosynthetic human insulin proteins, Science, 215:687-689.

Wong, H.C. and Chang, S., 1986, Identification of a positive retroregulator that stabilizes mRNA in bacteria, Proc. Natl. Acad. Sci. USA, 83:3233-3237.

Zabriskie, D.W. and Acuri, E.J., 1986, Factors influencing productivity of fermentations employing recombinant microorganisms, Enzyme Ferment. Technol. 8:706-717.

ACKNOWLEDGEMENTS

This work was partially supported by a grant from the Instituto Mexicano del Seguro Social. We are grateful to Enrique Merino for his helpful ideas and his assistance in the preparation of this manuscript. Oralia Ladrón de Guevara, Georgina Estrada, Leopoldo Güereca, Salvador Antonio and Fernando Zamudio developed and performed the protein purification procedures. Special acknowledgements to Dr. Victoria Valles, Dr. Josue Garza, Quim. Belia Wong and Quim. Ma. Concepción Rocha from the Instituto Nacional de la Nutrición and Dr. Salvador Zubirán for their participation in the development of the radioimmunoassay procedures. We also thank Dr. Enrique Hong and Mr. Julio Sánchez from CINVESTAV/IPN, as well as Dr. Alberto Chousleb, Dr. Enrique Foyo, Dr. Angélica Salas, Dr. Carlos Arámburo and Biol. Rocío Sánchez from U.N.A.M., for their invaluable help in the development of the biological assays for human insulin. X.A. is specially grateful to Dr. Enrique Hong and Dr. Carlos Arámburo for their continuous support and helpful discussions.

DIAGNOSIS AND CHARACTERIZATION OF NEW MUTATIONS IN MAN

Al Edwards[1] and C. Thomas Caskey[1,2,3]

[1]Department of Cell Biology, [2]Institute for Molecular Genetics, [3]Howard Hughes Medical Institute, Baylor College of Medicine, One Baylor Plaza, Houston, Texas 77030

INTRODUCTION

Complete deficiency of hypoxanthine phosphoribosyltransferase (HPRT) is associated with Lesch-Nyhan syndrome, a severe neurological deficiency characterized by choreoathetosis, mental retardation, and a compulsive behavior toward self mutilation, while a partial deficiency results in gouty arthritis (1). The HPRT enzyme is involved in the metabolic salvage of purines, and is encoded by an X-lined gene (1,2). As there is no selective advantage associated with Lesch-Nyhan syndrome, and the defect is effectively lethal since affected males do not reproduce, Haldane's principle predicts that the gene must be maintained in the population by new mutations (3).

The predominance of new mutations at the human HPRT locus and the availability of a cloned complementary DNA (cDNA) and well characterized genomic clones (4-8) has allowed for the (i) characterization of mutations in patients with Lesch-Nyhan syndrome, (ii) study of the mechanisms of mutation and reversion, and (iii) development of techniques for the accurate and rapid characterization and diagnosis of human mutations. The HPRT gene has several properties which have allowed investigators to pursue such studies. Since the gene is located on the X-chromosome, it is hemizygous in males and effectively hemizygous in females due to the process of Lyonization. Thus, mutational and reversional processes can be examined without the complications that can arise in the diploid state. The HPRT enzyme is not essential for cellular growth *in vitro*, thereby facilitating the isolation of mutants. Third, positive and negative selection procedures have been developed which allow for facile identification of mutants and revertants (reviewed in 9).

In this paper, we will describe studies from our laboratory on the (i) structure of the human HPRT gene, (ii) classes of mutations in the human HPRT gene of Lesch-Nyhan patients, (iii) mechanisms of mutations giving rise to major genomic rearrangements, and (iv) development of methods for the rapid characterization and diagnosis of all mutations classes.

Both human and mouse HPRT genes have been well characterized (reviewed in 9 and 10). Full length cDNA clones are available and the sequence is known (4,5). A summary of the exonic and genomic structures of the human HPRT gene is shown in Figures 1 & 2. Both the human and mouse genes consist of 9 exons distributed over 44 kilobase pairs (kb) and 33 kb, respectively (7,11). The difference in size is due to the length of the intervening sequences, which occur at identical positions in the coding sequence of the human and mouse genes. In addition to the X-linked HPRT gene, four independent HPRT related sequences are found in the human genome, which presumably represent processed pseudogenes (12). Two of the sequences have been localized to chromosome 11, and one each to chromosomes 3 and 5 (12). HPRT related sequences have also been identified in the mouse and hamster genomes (10).

The sequence of the human HPRT gene would greatly facilitate the rapid characterization and diagnosis of mutations utilizing both established and recently developed techniques. A detailed restriction map and localization of repeated elements would provide useful information about the structure of the gene and facilitate the development of unique probes for cloning and hybridization analysis. DNA sequence is indispensible for newer sequence based strategies, e.g., the polymerase chain reaction (PCR) in conjunction with methods for detecting mutations such as sequencing or ribonuclease (RNase) A cleavage. Further, the sequence of the HPRT gene in conjunction with such techniques will allow investigators to address questions about mutational rates, mechanisms, the DNA elements which may be involved in mutational events, and evolutionary issues.

Thus, we have initiated sequencing of the human HPRT gene in collaboration with Wilhelm Ansorge's group at the European Molecular Biology Laboratory (EMBL). The sequence is being determined by random or "shotgun" DNA sequencing and, at later stages, by more directed methods such as those utilizing specific oligonucleotide primers to fill in any remaining gaps or resolve sequence ambiguities (reviewed in 13). A sequencing library was constructed by sonicating recombinant lambda phage containing inserts from the human HPRT gene to 1-2 kb, size selection, and ligation into M13. The library was screened using radiolabelled total recombinant phage and arms to identify clones containing sequences from the lambda phage insert. Template was prepared using the glass fiber method (14). Dideoxy sequencing was performed using genetically modified T7 DNA polymerase graciously provided by Pharmacia and fluorescently tagged primers. The automated fluorescent sequencer developed by Ansorge's group is being used (15). The University of Wisconsin Genetics Computer Group (GCG) package is being used to assemble the overlapping sequence fragments. In collaboration with Peter Rice at EMBL, several new programs have been written to supplement the GCG gel assembly system. At this time 50% of the sequence of the human HPRT gene has been determined.

MUTATIONS AT THE HPRT LOCUS AND METHODS OF DETECTION AND STUDY

Using a full length human HPRT cDNA in Southern analyses to examine 28 independent Lesch-Nyhan mutations, 85% of the cases had blotting patterns indistinguishable from normal (16). Although Southern analysis detects mutations in only 15% of cases, it has proven a valuable method for the detection of major genomic arrangements, as illustrated in Figure 2. The laboratory is very interested in using these cases to describe in detail the lesions and use the mutations to ask basic biological questions about the mechanisms of mutation and reversion. Our progress in this area is summarized below. Of 15 cases examined by Northern analysis, 75% had a size and quantity of message which could not be distinguished from the normal controls (17). Even though 75% of cases had apparently normal mRNA levels, only 25% had levels of immunoreactive protein greater than or equal to 50% of normal levels. The levels (abnormal Southern, northern, or protein) at which the mutations in this set of patients are expressed are known in 15% of the cases and the sequence basis of the mutation is known fully in only 1 case (7%). At the time these studies were performed, existing methods for the characterization of mutations at the sequence level, i.e., cDNA and genomic cloning, were prohibitively slow. Thus, over the past two years our laboratory has been developing and applying techniques that will allow not only for the rapid characterization of point mutations and major genomic rearrangements, but also for facile diagnosis of mutations.

Ribonuclease A cleavage would allow a more direct examination of the 75% of Lesch-Nyhan cases which produce a message than would Southern analyses targeted at the level of genomic DNA. Therefore, RNase A cleavage has been applied to the detection of mutations in patients with Lesch-Nyhan syndrome (18). The technique is based upon the ability of RNase A to cleave some mismatches in RNA-RNA or RNA-DNA duplexes. RNase A cleavage has been adapted to the analysis of rare messages, such as HPRT which is estimated to represent 0.01% of mRNA, in our laboratory (18). The strategy involves separating free radiolabelled probe from probe hybridized to poly(A)$^+$ mRNA on messenger affinity paper (polyuridylic acid affinity paper). The method has been applied to a set of Lesch-Nyhan patients allowing for the detection of the mutation in 35% of Lesch-Nyhan cases in which a stable RNA is made. As shown in Figure 2, RNase A cleavage allowed for the identification of a possible point mutation in the mRNA of three patients (RJK 951, 906, and 894) and the detection of two deletion cases (RJK 855 and 888). Although RNase A cleavage allows for the detection of 35% of the mutations in mRNA and to some extent the position and type (e.g., point or deletion) of mutation, it does not provide exact positions or sequence information. This detailed information, albeit not necessary for diagnosis and carrier detection, is essential for understanding the mechanisms giving rise to the mutations and their influence on the function of the enzyme.

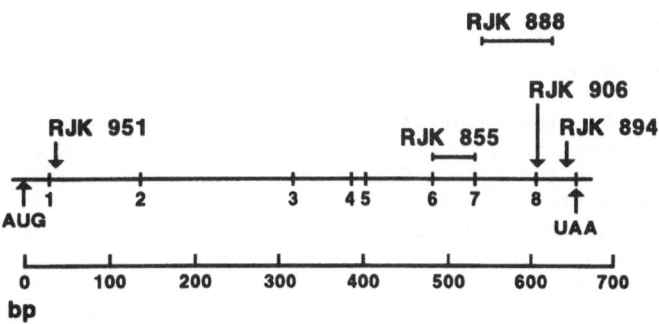

Figure 1. RNase A Cleavage of human HPRT mutants. The location of mutations in the HPRT coding region are shown.

The laboratory has identified 9 Lesch-Nyhan cases involving major genomic rearrangements detectable by Southern analysis (9, 16, and unpublished). Six of these are deletion mutants (Figure 2). Two cases (RJK 853 and a female Lesch-Nyhan case discussed below) have all nine HPRT exons deleted, and probably represent complete gene deletions. RJK853 is particularly useful as a control for HPRT-like autosomal sequences seen in Southern analysis with the cDNA. Three of the deletion mutants, namely RJK 3467, 984, and 849, have lost progressive amounts of the 3' end of the gene. It is tempting to speculate that the predominance of deletions in the 3' portion of the gene is due to the presence of a deletional "hot spot". The deletion in RJK 984 has been localized by detailed Southern mapping using unique intron and intron/exon probes to within less than 1 kb 5' to exon 6. We are currently cloning a fragment containing the deletional junction in this cell line towards understanding the deletional processes. The 3' deletion mutants RJK 3467 and RJK 849 have been localized to introns 6 and 3, respectively. RJK 1780 is a recently identified deletion case. A Southern performed using Pst I suggests that exons 1 and 2 are deleted, thus representing the first deletion extending 5' to the gene.

The banding pattern observed in Southerns using DNA derived from GM2227 was consistent with a translocation, inversion, or insertion. A cytogenetic analysis found that the region of the X chromosome containing the HPRT gene appeared normal, ruling out the possibility of a translocation accounting for the findings. There was, however, an unrelated translocation between the short arm of the X chromosome and the long arm of chromosome 13 in the population under study. Extensive Southern analysis of this mutant has been performed and the mutation has been localized to intron 5. The 5' junction of the mutation has been cloned and sequenced. The sequence is consistent with a topological mechanism involving illegitimate recombination between two similar, but not identical, Alu repeats. A repeat free probe isolated from sequences at the mutation suggests that the event is an inversion or, less likely, an insertion. This conclusion is based primarily upon the finding that the sequence involved in the mutation in GM2227 is also deleted in RJK 853. Furthermore, there are two additional sequences not localized to q24-qter of the X-chromosome which hybridize to the mutational probe. Additional molecular study will be required to delineate the precise mechanism.

Northern analysis of mRNA from GM1662 (fibroblast) or GM6804 (lymphocyte) identified a message with retarded migration consistent with a duplication or faulty initiation or termination of transcription (19). Southern analyses were consistent with a duplication of exons 2 and 3 (19). A cDNA was cloned from GM1662 and sequenced, revealing a precise endoduplication of exons 2 and 3. Southern analyses (shown in Figure 3) have localized the boundaries of the duplication to discreet regions within introns 1 and 3, and revealed that the duplication could not be accounted for by a simple unequal crossing-over event due to the presence of unknown rearrangements in the duplicated regions (19). Yang et. al. (19) also report the isolation of HPRT[+] revertants from GM6804 after selection in HAT. The revertants were found to produce a mRNA of size identical to normal, and to contain normal quantities and character of enzyme even though the mutant GM6804 cells lack enzyme activity and immunoreactive protein. Further, Southern analysis demonstrated the loss of one of the duplicated DNA segments in the revertants. This interesting observation has been extended by the observation that HPRT[+] revertants can be isolated in vitro in HAT after preselection in 6TG to remove any HPRT[+] cells (AE, CTC, unpublished). Several interesting questions remain unanswered: What mechanisms account for the inversion? For the reversion? Do the revertant cell lines arise by a similar mechanism? To answer these questions a genomic library has been constructed and is being screened, and independent revertants are being isolated. It is anticipated that the sequence of the normal HPRT gene will greatly facilitate the analyses.

An unusual case of a female Lesch-Nyhan patient has been described (25, 26). Separation of the X chromosomes in this patient by cell fusion allowed for examination of each separately by Southern analysis with a full length human HPRT cDNA (Ogasawara et al., in preparation). It was found that the maternal X, which could be distinguished from the paternal X by RFLP analysis with DXS-10, had suffered a complete deletion of the HPRT gene, while the paternal X contained an HPRT gene which gave a banding pattern indistinguishable from normal. These findings suggest that either (i) the paternal HPRT gene was nonfunctional, in spite of a normal Southern pattern or (ii) that the paternal X was selectively inactivated. Six independent hybrid clones containing the paternal X were shown not to express human glucose-6-phosphate dehydrogenase (G6PD) or phosphoglycerate kinase (PGK), while one hybrid containing the maternal X expressed both of the markers. The results support the latter hypothesis, i.e., that the paternal X had been selectively inactivated. A model consistent with these findings would hold that the development of Lesch-Nyhan syndrome in the female patient was due to a lethal mutation in the paternal X chromosome causing it to be selectively inactivated, in spite of the finding that it probably contains a functional HPRT gene. Alternatively, any mechanism which failed to reactivate the paternal X-chromosome could result in this event.

DIAGNOSIS OF NEW AND ACQUIRED MUTATIONS

New and acquired mutations pose a common diagnostic challenge. This is particularly the case for genes like HPRT, where all 18 mutations characterized thus far have been unique. The sporadic nature of mutations in the HPRT gene, and other genes, has made it difficult to identify mutations in more than 50% of cases (see above), and has prompted a search for more fruitful and specific methods. In the past, the detection of new mutations in probands, determination of carrier status, and identification of the origins of mutations have relied heavily upon assays for enzyme activity and restriction fragment length polymorphisms (RFLP).

Biochemical identification of affected males and carriers can be performed with the hair follicle assay or by selection in 6-thioguanine, to destroy cells containing HPRT activity (reviewed in 1). Despite an extensive search for RFLPs using a full length human cDNA, only one Bam HI polymorphism has been identified (20,21). This three allele RFLP is expressed phenotypically on Southern blots as three distinct pairs of fragments: (i) a 22 kb/25 kb pair; (ii) a 12 kb/25 kb pair; and (iii) a 22 kb/18 kb pair. Allele frequencies for a caucasian population were 0.77 for the 22 kb/25 kb allele, 0.16 for the 12 kb/25 kb allele, and 0.07 for the 22 kb/18 kb allele, giving an average heterozygosity of 38% in females of the population. Further, a Taq I RFLP has been reported with the anonymous probe DXS-10 (22). The mutation in a patient with gouty arthritis ($HPRT_{Toronto}$) was detected by Southern analysis using the restriction enzyme TaqI and an HPRT cDNA probe. The mutation was a C to G transversion giving rise to a new TaqI site in exon 3 as shown in Figure 2 (23).

Family studies of three Lesch-Nyhan patients has allowed for the identification of the origin of the mutations using Southern analyses to examine RFLPs and the exons. Study of the family of GM1662 identified an abnormal 4.1 kb BglII band associated with the mutation in the

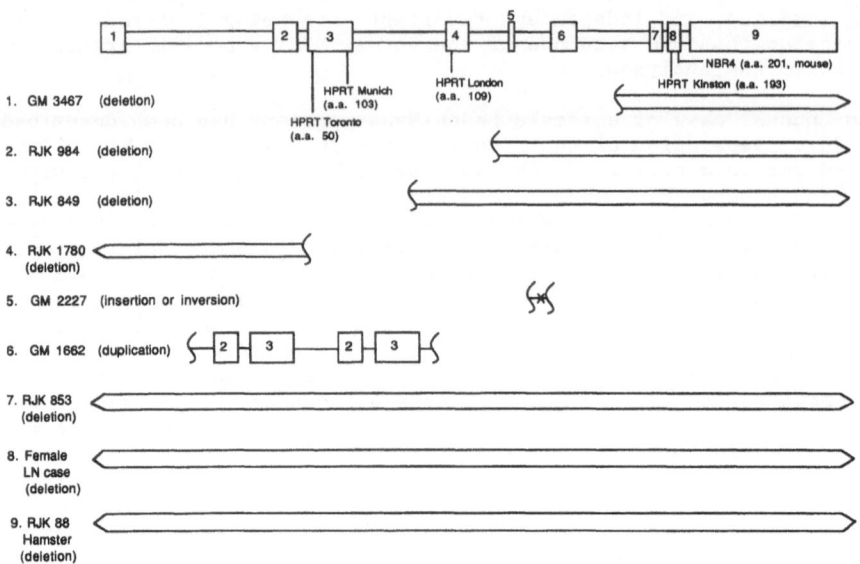

Figure 2. HPRT mutants. Point mutations identified are indicated within respective exons at the top while mutations involving major gene alterations are depicted in the lower half.

propositus, his mother, and two sisters (16). The abnormal band was, however, absent from the maternal grandmother, indicating that the mutation occurred in her germ line. Family study of two deletion cases (RJK 853 and 984) identified the maternal germ line as the source of the mutation (9). Although these carrier detection studies illustrate the ability to determine carrier status and decipher the origin of mutations, it should be noted that with one exception (23), the relative lack of information provided by the Southern analyses has allowed carrier studies only with families in which the mutation involved a major genomic rearrangement, such as a duplication or large deletion. Even so, RFLPs combined with Southern analysis with the cDNA and other enzymes to visualize all of the exons, has been the standard for mutational screening at the HPRT locus. As discussed above, the addition of RNase A cleavage to these methods has allowed for the identification of the mutation in 50% of cases. The laboratory is now focusing on the development and application of techniques utilizing the polymerase chain reaction (PCR) and direct sequencing.

A strategy for the rapid detection of nearly all mutations at the HPRT locus (and other loci) can be based upon three factors. First, the sequence of gene should be known, or, at a minimum, the coding regions, exon-intron boundaries, and other regions of the gene with regulatory functions. We feel that a complete sequence, with a detailed restriction map and the identification of potential problem areas (repeats, degenerative sequence tracts) will prove indispensible. Second, a knowledge of the classes, relative rates of each class, and origins of the mutations should be known. For example, 85% of mutations in the HPRT gene appear to involve point mutations or small genomic rearrangements not detectable my Southern analysis. What fraction of these mutations occur in coding regions (exons) versus 5' or 3' flanking regions of the cDNA? What

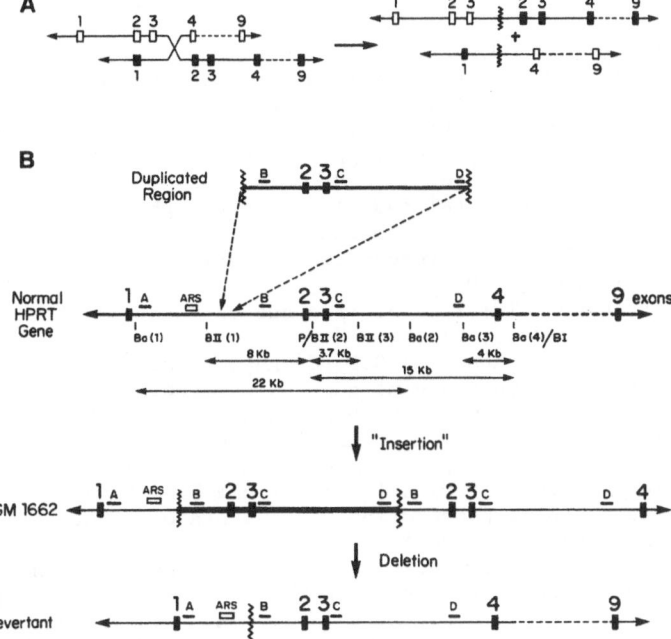

Figure 3. GM 1662 duplication mutant. Panel A shows an unequal crossover event giving rise to a duplication of exons 2 and 3. The duplication in GM 1662 cannot be accounted for completely by this mechanism (see
text). Panel B shows a detailed restriction map of the normal gene,
original duplication, and revertant cell lines.

fraction of the mutations occur outside of the coding region to influence
the mRNA, such as splicing mutations? Published (discussed above) and
unpublished studies suggest that while most of the point mutations are
located in the coding regions of the gene, the other classes also exist.
For example, while both RJK 855 and GM 2292 have Southern patterns indistinguishable from normal, RJK 855 produces a normal size and quantity of
message, whereas GM2292 fails to produce a detectable level of message
(16). Continued investigation of such mutants will allow for the identification of sequence elements important in the production of a mRNA and
active protein, even though the mutational events themselves appear to be
random. Third, the development and application of new technologies like
the PCR and direct sequencing of amplified products will allow for the
identification of the mutations rapidly at the level of sequence (24).

A possible strategy is to develop a set of PCRs based upon the known
set of mutants and to directly sequence the amplified products. For
example, in one or two tubes, one could amplify 500 bp to 2000bp surrounding each of the exons, the 5' and 3' noncoding regions, and the
promotor/enhancer/negative regulatory element regions. The amplified
products could be analyzed by RNase cleavage or preferably direct sequencing for deletions of exons and point mutations in exons, splice
donor, splice accepters, promoters, and other regions (24). A potential
problem is the error rate associated with the PCR; however, the error
rate can probably be avoided by direct sequencing of the amplified
products. Even doing each analysis twice to resolve ambiguities would
represent a significant time and cost advantage compared to Southern,
RFLP, Northern, cloning, and RNase A cleavage methods.

REFERENCES

1. W.N. Kelley, J.B. Wyngaarden. Clinical syndromes associated with hypoxanthine guanine phosphoribosyltransferase deficiency. *In The Metabolic Basis of Inherited Disease*, eds. J.B. Stanbury, J.B. Wyngaarden, D.S. Frederickson, J.L. Goldstein, M.S. Brown, pp. 1115-43. New York: McGraw Hill, 2032, pp. 5th ed., (1983).

2. G.S. Pai, J.A. Sprenkle, T.T. Do, C.E. Mareni, B.K. Migeon. Localization of loci for HPRT and glucose-6-phosphate dehydrogenase and biochemical evidence for non-random X-chromosome expression from studies of a human X-autosome translocation. *Proc. Natl. Acad. Sci. USA* 77:2810-13, (1980).

3. J.B.S. Haldane. The rate of spontaneous mutation of a human gene. *J. Genet.* 31:317-26, (1935).

4. J. Brennand, D.S. Konecki, C.T. Caskey. Expression of human and Chinese hamster hypoxanthine guanine phosphoribosyltransferase cDNA recombinants in cultured Lesch-Nyhan and Chinese hamster fibroblasts. *J. Biol. Chem.* 258:9593-96, (1983).

5. D. Jolly, H. Okayama, P. Berg, A.C. Esty, D. Filpula., *et al.* Isolation and characterization of a full-length expression cDNA for human hypoxanthine phosphoribosyltransferase. *Proc. Natl. Acad. Sci. USA* 80:477-81, (1983).

6. D.J. Jolly, A.C. Esty, H.V. Bernard, T. Friedmann. Isolation of a genomic clone partially encoding human hypoxanthine phosphoribosyltransferase. *Proc. natl. Acad. Sci. USA.* 79:5038-41, (1982).

7. P.I. Patel, P.E. Framson, C.T. Caskey, and A.C. Chinault. Fine structure of the human hypoxanthine phosphoribosyltransferase gene, *Mol. Cell Biol.* 6(2):393-403, (1966).

8. *S.H. Kim, J.C. Moores, D. David, J.G. Respess, D.J. Jolly, and T. Friedmann. The organization of the human HPRT gene, Nucleic Acids Research, 14(7):3103-3118, (1986).*

9. J.T. Stout, and C.T. Caskey. HPRT: gene structure, expression and mutation. In *Annual Review of Genetics, Vol 19* (A Campbell, I Herskowitz and L.M. Sandler Eds.) Palo Alto: Annual Review Inc., pp 127-148, (1985).

10. D.W. Melton. HPRT gene organization and expression. In *Oxford Surveys on Eukaryotic Genes* (N. Maclean Ed.) Vol. 4, (1987).

11. D.W. Melton, D.S. Konecki, J. Brennand, C.T. Caskey. Structure expression and mutation of the hypoxanthine phosphoribosyltransferase gene, *Proc. Natl. Acad. Sci. USA* 81:2147-51, (1984).

12. P.I. Patel, R.L. Nussbaum, P.E. Framson, D. Ledbetter, C.T. Caskey, A.C. Chinault. Organization of the HPRT gene and related sequences in the human genome, *Somat: Cell Mol. Genet.* 10:483-93, (1984).

13. W.M. Barnes. Sequencing DNA with Dideoxyribonucleotides as chain terminators: Hints and strategies for big projects, in *Methods in Enzymology* Vol. 152 Guide to Molecular Cloning Techniques (Shelby L. Berger, Ed.), Academic Press pp 538, (1987).

14. T. Kristensen, H. Voss and W. Ansorge. A simple and rapid preparation of M13 sequencing templates for manual and automated dideoxy sequencing, *Nucleic Acids Research* 15:5507-5516, (1987).

15. W. Ansorge, B. Sproat, J. Stegemann, C. Schwager, and M. Zenke. Automated DNA sequencing: ultrasensitive detection of fluorescent bands during electrophoresis, *Nucleic Acids Research* 15:4593-4602, (1987).

16. T.P. Yang, P.I. Patel, A.C. Chinault, J.T. Stout, L.G. Jackson. Molecular evidence for new mutation in the HPRT locus in Lesch-Nyhan patients, *Nature* 310:412-14, (1984).

17. J.M. Wilson, J.T. Stout, T.D. Palella, B.L. Davidson, W.N. Kelley, and C.T. Caskey. A molecular survey of hypoxanthine-guanine phosphoribosyltransferase deficiency in man, *J. Clin. Invest.* 77:188-195, (1986).

18. R.A. Gibbs, and C.T. Caskey. Identification and localization of mutations at the Lesch-Nyhan locus by ribonuclease A cleavage, **236**:303-305, (1987).

19. T.P. Yang, J.T. Stout, D.S. Konecki, P.I. Patel, R.L. Alford, and C.T. Caskey. Spontaneous reversion of a novel Lesch-Nyhan mutation by HPRT gene rearrangement, *Som. Cell Mol. Gen.* (in press).

20. R.L. Nussbaum, W.E. Crowder, W.L. Nyhan, and C.T. Caskey. A three-allele restriction-fragment-length polymorphism at the hypoxanthine phosphoribosyltransferase locus in man. *Proc. Natl. Acad. Sci. USA* **80**:4035-4039 (1983).

21. D.A. Gibbs, C.M. Headhouse-Benson, and R.W.E. Watts. Family studies of the Lesch-Nyhan syndrome: the use of a restriction fragment length polymorphism (RFLP) closely linked to the disease gene for carrier state and prenatal diagnosis. *J. Inher. Metab. Dis.* 9:45-58, (1986).

22. B.A. Boggs, and R.L. Nussbaum. Two anonymous X-specific human sequences detecting restriction fragment length polymorphisms in the region X126- qter. *Somat. Cell Mol. Genet.* 10:607-13, (1984).

23. J.M. Wilson, P. Frossard, R.L. Nussbaum, C.T. Caskey, and W.N. Kelley. Human hypoxanthine-guanine phosphoribysoltransferase; detection of a mutant allele by restriction endonuclease analysis. *J. Clin. Invest.* 72:767-772 (1983).

24. R.K. Saiki, D.H. Gelfand, S. Stoffel, S.J. Scharf, R. Higuchi, G.T. Horn, K.B. Mullis, and H.A. Erlich. Primer-directed enzymatic amplification of DNA with a thermostable DNA polymerase. *Science* **239**:487-491, (1988).

25. K. Hara, *et al*. A Female case of the Lesch-Nyhan syndrome. Tohoku J. Exp. *Med.* **137(3)**:275-282, (1982).

26. N. Ogasawara, *et al*. Hypoxanthine-guanine phosphoribosyl transferase (HGPRT) in a girl. *In Purine metabolism in man IV, Part A,* Plenum N.Y. 13-18 (1984).

CLINICAL ASPECTS OF INBORN ERRORS OF METABOLISM

Alessandra Carnevale

Jefe de la Div. de Investigación, Instituto Nacional de Pediatría, México, D.F.

The inborn errors of metabolism have caught the imagination of physicians and scientists due to their promise of insights into the mechanisms of both the pathogenesis of genetic disease and gene action itself. The etiologic agent of these disorders is the mutant gene. In this Symposium, attention is focused chiefly upon the mechanism by which the mutant gene produces clinical manifestations.

Since genes control the structure of polypeptides and their corresponding proteins, a gene mutation leads to a change in the amino acid chain. Depending on the nature of this amino acid change and its position in the molecule, the function of the protein may be altered (1).

The consequences of a genetic alteration in quality or quantity of a protein will depend on the normal role of that protein. When the protein involved is an enzyme the result is a metabolic block and the disorder is an inborn error of metabolism.

Several mechanisms can account for the reduction in enzymatic activity. First, in homozygotes for a mutant gene, the enzyme coded for by that gene may not be produced at all or be produced in an abnormal form with reduced activity. Second, if the mutation involves a gene that regulates the rate of production of the enzyme, an inadequate amount of normal enzyme will result. Third, the mutation may alter the rate of degradation of the enzyme leading to a deficiency of active enzyme. Fourth, normal activity may depend on association with a cofactor and mutations that interfere in some way with this association may reduce the activity of the enzyme. One useful way of classifying the diseases resulting from metabolic defects is according to the pathological consequences of the metabolic block. In this occasion we will present a brief inquiry into the potential effects and a few clinical examples of representative diseases.

Failure of formation of an end product

In these disorders the major clinical problems result from the absence of the end product of a metabolic pathway. One of the original errors of metabolism, albinism, is a good example. In the tyrosinase-negative form of albinism, because of the lack of tyrosinase activity in melanocytes, there is no melanin in skin, hair or eyes. Albinos from various racial back-grounds have similar phenotypic characteristics. All have snow-white hair, pink-white skin, gray to blue-gray irides in tangential illumination, a pink-eye appearance because of a red reflex from the unmelanized fundus, severe nystagmus and photophobia with a decreased visual acuity. Genetic heterogeneity exists, since there are several other ways of blocking the melanin production. At least seven different forms of albinism have been detected where tyrosinase-positive albinism and yellow mutant albinism are the most common (2). Another example of this type of metabolic defects is the recessive inherited goitrous cretinism. The major clinical manifestations are mental retardation, short stature and coarse face with macroglossia, all resulting from a lack of thyroid hormone (1).

Accumulation of precursors on the metabolic block

Accumulation of precursors on the metabolic block as in storage diseases, galactosemia, maple syrup urine disease or isovaleric acidemia.

Mucopolysaccharidoses are examples of storage diseases; they are hereditary, progressive disorders caused by the intralysosomal accumulation of glycosaminoglycans (acid mucopolysaccharides) in various tissues. Glycosaminoglycans are long-chain complex carbohydrates usually linked to proteins to form proteoglycans. Proteoglycans are major constituents of the ground substance of connective tissue. In the organism glycosaminoglycans are degraded by the sequential action of lysosomal enzymes leading to a stepwise shortening of the chain. The lack of a lysosomal enzyme results in the accumulation of partially degraded glycosaminoglycan molecules in lysosomes and interfere with normal cell function (3). Some clinical manifestations of the mucopolysaccharidoses such as coarse facial features, thick skin, corneal clouding and organomegaly are the direct expression of the molecule accumulation in tissues. Others, such as mental retardation, short stature, skeletal abnormalities are the result of a defective cell function. Joint contractures and herniae suggest an interference with other metabolic substance such as collagen or fibronectin.

There are at least 11 enzyme deficiencies leading to the accumulation of biochemically different glycosaminoglycan degradation products. As a general rule, the impaired degradation of heparan sulphate is more closed associated with mental deficiency and the impareid degradation of dermatan sulphate, chondroitin sulphate and keratan sulphate with mesenchymal abnormalities (3).

Another example is galactosemia, in which the defective enzyme is galactose-1-phosphate uridyl transferase, which normally converts galactose-1-phosphate to glucose-1-phosphate. In the mutant homozygote this step cannot occur and galactose-1-phosphate accumulates in blood cells, liver and other tissues. Untreated patients show distinctive manifestations early in life, when milk feedings are started. Food may be refused and vomiting is common; lethargy, hypotonia, jaundice, hepatomegaly and susceptibility to infection appear. Later, cataracts become evident and physical and mental retardation occur. It is believed that galactose-1-phosphate produces hepatic damage whereas galactitol accounts for the formation of cataracts (2).

Various disorders of amino acid metabolism fall in this category. Two examples of disorders of the branched chain amino acids are the maple syrup urine diseases (MSUD) and the isovaleric acidemia.

In the first, the patients inherit a deficiency of the branched chain α-keto acid dehydrogenase.

The enzyme is responsible for the oxidative decarboxylation of the α-keto acids formed by the deamination of the branched amino acids, leucine, isolencine and valine (3). A deficiency of the dehydrogenase prevents the further oxidation of the α-keto acids that are formed from their respective amino acids. Thus, patients accumulate α-ketoisocaproic acid, α-keto-β-methylcaproic acid and α-ketoisovaleric acid. These α-keto acids may be reaminated to form branched chain amino acids, which are also elevated in these patients.

Affected infants with the "classical" form of MSUD, in which the enzyme deficiency is almost complete, have symptoms of severe acidosis, lethargy and seizures soon after birth. If they survive, signs of brain damage and failure to thrive are usually present.

The characteristic odour of "maple syrup" can be detected early in the urine or even on the skin. Genetic heterogeneity exists and several variations with milder manifestations are known. The degree of severity has been related to the residual enzyme activity (4). Also, a variant of MSUD called "thiamine responsive form" has been described by Scriver (5).

In isovaleric acidemia, the defect is presumed to involve the enzyme isovaleryl-CoA-dehydrogenase, which normally converts isovaleric acid to β-methylcrotonic acid. Reduced activity of this enzyme leads to increased serum isovaleric acid.

During periods of decompensation, circulating isovaleric acid may reach levels of 3 to 10 m mol as compared with a normal value of 0.005 m mol or less. The clinical picture is that of intermittent acute attacks of vomiting acidosis, ataxia, progressing to lethargy and coma. They are commonly associated with acute infections. Mental retardation has been reported (6).

Production of excessive amounts of metabolites

The best example is Phenylketonuria (PKU) which is caused by a genetic deficiency of phenylalanine hydroxylase (7). Untreated patients develop mental retardation, eczema, hypopigmentation and neurological symptoms (8).

The characteristic "mousey" odour is due to the excretion of phenyla-cetic acid. The hypopigmentation is related to the competitive inhibition of tyrosine hydroxylase by the increased concentration of phenylalanine.

Finally, the exact mechanism that causes the neurological abnormalities such as hypertonicity, irritability, hyperactivity, autism and seizures, is unknown, but is believed to be related to interference in the functioning of metabolic pathways within the nervous system by phenylalanine and its accumulated by-products (9).

Interference with regulatory mechanisms

Interference with regulatory mechanisms as in adrenogenital syndromes, for instance, where there is a block at one of the several steps in the biosynthesis of cortisol by the adrenal cortex. This deficiency stimulates excessive production of ACTH that, in turn, stimula-tes the adrenal cortex to increase synthesis of cortisol precursors as far as the metabolic block. Breakdown of the accumulated precursors by alternative pathways leads to the production of androgens. The clinical effect is the virilization of a female fetus or the early virilization of affected boys. Clinical manifestations differ according to the specific enzyme deficiency and the precursors accumulated (10).

<u>Disorders of membrane transport, such as cystinuria, Hartnup disease, the</u>
<u>Fanconi syndrome</u>

From a clinical point of view, it has to be noted that there is a great variability in expression of different inherited metabolic defects. Some may be asymptomatic and may be classified as metabolic variants.

Examples are pentosuria and most cases of fructosoria. Some others become symptomatic only occasionally such as anemia associated with G-6-P-D deficiency that may be revealed only by exposure to certain drugs.

However, many of these conditions have severe consequences and may be lethal. Thus, it is becoming increasingly important for physicians to know how to diagnose genetic disease, what screening procedures should be used, and what treatments are available. Unfortunately, there are relatively few clinical clues other than mental retardation to help the physician develop an index of suspicion.

In recent year it has become apparent that inborn errors of metabolism may present in the newborn. The picture may be that of an infant in deep coma, with real or apparent sepsis, with massive ketosis, an unusual odour of urine or body or with very severe vomiting.

The signs and symptoms to be considered to suspect the presence of metabolic disease in childhood are mental retardation, hypopigmentation, dislocated lens, failure to thrive, osteoporosis, renal stones, coarse facies, liver damage, cataracts.

REFERENCES

1. J. J. Nora, F. C. Fraser, "Medical Genetics: Principle and Practice", 2nd. Ed. Lea and Febiger, Philadelphia, USA, (1981).
2. J. B. Stanbury, J. B. Wyngaarden, D. S. Fredrickson, "The metabolic basic of inherited disease", 4th. Ed., Mc Graw-Hill Book Co., New York, (1978).
3. A. E. Emery, D. L. Rimoin, "Principle and practice of medical genetics", Churchill Livngstone, Edinburgh, (1983).
4. A. Velázquez, F. Montiel, K. N. F. Shaw, A. Carnevale, V. del Castillo, Enfermedad de orina de jarabe de arce heterogeneidad genética, diagnóstico de heterocigotos y un nuevo enfoque terapéuti-co, <u>Rev. Invest. Clin.</u> (mex), 32:273, (1981).
5. C. R. Scriver, S. Mackenzie, C. L. Clow, E. Delvin, Thiamineresponsive maple syrup urine disease, <u>Lancet</u> 1:310, (1971).
6. I. Ando, W. G. Klinberg, K. ward An Rasnussen, W. Nyhan, Isovaleric acidemia presenting with altered metabolism of glycine, <u>Pediat. Res.</u>, 5:478, (1971).
7. C. Mitoma, R. M. Auld, S. Udenfriend, On the nature of enzymic defect in phenypyruvic oligophrenia, Proceding of the Society for, <u>Experimental Biology and Medicine</u>, 94:634, (1957).
8. A. Carnevale, A. Velázquez, F. Ruiz, V. del Castillo, El manejo en México de pacientes con fenilcetonuria, <u>Bol. Med. Hosp. Inf. Mex.</u>, 36:375, (1979).
9. W. L. Nyhan, Understanding inherited metabolic disease, Clinical Simposia, 32:3, (1980).
10. M. A. Sperling, <u>in</u>: "Birth defects compendium", 2nd. Ed., edited by Bergsma, New York, A. R. Liss, (1979).

FUNCTIONAL COMPONENTS OF CELL MEMBRANES

AS THE LOCALE OF PATHOLOGICAL PHENOMENA

Arnošt Kotyk

Institute of Physiology
Czechoslovak Academy of Sciences
142 20 Prague 4, Czechoslovakia

INTRODUCTION

Biomembranes are at the core of one of the most important structural-functional principles of cell organization. Each of the fundamental principles has one special type of macromolecule as its characteristic feature. The **chromosomes** have their nucleic acids, the **cytoskeleton** has its assortment of proteins, the **cell envelopes** have their carbohydrates, and finally the **biomembranes** have their lipids. However, it goes without saying that proteins are to a greater or lesser degree involved in the operation of all the principal organizational elements of a cell. Biomembranes are no exception.

Although the role of lipids in assisting, if not directly performing, various membrane functions is beyond any doubt (stabilization of membrane proteins, 'melting' or 'freezing' of distinct membrane domains, etc.) it is with the membrane-inserted proteins that the principal accomplishments of cell membranes are linked.

BASIC CATEGORIES OF MEMBRANE-LINKED FUNCTIONS

To attribute the development of a particular disease to a given membrane function it should be made clear what types of membrane functions or phenomena may be involved.

The oldest of all, evolution-wise, and the most ubiquitous even in modern membranes, is the **flow of matter**, represented by (1) membrane-associated enzymes (e.g., cytochrome P-450), translocating material in a scalar way, (2) membrane-spanning carriers (e.g., one of the ATPases), translocating substances vectorially across the membrane.

The second fundamental operation taking place in biomembranes is the **flow of energy**, represented by a variety of conversions, transformations, and transductions, involving radiation energy, free energy of oxidation, hydrolysis of ATP, and generation/dissipation of (electro)chemical potential gradients of various solutes.

The third basic function catalyzed by membranes is the **flow of information**. This involves first of all the perception of exogenous chemical or physical signals but also the interaction of membrane proteins with endogenous informational messengers, such as hormones and neurotransmitters.

DISEASES CAUSED BY MEMBRANE DEFECTS

Diseases involving the structure and/or function of membranes are as numerous as those that do not and it would hardly be feasible to present here an exhaustive account of them. Perhaps a selection classified according to a rather arbitrary scheme will be more helpful.

1. Pathological states in which some of the surface features of membranes are modified or destroyed.

2. Situations where one of the functional membrane proteins is missing.

3. Diseases based on misusing the normal function of a membrane protein with deleterious consequences.

4. Cases of direct membrane destruction.

The *first* group of diseases is best exemplified by neoplastic transformation. Various alterations within cells and their metabolism, such as an increased aerobic glycolysis, are accompanied here by a lack of contact inhibition which results in unhindered proliferation, unlike the behavior of somatic normal cells. One of the reasons for this defect, and perhaps the only major one, is the loss of some glycoprotein (and possibly glycolipid) sugars and hence inability of cognitive interaction between neighboring cells. A corollary of this is the appearance of surface features, mainly lectins, that contribute to the higher agglutinability of neoplastic cells.

According to the chalone theory, it is the production and membrane insertion of cell-produced glycoproteins called chalones which block cell proliferation in one of the phases of the cell cycle, either in the G-1 or in the G-2 phase and, at the same time, prolong the postmitotic maturation phase so that normal cells have a longer life expectancy than the tumor ones.

The *second* group of pathological states, those that affect specific membrane proteins, are more varied.

a. The most extensive subset of these diseases are those that affect transport proteins. In many of these cases the genetic link and the hereditary aspects of the disease are quite well understood.

i. Myopathic deposition of fats (resembling that after an exposure to diphtheria toxin) is caused by a genetic lack of carnitine, which impairs the function of the acyl group translocation by the carnitine cycle system of the inner mitochondrial membrane.

ii. Malfunction or dysfunction of the appropriate cation transport system in renal tubules bring about hyperkalemia, as well as hypokalemia, hypercalcemia, as well as hypocalcemia, and hypermagnesemia.

iii. Defects in enzymes of the intestinal (and renal) brush border cause blocks in the transport of the corresponding substrates. These may be directly localized in the membrane or precede the transport step. This applies to systems transporting glucose and galactose; neutral amino acids (Hartnup's disease); cystine and basic amino acids (cystinuria); proline, hydroxyproline and glycine (iminoglycinuria); tryptophan; methionine; vitamin B_{12} (megaloblastic anemia); chloride ions (metabolic alkalosis); as well as to hydrolytic, membrane-associated, enzymes, catalyzing the splitting of sucrose and isomaltose; lactose; and trehalose.

iv. Sickle-cell anemia in which the erythrocytes show a defective oxygen transfer caused by a structural change of their hemoglobin, is accompanied by changes in the calcium-transporting ATPase of their membranes. (In all fairness, this may have nothing to do with the manifestation of the disease, just like the 'low-K' sheep with a defective erythrocyte K^+ transport are to all appearances quite healthy.)

b. Membrane receptors for a variety of chemical signals are a sensitive group, whose lack or malfunction can cause disease. A few examples might be enlightening.

i. Familial hyper-β-lipoproteinemia is caused by the fact that fibroblasts lack the membrane receptors for plasma lipoproteins where the regulatory sequence for cholesterol synthesis begins. While in a normal subject high levels of β-lipoprotein and pre-β-lipoprotein cause a feedback inhibition of cholesterol synthesis, in the patient they canot enter cells and exert this inhibition. The consequence may be a 60 times higher cholesterol level in cells.

ii. One type of diabetes mellitus is caused by the lack of cell receptors for insulin which stimulates the entry of glucose into a number of organ cells. Obviously, the clinical picture is the same as with an insufficiency of insulin production by pancreatic B cells.

iii. A wide range of neurophysiological disturbances are caused by the lack of receptors for neurotransmitters. Parkinson's disease probably is due to either an insufficient production of dihydroxyphenylalanine or a lack of receptors for it. Huntington's chorea, resembling to a certain degree St. Vitus dance, is caused by a faulty phospholipid annulus at the membrane receptors for γ-aminoisobutyric acid, another neurotransmitter.

The *third* group of membrane diseases is characterized by the 'wrong' use of a membrane-bound functional protein.

i. A typical example is offered by the way cholera toxin damages a tissue cell. Its subunit L binds to the ganglioside (G_{M1}) part of an intestinal cell receptor which normally binds a mediator that activates adenylate cyclase. The cyclic AMP thus formed stimulates, among other things, the synthesis of ion-transporting enzymes, the ions being followed during transport by water. This water is under normal conditions reabsorbed but, due to a hyperstimulation of adenylate cyclase, the outflow of ions and hence of water is of such magnitude that it cannot be reabsorbed and the patient dies of dehydration.

ii. Another case is Graves' disease, during which a stable antibody against the thyreotropin receptor circulates in the bloodstream so that the receptor is continually stimulated and hyperthyreoidism ensues.

iii. Myasthenia gravis is also caused by a circulating antibody able to bind to and stimulate, acetylcholine receptors, bringing about enormous fatigue and muscle dystrophy.

The *fourth* group are diseases where membrane rupture or perforation is involved, even if the rupture itself need not be the cause but rather the objective symptom of the disease.

A case in point is gout. Whatever its biochemical cause may be (an insufficient excretion or uric acid from the body or a defect in the activity of hypoxanthine-guanine-phosphoribosyl transferase) the inflammation of joints itself is brought about by an influx of leukocytes phagocytosing sodium urate crystals. These crystals then bind by weak chemical bonds to cell membranes, cause damage to lysosomes and this results in complete cell lysis.

A genuine perforation of cell membranes, followed by leakiness for ions and, upon more severe exposure, even for larger molecules, can come about in the presence of various viruses and, although this is, strictly speaking, not a disease symptom, upon membrane insertion of the complement attack complex taking part in cellular immune response.

There is no doubt that every membrane function mediated by a gene-encoded protein is subject to a potential defect. It is, however, a special subclass of such defects that manifest themselves as a pathological state. There are some mutations that would be lethal even during embryo development (e.g., any major defect in the inner mitochondrial membrane H-ATPase or ATP synthase) while others may not be serious enough to warrant medical attention (possibly some functions in cells where they are of no importance for the organism, such as the above-mentioned ion disturbance of sheep erythrocytes, or defects involving mutually replaceable lipid constituents of membranes, such as various subspecies of a phospholipid class.

MOLECULAR AND PHYSIOLOGICAL PROPERTIES OF

PLASMA MEMBRANES: THE ROLE OF ION CHANNELS

M. Cereijido, M.S. Balda, A. Ponce
and J.J. Bolivar

Center of Research
and Advanced Studies
Apartado Postal 14-740
México 14, D.F.
México

INTRODUCTION

At the beginning of this century, once biologists convinced themselves that cells must be surrounded by a lipidic membrane, it became necessary to assume that this membrane has mechanisms to translocate ions and molecules which are not soluble in lipids. Among the first mechanisms proposed were water filled pores (*see* Cereijido and Rotunno, 1970). Years later, research with tracer fluxes and impaling ("classical") microelectrodes supported the concept that the cell membrane is perforated with a variety of water channels that are permeable to specific ionic species, may be opened and closed by gates which are sometimes sensitive to the electrical potential differences between the two sides of the membrane, and that may be influenced by signal molecules bound to nearby receptors.

However, the "modern era" of channel biology was inaugurated some twelve years ago by the introduction of techniques to study membrane patches that are small enough to contain only a few, or even a single channel (fig 1) (Neher and Sakmann, 1976; Horn and Patlak, 1980; Hamill et al, 1981). Powerful electronic techniques permit the measurement of the tiny currents of a few picoamperes flowing through these channels (fig 2). With these techniques, channels were found in almost all cell types, and even in the membrane of intracellular organelles. There are several types of channels for each one of the most important ions in biology (K^+, Na^+, Ca^{++}, Cl^-). A given cell may have more than one type of channel for the same ion species. Thus, Rae and Levis (1984) report at least six different K^+ channels in the cell membrane of a single lens epithelial cell. Channels are involved in biological processes as diverse as water reabsorption in the intestine, activation of lymphocytes by antibodies, excitation of neurons by acetylcholine, and insulin secretion by beta cells in the pancreas. They may not be continuously present on the cell membrane, but may be characteristic of a given period of the cell cycle. Unavoidably, these channels and the cellular mechanisms that modulate them are succeptible to pathological alterations, so that the number of diseases whose cause is tracked down to a particular ion channel, as well as the number of pharmaceutical drugs that act on them to alleviate human suffering, are continuously increasing.

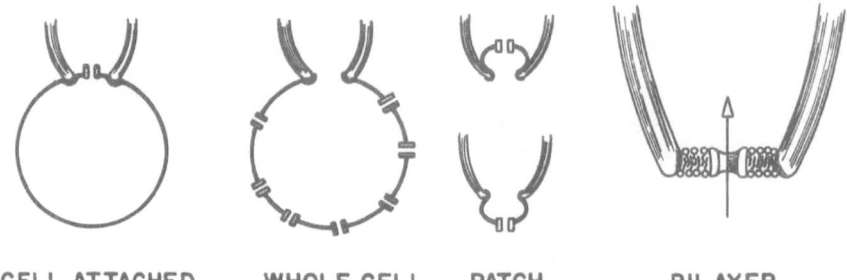

CELL ATTACHED   WHOLE CELL   PATCH   BILAYER

Fig 1. Technical approaches to study the biology of ion channels
       (*see   text*).

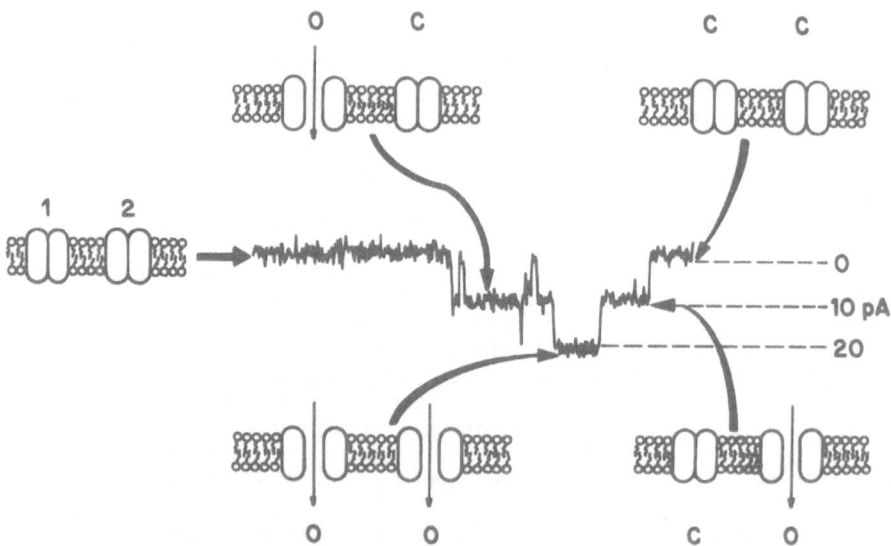

Fig 2. Interpretation of electrical activity in terms of functional
       states of two ion channels in the same patch. When no current
       flows, channels 1 and 2 are assumed to be closed. A negative
       deflection of 10 pA due to the outflux of K^+ is
       interpreded as the opening of one of the channels. This is
       followed by a second negative step that reaches 20 pA, and
       that is interpreted as the opening of both channels
       simultaneously.

The purpose of this presentation is to describe succinctly the field of channel biology. It refers briefly to the main techniques used, the variety of channels found, the diversity of biological functions in which channels participate, provides examples of diseases in which they are involved, and illustrates the use of drugs that act directly or indirectly on ion channels. However, the article does not intend to review these fields. References are only meant to serve as examples and as starting points for further inquiries.

WHAT IS AN ION CHANNEL?

The initial model of tiny water filled tubes that span the cell membrane and that allow the passage of ions was successfully confirmed by two crucial circumstances: (1) the development of "patch clamp techniques", that electrically isolate an area of membrane of 1-10 μm^2, and allow the recording of the small electric current carried by the ions flowing through the channels (Neher and Sakmann, 1976; Horn and Patlak, 1980; Hamill et al, 1981); and (2) biochemical purification of membrane proteins, that led to the isolation of channels (Hidalgo, 1986; Miller, 1986) and subsequent reconstitution in artificial bilayers (Latorre et al, 1982; Darszon et al, 1986), or in membrane vesicles (Racker, 1975; Garcia, 1986), where they may be subject to thorough structure/functional analysis.

Under an applied electrical potential gradient of 100 mV, ion channels may be traversed by an ion flow of 10^6 to 10^9 ions per second, ie a flow 1,000 to 10,000 larger than the ones translocated by pumps and carriers. Latorre and Miller (1983) have proposed to group the properties of channels under two headings: conduction and gating. Conduction properties are those which govern the rate of ion translocation through an open channel, while gating properties are those which determine whether the channel is in a conducting or a nonconducting state. The variety of channels already characterized with those criteria is really impressive: their conductance may be as small as 5 pS (Palmer and Frindt, 1986) or as large as 400 or more pS (Gray et al, 1984). Openings may appear in bursts separated by long nonconducting periods, as illustrated in fig 3. This is the case of the acetilcholine-activated cation channel of the end plate of frog skeletal muscle (Colquhoum and Sakmann, 1983). Sometimes, the current may double or triple, as illustrated in

CLOSED ——➤

OPEN ——➤

1 pA

1 msec

Fig 3. A K^+ channel undergoes a series
of openings and closings that, at
this time scale, are recorded as
spikes. The channel then becomes
dormant, until it exhibits a second
burst of activity on the extreme
right.

fig 2, suggesting that a single channel has more than one conductive state, or that there are several channels in the same patch. These channels may permit passage of only a single ion species, such as K^+ (Latorre and Miller, 1983; Hunter et al, 1986;) Na^+ (Salkoff and Tanouye, 1986; Papazian et al, 1988); Cl^- (Tank et al, 1982; Lotshaw et al, 1986; Madison et al, 1986); Ca^{++} (Fenwick et al, 1982; Cavalie et al, 1983; Nilius et al, 1985); or may allow the passage of other related species of the same electric charge, such as K^+ and Na^+ (Maruyama and Petersen, 1982), or Cl^- and I^- (Schneider et al, 1985). Finally, a channel may exhibit short-lived closings

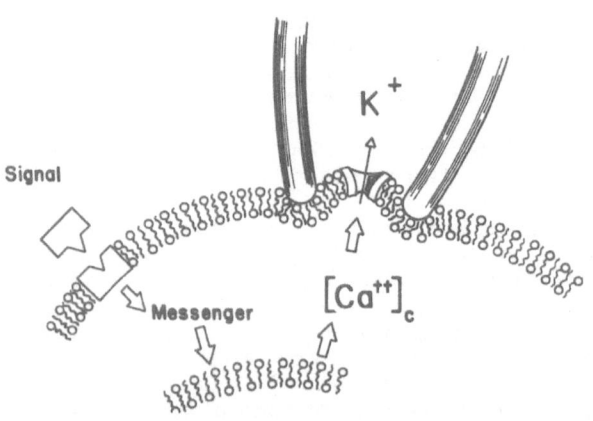

Fig 4. Detail of the cell attached patch illustrated in fig 1 (*left*). In this case, the border of the micropipette isolates a K^+ channel. The activation of a membrane receptor releases an intracellular messenger that stimulates the outflux of Ca^{++} from a cytoplasmic organelle. Ca^{++}, in turn, provokes the opening of the channel.

("flickerings"), originally described by Neher and Steinbach (1978) in the acetylcholine receptor channel under the effect of local anesthetics.

METHODS TO DETECT AND CHARACTERIZE ION CHANNELS

When a glass micropipette filled with the desired saline solution contacts the cell membrane as depicted in fig 1, it may seal a patch and insulate it electrically from the rest of the membrane, with resistances of 1-20 gigahoms, constituting the so called "cell attached patch". This configuration is particularly suited to study channels whose behaviour depends on an elaborate network of cellular events. Thus, fig 4 illustrates a K^+ channel activated by Ca^{++}. The concentration of this ion in the cytosol depends in turn on its release from internal reservoirs, and that is conditioned to the attachment of a molecule (eg a hormone, a neurotransmitter) to a membrane receptor. Signals are carried by a second messenger (eg Ca^{++}, cAMP, inositides) from the receptor to other organelles and molecules.

The patch at the mouth of the microelectrode may be ruptured to adopt the "whole cell clamp" configuration depicted in fig 1. In this case, all channels present in the cell membrane may be studied (Marty and Neher, 1983). Once the "cell attached" configuration is obtained, further manipulations may rip off a fraction of the plasma membrane, with its cytoplasmic side ("inside out" configuration) or its extracellular side ("outside out" configuration) exposed to the bathing solution. These configurations permit the study of factors that affect channels by acting on the extracellular or on the cytoplasmic side. Channels can be also isolated, reconstituted and studied with considerable experimental advantages in artificial systems, with lipid bilayers arranged as a flat sheet between two chambers or as vesicles (Darszon et al, 1986). These bilayers can be formed also at the tip of a microelectrode (Montal et al, 1983; Suarez-Isla et al, 1983; Coronado and Latorre, 1983) (fig 1). The saline solution inside the micropipette, and the one bathing the cells, are connected to sophisticated electronic devices that clamp the voltage across the membrane patch at a desired level and measure the amount of current crossing the channel.

Substances are used in two main ways: (1) the assay of a drug, a neurotransmitter or a hormone, to study its effects on an already known channel. Such is the case of the inhibition by serotonin of the K^+ channel in *Aplysia* sensory neurons (Siegelbaum et al, 1982). (2) In other cases, the effect of the substance is a very well known one, and it is mainly used to characterize a newly found channel. These drugs can be classified into: (a) <u>drugs that inhibit channel activity</u>: such as tetraethylammonium and quinidine that block high conductive Ca^{++} and voltage activated K^+ channels (Bolivar and Cereijido, 1987); lidocaine, a blocker of K^+ channels activated by cell swelling (Richards and Dawson, 1986); indanyloxyacetic and anthranilic acid derivatives, that inhibit epithelial Cl^- channels, (Landry et al, 1987). (b) <u>Toxins</u>: such as tetrodotoxin, a venom known to block Na^+ channels (Hartshorne et al, 1985) and apamin, a bee venom that specifically blocks a class of small (20 pS) Ca-activated K-channels (Romey et al, 1984). The field of toxins is an important one, because they not only allow to characterize, but also to purify the channel (Wu and Narahashi, 1988). (c) <u>Ions</u>: dwelling times in the open or closed states are stochastic quantities that may be highly sensitive to the concentration of Ca^{++} (Marty, 1981) (see fig 5), or Ba^{++} (Kirk and Dawson, 1983; Latorre and Miller, 1983; Palmer, 1986). (d)<u>Voltage-dependence</u>: when the electrical potential across the cell membrane decreases, or even reverses, some channels are stimulated to remain in the open configuration. A typical example is the large (220 pS) K^+ channel illustrated in fig 6. Its activation by depolarization increases the outflux of K^+, and the diffusion potential thus produced tends to restore the membrane potential. (e) <u>Substances that enhance or inhibit a specific regulatory step</u>: forskolin, a potent activator of adenylate cyclase, increases dramatically the rate of desensitization of the response to acetylcholine (Albuquerque et al, 1986; Middleton et al, 1986). However, in other systems such as the acetylcholine channel of rat myotubes, forskolin activity may be independent of cAMP phosphorylation (Grassi et al, 1987). Protein kinase C activators: substances such as phorbol esters and diacylglycerols decrease K-currents (Baraban et al, 1985; Malenka et al, 1986), Cl^-currents (Madison et

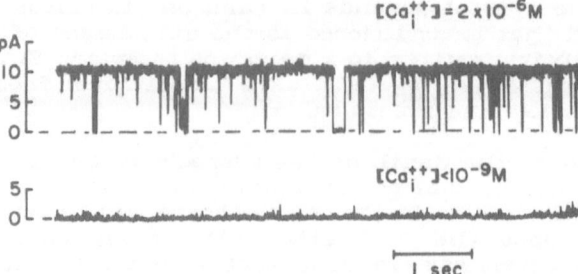

$[Ca_i^{++}] = 2 \times 10^{-6} M$

$[Ca_i^{++}] < 10^{-9} M$

|← 1 sec →|

Fig 5. Effect of Ca^{++} on a large K^+ channel found in the apical region of a renal epithelial cell. Above:in the presence of 2×10^{-6} M Ca^{++}, the channel undergoes an intense series of openings (*upward deflections*) and closings, thus allowing the passage of a K-current of about 10 pA. When Ca^{++} is removed (*below*), the channel remains closed (Taken with kind permission from Bolivar and Cereijido.1987).

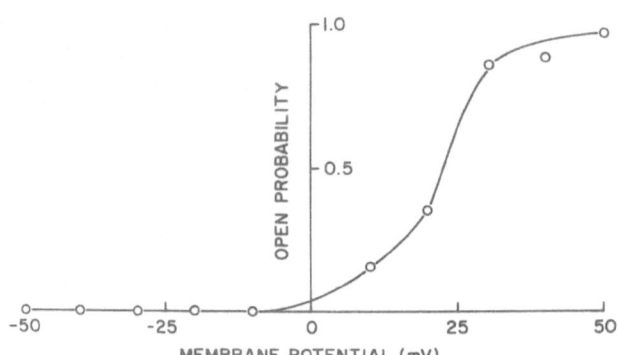

Fig 6. Probability of finding a channel in the open state, as a function of the electrical potential of the cell with respect to the bathing solution. At steady state potentials from -80 to -30 mV, the channel remains mainly closed, but it activates as the voltage decreasesand reverses (Taken with kind permission from Bolivar and Cereijido,1987).

al,1986) and Ca-currents (Rane and Dunlap, 1986). However, these are not universal effects, as the same activators may increase Ca-currents in other preparations (De Rimer et al, 1985). Cholera toxin causes activation of adenylate cyclase, and pertussis toxin may elicit an inhibition. These toxins are also useful to differentiate among several G proteins that affect channels (Miller, 1988).

Once a substance is known to act specifically on a given channel species, it becames of course a valuable tool to characterize newly found ones.

REGULATION OF CHANNEL ACTIVITY

Since ion channels are the crucial feature of an enormous number of physiological and pathological processes, it is not surprising that they are sensitive to a large number of substances and conditions interwoven in a complicated network of cellular controls: (1) Phosphorylation: this is one of the the ion channel modulatory mechanism most thoroughly investigated (Levitan, 1985, 1988). Binding of dihydropyridine to protein from skeletal muscle, can be used to purify and subsequently reconstitute a functional Ca^{++} channel that can be modulated by cAMP-dependent phosphorylation (Curtis and Catteral, 1985). Thus, in *Aplysia* sensory neurons, cAMP-dependent protein kinase closes the serotonin sensitive K^+ channel (Shuster et al, 1985). (2) ATP: glycolytic enzymes located in the membrane of adjacent cytoskeleton to K-channels, may supply ATP molecules that affect their function of (Noma,1983; Misler et al, 1986; Weiss and Lamp, 1987). Glucose may affect ATP-sensitive K^+ channels in insulin-secreting cells, probably through a change in intracellular metabolism (Ashcroft et al, 1984, Ashcroft, 1988). (3) cGMP and cAMP: cyclic nucleotides appear to interact directly with voltage-dependent cation channels in visual and olfactory receptor cells (Gold and Nakamura, 1987). (4) Calcium: the effect of calcium, usually elicited only from the cytoplasmic side of the channel (Latorre et al, 1982) is a widely distributed mechanism, indicating that this ion is one of the main intracellular regulators of channel activity (Bolivar and Cereijido, 1987). Cholecystokinin opens non-selective cation channels in mouse pancreatic acinar cells via calcium (Maruyama and Petersen, 1982) (5) Guanyl nucleotide-binding proteins: these proteins may interact directly with ion channels. One of the best studied examples is the cardiac K^+ channel stimulated by acetylcholine, which is directly regulated by a G-protein (Logothetis et al, 1987; Codina et al, 1987; Kurachi et al, 1986). (6) Inositol 1,4,5-triphosphate: this putative second messeger activates the conductance in excised patches of T-lymphocyte (Kuno and Gardner, 1987) and of a mast cell line (Jones et al, 1987). (7) Receptor site-ion channel: there is a direct binding of ligands to the receptor sites that are part of channel molecules. Such is the case of nicotinic acetylcholine activated channels (Changeaux et al, 1987) and aminoacid activated channels (Cull-Candy and Usowicz, 1987). (8) Growth factors and/or gene expression: treatment with nerve growth factor increases expression of the type II Na^+-channel gene, but has no effect on type I gene expression in cell derived from a rat pheocromocytoma (Mandel et al, 1988). Some channels are seldom found in actively dividing cells, and are mainly expressed in well differentiated stages. Thus, transfection with an specific oncogen expression vector temporarily supresses the formation of functional Ca^{++} and Na^+ channels in myocytes stimulated by mytogens. However, this is not a general effect, as under the same circumstances K^+ channels do not seem to be affected (Caffrey et al, 1987). (9) pH: some channels are inhibited by lowering of pH. Protons may act either as competitors for the use of the channel, or else may affect the dwelling times in the open or closed states (Cook et al, 1984). (10) Stretch: single ion channel currents in cultured embryonic chick skeletal muscle, were observed to increase upon stretching of the plasma membrane (Guharay and Sachs, 1984). Glycine-induced increase in Cl^- channel activity in Ehrlich ascites tumor cells is part of a volumen-regulatory response triggered by cell swelling (Hudson and Schultz, 1988).

PATHOLOGICAL CONDITIONS IN WHICH ION CHANNELS ARE INVOLVED

Information on the involvement of ion channels in human diseases, shows that channels may (a) be the central disorder causing the disease, (b) may be affected as a secondary consequence, or (c) may be responsible for the main symptoms. Thus an IgG from patients with Lambert-Eaton myasthenic syndrome, was shown to block voltage dependent Ca^{++} currents (Kim and Neher, 1988). However, the IgG does not appear to act directly on the physical properties of the channel itself, but through a reduction of the number of active channels present in the membrane.

Cl-permeability in sweat gland epithelia from patients suffering cystic fibrosis is abnormally low (Quinton, 1983). Studies with the "cell attached" configuration of apical Cl^- channels in sweat duct and tracheal epithelial cells show that they can be activated by the B-agonist isoproterenol. This stimulation is absent in cells from patients suffering cystic fibrosis (Welsh and Liedtke, 1986; Pedersen et al, 1986; Widdicombe, 1986). Nevertheless, when Cl^- channels obtained from patients with cystic fibrosis are studied in excised patches, they do not appear to be defective (Welsh, 1986; Frizzell et al, 1986). Therefore, in this disease not the channel, but its regulatory mechanism seems to be affected. In keeping with this interpretation, Schoumacher et al (1987) have demonstrated that phosphorylation does not activate Cl^- channels from cystic fibrosis airway cells.

Renaud et al (1986) have shown that muscles of patients with myotonic muscular dystrophy express a Ca^{++} activated K^+ channel that specifically binds apamin. This receptor is absent in normal human muscle as well as in muscles of patients with spinal anterior horn disorders.

Physical dependence on ethanol may also involve Ca^{++} channels, as antagonists of this ion are effective in the treatment of symptoms of alcohol withdrawal (Koppi et al, 1987).

Channels may be also involved in the massive intestinal secretion produced by cholera, and in diarrheas caused by enterotoxigenic bacteria (Field et al, 1972).

PHARMACOLOGY OF CHANNELS

There is a large number of drugs used therapeutically that appear to be involved in the modulation of K^+ channels (Cook, 1988). These drugs may be roughly classified in (a) Drugs that open K^+ channels: such as nicorandil (Furukawa et al, 1981; Karashima et al, 1982), pinacidil (Bray et al, 1987), and benzopyran compounds (Holzman, 1983; Hamilton et al, 1986). These drugs act as smooth muscle relaxants in the treatment of cardiovascular disorders, such as hypertension, angina pectoris, congestive failure and ischemia. They may be also related to asthma (Cook, 1988). Other openers of K^+ channels, such as diazoxide (Henquin and Meissner, 1984) and galanin (De Weille et al, 1988) are hyperglycemia inducing drugs. (b) Drugs that block K^+ channels: such as sotalol (Carmeliet et al, 1985), ISC 205-930 (Scholtysik, 1987), and nicorandil (Kakei et al, 1986) that are know to act as antiarrythmics. Bretylium tosylate and bethanidine sulphate are used in suppression of ventricular fibrilation (Bacaner et al, 1986), and tolbutamide is used as an hypoglycemia inducing drug (Henquin and Meissner, 1982, Turbe et al, 1986).

There are drugs that act on Ca^{++} channels and that can also be classified as activators and blockers. 1,4-D dihydropyridines activators and blockers differ only in minor structural aspects (Triggle and Janis, 1987). (a) Activators are used as vasoconstrictive drugs (Takenaka and Maeno, 1982; Schramm et al, 1983; Preus et al, 1985). (b) Blockers, such as flunarizine, verapamil and nimodipine are anticonvulsants, antagonizing seizures in experimental epilepsia; caroverine is effective in treatmet of alcoholism (Koopi et al, 1987); verapamil and diltiazem seem to be potentially useful in psychiatric disorders and as analgesics (Raeburn and Gonzales, 1988); dihydropyridines, phenylalkylamines and benzothiazepines are used in angina pectoris, in ischemic or hypoxic conditions and in hypertension (Godfraind et al, 1988; Triggle and Jains, 1987). The interaction of Ca^{++} antagonist with various receptors might be useful in treating or preventing migraine headache, cerebral vasospasm, excessive trombus formation, tumor cell lodgement, and related disorders (Defendis, 1987).

Na^+ channels may be modified pharmacologically by local anesthetics like lidocaine, procaine and its derivatives (Strichartz et al, 1976; Bean et al, 1983; Hondeghem and Katzung, 1984). Quinidine-like drugs inhibit votage-sensitive Na^+ channels in heart muscle and act as antiarrhythmics. Anticonvulsants like carbamazepine and phenytoin bind preferentially to Na^+ channels (Matsuki et al, 1984; Willow et al, 1985; Catteral, 1987) and can be used primarily against partial seizure and tonic-clonic (grand-mal) seizure.

CONCLUDING REMARKS

Models of ion channels are almost as old as the concept of the cell membrane itself. However, the last decade has seen a dramatic increase in our understanding of these channels, thanks to the introduction of techniques of patch and whole cell clamp. These in turn provide a deep insight into the biology of the plasma membrane, and the emerging picture is one of a variety of ion channels for each of the major physiological ions, that respond to the occupancy of specific receptors, to variations of the electrical potential, to changes in pH, and in the concentration of Ca^{++}, cAMP, cGMP, ATP and other intracellular messengers. Any of the numerous factors affecting the behaviour of a channel may eventually be liable to misfunction, thereby originating a pathological condition. Involvement of ion-specific channels were already detected in several human diseases. Fortunately, our information on the repertoire of drugs that operate directly or indirectly on ion channels is also rapidly increasing. On this basis, it may not constitute an exaggeration to affirm that we are witnessing the birth of a new branch of medicine: the one that attempts to cure or alleviate human suffering by acting on ion channels.

Acknowledgements: We should like to acknowledge the pleasant and efficient help of Mrs Elizabeth del Oso, and the economic support of research grants from COSBEL S.A. de C.V. and the NIH of the US.

REFERENCES

Albuquerque, E.X., Deshpande, S.S., Aracava, Y., Alkondon, M. and Daly, J.W. 1986. A possible involvement of cyclic AMP in the expression of desensitization of the nicotinic acetylcholine receptor: a study with forskolin and its analogs. FEBS Letters. 199:113.

Ashcroft, F.M. 1988. Adenosine 5'-triphosphate-sensitive potassium channels. Ann.Rev.Neurosci. 11:97.

Ashcroft, F.M., Harrison, D.E. and Ashcroft, S.J.H. 1984. Glucose induces closure of single potassium channels in isolated rat pancreatic β-cell. Nature 312:446.

Bacaner, M.B., Clay, J.R., Shrier, A. and Brochu, R.M. 1986. Potassium channel blockade: A mechanism for supressing ventricular fibrillation. Proc.Natl.Acad.Sci.USA. 83:2223.

Baraban, J.M., Snyder, S.H. and Alger, B.E. 1985. Protein kinase C regulates ionic conductance in hippocampal pyramidal neurons: electrophysiological effects of phorbol esters. Proc.Natl. Acad.Sci.USA. 82:2538.

Bean, B.P., Cohen, C.J. and Tsien, R.W. 1983. Lidocaine block of cardiac sodium channels. J.Gen.Physiol. 81:613.

Bolivar, J.J. and Cereijido, M. 1987. Voltage and Ca^{++}-activated K^+ channel in cultured epithelial cells (MDCK). J.Membr.Biol. 97:43.

Bray, K.M., Newgreen, D.T., Small, R.C., Southerton, J.S., Taylor, S.G., Weir, S.W. and Weston, A.H. 1987. Evidence that the mechanism of the inhibitory action of pinacidil in rat and guinea-pig smooth muscle differs from that of glyceryl trinitrate. Br.J.Pharmacol. 91:421.

Caffrey, J.M., Brown, A.M. and Schneider, M.D. 1987. Mitogens and oncogenes can block the induction of specific voltage-gated ion channels. Science 236:570.

Carmeliet, E. 1985. Electrophysiologic and voltage clamp analysis of the effects of sotalol on isolated cardiac muscle and purkinje fibers. J.Pharmacol.Exp.Ther. 232:817.

Catterall, W.A. 1987. Common modes of drug action on Na^+ channels: located anesthetics, antiarrhythmics and anticonvulsants. Trends Pharmacol.Sci. 8:57.

Cavalie, A., Ochi, R., Pelzer, D. and Trautwein, W. 1983. Elementary currents through Ca^{2+} channels in guinea pig myocytes. Pflügers.Arch. 398:284.

Cereijido, M. and Rotunno, C.A. 1970. Introduction to the study of biological membranes. Gordon and Breach Science Publishers, London

Changueux, J.P., Giraudat, J. and Denis, M. 1987. The nicotinic acetylcholine receptor: molecular architecture of a ligand-regulated ion channel. Trends Pharmacol.Sci. 8:459.

Codina, J., Yanati, A., Grenet, D., Brown, A.M. and Birnbaumer 1987. The α subunit of the GTP binding protein G_K opens atrial potassium channel. Science 236:442.

Colquhoun, D. and Sakmann, B. 1983. Burst of openings in transmitter-activated ion channels. In: Single-Channel Recording. B. Sakmann and E. Neher, editors. pp. 345-364. Plenum Press, New York, N.Y.

Cook, D.L., Ikeuchi, M. and Fuyimoto, W.Y. 1984. Lowering of pHi inhibits Ca^{2+}-activated K^+ channel in pancreatic β-cell. Nature 311:269.

Cook,N.S. 1988. The pharmacology of potassium channels and their therapeutic potential. Trends Pharmacol.Sci. 9:21.

Coronado, R. and Latorre, R. 1983. Phospholipid bilayers made from monolayers on patch-clamp pipettes. Biophys.J. 43:231.

Cull-Candy, S.G. and Usowicz, M.M. 1987. Patch-clamp recording from single glutamate receptor channels. Trends Pharmacol.Sci. 8: 218.

Curtis, B.M. and Catterall, W.A. 1985. Phosphorilation of the calcium antagonist receptor of voltage sensitive calcium channel by cAMP-dependent protein kinase. Proc.Nat. Acad.Sci.USA. 82:2528.

Darszon, A., Garcia-Soto, J., Lievano, A., Sanchez, J.A. and Islas-Trejo, A.D. 1986. Ionic channels in the plasma membrane of Sea Urchin

Sperm. In: Ionic channels in cells and model systems. R. Latorre, editor. pp. 291. Plenum Press, New York, N.Y.

De Rimer, S.A., Strong, J.A., Albert, K.A., Greengard, P. and Kaczmarek, L.K. 1985. Enhancement of calcium current in *Aplysia* neurons by phorbol ester and protein kinase C. Nature 313:313.

De Weille, J., Schmid-Antomarch, H., Fosset, M. and Lazdunski, M. 1988. ATP-sensitive K^+ channels that are blocked by hypoglycemia inducing sulfonylureas in insulin-secreting cells are activated by galanin, a hyperglycemia-inducing hormone. Proc.Natl.Acad. Sci.USA. 85:1312.

Defendis, F.V. 1987. Interaction of Ca^{++} antagonists at 5HT2 and H2 receptors and GABA uptake sites. Trends Pharmacol.Sci. 8: 200.

Fenwick, E.M., Marty, A. and Neher, E. 1982. A patch-clamp study of bovine chromaffin cells and of their sensitivity to acetylcholine. J.Physiol.(London). 331:577.

Field, M., Fromm, D., Al-Awqati, Q. and Greenough III, W.B. 1972. Effect of cholera enterotoxin on ion transport across isolated ileon mucosa. J.Clin.Invest. 51:796.

Frizzel, R.A., Rechkemmer, G. and Shoemaker, R.L. 1986. Altered regulation of airway epithelial cell chloride channels in cystic fibrosis. Science 233:558.

Furukawa, K., Itoh, T., Kajiwara, M., Kitamura, K., Suzuki, H., Ito, Y. and Kuriyama, H. 1981. Vasodilating actions of 2-Nicotiamidoethyl nitrate on porcine and guinea-pig coronary arteries. J.Pharmacol.Exp.Ther. 218:248.

Garcia, A.M. 1986. Methodologies to study channel-mediated ion fluxes in membrane vesicles. In: Ionic channels in cell and model systems. R. Latorre, editor. pp. 127. Plenum Press., New York, N.Y.

Godfraind, T., Morel, N. and Wibo, M. 1988. Tissue specificity of dihydrophyridine-type calcium antagonists in human isolated tissues. Trends Pharmacol. Sci. 9:37.

Gold, G.H. and Nakamura, T. 1987. Cyclic neucleotide-gated conductance: a new class of ion channels mediated visual and olfactory transduction. Trends Pharmacol.Sci. 8:312.

Grassi, F., Monaco, L. and Eusebi, F. 1987. Acetylcholine receptors channel properties in rat myotubes exposed to forskolin. Bioch.Biophys.Res.Comm. 147:1000.

Gray, P.T.A., Bevan, S. and Ritchie, J.M. 1984. High conductance anion-selective channels in rat cultured Schwann Cells. Proc.R. Soc.Lond.' B221:395.

Guharay, F. and Sachs, F. 1984. Stretch-activated single ion channel currents in tissue-cultured embryonic chick skeletal muscle. J.Physiol.(London) 352:658.

Hamill, O.P., Marty, A., Neher, E., Sakmann, B. and Sigworth, F.J. 1981. Improved patch clamp technique for high-resolution current recording from cell-free membrane patches. Pflügers Arch. 391:85.

Hamilton, T.C., Weir, S.W. and Weston, A.H. 1986. Comparison of the effects of BRL 34915 and verapamil on electrical and mechanical activity in rat portal vein. Br.J.Pharmacol. 88:103.

Hartshorne, R.P., Keller, B.U., Talvenheimo, J.A., Catterall, W.A. and Montal, M. 1985. Functional reconstitution of the purified brain sodium channel in planar lipid bilayers. Proc.Natl.Acad.Sci.USA. 82:240.

Henquin, J.C. and Meissner, H.P. 1982. Opposite effects of tolbutamide and diazoxide on 86Rb+ fluxes and membrane potential in pancreatic β-cells. Biochem.Pharmacol. 31:1407.

Henquin, J.C. and Meissner, H.P. 1984. Significance of ionic fluxes in membrane potential for stimulus-secretion coupling in pancreatic β-cells. Experentia. 40:1043.

Hidalgo, C. 1986. Isolation of muscle membranes containing functional ionic channels. In: Ionic channels in cells and model systems.

R. Latorre, editor. pp. 101. Plenum Press, New York, N.Y.

Holzman, S. 1983. Cyclic GMP as possible mediator of the coronary arterial relajation by nicorandil(SG-75).J.Cardiovasc.Pharmacol. 5:364.

Hodenghem, L.M. and Katzung, B.G. 1984. Antiarrhythmic agents: The modulated receptor mechanism of action of sodium and calcium channel-blocking drugs. Ann.Rev.Pharmacol.Toxicol. 24:387.

Horn, R. and Patlak, J.B. 1980. Single channel currents from excised patches of muscle membrane. Proc.Natl.Acad.Sci.USA. 77: 6930.

Hudson, R.L. and Schultz, S.G. 1988. Sodium-coupled glycine uptake by Ehrlich ascites tumor cells results in an increase in cell volume and plasma membrane channel activity. Proc.Natl.Acad.Sci.USA. 85:279.

Hunter, M., Lopes, A.G., Boulpaep, E. and Giebisch, G. 1986. Regulation of single potassium channel from apical membrane of rabbit collecting tubule.Am.J.Physiol. (Renal Fluid Electrolyte Physiol (20) 251:F725.

Jones, S.U.P., Cunha-Melo, J.R., Beaven, M.A. and Baker, J.L. 1987. Inositol-1,4,5,-triphosphate mimics antigen activation of membrane currents in a mast cell line. Fed.Proc. 46(3):A397

Kakei , M., Kelly, R.P., Aschroft, S.J.H. and Ashcroft, F.M. 1986. The ATP-sensitivity of K^+ channels in rat pancreatic β-cells is modulated by ADP. FEBS Letters 208:63.

Karashima, T., Itoh, T. and Kurimaya, H. 1982. Effects of 2-Nicotinamidoethyl nitrate on smooth muscle cells of the guinea-pig mesenteric and portal veins. J.Pharmacol.Exp.Ther. 221:472.

Kim, Y.I. and Neher, E. 1988. IgG from patients with Lambert-Eaton syndrome blocks voltage-dependent calcium channels. Science 239:405.

Kirk, K.L. and Dawson, D.C. 1983. Evidence for single-file ion flow. J.Gen.Physiol. 82:297.

Koppi, S., Eberhardt, C., Haller, R. and Konig, P. 1987. Calcium-channels-blocking agent in the treatment of acute alcohol withdrawal –Carovine versus meprobamate in a randomized double-blind study. Neuropsychobiology 17:49.

Kuno, M. and Gardner, P. 1987. Ion channels activated by inositol 1,4,5,-triphosphate in plasma membrane of human T-lymphocytes. Nature 326:301.

Kurachi, Y., Nakajima, T. and Sugimoto, T. 1986. On the mechanism of activation of muscarinic K^+ channels by adenosine in isolated atrial cells; involvement of GTP-binding proteins. Pflügers.Arch. 407:264.

Landry, D.W., Reitman, M., Cragoe, E.J.(Jr) and Al-Awqati, Q. 1987. Epithelial chloride channel: Development of inhibitory ligands. J.Gen.Physiol. 90:779.

Latorre, R. and Miller, C. 1983. Conduction and selectivity in potassium channels. J.Membr.Biol. 71:11.

Latorre, R., Vergara, C. and Hidalgo, C. 1982. Reconstitution in planar lipid bilayers of a Ca^{2+}-dependent K^+ channel from transverse tubule membranes isolated from rabbit skeletal muscle. Proc.Natl.Acad.Sci.USA 77:7484.

Levitan, I.B. 1985. Phosphorylation of ion channels. J.Membr.Biol. 87:177.

Levitan, I.B. 1988. Modulation of ion channels in neurons and other cells. Ann.Rev.Neurosci. 11:119.

Logothetis, D.E., Kurachi, Y., Galper, J., Neer, E.J. and Clapham, D.E. 1987. The beta-gamma subunit of GTP-binding proteins activated the muscarinic K^+ channel in heart. Nature 325:321.

Lotshaw, D.P., Levitan, E.S. and Levitan, I.B. 1986. Fine tuning of neuronal electrical activity: Modulation of several ion channels by intracellular messengers in a single identified nerve cells.

J.Exp.Biol. 124:307.

Madison, D.V., Malenka, R.C. and Nicoll, R.A. 1986. Phorbol esters block a voltage-sensitive chloride current in hippocampal pyramidal cells. Nature 321:695.

Malenka, R.C., Madison, D.V., Andrade, R. and Nicoll, R.A. 1986. Phorbol esters mimic some cholinergic actions in hippocampal pyramidal neurons. J.Neurosci. 6:475.

Mandel, G., Cooperman, S.S., Mane, R.A., Goodman, R.H. and Brehm, P. 1988. Selective induction of brain type II Na channels by nerve growth factors. Proc.Natl.Acad.Sci.USA. 85:924.

Marty, A. 1981. Ca^{2+}-dependent K^+ channels with large unitary conductance in chromaffin cell membranes. Nature 291:497.

Marty, A. and Neher, E. 1983. Tight-seal whole-cell recording. In: Single-channel recording. B. Sakman and E. Neher, editors. pp. 107-122. Plenum Press, New York, N.Y.

Maruyama, Y. and Petersen, O.H. 1982. Cholecystokinin activation of single-channel currents is mediated by internal messenger in pancreatic acinar cells. Nature 300:61.

Matsuki, N., Quandt, F.N., Ten Eik, R.E. and Yeh, J.Z. 1984. Characterization of the block of sodium channels by phenytoin in mouse neuroblastoma cells. J.Pharmacol.Exp.Ther. 228:523.

Middleton, P., Jaramillo, F. and Shuetze, S.M. 1986. Forskolin increase the rate of acetylcholine receptor desensitization at rat soleus endplates. Proc.Natl.Acad.Sci.USA. 83:4967.

Miller, C. 1986. Ion channel reconstitution: why bother?. In: Ionic channels in cell and model systems. R. Latorre, editor. pp. 257. Plenum, Press., New York, N.Y.

Miller, R.J. 1988. G proteins flex their muscles. Trends In Neuroscience 11(1):3.

Misler, S., Falke, L.C., Gillirs, K. and McDaniel, M.L. 1986. A metabolic-regulated potassium channel in rat pancreatic β- cells. Proc.Natl.Acad.Sci.USA 83:7119.

Montal, M., Suarez-Isla, B., Wan, K. and Lindstrom, J. 1983. Single-channel recordings from purifies acetylcholine receptors reconstituted in bilayers formed at the tip of patch pipettes. Biochemistry 22:2319.

Neher, E. and Sakmann, B. 1976. Single-channel currents recorded from membrane of denervated frog muscle fibres. Nature 260:799.

Neher, E. and Steinbach, J.H. 1978. Local anesthetics transiently block currents through single acetylcholine-receptor channels. J.Physiol.(London) 277:153.

Nilus, B., Hess, P., Lansman, J.B. and Tsien, R.W. 1985. A novel type of cardiac calcium channel in ventricular cells. Nature 316:443.

Noma, A. 1983. ATP-regulated K^+ channels in cardiac muscle. Nature 305:147.

Palmer, L.G. 1986. Patch-clamp technique in renal physiology. Am.J.Physiol.(Renal Fluid Electrolyte Physiol. 19) 250:F379.

Palmer, L.G. and Frindt, G. 1986. Amiloride-sensitive Na channels from the apical membrane of the rat cortical collecting tubule. Proc.Natl.Acad.Sci.USA. 83:2767.

Papazian, D.M., Schwarz, T.L., Tempel, B.L., Timpe, L.C. and Jan, L.Y. 1988. Ion channels in drosophila. Ann.Rev.Physiol. 50: 379.

Pedersen, P.S., Brandt, N.J. and Larsen, E.H. 1986. Qualitatively abnormal beta-adrenergic response in cystic fibrosis sweat duct cell culture. IRCS Med.Sci. 14:701.

Preuss, K.C., Gross, G.J., Brooks, H.L. and Warltiet, D.C. 1985. Slow channel calcium activators, a new groups of pharmacological agents. Life Science. 37:1271.

Quinton, P.M. 1983. Chloride impermeability in cystic fibrosis. Nature 301:421.

Racker, E. 1975. Reconstitution of membrane pumps. In: Proceedings of

10th FEBS Meeting; Biological Membranes. J. Montreuil and P. Mandel, editors. pp. 25.., North Holland, Amsterdam.

Rae, J.L. and Levis, R.A. 1984. Patch voltage clamp of lens epithelial cells: Theory and practice. Mol.Physiol. 6:115.

Raeburn, D. and Gonzales, R.A. 1988. CNS disorders and calcium antagonists. Trends Pharmacol.Sci. 9:117.

Rane, S.G. and Dunlap, K. 1986. Kinase C activator 1,2-oleoylacetylglycerol attenuates voltage dependent calcium current in sensory neurons. Proc.Natl.Acad.Sci. 83:184.

Renaud, J.F., Desneuelle, C., Schmid-Antomarchi, H., Hugwes, M., Serratrice, G. and Ladunski, M. 1986. Expression of apamin receptor in muscles of patients with myotonic muscular dystrophy. Nature 319:678.

Richards, N.W. and Dawson, D.C. 1986. Single potassium channels blocked by lidocaine and quinidine in isolated turtle colon epithelial cells. Am.J.Physiol. (Cell Physiol 20) 251:C85.

Romey, G., Hugues, M., Schmid-Antomarchi, H. and Lazdunski, M. 1984. Apamin: A specific toxin to study a class of Ca^{2+} -dependent K^+ channels. J.Physiol.(Paris) 79:259.

Salkoff, L.B. and Tanouye, M.A. 1986. Genetics of ion channels. Physiol.Rev. 66:301.

Schneider, G.T., Cook, D.I., Gage, P.W. and Young, J.A. 1985. Voltage sensitive, high-conductance chloride channels in the luminal membrane of cultured pulmonary alveolar (type II) cells. Pfluger.Arch. 404:354.

Scholtysik, G. 1987. Evidence for inhibition by ICS 205-930 and stimulation by 34915 of K^+ conductance in cardiac muscles. Naunym-Schmeed.Arch.Pharmacol. 335:692.

Schoumacher, R.A., Shoemaker, R.L., Halm, D.R., Tallant, E.A., Wallace, R.W. and Frizzell, R.A. 1987. Phosphorilaton fails to activate chloride channels from cystic fibrosis airway cells. Nature 330:752.

Schramm, M., Thomas, G., Towart, R. and Franckowiak, G. 1983. Novel diphydropyridines with positive inotropic action through activation of Ca^{++} channels. Nature 303:535.

Shuster, M., Camardo, J., Siegelbaum, S. and Kandel, E.R. 1985. Cyclic AMP-dependent protein kinase closes the serotonin-sensitive K^+ channels of Aplisia sensory neurones in cell-free membrane patches. Nature 313:392.

Siegelbaum, S.A., Camardo, J.S. and Kandel, E.R. 1982. Serotonin and cAMP close single K channels in *Aplysia* sensory neurons. Nature (London) 299:415.

Strichartz, G. 1976. Molecular mechanism of nerve block by local anesthetics. Anesthesiology 45:421.

Suarez-Isla, B.A., Wan, K., Lindstrom, J. and Montal, M. 1983. Single-channel recordings from purified acetylcholine receptors reconstituted in bilayers formed at the tip of patch pipettes. Biochemistry 22:2319.

Takenaka, T. and Maeno, H. 1982. A vasoconstrictive coumpound 1,4.-dihydropyridine derivative. Jpn.J.Pharmacol. 32:139

Tank, D.W., Miller, C.M. and Webb, W.W. 1982. Isolated-patch recording from liposomes containing functionally reconstituted chloride channel from Torpedo Electroplax. Proc.Natl.Acad.Sci.USA 79:7749.

Triggle, D.J. and Janis, R.A. 1987. Calcium channel ligans. Ann.Rev.Pharmacol.Toxicol. 27:347.

Turbe, G., Rorsman, P. and Ohno-Shosakv, T. 1986. Opposite effects of tolbutamide and diazoxide on the ATP-dependent K^+ channel in mouse pancreatic β-cells. Pflügers.Arch. 407:493.

Weiss, J.N. and Lamp, S.T. 1987. Glycolysis preferentially inhibits ATP-sensitive K^+ channels in isolated guinea pig cardiac myocytes.

Science   238:67.

Welsh, M.J. 1986. An apical-membrane chloride channel in human tracheal
     epithelial. Science 232:1648.

Welsh, M.J. and Liedtke, C.M. 1986. Chloride and potassium channels in
     cystic fibrosis airway epithelia. Nature 322:467.

Widdicombe, J.H. 1986. Cystic fibrosis and β-adrenergic response of
     airway epithelial cell cultures. Am.J.Physiol. 251:R818.

Willow, M., Gonoi, T. and Catterall, W.A. 1985. Voltage clamp analysis of
     inhibitory actions of diphenylhydantoin and carbamazepine on
     voltage-sensitive sodium channels in neuroblastroma cells.
     Mol.Pharmacol.   27:549.

Wu, C.H. and Narahashi, T. 1988.Mechanism of action of novel marine neuro
     toxins on ion channels. Ann.Rev.Pharmacol.Toxicol. 28: 141.

TRANSMEMBRANE CHANNELS AND DISEASE

C.A. Pasternak

Department of Biochemistry
St George's Hospital Medical School
Cranmer Terrace, London SW17 ORE

ABSTRACT

In this article, two examples of faulty transmembrane channels in disease are presented. The first concerns continuous and spontaneous firing of Na^+ channels in neurones, that may explain the pain associated with post-herpetic and other types of neuralgia. The second concerns the insertion of non-specific channels into the plasma membrane of certain cells, that appears to underly the action of toxins secreted by pathogenic bacteria and other organisms. The damage caused by such channels can be prevented by Zn^{2+} and other divalent cations.

INTRODUCTION

The plasma membrane is an important site at which cell function is regulated (Pasternak 1984a). Ion channels in the plasma membrane are often means by which extracellular signals (e.g. neurotransmitters) are sensed by the interior of the cell. Malfunction of transmembrane channels underlies several types of disease. We have studied two situations. The first concerns post-herpetic neuralgia, a condition associated with infection of sensory neurones by a herpes virus. The second concerns a whole variety of disease conditions, from bee stings to Staphylococcal infections, all of which are characterized by the creation in susceptible cells of non-specific channels, through which essential metabolites and nutrients leak. The fact that leakage can be prevented by divalent cations such as Zn^{2+}, may explain its beneficial effects in a number of situations.

SPONTANEOUS ELECTRICAL ACTIVITY IN SENSORY NEURONES

This is an example of a defect in a specific channel, namely the Na^+ channel, caused by a particular cellular event, namely infection with a herpes virus. It had been known for some time that infection with certain strains of herpes virus leads to spontaneous electrical activity in whole animals (Dempsher et al., 1955; Kiraly and Dolivo 1982). What we have been able to show is that one can study this phenomenon in single cells in culture (Mayer et al., 1985, 1986). This opens up the possibility of revealing the molecular mechanism by which it occurs.

The experimental system is to dissect out the dorsal root ganglia of neonatal animals and to plate out the dissociated cells under culture conditions such that cells capable of division (glia, fibroblasts, etc.) are killed by an inhibitor (e.g. cytosine arabinoside) that is present in the growth medium. Non-dividing cells, which are predominantly sensory neurones, are unaffected and become enriched on the culture dish. They may then be studied by suitable pharmacological and biochemical techniques.

When such neurones are infected with different strains of herpes simplex virus 1 (HSV1), different electrophysiological changes ensue after some time. These include a decreased membrane resistance, a loss of low threshold action potentials and altered inward rectification, as well as the appearance of spontaneous activity. In the latter case, impulses are fired without changing the membrane potential by electrical means or by the addition of specific neurotransmitters. It is this situation that is particularly relevant to neurological symptoms such as the neuralgia often associated with shingles, a disease caused by the chicken pox virus (varicella zoster) that is a member of the herpes family. Spontaneous electrical activity is due to the opening of Na$^+$ channels, since tetrodo toxin, an inhibitor that affects only Na$^+$ channels, blocks the activity; after its removal, spontaneous activity reappears (Fig. 1). Currently we are studying other drugs, such as phenytoin (Pasternak et al., 1988) and bradykinin (Dolphin et al., 1988) that may give information about the type of membrane lesion underlying spontaneous activity. This system obviously also provides a useful means for screening novel drugs that may be of potential use in the treatment of post-herpetic neuralgia (Schon et al., 1987) and other neurological disorders.

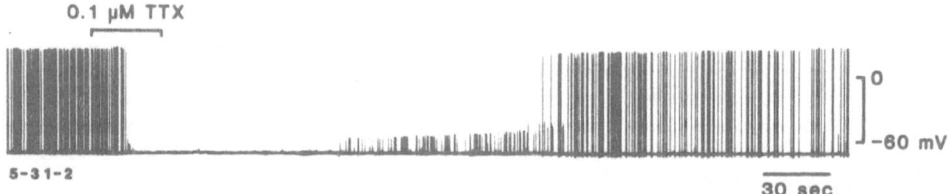

Fig. 1. Spontaneous electrical activity in sensory neurones blocked by tetrodotoxin. Adapted from Mayer et al. (1985), with permission.

It may be wondered whether the spontaneous activity in HSV1-infected neurones is due to cell damage caused by the entry of the virus or by the function of some of its associated viral proteins. The fact that the activity is seen only some hours after the infectious event argues against this possibility. Conclusive evidence that viral replication within the neurone is necessary is provided by the finding that acyclovir, - a drug that specifically prevents the replication of HSV1 (Elion et al., 1977), - blocks the induction of spontaneous firing (Fig. 2). Now we are examining what aspects of virus-infected cells, such as the formation of cell-cell syncytia by membrane fusion, and other cytopathic consequences dependent on the expression of particular viral genes, are necessary for the induction of spontaneous activity.

In conclusion, it is clear the **infection** of sensory **neurones** in culture by certain strains of **herpes virus** leads of **spontaneous** excitability that may underlie post-herpetic **neuralgia** and other neurological disorders.

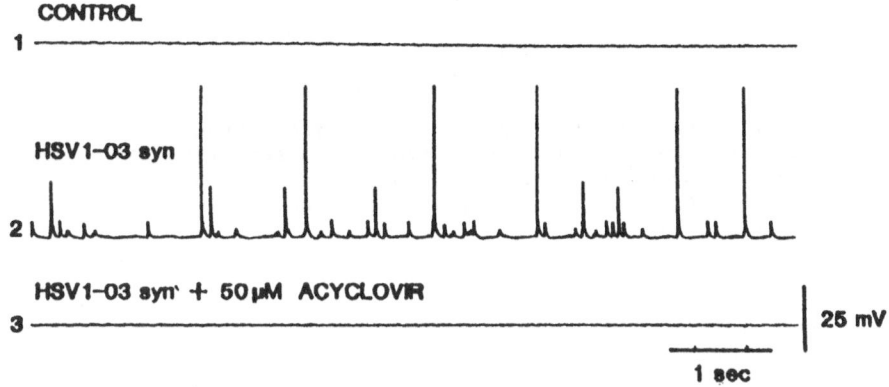

Fig. 2. Acyclovir prevents the induction of spontaneous electrical activity. Reproduced from Mayer et al. (1986), with permission.

CONSEQUENCES OF NON-SPECIFIC CHANNELS IN CELLS

In contrast to the kind of channel just described, which is specific to the transport of a particular ion (Na^+) and which is opened and shut as a result of specific events at the plasma membrane (action of neurotransmitters, infection by HSV1, etc.), there are situations in which non-specific channels become inserted into the plasma membrane of susceptible cells. These include the consequences of viral and bacterial infections, stings from bees and other animals, as well as the consequences of an immune response to some of these very infections. That is, the outcome of channel insertion may be beneficial as well as detrimental, - depending on the nature of the affected cell. Some agents that induce such channels in susceptible cells are shown in Table 1. Most of them are able to create holes in erythrocytes, i.e. to cause haemolysis, though in vivo this rarely occurs; instead their action is more local, and depends on the route of entry, - e.g. through the skin by a bee sting. Most cells in the body (e.g. nerve cells, heart cells, endocrine cells (Forda et al., 1982), endothelial cells (Suttorp et al., 1985) and so forth are sensitive, though the presence of specific receptors for parficular agents at the cell surface obviously increases their sensitivity.

Table 1. AGENTS THAT FORM DIVALENT CATION-SENSITIVE PORES

Viruses:	Sendai and NDV (paramyxovirus) Influenza (orthomyxovirus) and other viruses at low pH
Bacterial toxins:	Staphylococcus aureus α and δ toxin Streptolysin O Clostridium perfringens θ toxin
Animal toxins:	Melittin (honey bee) Cytolysin (sea anemone)
Immune proteins:	Activated complement Cytolysin (cytotoxic T cells)
Synthetic compounds:	Polylysine Triton X 100

All these agents induce non-specific channels in susceptible cells, that are in every case sensitive to inhibition by divalent cations. Based on previously-published (Bashford et al., 1986, 1987) and unpublished results.

Although the agents listed in Table 1 appear to have little in common, the lesions that they induce in cells share certain characteristic features (Bashford et al., 1986). The most striking of these is a sensitivity to divalent cations (and protons) that results in the prevention of leakage through the pore. The efficacy of different cations is $Zn^{2+} >$ $Ca^{2+} > Mg^{2+}$ in every case, with other cations being either Zn^{2+}-like (e.g. Cd^{2+}) or Ca^{2+}-like (e.g. Mn^{2+}) or rather ineffective (e.g. monovalent cations, and organic cations such as spermine, dimethonium, etc.) (Bashford et al., 1988). The mechanism(s) by which divalent cations act is not yet clear: at least part of it is explicable in terms of a 'gating' effect. That is, the channels that are induced in the membrane 'flicker' between an open and a closed state: it is the closed state that appears to be stabilized by divalent cations (Menstrina 1986).

What are the consequences of leakage through these pores? Because the lesions can be as large as 10nm in diameter, most low molecular weight metabolites present in cells, and even some cytoplasmic proteins, can leak out (Pasternak 1984b, 1988). But because the lesions are often quite transient, due to repair processes (Campbell and Morgan 1985; Carney et al., 1985) that act by an initial influx of Ca^{2+} (Morgan and Campbell 1985; Pasternak 1986), only minimal damage may occur. In other words, the consequences of this type of cell damage can vary from a completely reversible change in cell function (e.g. Forda et al., 1982) to irreversible cell death and tissue necrosis. The factors that affect the extent of cell damage include the amount of agent reaching a particular cell, and the concentration of protective extracellular ions such as Ca^{2+} and Zn^{2+} at that site.

Recent experiments with whole animals (Pasternak and Mahadevan 1988) have confirmed that Ca^{2+} and Zn^{2+} are indeed able to prevent cell damage resulting from the creation of membrane channels by toxins such as S.aureus α toxin (one of the major causes of infectious disease in Western hospitals) or by viruses such as Sendai virus (a virus similar to NDV which is a pathogen for many different types of farm animal). Because oral zinc supplement has been recommended to be of value against a number of infectious and other diseases, including the common cold (Eby et al., 1984), our results may provide new insights into the biochemical action of Zn^{2+} (Pasternak 1987), and encourage the research for other compounds having a similarly protective action on cells. They also offer an explanation for the long-observed beneficial action of Ca^{2+} on plant (True 1914) and animal (Frankenhauser and Hodgkin 1957) cells.

In conclusion, it appears that some cytotoxic pore-forming agents cause membrane leakage that is preventable by Ca^{2+} or Zn^{2+}

ACKNOWLEDGEMENTS

The author is grateful to his many colleagues for participating in the experiments described, and to the Cell Surface Research Fund and other grant-awarding agencies for financial support.

REFERENCES

Bashford, C.L., Alder, G.M., Menestrina, G., Micklem, K.J., Murphy, J.J. and Pasternak, C.A. (1986). Membrane damage by haemolytic viruses, toxins, complement and other cytotoxic agents: a common mechanism blocked by divalent cations. J.Biol.Chem. 261, 9300-9308.

Bashford, C.L., Alder, G.M., Graham, J.M., Menestrina, G. and Pasternak, C.A. (1988). Ion modulation of membrane permeability: effect of cations on intact cells and on cells and phospholipid bilayers treated with pore-forming agents. J.Membrane Biol. (in press).

Campbell, A.K. and Morgan, B.P. (1985). Monoclonal antibodies demonstrate protection of polymorphonuclear leukocytes against complement attack. Nature 317, 164-166.

Carney, D.F., Koski, C.L. and Shin, M.L. (1985). Elimination of terminal complement intermediates from the plasma membrane of nucleated cells: the rate of disappearance differs for cells carrying C5b-7 or C5b-8 or a mixture of C5b-8 with a limited number of C5b-9. J.Immun. 134, 1804-1809.

Dempsher, J., Larrabu, M.G., Bang, F.B. and Bodian, D. (1955). Physiological changes in sympathetic ganglia infected with pseudorabies virus. Am.J.Physiol. 182, 203-216.

Dolphin, A.C., McGuirk, S., Pasternak, C.A. and Vallis, Y. (1988). Bradykinin enhances excitability in rat sensory neurones by a GTP-dependent mechanism. J.Physiol. (in press).

Eby, G.A., David, D.R. and Halcomb, W.W. (1984). Reduction in duration of common colds by zinc gluconate lozenges in a double-blind study. Antimicrob.Agents and Chemother. 25(1), 20.

Elion, G.B., Furman, P.A., Fyfe, J.A., de Miranda, L., Beauchamp, L. and Schaeffer, H. (1977). Selectivity of action of an antiherpetic agent, 9-(2-hydroxyethoxymethyl)guanine. Proc.Nat.Acad.Sci. USA 74, 5716-5720.

Forda, S.R., Gillies, G., Kelly, J.S., Micklem, K.J. and Pasternak, C.A. (1982). Acute membrane responses to viral action. Neurosci.Lett. 29, 237-242.

Frankenhauser, B. and Hodgkin, A.L. (1957). The action of calcium on the electrical properties of squid axons. J.Physiol. 137, 218.

Kiraly, M. and Dolivo, M. (1982). Alteration of the electrophysiological Activity in sympathetic ganglia infected with a neurotropic virus. I. Presynaptic origin of the spontaneous bioelectric activity. Brain Res. 240, 43-54.

Mayer, M.L., James, M.H., Russell, R.J., Kelly, J.S., Wise, J.C.M. and Pasternak, C.A. (1985). Spontaneous electrical activity induced by herpes virus infection in rat sensory neuron cultures. Brain Research 341, 360-364.

Mayer, M.L., James, M.H., Russell, R.J., Kelly, J.S. and Pasternak, C.A. (1986). Changes in excitability induced by herpes simplex viruses in rat dorsal root ganglion neurons. J.Neurosci. 6, 391-402.

Menestrina, G. (1986). Ionic channels formed by Staphylococcus aureus alpha toxin: voltage-dependent inhibition by divalent and trivalent cations. J.Membr.Biol. 90, 177-190.

Morgan, B.P. and Campbell, A.K. (1985). The recovery of human polymorpho-nuclear leucocytes from sublytic complement attack is mediated by changes in intracellular free calcium. Biochem.J. 231, 205-208.

Pasternak, C.A. (1984a). Cell Surfaces. Interdisciplinary Science Rev. 10, 42-55.

Pasternak, C.A. (1984b). Virally-mediated changes in cellular permeability. In: Membrane Processes: Molecular Biological Aspects and Medical Applications. (Benga, G., Baum, H. and Kummerow, F., eds.) Springer-Verlag, N.Y. pp.140-166.

Pasternak, C.A. (1986). Effect of pore-formers on intracellular Ca^{2+}. Cell Calcium 7, 387-397.

Pasternak, C.A. (1987). A novel form of host defence: Membrane protection by Ca^{2+} and Zn^{2+}. Biosci.Rep. 7, 81-91.

Pasternak, C.A. (1988). Membrane-mediated cytotoxicity: measurement of changes and their prevention by divalent cations. In: Methodological Surveys in Biochemistry and Analysis (Reid, E., ed.) Vol.17, Plenum pp.189-198.

Pasternak, C.A. and Mahadevan, D. (1988). Novel role of extracellular calcium and zinc: protection against membrane damage induced by cyto-toxic agents. Ind.J.Biochem.Biophys. (in press).

Pasternak, C.A., Vallis, Y. and Dolphin, A.C. (1988). Spontaneous activity in virus-infected cells. In: Perspectives in Molecular Approaches to Human Disease (Gorrod, J., ed.) Ellis Horwood Ltd. (in press).

Schon, F., Mayer, M.L. and Kelly, J.S. (1987). Pathogenesis of post-herpetic neuralgia. Lancet II, 366-368.

Suttorp, N., Seeger, W., Dewein, E., Bhakdi, S. and Roka, L. (1985). Staphylococcal toxin-induced PG/2 production in endothelial cells: role of calcium. Am.J.Physiol. 248, C127-

True, R.H. (1914). The harmful action of distilled water. Amer.J.Bot. 1, 255.

PROTEIN KINASE C: STRUCTURE, FUNCTION AND MODULATION OF ITS
CATALYTIC ACTIVITY BY PHYSIOLOGICAL AND PHARMACOLOGICAL
AGENTS

Charles W. Mahoney and Angelo Azzi[*]

Institut für Biochemie und Molekularbiologie
Universität Bern
3012 Bern, Switzerland

Transmembrane Signal Transduction via Protein Kinase c
===

Extracellular agonist-cell surface receptor cell activation pathways can be mediated through the second messengers diacylglycerol and inositol-triphosphate which in turn can activate PKC kinase activity directly in the former case and indirectly in the latter by release of internally stored calcium.
The initial steps of physiological cell activation processes
are often of a transient nature and hence there have to be
mechanisms for turning off the activation process. Several
mechanisms of deactivation have been elucidated such as the
metabolism of DAG to either phosphatidic acid (re-generation of
phosphatidylinositol path) or further catabolism to monoacyl-
glycerol and glycerol thereby generating fatty acids such as
arachidonic acid, which can be further metabolized to a series
of potent biological effectors, the prostaglandins. Inositol-
1,4,5-triphosphate (IP_3) is also normally a transient message
being further phosphorylated to IP_4 IP_5, and IP_6 or being
dephosphorylated to IP_2, IP_1, and inositol (regeneration of
phosphatidylinositol path) (for reviews see Nishizuka, 1984,
1986; Berridge, 1987). The role of the other IP_x metabolites as
possible second messengers is currently being actively investigated (Berridge, 1987). In contrast to the transient signal of
DAG, the phorbol diesters, which also potently and directly
activate PKC kinase activity, have sustained action since they
are not readily metabolized. Because of this and the specificity of the phorbol diesters toward PKC, they are useful probes
of PKC function. Signal transduction across the cell membrane
can also be turned off by activated PKC which can down-regulate
phosphatidylinositol-bis-phosphate (PIP_2) phospholipase and up-
regulates a calcium pump (thereby reducing cytosolic calcium

---

[*]To whom correspondence should be addressed.
Abbreviations used: PKC, protein kinase C; PS, phosphatidylserine; PDB, phorbol 12, 13 diester; DAG, diacylglycerol; M_r,
apparent molecular weight; kDa, kilo-dalton; PKA, cAMP-dependent protein kinase; BME, 2-mercaptoethanol.

levels) (Berridge, 1987). The agonist/receptor/PI cycle/PKC
signal pathway can also regulate another signal pathway, ago-
nist/receptor/cAMP/PKA signal path, in both a negative or positive way depending on the cell type and experimental conditions
and the converse, the regulation of the agonist/receptor/PI
cycle/PKC signal pathway by cAMP/PKA path, has also been
documented in both a negative and positive sense, again
dependent on the cell type and experimental conditions
(Nishizuka, 1986). The elucidation of the complexities of these
signal/regulatory pathways coupled with spatial and temporal
resolution will eventually explain the specificities of various
extracellular ligands in inducing specific cell responses
through these common pathways.

PKC has been implicated in numerous biological processes
in various cell types, most notably in the regulation of secretion, growth, differentiation, tumor promotion, muscle contraction, platelet aggregation and secretion, the neutrophil respiratory burst, the allergic response, the immune response, ion
channels and membrane substrate transporters, and gap junctions
(Nishizuka, 1986). Various proteins have been proposed as substrates for PKC, some because of definitive evidence and others
because of suggestive evidence (Nishizuka, 1986). Most of these
proposed protein substrates are related to the above biological
processes although the function of many has not been identified
in vivo.

Activators of PKC

The tumor promoters, phorbol diesters, teleocidin, and
aplysiatoxin, as well as DAG are potent activators of PKC
kinase activity (Nishizuka, 1986; Brockerhoff, 1986; Jeffrey
and Liskamp, 1986). A tumor promoter is an agent that by itself
is not responsible for tumor induction but when applied after a
sub-threshold concentration of a carcinogen can induce tumor
development. These compounds activate PKC by lowering the Ca^{+2}
and phosphatidylserine requirement for activation of the kinase
activity (Castagna et al., 1982). Although these compounds have
very different structures, they all have functional groups with
similar three-dimensional coordinates that are critical for
their interaction with PKC (Brockerhoff, 1986; Jeffrey and
Liskamp, 1986). In the case of DAG, the carbonyl functional
groups in positions 1 and 2 and the 3 hydroxy group are
critical for binding to and activation of PKC. The carbonyl
groups provide hydrogen bond acceptor site with respect to a
hydrogen bond donor on PKC and the 3 hydroxy group provides a
hydrogen bond donor to an acceptor on the enzyme (Brockerhoff,
1986). In the case of the phorbol diesters, the 9 α-OH serves
as a hydrogen bond donor and the 12 and 13 carbonyl of the
esters serve as hydrogen bond acceptors (Brockerhoff, 1986). In
addition it has been demonstrated that DAG competes for the
phorbol diester binding site on PKC (Sharkey et al., 1984).
Other tumor promoters which activate PKC, teleocidin and
aplysiatoxin, also have analogous functional groups with
similar 3-dimensional coordinates (Brockerhoff, 1986). Jeffrey
and Liskamp (1986) through computer structural modeling deduced

that 4 oxygen groups in TPA (at C-3, C-4, C-9, and C-20) correspond to oxygen or nitrogen atoms in the 3- dimensional structures of teleocidin and aplysiatoxin suggesting that these atoms are critical in the interaction of these compounds with PKC. Hence protein-lipid and protein-tumor promoter hydrogen bonding interactions appear to be critical in the activation of PKC by neutral lipid and the tumor promoters.

The length and degree of saturation/unsaturation of the fatty ester side chains on DAG are critical in several respects for the binding to and activation of PKC kinase activity. Generally short saturated or unsaturated as well as long unsaturated fatty ester side chains are acceptable for these processes. The presence of a single long chain unsaturated fatty ester on DAG makes it possible for the second fatty ester to be a long chain saturated fatty ester in order to maintain binding and activation properties with respect to PKC (Go et al., 1987; Takai et al., 1979; Kishimoto et al., 1980; Lapetina et al., 1985; Davis et al., 1985).

Inhibitors of PKC

Many inhibitors of PKC kinase activity have been reported and most of these compete for either the PDB/DAG or the PS sites in the hydrophobic regulatory domain (30 kDa Mr) of PKC (Kuo et al., 1984; Mahoney et al., submitted). Some of the Ca^{+2} interacting drugs, such as trifluoperazine, and Ca^{+2} binding proteins (Kuo et al., 1984) can inhibit by competing for available free calcium. Several inhibitors interact at the active site in the 50 kDa catalytic domain: H-9 (Hidaka et al., 1984) and the K-252 compounds (Kase et al., 1987) inhibit by competing at the ATP binding site of the active site. TLCK (N-tosyl-L-lysine chloromethyl ketone) and TPCK (tosyl-L-phenylalanine chloromethyl ketone), the anti-serine protease compounds, inhibit PKC presumably by going to a lysine or phenylalanine recognition site in the active site of the enzyme (Solomon et al., 1985). This suggestion is consistent with the known structural requirement of serine-X-lysine in the substrate for phosphorylation of serine by PKC (Kondo et al., 1987). House and Kemp (1987) have elegantly demonstrated recently that an octadecapeptide corresponding to residues 19-36 of PKC ($alanine_{25}$ --> serine) containing $serine_{25}$-X-arginine could be phosphorylated by PKC suggesting a phosphorylation consensus sequence of serine-X-Z, where Z represents a basic residue. Almost all known inhibitors of PKC, with the exception of the staurosporines (Tamaoki et al., 1986; Mahoney et al., in press) and K-252 compounds (Nakanishi et al., 1986; Kase et al., 1986) which inhibit in nanomolar concentration ranges, inhibit PKC kinase activity in the micromolar concentration ranges.

The Gene and Deduced Amino Acid Sequences of PKC

Recently the nucleotide sequences for the α, β, γ forms of PKC and their deduced amino acid sequences from rat, bovine, human, and rabbit brain have been reported (Parker et al., 1986; Coussens et al., 1986; Kikkawa et al., 1987; Knopf et al., 1986; Ohno et al., 1987). The amino acid sequences of the α, β and γ forms of PKC contain four constant regions (C_1-C_4) and five variable regions (V_1-V_5). Animal species differences in each enzyme subspecies is much smaller than the enzyme subspecies differences found in a single animal species (Kikkawa et al., 1987). The first constant region (C_1) contains a cysteine rich region which corresponds to homologous sequences found in metalloproteins and in zinc finger structures of proteins that can bind to DNA. The third constant region (C_3) has a highly homologous sequence, G-G--G-----K, which is commonly found in ATP binding sites of various enzymes. The gamma isoform of PKC has only been found after birth in rat brain and only in central nervous tissue, suggesting an intimate and specific neural role for this isoform (Kikkawa et al., 1987). The α,β,γ forms of PKC in human brain have been localized to three distinct human chromosomes, 17, 16, and 19 (Coussens et al., 1986). Parker et al. (1986) have proposed a Ca^{+2} binding site, between the N-terminal cysteine rich area and the conserved ATP binding site area (residues 346-368), corresponding to residues 292-303 based on the abundance of acidic residues. On the other hand, Kikkawa et al. (1987), have found no obvious structure in the primary sequence of PKC to suggest a calcium binding site. These data suggest that PS in association with the enzyme may be necessary to form calcium binding sites. The differential activation properties of the three isoforms of PKC have been examined (for review see Mahoney and Azzi, in press, a). The differential expression, activation and inhibition properties, and substrate specificities of the three PKC isoforms will likely explain many of the specificities of signal transduction via PKC.

The Staurosporines Specifically Inhibit Purified PKC Kinase Activity

Staurosporine was first reported to potently inhibit PKC kinase activity by Tamaoki et al. (1986). We have confirmed these results and have synthesized several secondary amine derivatives of staurosporine in order to generate potent yet more specific inhibitors of PKC relative to other protein kinases. N-propan-2-ol- and N-phenyl- derivatives of staurosporine are approximately 2-4 fold less inhibitory toward PKC than the parent compound, staurosporine, yet both three derivatives show approximately 10 fold higher specificity toward PKC relative to PKA suggesting their usefulness in cell studies (Table I; Mahoney et al., submitted). In addition it was found that staurosporine and its derivatives (Table I) were all fluorescent and were characterized with respect to these properties (Mahoney et al., in press and submitted) suggesting their use in studying their interaction with PKC and the distribution of staurosporine binding sites in intact cells via their fluorescent properties.

Staurosporine Action with Intact Cells

Since PKC has been implicated in several biological processes, as discussed above, and since chemical compounds often

118

have different effects on purified enzymes as compared to intact cells we have examined the effects of staurosporine on several intact cell types in which PKC has a putative role. The respiratory burst of neutrophils which can be stimulated by the PKC-specific phorbol diesters or f-met-leu-phe is inhibited in a concentration-dependent manner by staurosporine (Fig. 2c and 2a; Mahoney, Wymann, et al., unpublished results) with the IC_{50} of this process (9 - 10 nM) being similar to the IC_{50} of purified PKC kinase activity (3.5 nM; Table I). PKC kinase activity appears to be essential throughout the respiratory burst, since addition of staurosporine during the up- , maximum- , or down- phase of the respiratory burst results in cessation of reactive oxygen species generation.

Table I

Specificity of Inhibition of PKC by the Secondary Amine Derivatives of Staurosporine

	IC_{50} (nM)		IC_{50}/catalytic unit $\dfrac{nM}{pmol/\ min\ mg/l}$		
	PKC	PKA	PKC	PKA	PKC/PKA[*]
staurosporine	3.5	14.7	0.11	17.9	6.48×10^{-3}
N-propanol-	8.3	260	0.27	317	0.86×10^{-3}
N-benzoyl-	166.4	606	5.5	739	7.40×10^{-3}
N-phenyl-	16.8	732	0.55	893	0.62×10^{-3}

PKC assay: 0.304 μg enzyme; final volume, 0.25 ml; specific activity (presence of Ca^{+2} and PS), 121 nmol/min mg; substrate, histone III-S.

PKA assay: 10 μg PKA; final volume, 0.1 ml; specific activity (presence of 2 μM cAMP), 8.2 nmol/min mg; substrate, histone III-S.

[*]PKC/PKA values are IC_{50}/catalytic unit ratios.

     Staurosporine inhibition of the respiratory burst during the above various phases is consistent with an inactivation of PKC kinase activity and a continual action of phosphatase(s) de-phosphorylating the activated phosphorylated product(s).

     PKC has been implicated in agonist-receptor coupled platelet aggregation which is a critical reaction in normal blood clotting as well as in pathological processes. We have tested the role of staurosporine in platelet aggregation by using the inhibitor staurosporine with intact platelets. Staurosporine is able to inhibit platelet aggregation in a concentration-dependent manner when stimulated by phorbol 12, 13 dibutyrate (Mahoney and Azzi, in press, b). Watson et al. (1988) have independently demonstrated that staurosporine is able to inhibit platelet aggregation as well as phorbol diester specific phosphorylation of proteins (47 and 20 kDa M_r) in intact platelets. These data provide further evidence for a regulatory role for PKC in platelet aggregation.

PKC has been implicated in tumor promotion by the phorbol diesters and related compounds, and in differentiation in

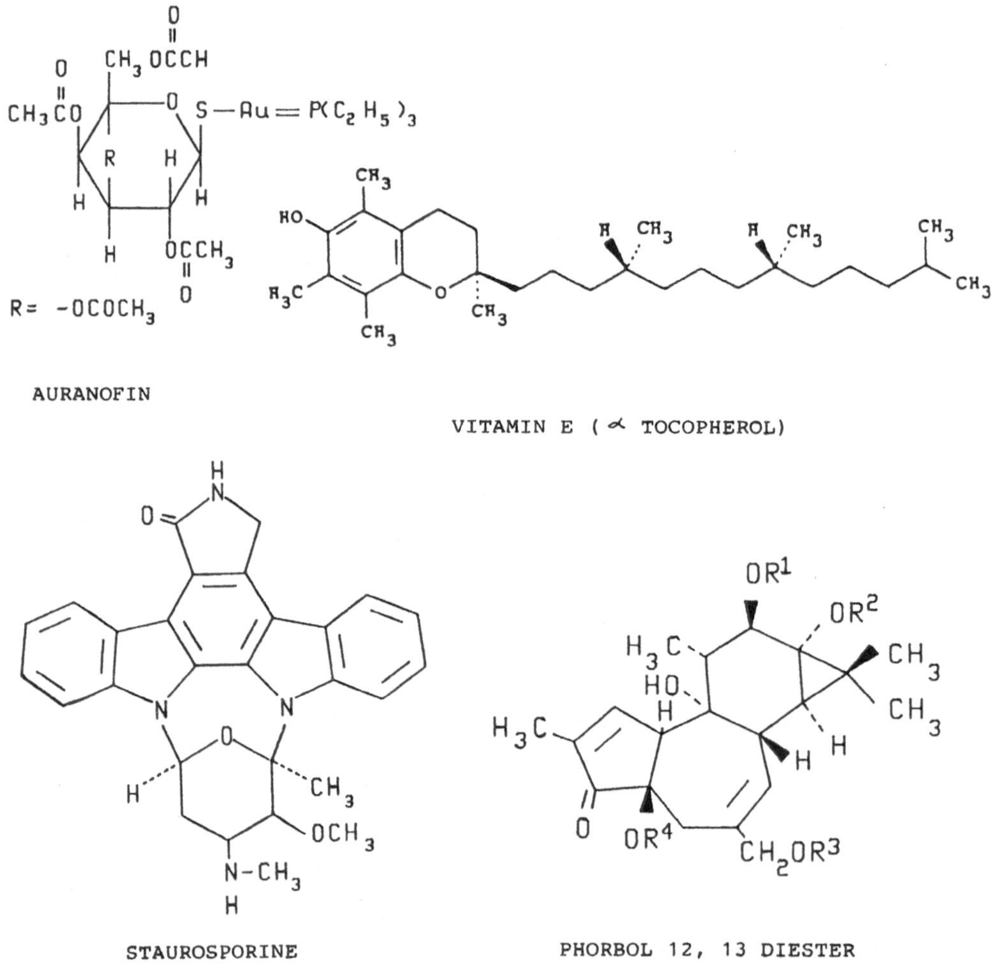

AURANOFIN

VITAMIN E (α TOCOPHEROL)

STAUROSPORINE

PHORBOL 12, 13 DIESTER

Fig. 1. Structures of Auranofin, vitamin E, staurosporine, and phorbol 12, 13 diester.

neuroblastoma (Liu and Chen, 1985; Ruusala et al., 1985). We have tested the role of PKC in differentiation using staurosporine as a differentiation-inducing agent with a neuroblastoma cell line as a model system (Mahoney and Azzi, in press, b).

At very low concentrations of staurosporine (4-5 nM) a neuroblastoma cell line was induced to differentiate as determined by dendrite outgrowth processes developing over a period of several days. Further studies are in progress to delineate the role of PKC in differentiation in neuroblastoma cell lines.

The Au(I)-thio- Anti-rheumatic Drugs Inhibit PKC Kinase Activity

Auranofin (Fig. 1) in the 10-100 μM concentration range inhibits the production of superoxide in neutrophils that have been stimulated by PKC-specific PDB or f-met-leu-phe (Hafstrom et al., 1984; Parente et al., 1986) which suggests a role for PKC in superoxide production and a role or co-role for PKC in the mechanism of Auranofin therapy as an anti-rheumatic.

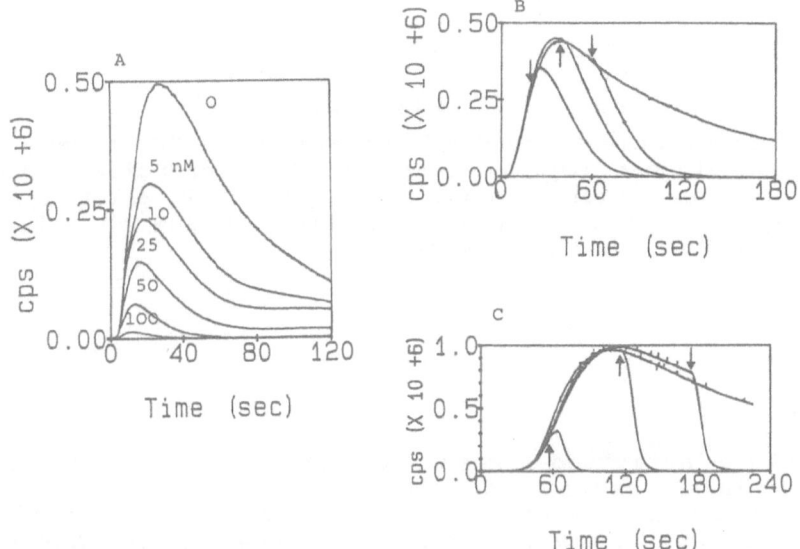

Fig. 2. Inhibition of the respiratory burst in human neutrophils. A) Neutrophils were pre-incubated with increasing concentrations of staurosporine and then stimulated with f-met-leu-phe at time 0. B) Neutrophils were stimulated with f-met-leu-phe at time 0 and then at various stages of the respiratory burst staurosporine was added. C) Neutrophils were stimulated with phorbol myristate acetate (PMA) at time 0 and then staurosporine was added at various stages of the respiratory burst. The respiratory burst was monitored by luminol-dependent chemiluminescence.

The Ca^{+2} - PS- activated kinase activity of purified PKC is similarly inhibited in a concentration dependent manner by Auranofin, Au-S-glucose, and Au-S-malate in the 1 μM - 1 mM concentration range and the control compounds HS-malate and malate show no inhibition of activity indicating that the Au(I) moiety is critical for the inhibition (Fig. 3; Mahoney and Azzi, 1988). Kinase activity of Ca^{+2} - PS- activated enzyme is much more sensitive to inhibition by Au-S-glucose (IC_{50} = 2.0 8 μM) and Au-S-malate (IC_{50} = 3.2 μM) than Auranofin (IC_{50} = 360 μM) (Fig. 3) (in the presence of BME).

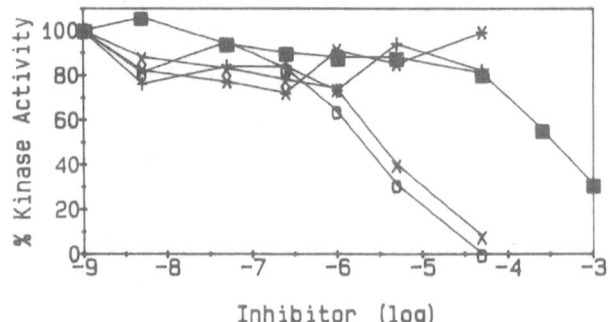

Fig. 3. Concentration-dependent inhibition of PKC kinase activity by Au-S-glucose (O), Au-S-malate (X), Auranofin (squares), and lack of inhibition by control compounds HS-malate (+), and malate (*). Assays were in the presence of 100 μM BME, 80 μM EGTA, 40 μg/ml PS, and 500 μM $CaCl_2$.

PDB- PS- activated kinase activity of purified PKC is also inhibited in a similar concentration dependent manner and control compounds lacking Au(I) showed no inhibition of activated kinase activity (data not shown). The Au(I)-thio-compounds have very complex chemistry (for reviews cf. Sadler, 1976; Shaw, 1979) and they have been shown to inhibit other enzymes (e.g. lysosomal) by several mechanisms: 1) complexation to N or S containing amino acid residues (e.g. cysteine, histidine, methionine, lysine); 2) redox reactions especially with the above residues; 3) complexation with nucleotides; and 4) thio exchange at the S-C bond. In order to further delineate the mechanism of inhibition of PKC by these compounds the effect of BME ($\pm$ 200 μM) and EGTA ($\pm$ 100 μM), which are normally present in our assay system, was examined with purified PKC (Figs. 4-5).

In all cases (Auranofin, Au-S-glucose, Au-S-malate) the presence or absence of 100 μM EGTA had little effect on the concentration dependent inhibition of kinase activity. In contrast to this, in the absence of BME, Auranofin (IC_{50} = 3.4-6.7 μM; Fig. 4) and Au-S-malate (IC_{50} = 0.66-0.87 μM, Fig. 5) were much more inhibitory than in the presence of BME (IC_{50} = 293 - 360 μM and 4.8-5.5 μM respectively). Inhibition of PKC kinase activity by Auranofin in the absence of BME occurs in the same order of magnitude concentration range as is the case with the inhibition of neutrophil superoxide production (IC_{50} = 23 μM) (these experiments were carried out in the absence of reducing agent).

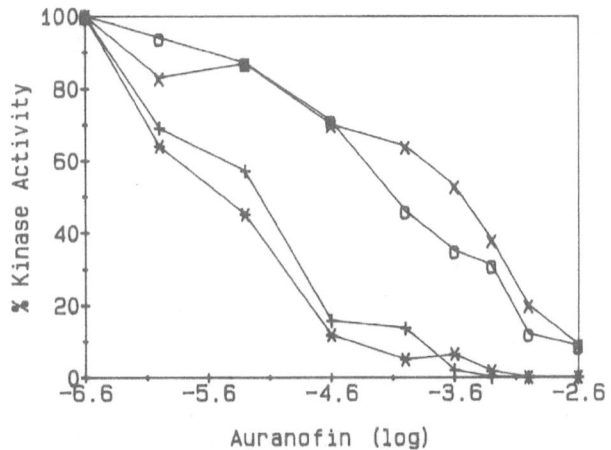

Fig. 4. Inhibition of PKC kinase activity by Auranofin and its dependence on 2-mercaptoethanol and EGTA. PKC was dialyzed into 20 mM Tris-HCl, pH 7.5, $\pm$ 2 mM BME , $\pm$ 1 mM EGTA to give final assay concentrations of presence of 200 μM BME (X,O); absence of BME (+,*); presence of 100 μM EGTA (X,+); absence of EGTA (O,*).

In contrast to the increased inhibitory potency of Auranofin and Au-S-malate in the absence of BME, Au-S-glucose showed the reverse effect, being more inhibitory in the presence of BME (IC_{50} = 2.0 μM) than in its absence (IC_{50} = 107 μM)(data not shown). These data suggest that the mechanism of inhibition of PKC kinase activity by Auranofin and Au-S-malate is different than that by Au-S-glucose. A reductive event or thio exchange between the Au(I)-thio- of Au-S-glucose and BME may be required for inhibitory action whereas in the case of Auranofin and Au-S-malate these events may inhibit the inhibitory interaction of these compounds with the enzymatic system.

Auranofin has been shown to not inhibit NADPH oxidase of neutrophils, the key enzyme in the respiratory burst (Parente et al., 1986) which further supports a regulatory role hypothesis for Auranofin/PKC in the respiratory burst. The concentrations of Auranofin required to inhibit PKC kinase activity and the respiratory burst of neutrophils are in the same order of magnitude as those found in the serum (30-40 μM) (Danpure et al., 1979) and synovium (17-22 μM) (Gerber et al., 1972) of patients undergoing Auranofin treatment. Hence these PKC inhibition studies are of physiological significance.

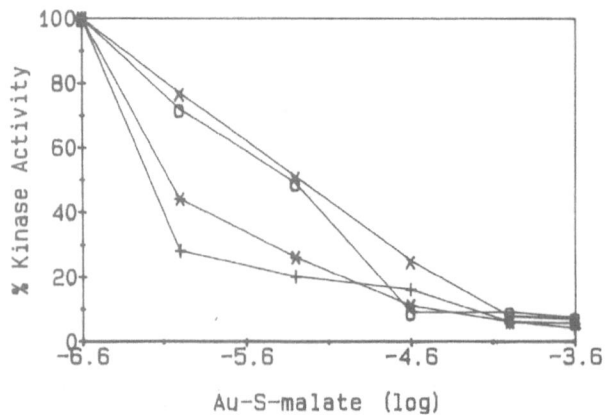

Fig. 5. Inhibition of PKC kinase activity by Au-S-malate and its dependence on 2-mercaptoethanol and EGTA. PKC was dialyzed into 20 mM Tris-HCl, pH 7.5, $\pm$ 2 mM BME, $\pm$ 1 mM EGTA to give final assay concentrations of 200 μM BME (X,O); absence of BME (*,+); presence of 100 μM EGTA (X,+); absence of EGTA (O,*).

In potential contrast to our results, Froscio et al. (1988) have recently reported that Auranofin stimulates phorbol diester-specific protein phosphorylation (40 and 20 kDa M_r) in intact platelets. It may be that in intact platelets a phosphatase is inhibited or other protein kinase(s) are activated resulting in increased phosphorylation of the 40 and 20 kDa M_r proteins. It is well known that most proteins which can be phosphorylated can be phosphorylated by different protein kinases, with some phosphorylation sites on a particular protein being phosphorylated by several kinases and other sites being uniquely phosphorylated by different kinases.

Vitamin E Inhibits PKC Kinase Activity

Vitamin E (dl-α-tocopherol) has been known for many years as an antioxidant (for reviews see Lubin and Machlin, 1982; Slater, 1960; Isler and Brubacher, 1982; Wasserman and Taylor, 1972); it has been shown that vitamin E can induce differentiation in human neuroblastoma cell lines (Helson et al., 1982), that H_2O_2 production can be inhibited in human neutrophils isolated from individuals whose diet was supplemented with 1600 units of Vitamin E/day (Baehner et al., 1982), and that it can act as platelet anti-aggregation agent (Steiner and Mower, 1982). Since PKC has been implicated in tumor promotion,

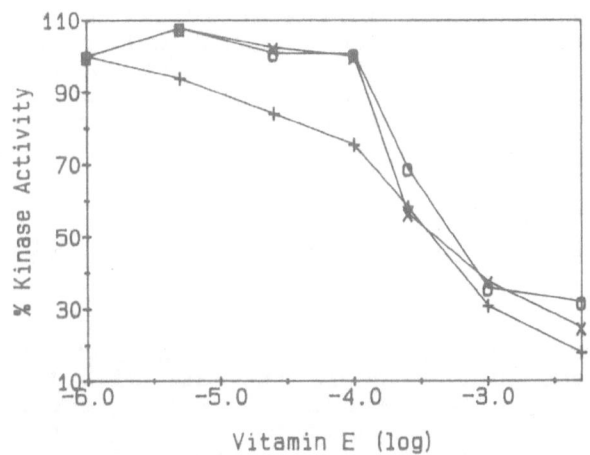

Fig. 6. Concentration dependent inhibition of PKC kinase activity by vitamin E (Hoffman-La Roche) and lack of an effect with increasing concentrations of PS. Assays were carried out in the presence of 40 (X), 100 (O), or 250 (+) µg/ml PS and activation was also by 500 µM CaCl$_2$.

differentiation, the granulocytes respiratory burst and platelet aggregation, and because of the hydrophobic nature of vitamin E and the regulatory domain of PKC we proposed and tested a direct modulatory role for vitamin E on PKC kinase activity. Ca^{+2} - PS- activated PKC kinase activity is inhibited in a concentration dependent manner by vitamin E with an approximate IC$_{50}$ of 456 µM (Fig. 6) (Mahoney and Azzi, in press, c). To further elucidate the mechanism of this inhibition increasing amounts of PS were added to the vitamin E inhibition assays. Under these conditions there was no significant effect on the concentration dependent inhibition curves (Fig. 6) suggesting that vitamin E is not competing for PS sites in the enzyme-phospholipid complex. PKC kinase activity, stimulated by PDB- PS- (in the presence and absence of added Ca^{+2}), yielded vitamin E concentration dependent inhibition curves essentially identical to each other as well as to the Ca^{+2} - PS- activated vitamin inhibition curves (Mahoney and Azzi, in press, c). These data indicate that Ca^{+2} and PDB also do not play a role in the interaction of vitamin E with PKC and its inhibition of the kinase activity. Inhibition studies with vitamin E, vitamin E acetate, and (+) vitamin E (all from Fluka AG) indicate little to no differences in the concentration dependent inhibition of PKC kinase activity, although the three vitamins from Fluka were less inhibitory than vitamin E from Hoffman-La Roche (data not shown).

The concentration of vitamin E in plasma from normal humans is the range of 12-35 μM (Stuart, 1982; Duggan, 1959) which is approximately 10-20 fold lower than the levels required here for 50% inhibition of purified PKC kinase activity yet inhibition of kinase activity in our assays starts at greater than 10 μM in some cases and it may be that in intact cells a local concentration effect of vitamin E in the membrane may allow for modulation of PKC activity. Of course intact cell studies should be pursued to either confirm or negate the ability of vitamin E to modulate PKC kinase activity in vivo.

Miscellaneous Pharmacological Agents which Inhibit or Have No Effect on PKC

In our screening program for detecting pharmacological and physiological agents that could interact with PKC and modulate its kinase activity, we were able to find several other compounds,i.e. azathioprine, suramin, antimony sodium gluconate, and 4-amino-fluorescein, which were able to inhibit the kinase activity of PKC in the micromolar concentration range (Table II). Antimony sodium gluconate (sodium stibogluconate) is used therapeutically as an anti-leishmania agent and its structure is related to that of the older therapeutic agent stibophen (no longer commercially available), which inhibits phosphvitin kinase I (K_i = 10 μM) purified from Ascaridia galli, a worm that can be found in the intestines of chicken (Ossikovski and Walter, 1984). In the same article it was reported that suramin also inhibited this protein kinase (K_i = 2 μM), but antimony potassium tartrate (10 mM) had no effect. Suramin (1 and 10 μM) and stibophen (10 and 100 μM) when tested against cAMP-dependent protein kinase had no effect on the kinase activity indicating the specificity of these two inhibitors toward phosphvitin kinase I.

In our case, antimony sodium gluconate inhibits PKC kinase activity (IC_{50} = 827 μM; Table II). Whether this agent can also inhibit other protein kinases remains to be seen.

Suramin, an anti-Trypanosoma gambiense agent (Walter, 1980), also inhibits HIV reverse transcriptase (for review see De Clercq, 1987) and was recently used in clinical trials for the treatment of AIDS (Collins et al., 1986), but was found to be too toxic and not as effective in vivo as in vitro (Baines, 1986). Suramin, an anionic dye which is mostly bound to serum

Table II

Other Inhibitors of PKC Kinase Activity

Compound	IC_{50}	Pharmacological Use
Azathioprine	1 mM	Immunosuppressant
Suramin	39 μM	Anti-HIV reverse transcriptase and Anti-trypanosome
Na stibogluconate	827 μM	Anti-leishmania
4-amino-fluorescein	546 μM	?

Kinase assays were carried out in the presence of 40 μg/ml PS and 500 μM CaCl2.

proteins in vivo (Collins et al., 1986), has been shown to also inhibit as mentioned above phosphvitin kinase but not cAMP-dependent protein kinase. We have found that suramin can inhibit (IC_{50} = 39 μM)Ca^{+2} - PS- activated PKC kinase activity (Table II). Four-amino-fluorescein, a hydrophobic dye, also inhibits Ca^{+2} - PS- activated PKC kinase activity (IC_{50} = 546 μM; Table II). How selective this agent is for protein kinases is not known.

Many compounds that might be expected to modulated PKC activity had no effect on Ca^{+2} - PS- activated PKC kinase activity (Table III) indicating structural specificity for the interaction with and modulation of PKC activity.

Differential Inhibition of PKC by Sulphydryl Group Reactive Compounds

PKC kinase activity is differentially sensitive to sulphydryl modifying agents. N-ethyl-maleimide (NEM) had no significant effect (0.1-5 μM) on the Ca^{+2} - PS- activated PKC kinase activity, whereas iodoacetic acid (IAA), iodoacetamide (IA), eosin-maleimide (EM), p-hydroxy-mercuric-benzoate (PMB), and $HgCl_2$ were increasing inhibitory in the 0.1-5 μM concentration ranges toward the kinase activity (Fig. 7). After this work was completed Kikkawa et al. (1987) reported inhibition of stimulated PKC kinase activity at greater than 1 mM NEM and that phorbol diester binding was only minimally inhibited.

These results indicate that a critical cysteine residue(s) is required for kinase activity and that this residue is not likely located at the phorbol diester binding site on the enzyme. The varying hydrophobic and charged natures of the SH modifying agents in conjunction with the inhibition data indicate that the critical cysteine residue required for activity is located in a hydrophobic pocket on the enzyme.

Table III

Pharmacological and Physiological Compounds Which Have No Effect on PKC Kinase Activity

Compound	Action
Quinine-HCl	Anti-malarial
Prednisolone	Anti-inflammatory
Phenylbutazone	Anti-inflammatory
Corticosterone-21-acetate	Anti-inflammatory
Progesterone	Pro-gestational
Amphotericin B	Anti-fungal
Halcinonide	Anti-inflammatory
Rifampicin	Anti-bacterial
Acetylsalicylic acid	Anti-inflammatory, analgesic
Ascorbic acid	Anti-oxidant
Lovastatin	HMG-co A Reductase Inhibitor

All kinase assays were carried out in the presence of 40 μg/ml PS and 500 μM CaCl2 and compounds were tested in the 1 nM - 1 mM concentration range.

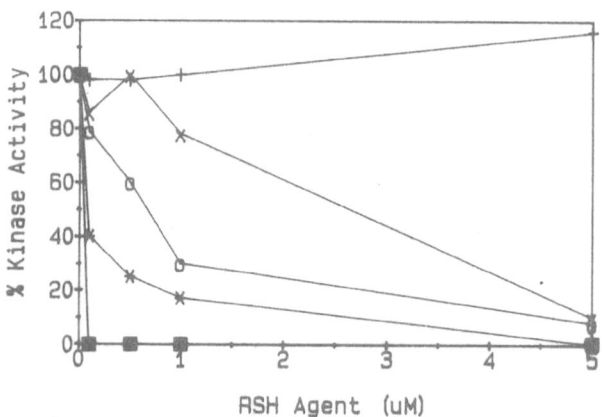

Fig. 7. Differential inhibition of PKC kinase activity by N-ethyl-maleimide (+); iodoacetic acid (X); iodoacetamide (O); eosin-maleimide (*); and HgCl$_2$ and para-hydroxy-mercuric-benzoate (squares).

Whether this critical cysteine residue is located in the active site of the enzyme or in a distal location, perhaps in the hydrophobic regulatory domain (and yet can still exert a distal inhibitory effect on the active site) remains unknown.

Experiments on the constitutively active catalytic fragment would shed further light on the mechanism of inhibition by these SH reagents. A reversibility study confirmed that inhibition by EM is by covalent modification of the enzyme and that by $HgCl_2$ is by complexation which can be reversed by increasing concentrations of BME (data not shown).

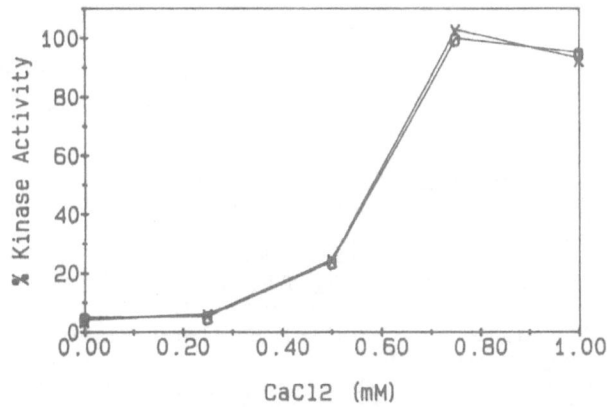

Fig. 8. Lack of an effect of Na_3VO_4 on PKC kinase activity. Assays were carried out in the presence of 40 μg/ml PS and 0-1 mM $CaCl_2$ (added). Na_3VO_4, 0 (O); 100 μM (X) were employed.

Lack of an Effect by Na_3VO_4 on Protein Kinase C

The VO_4^{-3} (vanadate, oxidation state (+5)) ion has many diverse chemical and biochemical properties (for reviews see Pope and Dale, 1968; Ramasarma and Crane, 1981; Simons, 1979; and Erdmann et al., 1984). 1) Vanadate potently inhibits the Na^+/K^+ ATPase (K_i = 4 nM) in the presence of MgATP, but other ATPases are not affected. Competitive inhibition experiments with the Na^+/K^+ ATPase indicate that vanadate competes at the ATP binding site (Cantley et al., 1978). 2) Vanadate in the presence of ADP is able to form a stable transition state complex with the enzyme myosin which can be isolated from free ADP and vanadate resulting in catalytically inactive enzyme

(Goodno, 1982). 3) Vanadate can form esters with glucose (glucose-6-vanadate; Nour-Eldeen, 1985) and phenol and tyrosine (Tracey and Gresser, 1986). 4) Vanadate can act as an insulin-mimetic agent and induce phosphorylation of the insulin receptor at a tyrosine residue (Heyliger et al., 1985; Ramasarma and Crane, 1981; Tamura et al., 1984).

Because of the diverse effects that vanadate can have on phosphotransferase enzymes we examined the effects of Na_3VO_4 (5 μM - 1 mM) on the PKC kinase activity. Vanadate (100 μM) has no effect on the activation properties of PKC in the presence of 40 μg/ml PS in the Ca^{+2} activation range of 0-1 mM added $CaCl_2$ (Fig. 8). In a second experiment, vanadate (5 μM- 1 mM) had no significant effect on the activation properties of PKC (0 - 5 mM $CaCl_2$) in the presence of 40 μg/ml PS (data not shown). These data suggest that vanadate is not able to interact with and modulate PKC kinase activity.

References

Baehner, R. L., Boxer, L. A., Ingraham, L. M., Butterick, C., and Haak, R. A., 1982, Ann. N.Y. Acad. Sci., 393:235.

Baines, D.M., 1986, Science, 233:419.

Berridge, M.J., 1987, Ann. Rev. Biochem., 56:159.

Brockerhoff, H., 1986, FEBS Lett., 201:1.

Cantley, L.C., Cantley, L.G., and Jospephson, L., 1978, J. Biol. Chem., 253:7361.

Castagna, M., Takai, Y., Kaibuchi, K., Sano, K., Kikkawa, U., and Nishizuka, Y., 1982, J. Biol. Chem., 257:7847.

Collins, J.M., Klecker, R.W., Yarchoan, R., Lane, H.C., Fauci, A.S., Redfield, R.R., Broder, S., and Meyers, C.E., 1986, J. Clin. Pharmacol., 26:22.

Coussens, L., Parker, P.J., Rhee, L., Yang-Feng, T.L., Chen, E., Waterfield, M.D., Francke, U., and Ullrich, A., 1986, Science, 233:859.

Danpure, C.J., Fyfe, D.A., and Gumpel, J.M., 1979, Ann. Rheum. Dis., 38:364.

Davis, R.J., Ganong, B.R., Bell, R.M., and Czech, M.P., 1985, J. Biol. Chem., 260:1358.

De Clercq, E., 1987, Antiviral Res., 7:1.

Duggan, D.E., 1959, Arch. Biochem. Biophys., 84:116.

Erdmann, E., Werdman, K., Krawietz, W., Schmitz, W., and Scholz, H., 1984, Biochem. Pharmacol., 33:945.

Froscio, M., Solanki, U., Murray, A.N., and Hurst, N.P., 1988, Biochem. Pharmacol., 37:366.

Gerber, R.C., Paulus, H.E., Bluestone, R., and Lederer, M., 1972, Arth. Rheum., 15:622.

Go, M., Sekiguchi, K., Nomura, H., Kikkawa, U., and Nishizuka, Y., 1987, Biochem. Biophys. Res. Commun., 144:598.

Goodno, C.C., 1982, Methods Enzymol., 85:116.

Hafstrom, I., Seligman, B.E., Friedman, M.M., and Gallin, J.I., 1984, J. Immunol., 132:2007.

Helson, L., Verma, M., and Helson, C., 1982, Ann. N.Y. Acad. Sci., 393:226.

Heyliger, C.E., Tahiliani, A.G., and McNeill,J.H., 1985, Science, 227:1474.

Hidaka, H., Inagaki, M., Kawamoto, S., and Sasaki, Y., 1984, Biochemistry, 23:5036.

House, C., and Kemp, B.E., 1987, Science, 238:1726.

Isler, O., and Brubacher, G., 1982, in: "Vitamine", Vol. 1, Georg Thieme Verlag, NY. p. 126.

Jeffrey, A.M., and Liskamp, M.J., 1986, Proc. Natl. Acad. Sci., USA,83:241.

Josephson, L., and Cantley, L.C., 1977, Biochemistry, 16:4572.

Kase, H., Iwahashi, K., Nakanishi, S., Matsuda, Y., Yamada, K., Takahshi, M., Murakata, C., Sato, A., and Kaneko, M., 1987, Biochem. Biophys. Res. Commun., 142:436.

Kase, H., Iwahashi, K., and Matsuda, Y., 1986, J. Antibiotics, 39:1059.

Kawahara, Y., Takai, Y., Minakuchi, R., Sano, K., and Nishizuka, Y., 1980, Biochem. Biophys. Res. Commun., 97:309.

Kikkawa, U., Ogita, K., Ono,Y., Asaoka, Y., Shearman, M.S., Fujii, T., Ase,K., Sekiguchi, K., Igarashi, K., and Nishizuka, Y., 1987, FEBS Lett., 223:212.

Kishimoto, A., Takai, Y., Mori, T., Kikkawa, U., and Nishizuka, Y., 1980,, J. Biol. Chem., 255:2273.

Knopf, J.L., Lee, M., Sultzman, L.A., Kriz, R.W., Loomis, C.R., Hewick, R.M., and Bell, R.M., 1986, Cell, 46:491.

Kondo, H., Baba, Y., Takai, K., Kondo, K., and Kagamiyama, H., 1987, Biochem. Biophys. Res. Commun., 142:155.

Kuo, J.F., Schatzman, R.C., Turner, R.S., and Mazzei, G.J., 1984, Mol. and Cell. Endocrinol., 35:65.

Lapetina, E.G., Reep, B., Ganong, B.R., and Bell, R.M., 1985, J. Biol. Chem., 260:1358.

Liu, A.Y., and Chen K.Y., 1985, J. Cell Physiol., 125:387.

Lubin, B., and Machlin, L.J., 1982, Ann. N.Y. Acad. Sci., 393:1.

Mahoney, C.W., and Azzi, A., 1988, Experientia, 44:A82.

Mahoney, C.W., and Azzi, A., a) in: "Dynamics of Membrane Proteins and Cellular Energetics", N. Latruffe, Y. Gaudemer, P. Vignais, A. Azzi, eds., Springer Verlag, Berlin, in press.

Mahoney, C.W., and Azzi, A., b) in: "Perspectives in Molecular Approaches to Human Disease", S. Papa, J. Tager, J. Jaz, eds., Ellis Horwood, London, in press.

Mahoney, C.W., and Azzi, A., in press, c) Biochem. Biophys. Res. Commun.

Mahoney, C.W., Fredenhagen A., Peter, H., and Azzi, A., Biological Chem. Hoppe-Seyler, in press.

Mahoney, C.W., Fredenhagen, A., Peter, H., and Azzi, A., (submitted).

Naka, M., Nishikawa, M., Adelstein, R.S., and Hidaka, H., 1983, Nature, 306:490.

Nakanishi, S., Matsuda, Y., Iwahashi, K., and Kase, H., 1986, J. Antibiotics, 39:1066.

Nishizuka, Y., 1984, Nature, 308:693.

Nishizuka, Y., 1986, Science, 233:305.

Nour-Eldeen, A.F., Craig, M.M., and Gresser, M.J., 1985, J. Biol. Chem., 260:6836.

Ohno, S., Kawasaki, H., Imajoh, T., Suzuki, K., Inagaki, M., Yokokura, H., Sakoh, T., and Hidaka, H., 1987, Nature, 325:161.

Ossikovski, E., and Walter, R.D., 1984, Mol. and Biochem. Parasitol., 12:299.

Parente, J.E., Wong, K., and Davis, P., 1986, Inflammation, 10:303.

Parker, P.J., Coussens, L., Totty, N., Rhee, L., Young, S., Chen, E., Stabel, S., Waterfield, M.D., and Ullrich, A., 1986, Science, 33:853.

Pope, M.T., and Dale, B.W., 1968, <u>Quart. Rev. Chem. Soc.</u>, 22:527.

Ramasarma, T., and Crane, F.L., 1981, <u>Curr. Cont. Cell. Regul.</u>, 20:247.

Ruusala, A.I., Mattsson, M., Esscher, T., Abrahamsson, L., Jergil, B., and Pahlman, S., 1985, <u>Develop. Brain Res.</u>, 18:27.

Sadler, P.J., 1976, <u>Struct. Bonding</u>, 29:171.

Sano, K., Takai, Y., Yamanashi, J., and Nishizuka, Y., 1983, <u>J. Biol. Chem.</u>, 258:2010.

Sharkey, N.A., Leach, K.L., and Blumberg, P.M., 1984, <u>Proc. Natl. Acad. Sci., USA</u>, 81:607.

Shaw, C.F., 1979, <u>Inorg. Persp. in Biol. and Med.</u>, 2:287.

Simons, T.J.B., 1979, <u>Nature</u>, 281:337.

Slater, E.C., 1960, <u>in</u>: "4th International Congress of Biochem.", Vienna; Pergamon, London, p. 316.

Solomon, D.M., O'Brian, C.A., and Weinstein, I.B., 1985, <u>FEBS Lett.</u>, 190:342.

Steiner, M., and Mower, R., 1982, <u>Ann. N.Y. Acad. Sci.</u>, 393:289.

Stuart, M.J., 1982, <u>Ann. N.Y. Acad. Sci.</u>, 393:277.

Takai, Y., Kishimoto, A., Kikkawa, U., Mori, T., and Nishizuka, Y., 1979, <u>Biochem. Biophys. Res. Commun.</u>, 91:1218.

Tamaoki, T., Takahashi, I., Kato, Y., Morimoto, M., and Tomita, F., 1986, <u>Biochem. Biophys. Res. Commun.</u>, 135:397.

Tamura, S., Brown, T.A., Whipple, J.H., Fujita-Yamaguchi, Y., Dubler, R.E., Cheng, K., and Larner, J., 1984, <u>J. Biol. Chem.</u>, 259:6650.

Tracey, A.S., and Gresser, M.J., 1986, <u>Proc. Natl. Acad. Sci., USA</u>, 83:609.

Walter, R.D., 1980, <u>Mol. and Biochem. Parasitol.</u>, 1:139.

Wasserman, R.H., and Taylor, A.N., 1972, <u>Ann. Rev. Biochem.</u>, 41:182.

Watson, S.P., McNally, J., Shipman, L.J., and Godfrey P.P., 1988, <u>Biochem. J.</u>, 249:345.

THE SIGNALING FUNCTION OF CALCIUM AND ITS REGULATION

Ernesto Carafoli

Laboratory of Biochemistry
Swiss Federal Institute of Technology (ETH)
8092 Zürich, Switzerland

INTRODUCTION

In the course of evolution Ca has become a signaling agent of universal significance, which controls a large number of cellular functions: prominent among them are the synthesis and release of hormones, muscle and non-muscle motility, and a multiplicity of membrane-linked processes (see Carafoli, 1987, for a recent comprehensive review). It is self evident that the signaling function of Ca demands its maintenance within cells at a very low free concentration, and mechanisms to efficiently modulate it in the cell compartments where the targets of the signaling function are located. Other signaling agents are regulated within cells by biosynthesis and breakdown. Since this is impossible in the case of Ca, evolution has selected an entirely different control mechanism, i.e., the reversible complexation by specific proteins, which are either soluble, organized in non membranous structures, or intrinsic to membranes. These proteins "buffer" intracellular Ca at a concentration which is at least 10,000 fold lower than in the external spaces. The functional cycle of cells requires both short and long term regulation of Ca. The rapid and precise modulation is performed by intracellular Ca binding proteins but also (in fact mostly, see below) by high Ca affinity membrane intrinsic proteins. The Ca-filtering function of the plasma membrane, which depends on the operation of membrane-intrinsic Ca binding proteins, is responsible for the long term maintenance of the Ca gradient between cells and medium. The trans-plasma membrane gradient is convenient to the signaling function of Ca, since the large inwardly directed Ca pressure ensures that even minor increases in the Ca permeability of the plasma membrane results in significant swings in its intracellular concentration, and thus in turn in the modulation of the signaling function.

The high affinity intracellular Ca binding proteins

The solution of the crystal structure of the muscle Ca-binding protein parvalbumin by Kretsinger and his associates (1973) has led to the proposal of a set of structural guidelines for the complexing of Ca by parvalbumin and other high affinity Ca binding proteins (see Kretsinger and Nelson, 1977, for a review). The parvalbumin principles demand that Ca-binding proteins of this type contain repeat units made of two perpendicular α-helices, flanking a non-helical loop of 10-12 amino acids where Ca coordinates to 6-8 oxygen atoms of carboxylic side chains

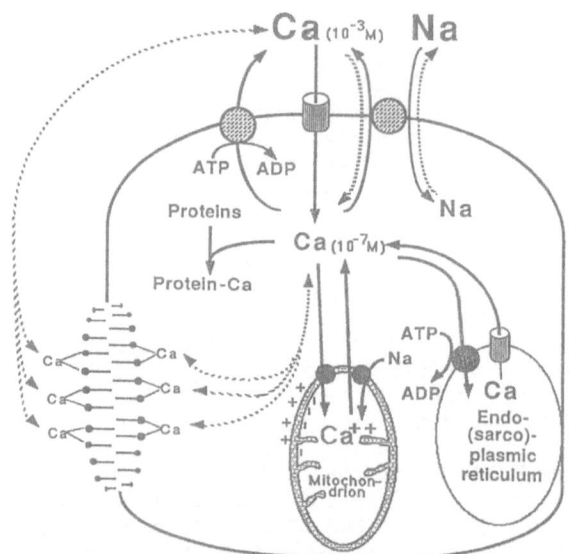

Fig. 1. Ca transporting systems in eucaryotic cells. The Figure shows the transporting systems which have been characterized so far. The systems in the Golgi membranes and in the lysosomes are not shown because they are still very poorly known. Three in the plasma membrane (a Ca-ATPase, a Na/Ca exchanger, which normally imports Ca but sometimes also exports it, and a Ca channel); two in endo(sarco)plasmic reticulum (a Ca-ATPase and a still poorly characterized release channel); two in the inner membrane of mitochondria (an electrophoretic uptake uniporter and a Ca releasing Na/Ca exchanger. The soluble Ca binding proteins of the cytosol and the acidic phospholipids of the plasma membrane have been included in the Figure as participants in the Ca-buffering function (see text).

and, less frequently, to carbonyl oxygens of the peptide backbone. This Ca binding site has been optimized in the course of evolution to a point where only minimal variability is now tolerated. The parvalbumin model has become known as the EF-hand model, and has been shown to be compatible with the amino acid sequences of a number of other Ca-binding proteins; recently, it has been directly validated by the crystal structures of three other Ca-binding proteins (Szebenyi et al., 1981; Herzberg and James, 1985; Babu et al., 1985).

The soluble high affinity Ca-binding proteins complex Ca and thus contribute to its buffering. But they have an important additional function: upon binding Ca they express hydrophobic domains on the surface, and the conformational change enables them to interact with (enzyme) targets. Therefore, it is convenient to consider these proteins more as processors of the Ca signal, rather then as intracellular Ca buffers. That their Ca buffering role is accessory seems plausible considering that their buffering function would be quantitatively limited by their amount in the cell: one can nevertheless attempt to evaluate the total theoretical Ca buffering capacity of the two most important Ca-binding proteins of this type, calmodulin and troponin C. The concentration of the former is lower in muscles than in other tissues, where its total Ca buffering capacity corresponds to approximately 8 μM Ca (Carafoli, 1987). Troponin-C, on the other hand, is more concentrated in muscles than in non muscle tissues; based on a measured content of 2 in heart, and 20 μM in skeletal muscles (Grand et al.,1979),it can be calculated that troponin-C could handle up to 6 μM and 80 μM Ca in heart and skeletal muscles, respectively (Carafoli, 1987).

Membrane transport of calcium

The limitations in the Ca buffering function described above do not apply to other Ca-binding proteins, intrinsic to membranes. They complex Ca at one membrane side, deliver it across, and repeat the operation continuously, thus eliminating the problems arising from the possibly insufficient amount of protein available. Therefore, membrane-intrinsic Ca binding (and transport) proteins play the dominant role in the buffering of intracellular Ca. Eucaryotic cells possess several Ca transporting systems in the plasma membrane and in the membranes of the organelles (Figure 1). The various transport systems operate with different kinetic properties, responding to the varying demands of the functional cycle of cells: sometimes, rapid, precise, and high affinity

Table 1. Ca transporting systems in cell membranes

Transporting mode	Membrane system	Ca affinity
ATPases	Plasma membrane Endo(sarco)plasmic reticulum (Golgi? lysosomes?)	high
Exchangers (Na/Ca)	Plasma membrane Mitochondria	low
Channels	Plasma membrane Endo(sarco)plasmic reticulum	low
Electrophoretic uniporters	Mitochondrial inner membrane	low

regulation of Ca is required, at some other times lower affinity and less rapid regulation (i.e., transport) may suffice. As figure 1 indicates, eucaryotic cell membranes contain several (proteinaceous) Ca transporting systems, corresponding to four basic transporting mechanisms: ATPases, exchangers, channels, electrophoretic uniporters (Table 1). High affinity Ca regulation depends obligatorily on ATPases, since only this transporting mode confers to the system high affinity for Ca. Lower affinity regulation may on the other hand choose from amongst the other three modes listed in Table 1.

Ca transport in the plasma membrane

Figure 2 shows the three Ca transporting systems known to exist in the plasma membrane: the Ca channel, the Na/Ca exchanger, and the Ca pump. The Ca channel is responsible for the penetration of Ca into cells. Although known for a long time, the channel has been dissected at a molecular level only recently: the introduction of the patch-clamp technique has permitted the study of single channel activity, and has led to the definition of some of its kinetic parameters (see Reuter, 1984, for a review). Recordings of single channel currents have revealed a conductance of 15-25 pS, corresponding to the passage of about 3×10^6 Ca ions per sec. (Reuter et al., 1982). The density of the Ca channel is of the order of 0.1 to 1.0 per M^2 of membrane surface in heart, but is higher in the T-tubular membranes of skeletal muscles. The channels are controlled by the electrical potential across the plasma membrane, and begin to open as the transmembrane potential increases from the resting level of -70 mv to about -40 mv (maximal Ca currents are seen at about 0 mv). The channels are blocked by several "Ca antagonists" (Fleckenstein, 1983), the most important of which are the dihydropyridines. Work on Ca antagonists has greatly advanced the understanding of the Ca channel: one important recent development has been the solution of the primary structure of the dihydropyridine receptor (Tanabe et al., 1987). Another interesting aspect of the Ca-channel is its stimulation by adrenergic neurotransmitters: patch-clamp experiments on heart plasma membranes have shown that cAMP enhances the opening probability of the channel (Reuter et al., 1982), a finding that has obvious therapeutic implications. Several Ca channel types different in properties have been recently recognized. Two have been identified by patch-clamp experiments in mammalian hearts (Nilius et al., 1985), the most common being the dihydropyridine sensitive L-type, whose openings produce currents of long duration. The T-type opens at more negative transmembrane potentials, is insensitive to dihydropyridines, and produces currents of shorter duration.

The Na/Ca exchanger is one of the two Ca exporting systems of the plasma membrane (under some conditions it can also mediate the influx of Ca). It is a large-capacity, low affinity system, particularly active in excitable tissues like heart. The system has been studied essentially on heart and the giant axon of the squid (see Philipson, 1985 for a review). Early electrophysiological work has established that the system operates electrogenically, exchanging 3 Na for 1 Ca; thus, it does not only respond to the Na and Ca gradients, but also to the transmembrane electrical potential. Recent work on heart sarcolemmal vesicles (Reeves and Sutko) has shown that the system has low affinity for Ca (K_m, 1-20 M) but high maximal rate of transport, corresponding in heart to 20 nmoles per mg of sarcolemmal protein per sec. The low affinity of the system for Ca is puzzling, since the free Ca concentration in the cytosol presumably never increases to 10 µM. Possibly, the kinetic parameters of the exchanger in isolated vesicles are different from those in the intact tissue, but other physiological mechanisms to increase the Ca affinity of the system may also exist. The activation of the exchanger by a kinase-linked phosphorylation process (Caroni and Carafoli, 1983), which decreases the K_m(Ca) of the system to about 1 µM may be relevant in this context.

The Ca ATPase of the plasma membrane interacts with Ca with high affinity (K_m, about 0.5 µM), but has low total transport capacity: (about 0.5 nmol per mg of membrane protein per sec in heart). The high Ca affinity of the enzyme qualifies it to export Ca from cells even when its concentration in the cytosol is at the resting sub- Molar level. Thus, the Ca ATPase probably plays the most important role in maintaining the gradient of Ca between cells and medium. The enzyme is an ATPase of the P-class, i.e., it forms an spartyl phosphate during the reaction mechanism and is inhibited by vanadate (see Schatzmann, 1982, for a

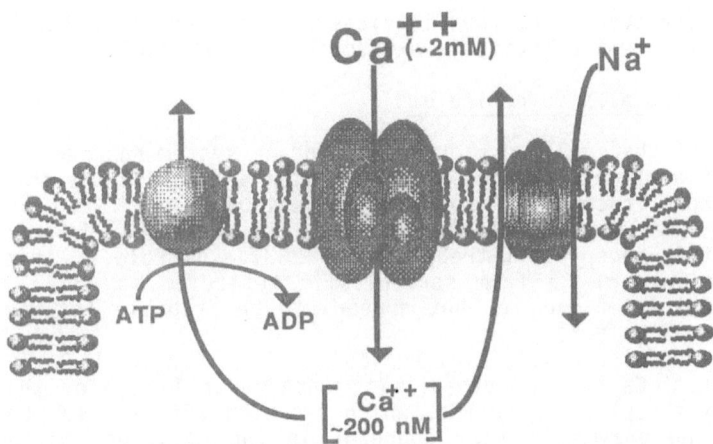

Fig. 2. A scheme of the three plasma membrane Ca transporting systems. The subunit composition of the Ca channel and of the Na/Ca exchanger is based on the isolation work of several workers for the former (see for example Curtis and Catterall, 1983, and Borsotto et al., 1984, and on a proposal by Longoni and Carafoli (1987) for the latter).

review). In addition, it is a target of calmodulin stimulation: the interaction with calmodulin has permitted the purification of the ATPase on calmodulin columns as a single polypeptide of 138 kDa, which reconstitutes in liposomes with optimal transporting efficiency (Niggli et al., 1981). The liposomal system has permitted to establish that the ATPase transports Ca with a 1:1 stoichiometry to ATP hydrolysis, as compared to a figure of 2 for the analogous enzyme of sarcoplasmic reticulum (see below). Caroni and Carafoli (1981) have found that a cAMP-linked phosphorylation process stimulates the ATPase and the associated pumping of Ca. Since cAMP also stimulates the Ca channel activity (see above), the role of cAMP is the overall increase of the plasma membrane Ca flux, rather than the stimulation of a single Ca transport reaction.

Work on the purified Ca ATPase has established that the enzyme can be stimulated by several treatments alternative to calmodulin, among them the exposure to acidic phospholipids and polyunsaturated fatty acids, and the controlled proteolysis by a number of proteases. The proteolysis work has been instrumental in a recent comprehensive attempt, based on chemical sequencing and DNA recombination technology, to establish the primary structure of the enzyme. At the present stage of development the work has solved the sequence of approximately 90% of the ATPase. It has mapped its essential functional domains, among them the calmodulin binding domain, which is located near the C-terminus.

Ca transport across intracellular membranes

The long-term maintenance of the Ca gradient between cells and medium is thus the result of the action of the plasma membrane transport systems. However, the plasma membrane fluxes are quantitatively minor when compared to the total amount of Ca involved in the functional cycle of cells. The Ca crossing the plasma membrane triggers important intracellular events and thus is vital to cell function, but most of the Ca needed for cell activity is extracted from intracellular stores.

The endo(sarco)plasmic reticulum

Most of the early work on this membrane system has been performed on the easily available sarcoplasmic reticulum and on its key enzyme, the Ca ATPase. More recently, endoplasmic reticulum has seen considerable activity, due to the discovery that its Ca pool is sensitive to inositol-tris-phosphate (Streb et al., 1983). The role of this messenger in the release of Ca from sarcoplasmic reticulum is still unclear in heart and skeletal muscle, but appears to be probable in smooth muscles (Somlyo et al., 1985).

The endo(sarco)plasmic reticulum is responsible for the fine regulation of Ca in the cytosol. It contains an ATPase of the same type of that of the plasma membrane, which has high affinity for Ca (K_m below 0.5 µM). The enzyme is very abundant in the membranes of sarcoplasmic reticulum, which thus has a large total Ca transporting capacity. This is particularly so in fast skeletal muscles where it may reach 70 nmol per mg of membrane protein per sec. The sarcoplasmic reticulum ATPase has been purified by MacLennan in 1970 as a single polypeptide of about 100 kDa, and has been characterized extensively in several Laboratories. It belongs to the same ATPase class of the plasma membrane Ca pump: it forms an aspartyl-phosphate, is inhibited by vanadate, and can be reconstituted into liposomes, where it transports Ca with a 1:1 stoichiometry to the hydrolyzed ATP. Recently, its primary structure has been determined (MacLennan et al., 1985) on both heart and fast skeletal muscles reticulum. The ATPase in heart, smooth, and slow (but not fast) skeletal muscles is modulated by the acidic proteolipid phospholamban (Tada et al., 1975). This proteolipid is a pentamer of five identical subunits of M_r about 6 kDa, and is phosphorylated by both the cAMP and the calmodulin dependent protein kinase. Phosphorylated phospholamban activates the ATPase-linked transport of Ca, thus transmitting to sarcoplasmic reticulum hormonal messages from the plasma membrane.

The release of Ca from heart sarcoplasmic reticulum is now the subject of very intensive investigations, also as the result of the great interest in the function of inositol tris phosphate in endoplasmic reticulum and in sarcoplasmic reticulum of smooth muscles. The favored mechanism for the Ca release phenomenon in heart is the Ca-induced Ca release (Fabiato and Fabiato, 1975), in which the liberation of massive amounts of Ca from the vesicles of sarcoplasmic reticulum is triggered by Ca added to the medium at concentrations below 0.1 µM. The channel that mediates the release phenomenon now also intensively studied. Recent work

(Meissner and Henderson, 1987) has indicated that it is activated by Ca, adenine nucleotides, and caffeine, and inhibited by ruthenium red, Mg, protons, and calmodulin.

Much less is known on the mechanism of Ca release induced by inositol tris phosphate. One interesting recent development is the proposal (Volpe et al., 1987) that the organelle that responds to inositol-tris phosphate in non-muscle tissues is not the endoplasmic reticulum, but a separate vesicular system termed the calciosome.

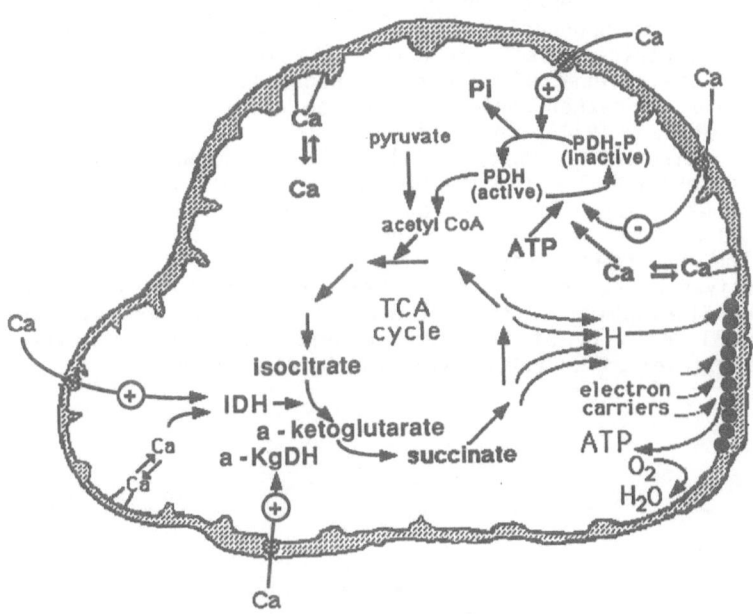

Fig. 3. The regulation of matrix dehydrogenases by Ca. The information in the Figure comes from the work of Denton and his coworkers (1972, 1978, 1979). PDH = pyruvic acid dehydrogenase; IDH = NAD-linked isocitric acid dehydrogenase; Kg-DH = α-ketoglutaric acid dehydrogenase.

Mitochondria

Mitochondria have been traditionally considered as important in the regulation of both cytosolic and intramitochondrial Ca (see Carafoli, 1982, for a review). The demonstration that they only handle Ca with low affinity (Crompton et al., 1976) and the finding that they contain much less Ca in situ than generally assumed based on experiments on isolated mitochondria (Somlyo et al., 1979), have forced a re-evaluation of their importance as cytosolic Ca regulators, a function which is now considered of minor importance. Their main Ca-related task is now accepted to be the regulation of matrix Ca. This is an important function, due to the existence in the matrix of dehydrogenases which are precisely controlled by Ca (Fig. 3). Mitochondria accumulate Ca by means of an electrophoretic route which is energized by the electrical component of the proton-motive force across the inner membrane and is inhibited by ruthenium red. The route has low Ca affinity (K_m, 1-10 μM) but could accumulate Ca at an optimal rate of up to 10 nmoles per mg of protein per sec. In vivo, however, the sub-optimal concentration of Ca in the cytosol, and the presence of Mg which also inhibits the route reduce the uptake rate to a fraction of the optimal. Interestingly, mitochondria also accumulate inorganic phosphate to precipitate Ca as an insoluble salt, probably hydroxyapatite, in the matrix. This results in the damping of the changes in ionic Ca in the latter compartment, limiting the disturbance to the Ca-modulated dehydrogenases by newly accumulated Ca.

Since the mitochondrial Ca uptake process is electrophoretic, and since the electrical potential across the inner membrane is unlikely to fluctuate, the release of Ca can not occur through the reversal of the uptake route. It rather occurs through an electroneutral Na/Ca exchanger (Carafoli et al., 1974), which is particularly active in heart and other excitable tissues (Crompton et al., 1978), and which releases Ca at a slow rate (about 0.2 - 0.3 nmoles per mg of mitochondrial protein per sec). The Na-promoted release route is insensitive to ruthenium red but is blocked by some of the inhibitors of the plasma membrane Ca channel, e.g., the benzothiazepine diltiazem (Vaghy et al., 1982). The energy-driven uptake route and the Na/Ca exchanger can be integrated into an energy-dissipating Ca cycle (Carafoli, 1979, Figure 4). However, the low overall activity of the mitochondrial Ca transport system in vivo limits the level of energy dissipation by the cycle.

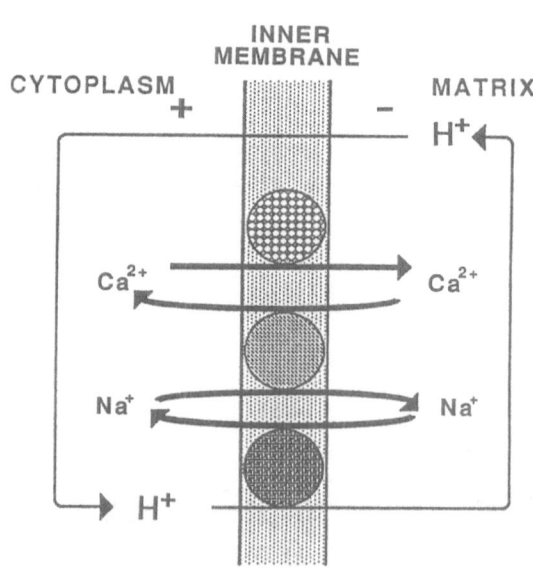

Fig. 4. The mitochondrial Ca cycle (Carafoli, 1979)

The minor role of mitochondria in the regulation of cytosolic Ca under physiological conditions is thus essentially determined by the insufficient concentration of Ca in cells. However, injuring conditions may alter the permeability properties of the plasma membrane permitting the influx of excess Ca into the cytosol. If this occurs, the mitochondrial uptake system may become activated, leading to the storage of large amounts of precipitated Ca-phosphate in the matrix. Deposits of Ca-phosphate in mitochondria of injured cells, including heart cells, have been repeatedly documented (see Carafoli, 1982, for a review). If the injuring condition disappears, mitochondria slowly release the stored Ca at a rate compatible with the exporting ability of the plasma membrane systems. Thus, they play a decisive protective role against cytosolic Ca overload.

CONCLUSIONS

The control of cell Ca is essentially performed by the reversible complexation to specific proteins. Soluble proteins contribute to Ca buffering but are more important in the transmission of Ca signal to targets. Membrane-intrinsic proteins play the main role in the buffering of cell Ca. They may control Ca very precisely, with high affinity (ATPases) or with lower affinity (channels, exchangers. the electrophoretic uniporter of mitochondria). Endo(sarco)plasmic reticulum is responsible for the fine regulation of Ca, whereas the mitochondrial transporting systems essentially control intramitochondrial Ca. The mitochondrial system, however, also protects the cytosol against pathological Ca increases.

REFERENCES

Babu, Y.S., Sack, J.S., Greenhough, T.J., Bugg, C.E., Means, A.R., and Cook, W.J., 1985, Three-dimensional structure of calmodulin, Nature, 315:37.
Borsotto, M., Norman, R.I., Fosset, M., and Lazdunski, M., 1984, Eur. J. Biochem., 14:449.
Carafoli, E., 1979, The calcium cycle of mitochondria, FEBS Letters, 104:1.
Carafoli, E., 1982, The transport of calcium across the inner membrane of mitochondria, in: "Membrane Transport of Calcium", E. Carafoli, ed., Academic Press, London, pp 109.
Carafoli, E., 1982, Membrane transport and the regulation of the cell calcium levels, in: "Pathophysiology of Shock, Anoxia, and Ischemia", R.A. Cowley, and B.F. Trump, eds., Williams and Wilkins, pp 95.
Carafoli, E., 1987, Intracellular calcium homeostasis, Ann. Rev. Biochem., 56:395.
Carafoli, E., Tiozzo, G., Lugli, F., Crovetti, F., and Kratzing, C., 1974, The release of calcium from heart mitochondria by sodium, J. Molec. Cell. Cardiol., 6:361.
Caroni, P., and Carafoli, E., 1981, Regulation of Ca^{2+}-pumping ATPase of heart sarcolemma by a phosphorylation-dephosphorylation process, J. Biol. Chem., 256:9371.
Caroni, P., and Carafoli, E., 1983, The regulation of the Na^+/Ca^{2+} exchanger of heart sarcolemma, Eur. J. Biochem., 132:451.
Crompton, M., Moser, R., Lüdi, H., and Carafoli, E., 1978, The interrelations between the transport of sodium and calcium in mitochondria of various mammalian tissues, Eur. J. Biochem., 82:25.
Crompton, M., Sigel, E., Salzmann, M., and Carafoli, E., 1976, A kinetic study of the energy-linked influx of Ca^{2+} into heart mitochondria, Eur. J. Biochem., 69:429.

Curtis, B.M., and Catterall, W.A., 1985, Phosphorylation of the calcium antagonist receptor of the voltage-sensitive calcium channel by cAMP-dependent protein kinase, Proc. Nat. Acad. Sci., USA, 82:2528.

Denton, R.M., Randle, P.J., and Martin, B.R., 1972, Stimulation by calcium ions of pyruvate dehydrogenase phosphate phosphatase, Biochem. J., 128:161.

Denton, R.M., Richards, D.A., and Chin, J.G., 1978, Calcium ions and the regulation of NAD^+-linked isocitrate dehydrogenase from the mitochondria of rat heart and other tissues, Biochem. J., 176:899.

Fabiato, A., and Fabiato, F., 1975, Contractions induced by a calcium triggered release of calcium from the sarcoplasmic reticulum of single skinned cardiac cells, J. Physiol., 249:457.

Fleckenstein, A., 1973, Calcium antagonism in heart and smooth muscle, John Wiley, New York.

Grand, R.J.A., Perry, S.V., and Weeks, R.A., 1979, Troponin-C like proteins (calmodulin) from mammalian smooth muscle and other tissues, Biochem. J., 177:521.

Herzberg, O., and James, M.N.G., 1985, Structure of the calcium regulatory muscle protein troponin-C at 2.8. A resolution, Nature, 313:665.

Krause, K.H., Volpe, P., Zorzato, F., Hashimoto, S., Pozzan, T., Meldolesi, J., and Lew, P.D., 1987, Calciosomes: evidence for a new type of organelle regulating intracellular Ca^{2+}, Seventh Intern. Washington Spring Symposium, Abstract 64.

Kretsinger, R.H., and Nelson, D.J., 1977, Calcium in biological systems, Coord. Chem. Rev., 18:29.

Kretsinger, R.H., and Nockolds, C.E., 1973, Carp muscle calcium-binding protein, J. Biol. Chem., 248:3313.

Longoni, S. and Carafoli, E., 1987, Identification of the Na^+/Ca^{2+} exchanger of calf heart sarcolemma with the help of specific antibodies, Biochem. Biophys. Res. Commun., 145:1059.

MacLennan, D.H., 1970, Purification and properties of an adenosine triphosphatase from sarcoplasmic reticulum, J. Biol. Chem., 245:4508.

MacLennan, D.H., Brandl, C.J., Korczak, B., and Green, N.M., 1985, Amino-acid sequence of a Ca^{2+} + Mg^{2+}-dependent ATPase from rabbit muscle sarcoplasmic reticulum, deduced from its complementary DNA sequence, Nature, 316:696.

McCormack, J.G., and Denton, R.M., 1979, The effects of calcium ions and adenine nucleotides on the activity of pig heart 2-oxoglutarate dehydrogenase complex, Biochem. J., 180:533.

Meissner, G., and Henderson, J.S., 1987, Rapid calcium release from sarcoplasmic reticulum vesicles is dependent on calcium and is modulated by Mg^{2+}, adenine nucleotide, and calmodulin, J. Biol. Chem. 262:3065.

Niggli, V., Adunyah, E.S., Penniston, J.T., and Carafoli, E., 1981, Purified $(Ca^{2+}-Mg^{2+})$-ATPase of the erythrocyte membrane; reconstitution and effect of calmodulin and phospholipids, J. Biol. Chem., 256:395.

Nilius, B., Hess, P., Lansmann, J.B., and Tsien, R.W., 1985, A novel type of cardiac calcium channel in ventricular cells, Nature, 316:443.

Philipson, K.D., 1985, Sodium-calcium exchange in plasma membrane vesicles, Ann. Rev. Physiol., 47:561.

Reeves, J.P., and Sutko, J.L., 1979, Sodium-calcium exchange in cardiac membrane vesicles, Proc. Nat. Acad. Sci. USA, 76:590.

Reuter, H., 1984, Ion channels in cardiac cell membranes, Ann. Rev. Physiol., 46:473.

Reuter, H., Stevens, C.F., Tsien, R.W., and Yellen, G., 1982, Properties of single calcium channels in cardiac cell culture, Nature, 297:501.

Schatzmann, H., 1982, The calcium pump of erythorcytes and other animal cells, in "Membrane Transport of Calcium", E. Carafoli, ed., Academic Press, London, pp 41.

Somlyo, A.V., Bond, M., Somlyo, A.P., and Scarpa, A., 1985, Inositol tris phosphate-induced calcium release and contraction in vascular smooth muscle, Proc. Nat. Acad. Sci., 82:5231.

Somlyo, A.P., Somlyo, A.V., and Shuman, H., 1979, Electron probe analysis of vascular smooth muscle, composition of mitochondria, nuclei, and cytoplasm, J. Cell Biol., 81:316.

Streb, H., Irvine, R.F., Berridge, M.J., and Schulz, I., 1983, Release of Ca^{2+} from a non-mitochondrial intracellular store in pancreatic acinar cells by inositol-1,4,5-trisphosphate, Nature, 306:66.

Szebenyi, D.M.E., Obendorf, S.K., and Moffat, K., 1981, Structure of vitamin D-dependent calcium binding protein from bovine intestine, Nature, 294:327.

Tada, M., Kirchberger, M.A., and Katz, A.M., 1975, Phosphorylation of 22.000-Dalton component of the cardiac sarcoplasmic reticulum by adenosine 3":5"-monophosphate-dependent protein kinase, J. Biol. Chem., 250:2640.

Tanabe, T., Takeshima, H., Mikami, A., Flockerzi, V., Takahashi, H., Kangaura, K., Kojima, M., Matsuo, H., Hirose, T., and Numa, S., 1987, Primary structure of the receptor for calcium channel blockers from skeletal muscle, Nature, 328:313.

Vaghy, P.L., Johnson, J.D., Matlib, M.A., Wang, J., and Schwarz, A., 1982, Selective inhibition of Na^+-induced Ca^{2+} release from heart mitochondria by diltiazem and certain other Ca^{2+} antagonist drugs, J. Biol. Chem., 257:6000.

Volpe, P., Krause, K.H., Hashimoto, G., Zorzato, F., Pozzan, T., Meldolesi, J., and Lew, D.P., 1988, "Calcisome", a cytoplasmic organelle: the inositol 1,4,5-trisphosphate-sensitive Ca^{2+} store of non-muscle cells? Proc. Nat. Acad. Sci., U.S.A., 85:1091.

pH HOMEOSTASIS AND CELL FUNCTIONS AND DISEASES

Sergio Papa, Michele Lorusso and Ferdinando Capuano

Institute of Medical Biochemistry and Chemistry
University of Bari, Bari, Italy

The pH homeostasis in the cell appears to be a critical element for its functions and growth. Hydrogen ions besides regulating or optimizing biological activities of enzymes, carriers, receptors and nucleic acids, mediate energy transfer in the cell in the form of protonmotive force (PMF, $\Delta\mu H^+$) across membranes.

There are, furthermore, systems which utilize protons to transmit signals for cell growth (Pouyssegur, 1985) or to direct cellular distribution of solutes and macromolecules (Mellman et al., 1986; Rudnick, 1986; Sze, 1985).

In Fig. 1 a general picture of systems involved in pH homeostasis and protonmotive energy transfer in eukaryotic cell is reported. The mitochondrial transmembrane protonmotive force generated by respiratory chain activity is utilized by the H^+-ATPase to synthesize most of the ATP produced by the cell. Different types of ATPases are operating in the plasma membrane and organelles of the vacuolar system, where ATP is utilized to acidify the interior of organelles. Important for both iron uptake and cell growth is the transplasma membrane NAD(P)H dependent diferric-transferrin reductase. Reduction of Fe^{3+}-transferrin is followed by the receptor-mediated internalization of the complex into acidic endosomes where iron is released (Low et al., 1987).

A large body of observations, also deriving from studies with human tissues, indicate that disturbances of the pH homeostasis and protonmotive energy transfer systems are involved in hereditary and/or acquired metabolic dysfunctions (Scholte, 1988; Di Mauro et al., 1985; De Vivo et al., 1983; Rebouche and Engel, 1983) cardiovascular diseases (Rouslin, 1983) malpractice in physical exercise (Packer, 1987). Definite progress is being made in the elucidation of the molecular biology, functions and dysfunctions of proteins which transport H^+, adjust and/or respond to pH in the different cell compartments, generate and utilize proton current and gradient.

This paper will deal with certain aspects of: i) the Na^+/H^+ antiport of the plasma membrane and of the inner mitochondrial membrane; ii) the ΔpH driven anion uptake into the mitochondrial matrix; iii) the protonmotive cytochrome system of mitochondria.

145

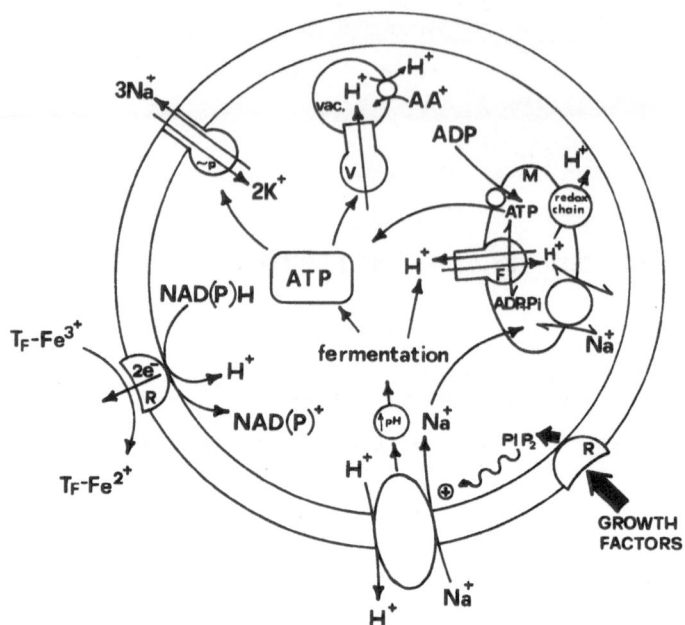

Fig. 1. Proton and cation translocators in animal cell membranes.
M and Vac represent mitochondrial and vacuolar compartments
respectively. F, P and V represent the three categories of
ATPases according to Pedersen and Carafoli (1987). PIP_2,
phosphatidylinositol-4,5-biphosphate.

Na^+/H^+ ANTIPORTER

 Na^+/H^+ antiport systems, which mediate 1/1 exchange of Na^+ with H^+,
have been identified in the plasma membrane of all the animal cells so far
described as well as in the inner mitochondrial membrane (Pouyssegur, 1985;
Krulwich, 1983; Aronson, 1985).
 The plasma membrane Na^+/H^+ antiporter of animal cells, which is specifically inhibited by amiloride, has been shown to be involved in the regulation of cytosolic pH, cell volume, transepithelial transport of Na^+ and
HCO_3^- and in the initiation of cell growth and proliferation in response to
external stimuli. Insulin, tumor promoters and growth factors (serum, epidermal growth factor, platelet-derived growth factor, α-thrombin, vasopres-
sin, bradykinin, concanavalin A etc.) stimulate the Na^+/H^+ antiporter in
resting cell lines with Na^+ uptake and alkalinization of the cytosol (Pouyssegur, 1985; Aronson, 1985). Activation by growth factors seems to involve the phosphatidylinositol system, protein c kinase, phosphorylation of
the Na^+/H^+ antiporter. This results in enhancement of cytosolic pH, which
in turn promotes DNA synthesis and growth. The concept emerges from these
interesting observations that cytosolic pH may act as a second messenger
in the action of growth promoting agents in normal and neoplastic cells.
This would suggest that substances capable of inducing cytosolic acidification may depress growth of neoplastic cells, thus serving as cytostatic a-
gents (Margolis et al., 1987).
 The $Na^+ (K^+)/H^+$ antiporter of the inner mitochondrial membrane, parti-

cularly active with Na^+ and at alkaline pH's, has been proposed to play a role in regulating intracellular pH (Rosen and Futai, 1980) and Na^+ concentration (West and Mitchell, 1974). It has been found, in our laboratory, that tumor mitochondria exhibit increased activity of Na^+ $(K^+)/H^+$ exchange system and enhanced passive permeability to Na^+ when compared to normal mitochondria (Papa et al., 1983; Capuano et al., 1982). As a consequence in tumor mitochondria a large amount of the ΔpH component of transmembrane $\Delta\mu H^+$ set up by respiration, is lost via Na^+/H^+ exchange. It can be noted, in fact, from Fig. 2 that at alkaline pH, replacement in the medium of K^+ with Na^+ results in a ΔpH decrease which is more marked in tumor than in rat liver mitochondria. The decrease of aerobic ΔpH is accompanied by a parallel, small increase of Δψ component which compensates for the drop of ΔpH.

The critical role of the mitochondrial Na^+/H^+ antiporter in regulating the extent of aerobic transmembrane ΔpH is shown by the effect of $MgCl_2$ (Table 1) which inhibits the antiporter (Douglas and Cockrell, 1974) and restores aerobic ΔpH without affecting significantly the aerobic Δψ.

The ΔpH generated by mitochondrial respiration drives, by means of specific proton symporters, the accumulation of anionic substrates into mitochondria (Chappel, 1968; Papa et al., 1970; Paradies and Papa, 1977). Thus, ΔpH dissipation would result in inhibition of the uptake of substrates like phosphate and pyruvate by mitochondria with consequent depression of oxidative

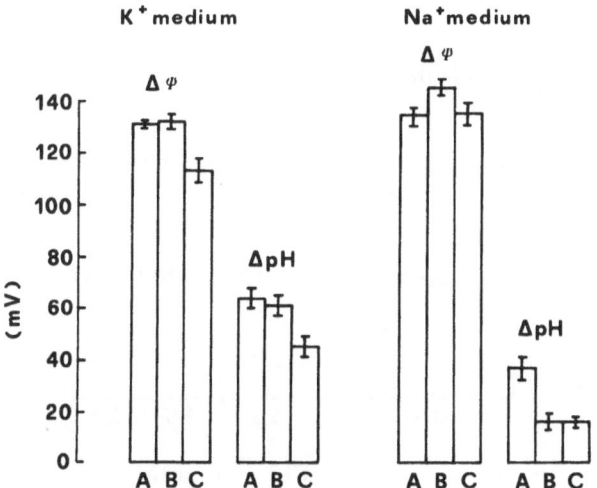

Fig. 2. Membrane potential, Δψ, and transmembrane pH difference, ΔpH, in tumor and in rat liver mitochondria.
Mitochondria (1.5 mg/ml) were suspended at 25°C in: 100 mM sucrose, 1 μg/mg prot. rotenone, 2 μg/mg prot. oligomycin, 30 nmol/mg prot. N-ethylmaleimide, 20 mM Hepes (pH 8.0), 0.5 mM EGTA, 20 mM KCl or NaCl. After 10 min preincubation, respiration was started by addition of 10 mM potassium or sodium succinate. Incubation was interrupted by rapid centrifugation 90 s after succinate addition. The values determined as described by Papa et al. (1983) are the means ± S.E.M. for 6-16 determinations. A, rat liver; B, Ehrlich ascites cells; C, Morris hepatoma 3924A.

Table 1. Effect of Magnesium on transmembrane $\Delta\mu H^+$ in respiring mitochondria from rat liver and Ehrlich Ascites tumor cells.

		$\Delta\psi$ (mV)	$-59\ \Delta pH$ (mV)
Rat liver	--	147	36
	+MgCl$_2$	158	66
Ascites cells:	--	132	20
	+MgCl$_2$	143	44

Mitochondria (1.5 mg/ml) were suspended in a reaction mixture containing: 80 mM sucrose, 30 mM NaCl, 1 µg/mg prot. rotenone, 2 µg/mg prot. oligomycin, 30 nmol/mg prot. N-ethylmaleimide, 20 mM Hepes, 0.5 mM EGTA. pH was adjusted to 8.0 with choline. Where indicated 25 mM MgCl$_2$ was included in the medium. Respiratory substrate, 10 mM succinate (see Papa et al., 1983).

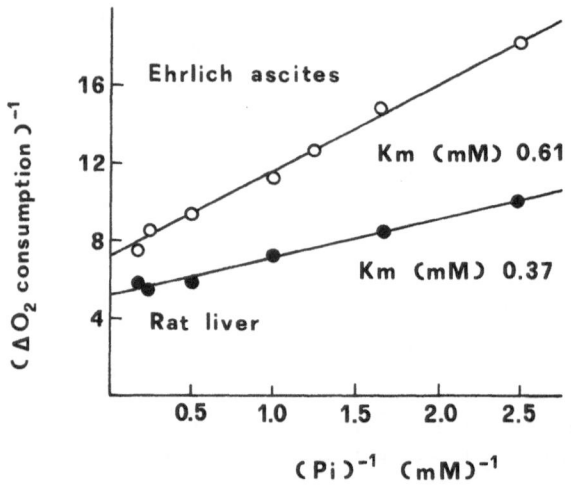

Fig. 3. Double reciprocal plot of the Pi-stimulated oxygen uptake in normal and tumor mitochondria. Mitochondria (2.5 mg/ml) were incubated at 25°C in: 75 mM sucrose, 30 mM Tris-HCl, 50 mM KCl, 1 mM MgCl$_2$, 1 mM EDTA, 50 mM glucose, 0.3 mM ADP, 10 mM glutamate, 1 U/ml hexokinase, pH 7.4. After 5 min preincubation 10 mM succinate was added. Changes in the respiratory rate, as a function of Pi additions to the respiring mitochondria, are reported.

Table 2. Phosphate uptake during respiratory steady-state in normal and in tumor mitochondria.

	n° of expt.s	(nmol/mg prot.) pH 7	pH 8
K+ medium			
Rat liver	3	30.5 ± 1.3	26.4 ± 1.7
Ehrlich Ascites cells	4	27.3 ± 1.2	20.7 ± 1.1
Na+ Medium			
Rat liver	3	28.1 ± 2.5	23.2 ± 2.1
Ehrlich Ascites cells	4	17.4 ± 0.3	11.4 ± 0.5

Mitochondria (5 mg/ml) were suspended at 25°C in: 100 mM sucrose, 1 µg/mg prot. rotenone, 2 µg/mg prot. oligomycin, 20 mM Hepes, 0.5 mM EGTA, 20 mM KCl or Na Cl. After 2 min preincubation, respiration was started by addition of 10 mM potassium or sodium succinate. 2 min later 2 mM Pi (0.2 µCi//ml) were added. Incubation was interrupted by centrifugation 2 min after Pi addition. Pi was measured in $HClO_4$ extracts of the mitochondrial pellet and in the supernatant. The substrate content of the matrix space was calculated by correcting the amount in the mitochondrial extract with that in sucrose-permeable space plus adherent supernatant (see Papa et al., 1970).

phosphorylation. Measurements of phosphate uptake during respiratory steady-state (Table 2) shows that, in a Na+ medium and at alkaline pH, tumor mitochondria show reduced capacity to take up phosphate when compared to normal mitochondria. As a consequence tumor mitochondria exhibit a deficiency of the phosphorylating capacity (Fig. 3), measured as stimulation of oxygen uptake induced by phosphate addition to respiring mitochondria. It can be seen that the Km for activation of oxygen uptake is increased as compared to that in rat liver mitochondria and Vmax decreased in tumor mitochondria. No difference, however, in the activity of the phosphate translocator between normal and tumor mitochondria was found (Capuano et al., 1985). As far as pyruvate translocator is concerned, measurements of kinetic parameters of pyruvate uptake in three different tumors (Table 3) (see also Paradies et al., 1983) showed a decreased activity of the translocator, and of the pyruvate supported oxygen consumption, in tumor mitochondria as compared to normal mitochondria. Interestingly the observed variations seem to be correlated with the growth rate of tumor cells.

Fig. 4 summarizes the relationships between transport processes of protons, monovalent cations and anionic substrates across the plasma membrane and inner mitochondrial membrane. Enhanced entry of Na+ in the cytosol of tumor cells, followed by Na+ passive diffusion into mitochondria and its extrusion in exchange with H+, via the mitochondrial Na+/H+ antiporter, may be primarily responsible for alkalinization of the cytosol with stimulation of DNA synthesis and depression of mitochondrial utilization of pyruvate and Pi with impairment of oxidative phosphorylation. The latter, as well as enhanced ATP expenditure for extrusion of cytosolic Na+ by the Na+-pump will lo-

Table 3. Pyruvate uptake and pyruvate-supported respiratory rates in normal and tumor mitochondria.

| | Pyruvate uptake | | Oxygen uptake |
| | Km | Vmax | |
	(mM)	(nmol min^{-1} mg prot.$^{-1}$)	
Rat liver	0.64 ± 0.01	19.8 ± 0.5	23.5 ± 1.8
Ehrlich Ascites	0.63 ± 0.02	11.9 ± 0.6	14.5 ± 2.2
MH 44	0.74 ± 0.03	11.6 ± 0.8	13.3 ± 2.5
MH 3924A	1.1 ± 0.07	4.7 ± 0.5	7.5 ± 2.7

Pyruvate uptake was measured as described (Paradies et al., 1983). The rate of oxygen uptake (natom min^{-1} mg prot.$^{-1}$) was measured in the presence of 0.5 mM pyruvate. MH, Morris hepatoma. The values are means ± S.E.M. for 3-8 determinations.

wer the phosphate potential in the cytosol and enhance the rate of aerobic glycolisis as observed in rapidly growing tumor cells (Pedersen, 1978; Papa and Del Pesce, 1979).

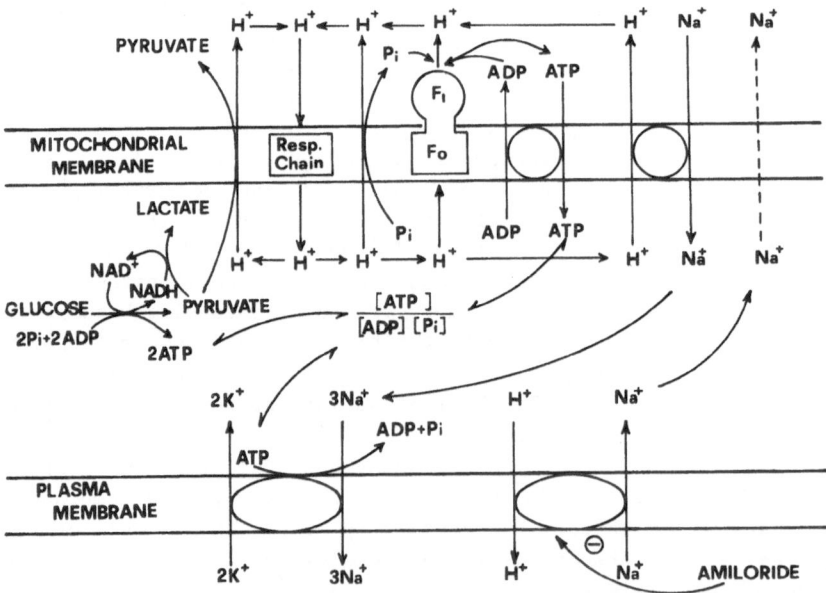

Fig. 4. Transport processes of protons, cations and anionic substrates across the plasma membrane and the inner mitochondrial membrane. Dotted line indicates passive diffusion pathway of Na^{+} into mitochondrial matrix.

Table 4. Properties of Energy Transfer Complexes

Complex	Mw(kDa)	Polypeptides N-DNA	Mt-DNA	Prosthetic groups
I NADH–UQ reductase	800	25	7	FMN, Fe-S Clusters SQ_N
II Succinate–UQ reductase	140	4-5		FAD, Fe-S Clusters SQ_S
III Ubiquinol–cyt.c reductase	250	10	1 apocyt.b	b_{562}, b_{566}, c_1 hemes Fe-S cluster, SQ_c
IV Cytochrome c oxidase	162	10	3 Co I,II,III	a, a_3 hemes Cu_A, Cu_B
V ATP Synthase	500	11-13	2 ATPase 6, A6L	

Data from Hatefi (1985) Chomyn et al. (1985, 1986) Shagger et al. (1985), Kuhn-Nentwig and Kadenbach (1985), Walker (1987). Complexes I, III and IV exhibit a coupled outwards proton translocation. ATP synthesis is coupled to inwards proton translocation.

THE MITOCHONDRIAL CYTOCHROME SYSTEM

Detailed knowledge is now available of the polypeptide composition of the cytochrome chain, the amino acid sequence of the constituent proteins, their three-dimensional folding, membrane topology and possible involvement in the redox and protonmotive activity. Table 4 summarizes properties of the complexes involved in electron transfer and energy transduction. Cytochrome c reductase and cytochrome c oxidase represent, like NADH-ubiquinone reductase and the H^+-ATP synthase, the product of the concerted expression of nuclear and mitochondrial genes (Table 4). The interaction of the nuclear and the mitochondrial genomas is likely to represent a critical process in the functional organization of eukaryotic cells and disturbance of its normal course could be involved in dysfunctions of cells metabolism and growth.

Cytochrome b of the mammalian cytochrome c reductase is the only subunit encoded by a mitochondrial gene, the other two catalytic subunits, cytochrome c_1 and the Fe-S protein being encoded by nuclear genes. Cytochrome c oxidase of animal cells has three subunits encoded by mitochondrial genes. Of these, subunit I seems to be the essential unit involved in redox and protonmotive activity. In addition to the above subunits, both cytochrome c reductase and cytochrome c oxidase have additional subunits all encoded by nuclear genes (Papa et al., 1987b). The function of the supernume-

Table 5. Observed characteristics of the protonmotive activity of cytochrome c reductase.

---

1. H^+ pumping is effected by an electrongenic process located between cytochrome b_{562} and c_1.

2. Protein stabilized $\dot{Q}_B^-$, acting between b_{562} and c_1, is essential for activity.

3. Hydrophylic segments of core protein II and 14 kDa subunit, provide H^+ conducting path from the matrix space to the Q_B center.
The 8 kDa, DCCD-binding subunit also contributes H^+ conducting path to the Q_B center.

---

rary subunits present in the eukaryotic enzyme is as yet unclear. Besides a possible control function of these subunits, there is evidence indicating that some of them in cytochrome oxidase are tissue specific and there could be structural differences and/or different rate of synthesis depending on the developmental stage, age and functional activity (Kuhn-Nentwig and Kadenbach, 1985).

Cytochrome c reductase exhibit a well characterized protonmotive activity with an overall stoicheiometry of $2H^+/e^-$ for proton pumping (see Papa et al., 1983; Papa and Lorusso, 1984) (Table 5). Studies in our laboratory indicate a role of the supernumerary subunits in proton pumping by the reductase. In fact selective enzymatic digestion of core protein II and the subunit of 15 kDa results in decoupling of proton pumping from electron flow (Lorusso et al., 1985). A protein-stabilized ubisemiquinone species (UQ_c^-) (Ohnishi and Trumpower, 1980) appears to play an essential role in the redox and protonmotive activity of the complex. Analogous prosthetic forms of ubiquinone seem to be associated to NADH and succinate ubiquinone reductase (Yu and Yu, 1981). In the case of cytochrome c reductase the subunit of 15 kDa (Wang and King, 1982) and the Fe-S protein (Cocco et al., 1987) are apparently involved in the binding of ubiquinone. Defects of the ubiquinone-binding proteins may be involved in dysfunctions of the respiratory chain also in the presence of normal biosynthesis and overall content of ubiquinone.

The reduction of dioxygen by ferrocytochrome c catalyzed by the oxidase is anisotropically organized in the mitochondrial membrane so to result directly in the generation of transmembrane $\Delta\mu H^+$ (H^+ consumed in the formation of H_2O derive from the matrix space) (Papa, 1976; Mitchell and Moyle, 1970; Papa et al., 1974). In addition, the oxidase appears capable to couple electron flow with proton ejection in the outer space (Wikström et al., 1981; Papa et al., 1987a). The H^+/e^- stoicheiometry of this process, which can amount up to 1 (Wikström et al., 1981; Papa et al., 1987a) appears, however, to be variable among species. Furthermore, in the case of the mammalian oxidase the H^+/e^- ratio can be affected by a number of factors (Table 6) (Papa et al., 1987a; Papa, 1988). In particular it seems that at the

Table 6. Characteristics of the protonmotive activity of cytochrome c oxidase.

---

1. The reduction of dioxygen to H_2O in cytochrome c oxidase of prokaryotes and eukeryotes is <u>anisotropically organized</u> in the membrane so to generate <u>per se</u> a $\Delta\mu H^+$ with effective outward translocation of 1 q^+/e^-.

2. In eukaryotic and various prokaryotic oxidases the redox reaction is also associated with H^+ ejection.
   In higher eukaryotes the data produced seem to qualify this process as electrogenic H^+ pumping.
   There are, however, prokaryotic oxidases which do not exhibit H^+ ejection.

3. The H^+/e^- ratio with the mammalian oxidase appears to vary from 1 to around 0 being affected by factors like: the actual concentration of cyt. c^{2+} and cyt. c^{3+}, the rate of electron flow, the presence of divalent cations, the actual pH and magnitude of $\Delta\mu H^+$, the state of the enzyme (Papa et al., 1987a; Proteau et al., 1983).

4. In intact mitochondria under physiological conditions of dehydrogenase activity and $\Delta\mu H^+$ generation by the respiratory chain, cytochrome oxidase does not appear to contribute significant H^+ pumping (Papa et al., 1988; Murphy and Brand, 1987).

---

rate of electron flow supported by physiological concentrations of dehydrogenase substrates and in the presence of $\Delta\mu H^+$ generated by the first two energy conserving sites little if any proton pumping is exhibited by cytochrome oxidase (Papa, 1988; Papa et al., 1987a; Murphy and Brand, 1987).

It is possible that the supernumerary subunits of mammalian cytochrome oxidase regulate the kinetics of redox activity and the efficiency of energy linked proton pumping thus adapting the function of the oxidase to the highly variable physiological conditions meet in the various human tissues (see also Kuhn-Nentwig and Kadenback, 1985).

Recently different human diseases have been identified which are characterized by defects of enzymes of the respiratory chain (Table 7) and more of these diseases seem to be continuously discovered (Di Mauro et al., 1985; Darley-Usmar et al., 1983; Rouslin, 1983; Tanaka et al., 1986; Scholte, 1988). Some of these mitochondrial defects result to be tissue specific, other are systemic. The less severe defects appear to spontaneously reverse with age, (benigne cytochrome c oxidase deficiency), others are lethal.

According to a clasification of mitochondrial diseases recently introduced (Scholte, 1988) primary dysfunctions are those resulting from genetic defects of one or more mitochondrial enzymes or translocators. Secondary mitochondrial dysfunctions can directly derive from genetic defects (i.e., phenylpyruvate overproduction in phenylketonuria which inhibits pyruvate

Table 7.   Respiratory chain and oxidative phosphorylation defects.

NADH–UQ reductase deficiency	–Deficiency of Fe–S clusters –high Km for NADH	Riboflavin – responsive patients
Defect of Q-binding protein	–Oxygen radical production	UQ-responsive
UQH_2 –cyt. c reductase deficiency	–cyt. b or c_1 deficiency –Fe–S cluster deficiency	Menadione-responsive patients
Cytochrome c oxidase deficiency	–lack of reducible cyt. a, a_3	
Defects in energy transduction	–ATP synthase deficiency –dysfunction of ANT –loose coupling –defective proton pumping –alteration in the binding of the ATPase inhibitor	

ANT: Adenosine Nucleotide Translocator.

and ketone bodies translocator) (Paradies and Papa, 1977) or from a series of dysmetabolic or pathological situations including defects in circulation (ischemia and anoxia), viral infections, hormone and/or neurotrasmitter defects (untreated diabetes mellitus, hypothyroidism), poisoning antitumor or antibacterial drugs (adriamycin, tetracyclines, chloramphenicol), malnutrition.

Clearly a better understanding of the molecular structure and mechanism of action of the enzymes of the respiratory chain and of their genetics will help to better characterize dysfunctions of the protonmotive respiratory chain and provide useful indications for their therapeutical treatment.

REFERENCES

Aronson, P.S., 1985, Kinetic properties of the plasma membrane Na^+-H^+ exchanger. Ann. Rev. Physiol., 47: 545-560.
Capuano, F., Carrieri, E.and Papa, S., 1985, Proton coupled transport of phosphate and oxidative phosphorylation in tumour cells mitochondria, in: "Cell Membranes and Cancer" T. Galeotti et al. eds, Elsevier Publishers B.V., Amsterdam, 265-268.

Capuano, F., Paradies, G., Capitanio, N. and Papa, S., 1982, Proton-cation translocation and transport of anionic substrates in tumour cells mitochondria. In: "Membranes in Tomour Growth", T. Galeotti et al., eds., Elsevier Biomedical Press, Amsterdam, 345-352.

Chappell, J.B., 1968, Systems used for the transport of substrates into mitochondria, Br. Med. Bull., 24: 150-171.

Chomyn, A., Cleeter, M.W.J., Ragan, C.I., Riley, M. Doolittle, R.F. and Attardi, G., 1986, URF6, Last Unidentified Reading Frame of Human mtDNA, Codes for an NADH Dehydrogenase Subunit, Science, 234: 614-618.

Chomyn, A., Mariottini, P., Cleeter, M.W.J., Ragan, C.I., Matsuno-Yagi, A., Hatefi, Y., Doolittle, R.F. and Attardi, G., 1985, Six unidentified reading Frames of Human mitochondrial DNA encode components of the respiratory-chain NADH dehydrogenase, Nature, 314: 592-597.

Cocco, T., Meinhardt, S., Gatti, D., Lorusso, M., Ohnishi, T. and Papa, S., 1987, Effects of protheolytic digestion and chemical modification of b-c$_1$ complex on the Fe-S centre and the stable ubiquinone (SQ$_c$), in: "Cytochrome System: Molecular Biology and Bioenergetics", S. Papa, B. Chance and L. Ernster, eds., Plenum Press, New York, 559-560.

Darley-Usmar, V.M., Kennaway, N.G., Buist, N.R.M. and Capaldi, R.A., 1983, Deficiency in ubiquinone cytochrome c reductase in a patient with mitochondrial myopathy and lactic acidosis, Proc. Natl. Acad. Sci. USA, 80: 5103-5106.

De Vivo, D.C., Uziel, G., 1983, Disturbances of pyruvate metabolism in neuromuscolar diseases, in: "Mitochondrial Pathology in Muscle Disease", G. Scarlato, C. Cerri, eds., Piccin Medical Books, 57-70.

Di Mauro, S., Bonilla, E., Zaviani, M., Nakagawa, M. and De Vivo, D.C., 1985, Mitochondrial myopathies, Ann. Neurology, 17: 521-538.

Douglas, M.G. and Cockrell, R.S., 1974, Mitochondrial cation-hydrogen ion exchange, J. Biol. Chem., 249: 5464-5471.

Hatefy, Y., 1985, The mitochondrial electron trnsport and oxidative phosphorylation system, Ann. Rev. Biochem., 54: 1015-1069

Krulwich, T.A., 1983, Na$^+$/H$^+$ antiporters, Biochim. Biophys. Acta, 726: 245-264.

Kuhn-Nentwig, L. and Kadenbach, B., 1985, Isolation and properties of cytochrome c oxidase from rat-liver and quantification of immunological differences between isoenzymes from various rat tissues with subunit specific antisera, Eur. J. Biochem., 149: 147-158.

Lorusso, M., Gatti, D., Marzo, M. and Papa, S., 1985, Effect of papain digestion on redox-linked proton translocation in b-c$_1$ complex from beef heart reconstituted into liposomes. FEBS Lett., 182: 370-374.

Low, H., Grebing, C., Lindgren, A., Tally, M., Sun, I.L. and Crane, F.L., 1987, Involvement of transferrin in the Reduction of Iron by the Transplasma Membrane Electron Transport System, J. Bioenerg. Biomembr. 19: 535-549.

Margolis, L.B., Rozovskaja, I.A. and Skulachev, V.P., 1987, Acidification of the interior of Ehrlich ascites tumor cells by nigericin inhibits DNA synthesis. FEBS Lett., 220: 288-290.

Mellmann, I., Fuchs, R. and Helenius, A., 1986, Acidification of the endocytic and exocytic pathways, Ann. Rev. Biochem., 55: 663-700.

Mitchell, P. and Moyle, J., 1970, The intrinsic anisotropy of the cytochrome oxidase region of the mitochondrial respiratory chain and the consequent vectorial property of respiration, in: "Electron Transport and Energy Conservation", J.M. Tager, S. Papa, E. Quagliariello and

E.C. Slater, Adriatica Editrice, Bari, 575-587.

Murphy, M.P. and Brand, M.D., 1987, Variable stoicheiometry of proton pumping by the mitochondrial respiratory chain, Nature, 329: 170-171.

Ohnishi, T. and Trumpower, B.L., 1980, Differential effects of antimycin on ubisemiquinone bound in different environments in isolated succinate-cytochrome c reductase complex, J.Biol. Chem., 255: 3278-3284.

Packer, L., 1987, Exercise, aging and antioxidant, in: "Advances in Myochemistry" G. Benzi ed., J. Libbey, London, 37-50.

Papa, S., 1976, Proton translocation reactions in the respiratory chain, Biochim. Biophys. Acta, 456: 39-84.

Papa, S., 1988, Cytochrome c oxidase and its protonmotive activity, an overview, in: "Oxidases and Related Redox Systems", H.S. Mason, ed., Alan R. Liss, New York, in press.

Papa, S., Capitanio, N. and De Nitto E., 1987a, Characteristics of the redox-linked proton ejection in beef-heart cytochrome c oxidase reconstituted in liposomes, Eur. J. Biochem., 164: 507-516.

Papa, S., Capuano, F., Capitanio, N., Lorusso, M. and Galeotti, T., 1983, Proton/cation translocation in tumour cell mitochondria, Cancer Res., 43: 834-838.

Papa, S., Chance, B. and Ernster, L., 1987b, "Cytochrome Systems: Molecular Biology and Bioenergetics", Plenum Press, New York.

Papa, S. and Del Pesce, C., 1979, Bioenergetica delle cellule tumorali, Folia Oncologica, 2: 3-24.

Papa, S., Guerrieri, F. and Lorusso, M., 1974, Mechanism of respiration-driven proton translocation in the inner mitochondrial membrane. Analysis of proton translocation associated to oxidoreductions of the oxygen-terminal respiratory carriers, Biochim. Biophys. Acta, 357: 181-190.

Papa, S., Lofrumento, N.E., Quagliariello, E., Meijer, A.J. and Tager, J.M., 1970, Coupling mechanisms in anionic substrate transport across the inner membrane of rat-liver mitochondria, Bioenergetics, 1: 287-307.

Papa, S. and Lorusso, M., 1984, The cytochrome chain of mitochondria: Electron transfer reactions and transmembrane proton translocation, in: "Biomembranes: Dynamics and Biology", R.M. Burton and F. Carvalho Guerra, eds., Plenum Press, New York, 257-290.

Papa, S. Lorusso, M., Boffoli, D. and Bellomo E., 1983, Redox-linked proton translocation in the b-c$_1$ complex from beef-heart mitochondria reconstituted into phospholipid vesicles. General characteristics and control of electron flow by $\Delta\mu H^+$, Eur. J. Biochem., 137: 405-412.

Paradies, G., Capuano, F., Palombini, G., Galeotti, T. and Papa, S., 1983, Transport of pyruvate in mitochondria from different tumour cells, Cancer Res., 43: 5068-5071.

Paradies, G. and Papa, S., 1977, On the kinetics and substrate specificity of the pyruvate translocator in rat liver mitochondria, Biochim. Biophys. Acta, 462: 333-346.

Pedersen, P.L., 1978, Tumour mitochondria and the bioenergetics of cancer cells, Progr. Exp. Tumor Res., 22: 190-274.

Pedersen, P.L. and Carafoli, E., 1987, Ion motive ATPases.I. Ubiquity, properties and significance to cell function, Trends in Biochemical Sciences, 12: 146-150.

Pouyssegur, J, 1985, The growth factor-activatable Na^+-H^+ exchange system: a genetic approach, Trends in Biochemical Sciences, 11: 453-455.

Proteau, G., Wrigglesworth, J.M. and Nicholls, P., 1983, Protonmotive functions of cytochrome c oxidase in reconstituted vesicles, Biochem. J., 210: 199-205.

Rebouche, C.J., Engel, A.G., 1983, Carnitine metabolism and deficiency syndrome, Mayo Clin. Proc., 58: 533-540.

Rosen, B.P. and Futai, M., 1980, Sodium/proton antiporter of rat liver mitochondria, FEBS Lett., 17: 39-43.

Rouslin, W., 1983, Mitochondrial complexes I,II,III,IV and V in myocardic ischemia and autolysis, Am. J. Physiol., 252: H622-H627.

Rudnick, G., 1986, ATP-driven H^+ pumping into intracellular organelles, Ann. Rev. Physiol., 48: 403.

Schägger, H., Borchart, U., Aquila, H., Link, T.A. and von Jagow, G., 1985, Isolation and amino acid sequence of the smallest subunit of beef heart $b c_1$ complex, FEBS Lett., 190: 89-94.

Scholte, H.R., 1988, The biochemical basis of mitochondrial diseases, J. Bioenrg. Biomembr., 20: 161-191.

Sze, H., 1985, H^+-translocating ATPases: Advances using membrane vesicles, Ann. Rev. Plant. Physiol., 36: 175-208.

Tanaka, M., Nishikimi, M., Suzuki, H., Ozawa, T., Nishizawa, M., Tanaka, K. and Miyatake, T., 1986, Deficiency of subunits in heart mitochondrial NADH-ubiquinone oxidoreductase of a patinet with mitochondrial encephalomyopathy and cardiomyopathy, Biochem. Biophys. Res. Commun., 140: 88-93.

Walker, J.E., Runswick, M.J. and Poulter, L., 1987, ATP synthase from bovine mitochondria. The characterization and sequence analysis of two membrane-associated subunits and the corresponding cDNAs, J. Mol. Biol., 197: 89-100.

Wang, T. and King, T.E., 1982, Isolation of QP-C and reconstitution of the QH_2-c reductase, Biochim. Biophys. Res. Commun., 104: 591-596.

West, J.C. and Mitchell, P., 1974, Proton/sodium ion antiport in Escherichia coli, Biochem. J., 144: 87-90.

Wikström, M., Krab, K. and Saraste, M., 1981, Proton-translocating cytochrome complexes, Ann. Rev. Biochem., 50: 623-655.

Yu, C.A. and Yu, L., 1981, Ubiquinone-binding proteins, Biochim. Biophys. Acta, 639: 99-128.

THE "CYTOSOL": A NEGLECTED AND POORLY UNDERSTOOD COMPARTMENT OF EUKARYOTIC CELLS

J.S. Clegg
University of California
Bodega Marine Laboratory
Bodega Bay, CA 94923 USA

and

M.B. Barrios
Universidad de Carabobo
Valencia, Venezuela

INTRODUCTION

The study of cellular components obtained by cell fractionation has been a very profitable approach to the study of cell structure and function. However, these tactics have provided no reliable information about the physical properties and metabolic activities of the "aqueous compartments" of animal cells: nucleoplasm, the fluid interiors of cytoplasmic membrane bound organelles, and the intervening "cytosol." Although somewhat superficial we can consider these compartments to be the aqueous volumes, including dissolved solutes, located between all presently described ultrastructural components. They collectively make up about 75% of the mass and volume of living cells so it is important that we understand them. That requires detailed knowledge about their solute composition, aqueous phase properties, interrelationships with the ultrastructure with which they are in contact, and the functions (metabolism) that occur there, matters about which we know relatively little at present.

The most extensive aqueous compartment of animal cells is the "cytosol," whose study is the central focus of this paper. The origin of this widely used term is attributed to Henry Lardy (1965) who first defined it as that portion of the cell which is found in the supernatant after centrifuging an homogenate at 105,000xg for 1 hour. Thus, the "cytosol" was originally defined as a solution obtained by cell disruption in a dilution buffer. While a perfectly useful term in that context it soon was also applied to the *intact* cell, a practice widely used at present. It is important to ask whether or not the *in vitro* cytosol tells us anything reliable about the *in vivo* cytosol, which I will refer to here as the "aqueous cytoplasm" to lessen ambiguity (Clegg, 1984). It is well known that the cytosol varies widely in composition depending on the method used to disrupt the cells but it always contains a very large proportion of cellular macromolecules, especially proteins (Anderson and Green, 1967; Clegg, 1984). Similarly, a wide variety of metabolites, coenzymes and inorganic ions are found through cell fractionation to be "cytosolic" or "soluble." If one is willing to extrapolate to the intact cell then the aqueous cytoplasm is viewed as a crowded solution, whose total solute concentration is in the range of 20-30 wt.%., within which the various organelles and other cell components are presumably suspended, and an assigned locus of much metabolic activity (the "soluble" enzymes). Of course, no one considers the cytosol to be an *accurate* reflection of the aqueous cytoplasm; nevertheless, the foregoing description is a widely used paradigm for experimental design, data interpretation and metabolic model building, and is solidly entrenched in current textbooks. The distinction between cytosol and aqueous cytoplasm is not trivial since so much of current thought in cell physiology and biochemistry has, from "cytosolic inference," been constructed on events that are believed to take place in solution in this major cellular compartment.

THE AQUEOUS CYTOPLASM

Recent reviews and books summarize the abundant evidence that the aqueous cytoplasm is very dilute and bears little if any resemblance to the cytosol (Clegg, 1984; Welch, 1985; Bhargava, 1985; Welch and Clegg, 1987; Clegg, 1988 a,b). If the aqueous cytoplasm is not a crowded chaotic solution then most of the macromolecules that are released by cell disruption must exist in an organized but fragile state within the intact cell and, by inference, intimately connected with function and regulation. The idea that such extensive organization exists in cells can be traced to the early part of this century (see Porter, 1984), and isolated studies subsequently reminded us that relatively few macromolecules were free to diffuse in intact cell (for example, see Chambers, 1940; Kempner and Miller, 1967). I will summarize next some examples of more recent evidence on this issue.

Using an ingenious "reference phase technique" (Horowitz and Miller, 1984), Paine and colleagues have examined the diffusion of intracellular (native) proteins in amphibian oocytes (see Paine, 1984; 1987). They microinjected droplets of gelatin (the "reference phase"), caused them to gel by placing the egg at low temperature, waited for diffusible proteins to enter and equilibrate with the reference phase, and then cryomicrodissected the reference phase, nucleus, and fragments of cytoplasm. Analysis of the protein content of these components by gel electrophoresis showed that some proteins do indeed diffuse into the reference phase but that many do not enter it at all. Fulton (1982) analyzed reference phase data and concluded that over 80% of the non-yolk proteins of the oocyte do not enter the reference phase over very long time scales.

Dabauville and Franke (1986) more recently used a variation on the reference phase theme. They punched small holes into *Xenopus* oocytes and introduced gel filtration beads of different exclusion limits. After waiting for diffusion "equilibrium" they dissected the beads, "rinsed" them to remove surface-bound material, and extracted the proteins. Gel electrophoresis then provided information on the relative amount and distribution of diffusible proteins. Once again, the results provided evidence for the existence of diffusing and non-diffusing proteins in these cells. The authors did not attempt an evaluation of the relative amounts of diffusible protein (compared to the total) but inspection of their gels suggests an outcome similar to that of Paine and colleagues.

While these excellent studies are not free from interpretive difficulty they do have the significant virtue of studying the intracellular mobility of proteins already present in the cell. One shortcoming of these techniques is their restriction to giant cells. Other studies have examined the diffusion of "foreign" proteins introduced into mammalian cells by microinjection or cell-fusion techniques and followed by fluorescence methods (see Jacobsen and Wojcieszyn, 1984). The evident caveat to such work is that the "foreign" macromolecules may not reflect the diffusive behavior of native ones. Nevertheless, the general outcome of these studies is that the diffusion of introduced macromolecules is greatly retarded compared to behavior in water.

The group at Carnegie-Mellon University has contributed greatly to our understanding of the organization of cytoplasm (for review see Luby-Phelps *et al.*, 1988). They introduced fluorescein isothiocyanate (FITC) labelled dextrans and Ficoll beads into mammalian cells by cell-fusion and measured their motion by fluorescence recovery after photobleaching. Among other conclusions they proposed that this evidence supports the existence of a structural network in the cytoplasmic ground substance and cannot be explained by the presence of a concentrated protein solution. Their results provide strong evidence that the degree of cytoplasmic organization is far greater than the crowded solution paradigm of the cytosol infers.

Mastro, Keith and colleagues examined the motional behavior of spin-label probes in mammalian cells by electron spin resonance with results that are important to cytoplasmic organization (Mastro and Hurley, 1987). They studied the effect of osmotic shrinkage on the *translational* and *rotational* motion of the spin label, tempone, in BHK cells. They predicted that, if the "solution paradigm" of the cytosol were correct then both parameters should show similar dependence when the cell volume was reduced. But that was not observed: "rotation" of the probe was hardly effected whereas "translation" was profoundly dependent on cell volume, probably reflecting the existence of a structural network in cells, surrounded by a dilute aqueous phase of relatively low viscosity. That interpretation is similar to ours, arrived at from metabolic studies on osmotically perturbed cells (Mansell and Clegg, 1983; Clegg and Gordon, 1985; Clegg, 1986 and 1988 a,b) and that of the Carnegie-Mellon group mentioned above. Moreover, it is in accord with images produced from high voltage electron microscopy (HVEM) of mammalian cells.

Porter, his students and associates have proposed from well known HVEM observations (see Porter, 1987 for recent review) that an extensive highly branched network exists in the cytoplasm of animal cells, the "microtrabecular lattice" (MTL). Figure 1 illustrates some HVEM images of the MTL and a speculative diagram of a trabecula. It is obvious that the MTL could represent the ultrastructural equivalent of the organized network postulated to exist from the work previously considered in this paper. Penman and his associates (see Fey and Penman, 1987) from an extensive series of studies arrived at a similar position on intracellular organization, although their views do differ in several respects from the MTL as described by Porter. I hasten to recognize that the MTL has not been accepted by all as "real" and alternative interpretations have been proposed (see Bridgman and Reese, 1984; Kondo, 1984). Be that as it may, the HVEM images are nicely consistent with a rather large body of evidence derived from independent studies on intact living cells, some of which I have sampled in this paper.

Although I have not considered here the aqueous compartments of the nucleus (nucleoplasm) or interiors of cytoplasmic membrane bounded organelles it seems likely that these also exhibit extensive macromolecular organization. The excellent recent review by Srere (1987) summarizes compelling evidence in the case of the

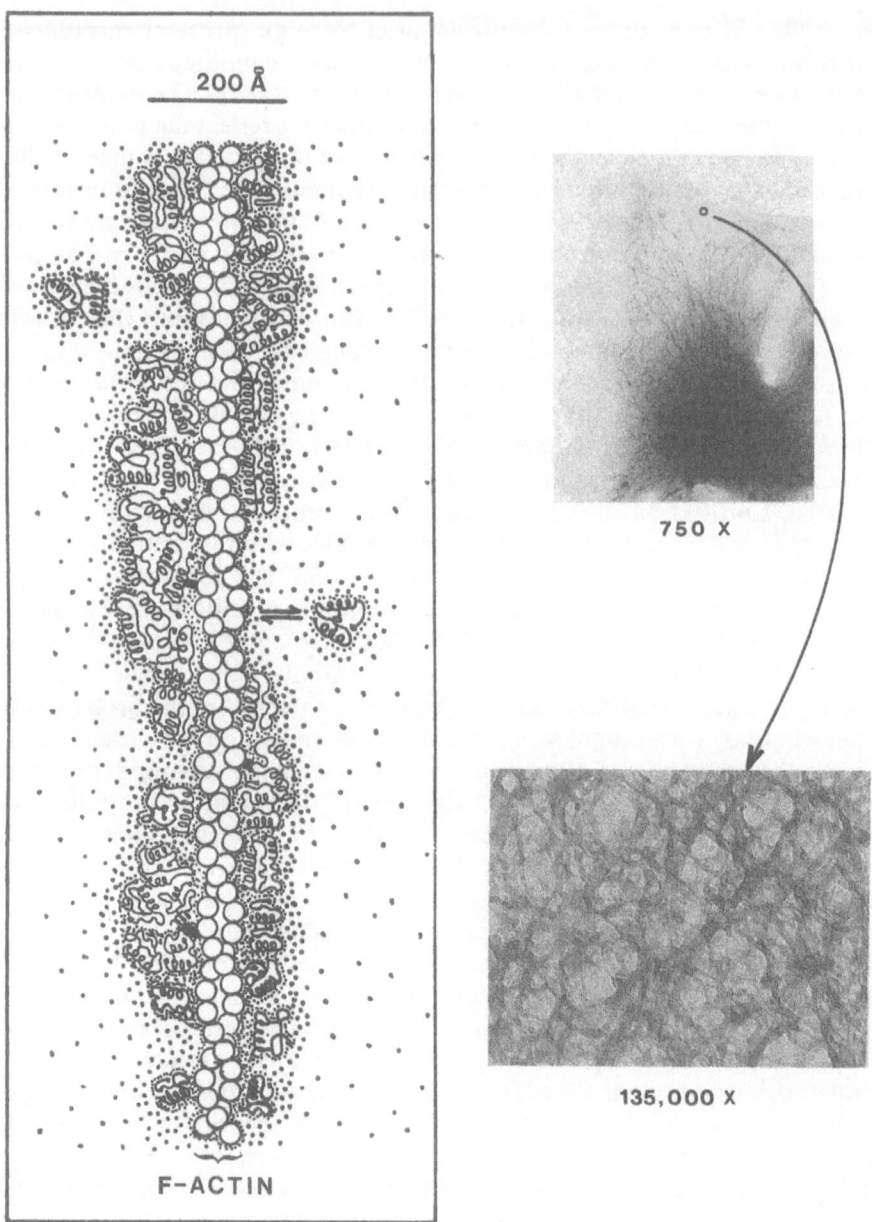

Figure 1. The high voltage electron micrographs shown to the right, generously supplied by Prof. Keith Porter, illustrate the microtrabecular lattice as described by Porter and colleagues. The panel to the left represents a speculative rendition (Clegg, 1984) of one possible arrangement for the composition of part of a trabeculum (see text for discussion).

mitochondrial matrix, and can be consulted with benefit on the general issue of intracellular enzyme organization.

The broad image drawn here of the aqueous cytoplasm infers that most macromolecules released by cell disruption are, in the intact cell, in highly organized states. However, the extent of that organization remains to be determined, and we know little about the details of its origin and control, and significance. The best documentation comes from the study of glycolytic enzymes for which there is now overwhelming evidence for their association with F-actin and associated proteins. Most of the recent evidence comes from Australian laboratories (Clarke et al., 1985; Stephen et al., 1986; Masters et al., 1987) and the work of Knull et al., (1987); however, some of it goes back over 20 years to the pioneering studies of Amberson et al., (1965) and Pette (Arnold and Pette, 1968) among others. We have also added evidence for extensive glycolytic enzyme organization using permeabilized mouse fibroblasts (Clegg, 1987 and 1988).

It appears from this work and that of Schliwa et al. (1981) that F-actin is a major "scaffold" upon which many "soluble" proteins may be located in intact animal cells, generating at least part of the MTL (Figure 1). Actin does appear to be a critical component in the emerging picture of intracellular organization (Stossel et al. 1985; Pollard and Cooper, 1986), in addition to its well documented roles in cell motion and locomotion. Of course, microtubules (Olmstead, 1986) and intermediate filaments (Steinert and Parry, 1985) are also strong candidates.

CELL WATER, IONS AND METABOLITES

The preceeding sections have focused on macromolecular organization. However, there is also evidence that at least some of the smaller molecules present in the *in vitro* cytosol are not undergoing free diffusion in the intact cell, implying their organization as well.

It has long been known from radioactive tracer studies on "intracellular pools" that a number of amino acids do not behave as though they existed in freely diffusing forms in a well-mixed pool (Holden, 1962). The more recent work of Wheatley and colleagues (reviewed by Wheatley et al., 1986) presents convincing evidence that most, and possibly all amino acids exist in both free and complexed states. Srivastava and Bernhard (1987) reviewed the data on intermediates of the glycolytic pathway and suggested that these are enzyme bound. That is to be expected if these enzymes are organized in the cell and channeling of these metabolites occurs between enzymes. Indeed, it can be suggested that few metabolic intermediates from any metabolic pathway escape into the surrounding aqueous phase.

The major intracellular inorganic ions (Na^+, K^+, Cl^-) have also generally been assumed to exist in freely diffusing forms in cells. However, a growing body of evidence indicates that substantial fractions of these ions are not undergoing random thermal motion, being "bound" to various intracellular sites (Negendank, 1982 and 1986; Ling, 1984; Edelmann, 1986; Kellermayer et al., 1986; Cameron et al., 1986; Dawson and Smith, 1986; Adam et al., 1987). Indeed, the "association-induction hypothesis (AIH) proposed by Ling (1984) considers the specific binding of Na^+ and K^+ as major governing mechanisms in cell physiology, and offers a completely different paradigm of the cell compared to the one adopted by most cell biologists. However, the AIH has not yet gained wide acceptance.

The physical properties of intracellular water also remain topics of controversy and variable data interpretation. Views range between extremes, that practically all

163

or none of cell water exhibits properties like those of ordinary aqueous solutions: someone is obviously quite wrong. The interested reader can consult the contrasting evidence and views in a number of books (Keith, 1979; Drost-Hansen and Clegg, 1979; Franks and Mathias, 1982; Ling, 1984; Pullman, *et al.*, 1985; Welch and Clegg, 1987). Additional access to this literature can be obtained from recent papers (Merta *et al.*, 1986; Kasturi *et al.*, 1987; Wheatley *et al.*, 1987; Watterson, 1987). My own bias is that most of the aqueous volume in animal cells, and possibly all of it, exhibits properties that do differ importantly from those of ordinary dilute aqueous solutions (Clegg, 1987). Part of the reason for this opinion is the vast surface area that exists in cells (Gershon *et al.*, 1985), possibly in the form of the MTL, and the emerging evidence that the physical properties of water adjacent to surfaces can be influenced over distances of 30 Å or so (see Parsegian and Rau, 1984; Evans and Ninham, 1986; Pashley and Israelachvilli, 1984) and possibly much farther (Drost-Hansen, 1982). Thus, although I have argued that prevailing evidence indicates that the aqueous compartments are dilute, that should not be taken to indicate that their properties are like those of dilute solutions.

SIGNIFICANCE AND CONCLUDING COMMENTS

It appears to me that we should abandon completely the idea that studies on the cytosol can tell us anything reliable about the nature of cytoplasmic organization and function. Extrapolation of this sort is worse than wrong, for it leads us to believe that we might understand cells by this approach. The organized paradigm presented here, even though very briefly and incompletely, considers the cell to be a unit structure in which the whole cannot be unambiguously described through a study of its isolated parts. The widespread idea of "levels or organization" is, in my view, a fictional and deceptively useful concept, being more a construction of cell biologists than that of the cell. To me a continuum exists, and it is that which sooner or later requires study.

I view the MTL as the most reliable visual representation of this continuum that we have available to us at the present time. Those images are, of course, static snapshots and lack all the dynamics we know must exist. These methods are also prone to artifact and misrepresentation, but we can hope that future developments will correct these problems, and provide more accurate images. Given that caveat, it seems warranted from what we know now to envision the cytoplasm as an intricate interconnected network of microfilaments (F-actin) and probably other classical cytoskeletal structures (Olmstead, 1986; Steinert and Parry, 1985) upon which the enzymes of intermediary metabolism, as well as other cytoplasmic macromolecules and organelles, are associated into functional units. The intervening aqueous volume appears to contain little of consequence in terms of solutes, but is probably of vital importance to homestasis. I will consider one example here (Figure 2).

Consider the evidence summarized earlier that a large fraction of inorganic ions and small metabolites in cells are associated with cytoplasmic structure and not undergoing free diffusion. Consequently, these are not available to exert osmotic pressure. However, the conventional paradigm considers these solutes to be the major means by which cell volume and its regulation are determined. Others have pointed out this problem (Ling, 1984; Negendank, 1982 and 1986) and proposed that one explanation involves the altered properties of cell water. Thus, the lower concentration of total solute that is dissolved in cell water, compared to that in the solution surrounding the cell, nevertheless exerts an osmotic pressure equivalent to the exterior because of the altered properties of the cell water in which the solutes are

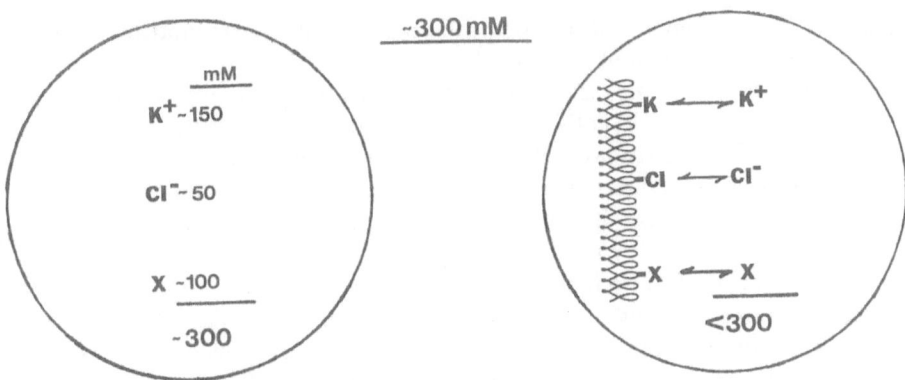

Figure 2. Highly simplified version of two views on the contribution of freely dissolved ions and other small solutes (X) to the determination of cell volume in mammalian cells. The representation to the left, the conventional paradigm, considers all these solutes to exist free in solution, the total concentration of which is close to that of the extracellular solution (about 300 mM). The alternative, shown to the right, considers these solutes to exist in free and bound form; however, the total concentration of the free species is considered to be much less than 300 mM. See text for further discussion.

dissolved. Other examples of how intracellular water may participate in cell structure and function, other than as a bulk dielectric, have been suggested (see Drost-Hansen and Clegg, 1979; Franks and Mathias, 1982; Ling, 1984; Pullman *et al.* 1984; Watterson, 1987).

Much of what has been presented here is not well understood, and I am well aware of its speculative nature. What seems to be established, in my opinion, is the existence of an intimate connection between water, cellular architecture and most, probably all of the metabolic machinery of animal cells. These relationships appear to be dynamic, and their disruption can be deduced to lead to malfunction and cellular disease. It certainly seems likely that viral and malignant transformation processes in cells might be intimately related to the disruption of such an organization. A few recent examples are worth noting.

Alterations in the glycolytic pathway have been known to occur in tumor cells since the time of Warburg, and many workers have demonstrated changes in the level of glycolytic enzymes in tumor cells (see Hennipman *et al.*, 1987). Most interesting is the work of Cooper *et al.*, (1983) who found that three glycolytic enzymes are phosphorylated at tyrosine residues in cells transformed by viruses whose transforming proteins are tyrosine protein kinases. In pondering the role of this phosphorylation in the transformation process they considered several possibilities, one of which was a phosphorylation-induced "modification in the intracellular organization of the glycolytic enzymes." Another example comes from Naharro *et al.*, (1984) who demonstrated that the primary translation product of a sarcoma transforming virus is a hybrid protein containing a portion of actin as well as a tyrosine-specific protein kinase. These authors point out that their findings bear on the generally valid correlation between alterations in the actin-cytoskeleton (usually disruption) and the transformation process. We might add a third component to the correlation -- the changes that often occur in metabolic pathways during transformation, commonly involving "soluble" enzymes and their regulation.

Eigenbrodt *et al.*, (1983) also found that the glycolytic enzyme enolase was phosphorylated at tyrosine in cells transformed by a sarcoma virus. They noted that "Changes in [enzyme] partitioning may be another mechanism of metabolic control regulated by the phosphorylation state of an enzyme." They also take the position that to understand these changes it will be necessary to ultimately carry out studies on intact cells. Finally, it is of some interest that the src-gene product (tyrosine protein kinase) has been found to be localized in the same regions as vinculin, one of the cytoskeletal proteins, in transformed rat, chicken and mouse cells. The authors point out that their methods could not distinguish between the existence of the kinase as a soluble protein, or as one associated with a "diffusely distributed cytoplasmic structure."

This brief excursion into the vast literature on transformation does not, of course, do justice to all the complexities involved. Nor do I pretend to have arrived at some unique position on the matter: it is obvious to everyone that relationships between the cytomatrix, transforming viruses (and their translation products) and metabolism are of great importance. I do suggest that these relationships may perhaps be better appreciated if viewed within the holistic description of the animal cell advocated here.

A final consideration relates to the nature of intracellular water in tumor and other transformed cells. As long ago as 1916 it was documented that tumor cells contained more water than the cells from which they were derived (Cramer, 1916). That generality has stood the test of time, and it seems likely that alterations in cell water may be involved with a variety of cellular diseases (Szent-Gyorgyi, 1971). For example, it has been shown that tumor cell water exhibits motional properties that more closely resemble those of ordinary dilute solutions than those of normal cells (see Hazlewood, 1979, 1986 and the present volume). In the context of the present article it can be suggested that such a result is due, at least partly, to the loss of cytomatrix surface area that also accompanies transformation, and that some of us believe is responsible for generating the unusual properties of water in healthy cells. These correlations between transformation-associated changes in the nature of cytoplasmic organization, metabolic activity and cell water lend further credibility to the holistic, continuum nature of cells that has been the central theme of this paper.

ACKNOWLEDGEMENTS

I thank Professor Keith Porter for the high voltage electron photomicrographs shown in Figure 1. Vicki Hoffman and Diane Cosgrove prepared the manuscript with skill and good cheer under duress. Supported by grants from the U.S. National Science Foundation.

REFERENCES

Adam, W.R., Koretsky, A.P., and Weiner, M.W., 1987, Measurement of tissue potassium *in vivo* using ^{39}K nuclear magnetic resonance, *Biophy. J.*, *51*: 265.

Anderson, W.C., Rosen, F.J., and Bauer, A.C., 1965, The attachment of glycolytic enzymes to muscle ultrastructure, *J. Cell. Comp. Physiol.*, *66*: 71.

Anderson, N.G., and Green, J.G., 1967, The soluble phase of the cell, in: "Enzyme Cytology," D.B. Roodyn, ed., Academic Press, New York, p. 475.

Arnold, H. and Pette, D., 1968, Binding of glycolytic enzymes to structural proteins of muscle, *Eur. J. Biochem.*, *6*: 163.

Bhargava, P., 1985, Is the "soluble" phase of cells structured? *Biosystems*, *18*: 135.

Bridgeman, P.C. and Reese, T.S., 1984, The structure of cytoplasm in directly frozen cultured cells. I. Filamentous meshworks and the cytoplasmic ground substance, *J. Cell Biol.*, *99*: 1655.

Cameron, I.L., Hansen, J.T., Hunter, K.E., and Padilla, G.M., 1986, Elemental concentration gradients between subcellular compartments, *J. Cell Sci.*, *81*: 283.

Chambers, R., 1940, The micromanipulation of living cells, in: "The Cell and Protoplasm," AAAS, No. 14., F.R. Moulton, ed. Science Press, Lancaster, p. 20.

Clark, F.M., Morton, D.J., Stephan, P., and Wiedemann, J., 1985, The functional duality of glycolytic enzymes: potential integrators of cytoplasmic structure and function, in: "Cell Molility: Mechanism and Regulation," H. Ishikawa, S. Hatano, and H. Sato, eds., University of Tokyo Press, Tokyo, p. 235.

Clegg, J.S., 1984, Properties and metabolism of the aqueous cytoplasm and its boundaries, *Amer. J. Physiol.*, *246*: R133.

Clegg, J.S., 1986, L-929 cells under hyperosmotic conditions: volume changes, *J. Cell. Physiol.*, *129*: 367.

Clegg, J.S., 1988a, Reversible dehydration and the aqueous compartments of cells, in: "Water Transport in Biological Membranes," G. Bhenghe, ed., CRC Press, Boca Raton, in press.

Clegg, J.S., 1988b. On the internal environment of animal cells, in: "Microcompartmentation," D. Jones, ed., CRC Press, Boca Raton, in press.

Clegg, J.S., 1987a, Evidence for causal connections between the cytomatrix, the properties of cell water and metabolic compartmentation, in: "Modern Cell Biology," C. Waymouth, ed., A.R. Liss, Inc., New York, p. 151.

Clegg, J.S., 1987b, On the physical properties and potential roles of intracellular water, in: "Organization of Cell Metabolism," G.R. Welch and J.S. Clegg, eds., Plenum Press, New York, p. 41.

Clegg, J.S. and Gordon, E.P., 1985, Respiratory metabolism of L-929 cells at different water contents and volumes, *J. Cell Physiol.* *124*: 299.

Cooper, J.A., Reiss, N.A., Schwartz, R.J., and T. Hunter, 1983, Three glycolytic enzymes are phosphorylated at tyrosine in cells transformed by Rous sarcoma virus, *Nature, Lond.*, *302*: 218.

Cramer, W., 1916, On the biochemical mechanism of growth, *J. Physiol.*, *50*: 323.

Dabauvalle, M.C. and Franke, W.W., 1986, Determination of the intracellular state of soluble macromolecules by gel filtration *in vivo* in the cytoplasm of amphibian oocytes, *J. Cell Biol.*, *102*: 2006.

Dawson, W.D. and Smith, C.T., 1986, Intracellular Na^+, K^+ and Cl^- activities in Ehrlich ascites cells, *Biochim. Biophys. Acta*, *860*: 293.

Drost-Hansen, W. and Clegg, J.S., 1979, "Cell-Associated Water," Academic Press, New York, p. 1.

Drost-Hansen, W., 1982, The occurrence and extent of vicinal water. in: "Biophysics of Water," F. Franks and S. Mathias, eds., Wiley, New York, p. 163.

Edelmann, L., 1986, Two opposing theories of the cell: experimental testing by cryomethods and electron microscopy, in: "The Science of Biological Specimen Preparation," M. Muller, R.P. Becher, A. Boyde, and J.J. Wolosewick, eds., Scanning Electron Microscopy, Inc., AMF O'Hare, Chicago, p. 33.

Eigenbrodt, E., Fister, P., Rubssmen, H., and Friis, R.R., 1983, Influence of transformation by Rous sarcoma virus on the amount, phosphorylation and enzyme kinetic properties of enolase, *EMBO Jour.*, 2: 1567.

Evans, D.F. and Ninham, B.W., 1986, Molecular forces in self-organization of amphiphiles, *J. Phy. Chem.*, 90: 226.

Fey, E.G. and Penman, S., 1987, New views of cell and tissue cytoarchitecture: embeddment-free electron microscopy and biochemical analysis, in: "Organization of Cell Metabolism," G.R. Welch and J.S. Clegg, eds., Plenum Press, New York, p. 89.

Franks, F. and Mathias, S., 1982, "Biophysics of Water," Wiley, New York, p. 1.

Fulton, A.B., 1982, How crowded is the cytoplasm? *Cell*, 30: 345.

Garlid, K.D., 1979, Aqueous phase structure in cells and organelles, in: "Cell-Associated Water," W. Drost-Hansen and J.S. Clegg, eds., Academic Press, New York, p. 393.

Gershon, N.D., Porter, K.R., and Trus, B.L., 1985, The cytoplasmic matrix: its volume, surface area, and the diffusion of molecules through it, *Proc. Natl. Acad. Sci. USA 82*: 5030.

Hazlewood, C.F., 1979, A view of the significance and understanding of the physical properties of cell water, in: "Cell-Associated Water," W. Drost-Hansen and J.S. Clegg, eds., Academic Press, New York, p. 165.

Hazlewood, C.F., 1986, Nuclear magnetic resonance parameters of water in biological tissues, *Microcirc. Endothel. Lymphat.*, 2: 597.

Hennipman, A., Smits, J., Rijksen, G., Neyt, J.P. and Staal, G.E.J., 1987, Glycolytic enzymes in breast cancer, benign breast disease and normal breast tissue, *Tumor Biol.*, 8: 251.

Holden, J.T., 1962, "Amino Acid Pools," Elsevier, Amsterdam, p. 840.

Horowitz, S.B. and Miller, D.S., 1984, Solvent properties of ground substance studied by cryomicrodissection and intracellular reference phase techniques, *J. Cell Biol.*, 99: 172s.

Jacobson, K., and Wojcieszyn, J.W., 1984, The translational mobility of substances within the cytoplasmic matrix, *Proc. Natl. Acad. Sci. USA*, 81: 6747.

Kasturi, S.R., Hazlewood, C.F., Yamanashi, W.S., and Dennis, L.W., 1987, The nature and origin of chemical shift for intracellular water nuclei in *Artemia* cyts. *Biophys. J.*, 52: 249.

Keith, A.D., 1979, "The Aqueous Cytoplasm," Marcel Dekker, Inc., New York, p. 1.

Kellermayer, M., Ludany, A., Jobst, K., Szucs, G., Trombitas, K., and Hazlewood, C.F., 1986, Cocompartmentation of proteins and K^+ within the living cell, *Proc. Natl. Acad. Sci. USA*, 83: 1011.

Kempner, E.S. and Miller, J.H., 1968a, The molecular biology of *Euglena gracilis*. IV. Cellular stratification by centrifuging, *Exp. Cell Res.*, 51: 141.

Kempner, E.S. and Miller, J.H., 1968b, The molecular biology of *Euglena gracilis*. V. Enzyme localization, *Exp. Cell Res.*, 51: 150.

Knull, H.R., 1987, Glycolytic enzyme-cytomatrix interactions, in: "Modern Cell Biology," volume 5, C. Waymouth, ed., Alan R. Liss, Inc., New York, p. 157.

Kondo, H., 1984, Reexamination of the reality or artifact of the microtrabeculae, *J. Ultrastr. Res.*, 87: 124.

Lardy, H.A., 1965, On the direction of pyridine nucleotide oxidation-reduction reactions in gluconeogenesis and lipogenesis, in: "Control of Energy Metabolism," B. Chance, R. Estabrook, and J.R. Williamson, eds., Academic Press, New York, p. 245.

Ling, G.N., 1984, "In Search of the Physical Basis of Life," Plenum Press, NY, p. 1.

Luby-Phelps, K., Lanni, F., and Taylor, D.L., 1988, The submicroscopic properties of cytoplasm as a determinant of cellular function, *Ann. Rev. Biophys. Biophys. Chem.*, *17*: 369.

Mansell, J.L. and Clegg, J.S., 1983, Cellular and molecular consequences of reduced cell water content, *Cryobiology*, *20*: 591.

Masters, C.J., 1984, Interactions between glycolytic enzymes and components of the cytomatrix, *J. Cell Biol.*, *99*: 222s.

Masters, C.J., Reid, S., and Don, M., 1987, Glycolysis - new concepts in an old pathway, *Mol. Cell. Biochem.*, *76*: 3.

Mastro, A.M. and Hurley, D.J., 1987, Diffusion of a small molecule in the aqueous compartments of mammalian cells, in: "Organization of Cell Metabolism," G.R. Welch and J.S. Clegg, eds., Plenum Press, New York, p. 57.

Merta, P.J., Fullerton, G.D., and Cameron, I.L., 1986, Characterization of water in unfertilized and fertilized sea urchin eggs, *J. Cell. Physiol.*, *127*: 439.

McConkey, E.H., 1982, Molecular evolution, intracellular organization, and the quinary structure of proteins, *Proc. Natl. Acad. Sci. USA*, *79*: 3236.

Morton, D.J., Wiedemann, J.F., Clarke, F.M., Stephan, R., and Stewart, M., 1982, A cytoskeletal role for glycolytic enzymes, *Micron*, *13*: 377.

Mowbray, J. and Moses, V., 1976, The tentative identification in *Escherichia coli* of a multienzyme complex with glycolytic activity, *Eur. J. Biochem.*, *66*: 25.

Naharro, G., Robbins, K.C. and Reddy, E.P., 1984, Gene product of v-fgr onc: hybrid protein containing a portion of actin and a tyrosine-specific protein kinase. *Science*, *223*: 63.

Negendank, W., 1982, Studies of ions and water in human lymphocytes, *Biochem. Biophys. Acta*, *694*: 123.

Negendank, W., 1986, The state of water in the cell, in: "The Science of Biological Specimen Preparation," M. Muller, R.P. Becker, A. Boyde, and J.J. Wolosewick, eds., SEM, Inc. AMF O'Hare, Chicago, p. 21.

Nelson, W.G., Pienta, K.J., Barrack, E.R., and Coffey, D.S., 1986, The role of the nuclear matrix in the organization and function of DNA, *Ann. Rev. Biophys. Chem.*, *15*: 457.

Nicolini, C., 1986, "Biophysics and Cancer," Plenum Press, New York, p. 463.

Olmstead, J.B., 1986, Microtubule-associated proteins, *Ann. Rev. Cell Biol.*, *2*: 421.

Paine, P.L., 1984, Diffusive and nondiffusive proteins *in vivo*, *J. Cell Biol.*, *99*: 188s.

Paine, P.L., 1987, The *in vivo* cytomatrix: minimally disturbed systems, in: "Modern Cell Biology," vol. 5, C. Waymouth, ed., A.R. Liss, Inc. New York, p. 169.

Parsegian, V.A. and Rau, D.C., 1984, Water near intracellular surfaces, *J. Cell Biol.*, *99*: 196s.

Pashley, R.M. and Israelachvilli, J.N., 1984, Molecular layering of water in thin films between mica surfaces and its relation to hydration forces, *J. Coll. Interf. Sci.*, *101*: 511.

Pollard, T.D. and Cooper, J.A., 1986, Actin and actin-binding proteins, *Ann. Rev. Biochem.*, *55*: 987.

Porter, K.R., 1984, The cytomatrix: a short history of its study. *J. Cell Biol.*, *99*: 3s.

Porter, K.R., 1987, Structural organization of the cytomatrix, in: "Organization of Cell Metabolism," G.R. Welsh and J.S. Clegg, eds., Plenum Press, New York, p. 9.

Pullman, A., Vasilescu, V., and Packer, L., 1985, "Water and Ions in Biological Systems." Plenum Press, New York, p. 1.

Schliwa, M., van Blerkom, J., and Porter, K.R., 1981, Stabilization of the cytoplasmic ground substance in detergent opened cells, *Proc. Natl. Acad. Sci. USA.*, *78*: 4329.

Srere, P.A., 1987, Complexes of sequential metabolic sequences, *Ann. Rev. Biochem.*, *56*: 21.

Srivastava, D.K. and Bernhard, S.A., 1986, Enzyme-enzyme interactions and the regulation of metabolic reaction pathways, *Curr. Top. Cell Reg.*, *28*: 1.

Steinert, P.M. and Parry, A.D., 1985, Intermediate filaments, *Ann. Rev. Cell Biol.*, *1*: 41.

Stephen, P., Clarke, F. and Morton, D., 1986, The indirect binding of triose-phosphate isomerase to myofibrils to form a glycolytic mini-complex, *Biochim. Biophys. Acta*, *873*: 127.

Stossel, T.P., Chaponnier, C., Ezzell, R.M., Hartwig, J.H., Janmey, P.A., Kwiatkowski, D.J., Lind, S.E., Smith, D.E., Southwick, F.S., Yin, H.L. and Zaner, K.S., 1985, Nonmuscle actin-binding proteins, *Ann. Rev. Cell Biol.*, *1*: 353.

Szent-Gyorgyi, A., 1971, Biology and the pathology of water, *Persp. Biol. Med.*, *14*: 239.

Walsh, J.L. and Knull, H.R., 1988, Heteromerous interactions among glycolytic enzymes and of glycolytic enzymes with F-actin, *Biochim. Biophys. Acta*, *952*: 83.

Watterson, J.G., 1987, A role for water in cell structure, *Biochem. J.*, *248*: 615.

Welch, G.R., ed., 1985, "Organized Multienzyme Systems," Academic Press, New York, p. 457.

Welch, G.R. and Clegg, J.S., 1987, "Organization of Cell Metabolism," Plenum Press, New York, p. 389.

Wheatley, D.N., Inglis, M.S., and Malone, P.C., 1986, The concept of the intracellular amino acid pool, *Curr. Top. Cell. Reg.*, *28*: 107.

Wheatley, D.N., Inglis, M.S., and Foster, M.A., 1987, Hydration, volume changes and NMR proton relaxation times of HeLa S-3 cells in M-phase and subsequent cell cycle. *J. Cell Sci.*, *88*: 13.

MOLECULAR BASIS OF RADIOPROTECTION BY AMINOTHIOLS

Dan Vasilescu

Biophysics Laboratory, Nice University, Parc Valrose 06034 Nice Cedex, France

INTRODUCTION

It is now well established that aminothiols are radioprotector agents, and in some cases anticancer drugs[1-6]. Aminothiols which belong to the radioprotector family are able to reduce the radiation damage when administrated to animals or cellular cultures before irradiation with ionizing radiation. These drugs are characterized by their dose reduction factor (DRF = ratio of lethal dose of irradiation for 50% of animals treated with radioprotector to lethal dose of irradiation for 50% of control animals). On another hand, it was shown that both in vivo and in vitro conditions, ionizing radiation (UV, X or γ rays) induce, by intermediate radicals or hydrated electrons, single breaks and double breaks of DNA backbone and base modifications[7-10].

We have represented in figure 1 some aminothiols derived from cysteamine and in figure 2 some new drugs derived from the naturally occurring molecules cysteine (DRF = 1.4) and glutathione (GSH) which is an endogenous radioprotector. Cysteamine is a good radioprotector (DRF = 1.6) but introduces an unacceptable toxicity. The best known radioprotector is WR-2721, which possesses the highest dose reduction factor (DRF = 2.7). The toxicity of this molecule is reduced by S-phospho-rylation of WR-1065. Actually, this free sulfhydryl molecule is the active metabolite of WR-2721 which is able to interact with the nuclear material.

Purdie has demonstrated that, in vitro, cysteamine and WR-1065 are the metabolic fractions able to interact with the DNA in cultured human cells[11]. Yuhas et al. have shown firstly that WR-2721 is an anticancer drug[4]. This aminothiol can protect solid tumors and normal

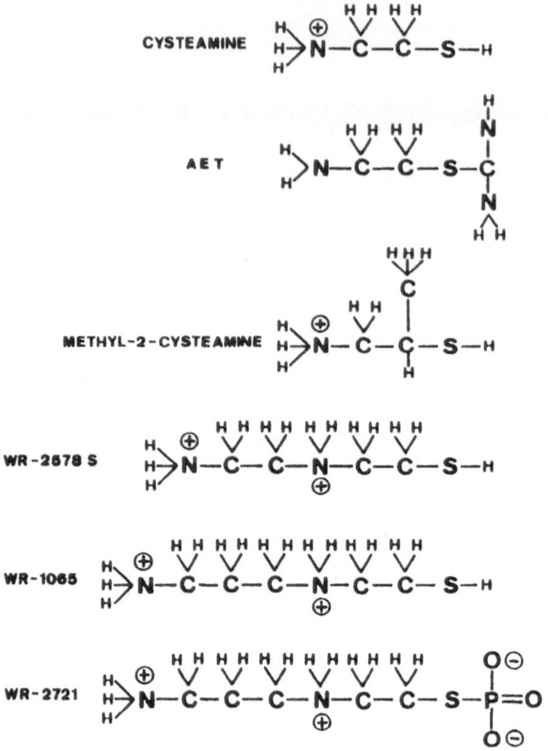

CYSTEAMINE

AET

METHYL-2-CYSTEAMINE

WR-2678 S

WR-1065

WR-2721

Fig. 1. Some aminothiols derived from
cysteamine.

tissues in a differential manner, in radiotherapy and chemotherapy; it can also reduce the nephrotoxicity of cisplatin without antitumoral activity alteration[12-14]. Recently, new aminothiols containing a pseudo-peptide group were synthesized by Imbach et al[15-16]. These molecules are derived from the γ-Glu-Cys branch of GSH which is the most abundant thiol inside cells and possesses multifunctional activities in biology and cancer therapy[17].

Figure 2 shows three of these molecules : I-102, I-102 S which is the free sulfhydryl form of I-102 (DRF = 1.4) and I-143 which is a dithiol.

Comparative studies were carried out with I-102 and WR-2721 by Imbach et al., and led to the conclusion that I-102 is less toxic than WR-2721. I-102, like WR-2721, is an anticancer drug by preferential protection of normal tissues against ionizing radiation. Nevertheless, an important metabolic fraction of I-102 S is cysteamine and this indicates that this molecule appears as a "prodrug".

So, many questions are involved in the different mechanisms of radioprotection, and then it is necessary to distinguish different levels in the field of numerous studies. For instance, at the clinical

I-102

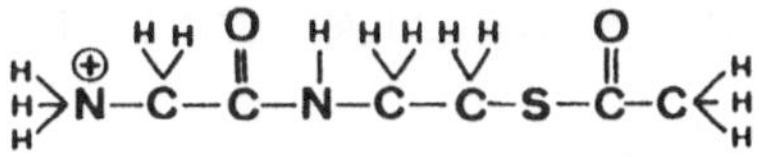

I-102 S

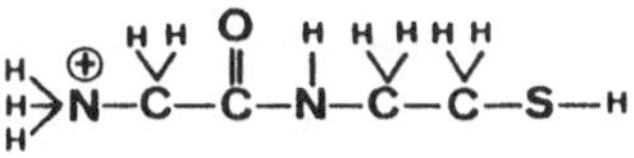

I-143

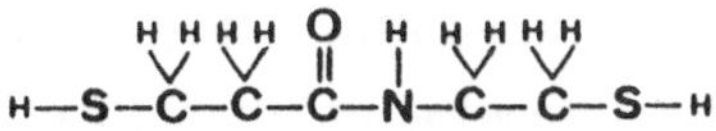

GSH

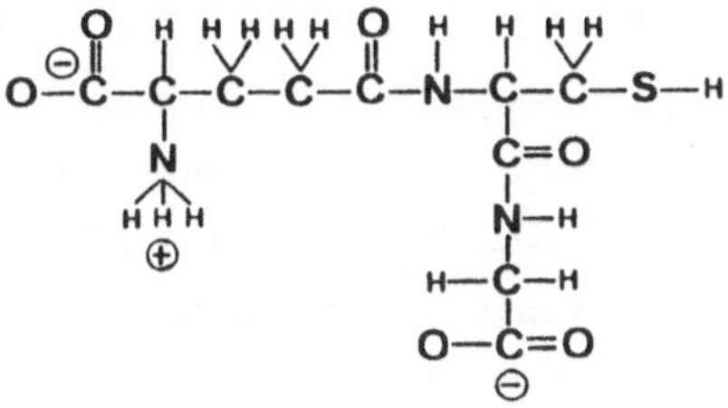

Cysteine

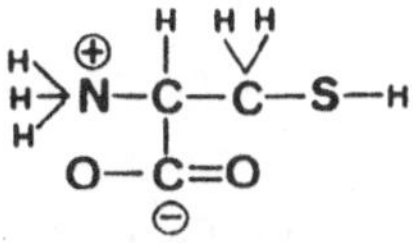

Fig. 2. Aminothiols containing a peptide or a
pseudopeptide group. Cysteine and GSH
are naturally occurring thiols.

level there are pharmacokinetic problems which are not fully resolved[18].
At the molecular level, the mechanism of repair of DNA simple and
double strands breaks involves the aminothiols as radical scavengers
(H_2O radiolysis) and as hydrogen donators which are able to chemically
repair sugar damage induced by irradiation (i.e. : a C4' radical is
involved on the DNA deoxyribose[19,20]). When the metabolic fraction
penetrating cell nucleus is considered it appears that - in the absence
of ionizing radiation - the interaction between acting drug and the
accessible portions of DNA belonging to genetic material, is an important
factor. We have demonstrated previously by using spectrophotometric
and dielectric studies that, in vitro, this interaction is essentially
electrostatic and involves the anionic DNA phosphate sites and the
cationic groups of aminothiols[21-23].

   The aim of this paper is to point out :

   - firstly, the molecular similarity in aminothiol radioprotectors
by using topological and quantum arguments ;

   - secondly, the mechanism of depolarization of DNA, when an aminothiol is approaching the counterionic atmosphere around the phosphate
sites, with the help of ^{23}Na NMR and Electric Birefringence techniques.

TOPOLOGICAL AND QUANTUM MOLECULAR SIMILARITIES IN AMINOTHIOLS FAMILY

Topological Similarities

When we observe figures 1 and 2, it is obvious that similarities exist between the molecules derived from cysteamine or cysteine. If we start with cysteamine (which is obtained by decarboxylation of cysteine) as a substructure, we can build the WR-2578 S which appears as a fundamental structure, then an hyperstructure GSH. All these arguments were developed in a recent study, by using topological descriptors introduced by Randic[24]. Topological descriptors and indexes describe a molecule by a graph invariant[25-28]. For each molecule (represented only by its heavy atoms), we can build the atomic path sequences by counting the number of paths : P_0, P_1, ... P_k ... P_n. P_0 is the number of heavy atoms, P_1 is the number of paths of length 1 (1 bond), P_k is the number of paths of length k (k bonds). It is also possible to introduce a variation in the count of path, by adopting different weights for the heterobonds (For instance, we can adopt the bond weights : C-C = 1 ; C-O = x ; C-N = y and C-S = z).

Then, a quantitative discussion of molecular similarity - in a class of molecules possessing similar properties - may be conducted by using similarity matrices. Each element of a similarity matrix represents the Euclidean distance $d_{\alpha\beta}$ between two molecules α and β in an n-dimensional vector space of paths (P_k). In other words, in the n-dimensional vector space of paths :

- the molecule α is represented by a vector $\vec{V}_\alpha$ of components $(P_{1\alpha}, P_{2\alpha} ... P_{n\alpha})$, and the molecule β is represented by a vector $\vec{V}_\beta$ $(P_{1\beta}, P_{2\beta}, ... P_{i\beta}, ... P_{m\beta})$;
   - the Euclidean distance between $\vec{V}_\alpha$ and $\vec{V}_\beta$ is :

$$d_{\alpha\beta} = \left\{ | \vec{V}_\alpha - \vec{V}_\beta |^2 \right\}^{1/2} = \left\{ \sum_{i=0}^{n,m} | P_{i\alpha} - P_{i\beta} |^2 \right\}^{1/2}$$

When a similarity matrix is constructed the more similar couples of molecules α and β are those for which the $d_{\alpha\beta}$ is the shortest.

By varying the numerical values of heterobonds x, y, z parameters, we have found :

- a remarkable similarity between WR-1065 and I-102 S and between WR-1065 and WR-2578 S (DRF = 2.0) which are the most potent known radio-protectors.

Other similarities were found, by using an index $(mn)^{-1/2}$ which is defined, as a weight $(mn)^{-1/2}$ for each bond between two atoms, with m and n neighbors for the terminal atoms of the bond.

Analysis of the similarity, when using similarity matrices built with the $(mn)^{-1/2}$ index, leads to logical couples of molecules, when considering the cysteamine main chain or the WR-2578 S main chain. For instance, the new synthesized molecules I-102 S and I-143 are very similar ; it is also notable to point out that for the couple involving GSH and I-143, we obtain an association between two agents possessing a weak radioprotective effect.

For short molecules derived from cysteamine main chain it is to be noted that Me2 cysteamine is the more isotopologous neighbour of cysteamine ; this result is very interesting when considering that Me2 cysteamine possesses a DRF = 1.8 > DRF of cysteamine, with the advantage of reducing the toxic effects of cysteamine. We also observe a good similarity between the pairs: Me2 cysteamine-AET (DRF = 1.6) and cysteamine-cysteine.

Electrostatic Similarities

An another important characteristic aspect concerns the intrinsic electrostatic properties of these molecules. When dissolved in water or aqueous electrolytes, these aminothiols are generally ionized : cysteamine and I-102 S are monocationic and WR-2578 S and WR-1065 are bicationic ; GSH is globally monoanionic at neutral pH.

We have previously measured the intrinsic ionic conductivity of cysteamine and WR-1065 and deduced their electrical and mechanical mobility[29]. These intrinsic electrostatic properties of aminothiols may also be emphasized by the help of quantum mechanical studies[30,31].

We present in figure 3 some results obtained by using PCILO and ab initio quantum methods on cysteamine, I-102 S, WR-2578 S and WR-1065. The minimal energy conformation of these molecules was obtained by using PCILO method ; then the charge distribution was obtained with the help of an ab initio computation. On the basis of the Mulliken atomic net charges (in proton unit) centered on each nucleus of a molecule, we can introduce the charges of different groups like : NH_3, NH_2, NH, CH_2 and SH.

When we examine the charge group distribution on the four molecules of figure 3, we can show-up a "quantum similarity" and particularly we can observe :

- In all cases, there is a large concentration of the positive charge on the ammonium group. The value of this charge is reinforced in the case of monocation I-102 S and in the case of bications WR-2578 S and WR-1065. Nevertheless, it is to be noted that the effect of methyl substitution on the CH_2 group near sulfhydryl in cysteamine (Me2 cysteamine

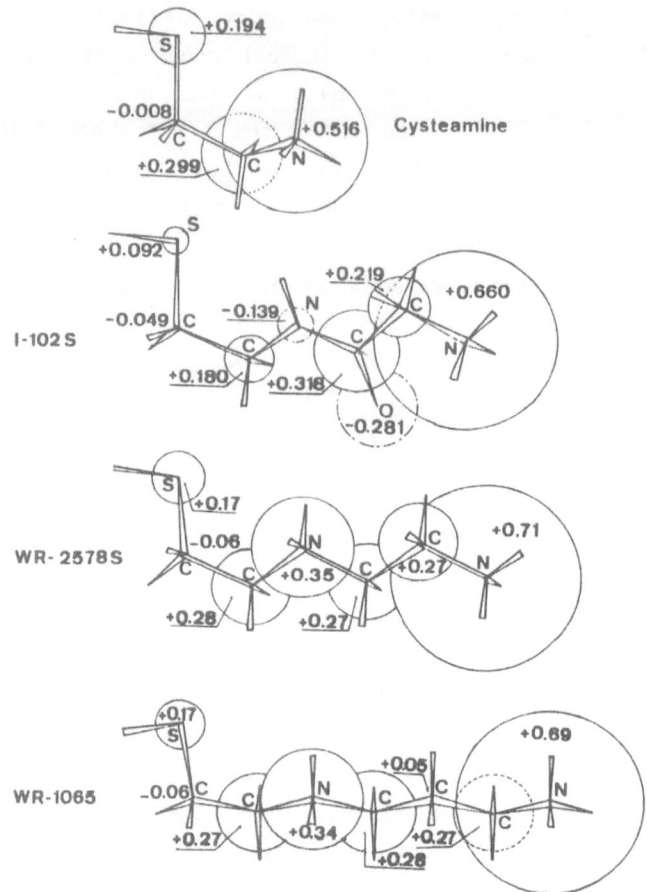

Fig. 3. Charge group distribution in monocations cysteamine and I-102 S, and in bications WR-2578 S
and WR-1065. Molecules are represented in their
minimal energy conformation. The charges of SH,
CH_2, NH, NH_2 and NH_3 groups are represented by
spheres centered on S, C, N atoms with a radius
proportional to the charge group value. In the
case of I-102 S, the C and O atoms of the pseudopeptide group are isolated.

molecule) reinforce the ammonium positive value (+0.64) to the detriment
of the SH group charge (+0.14).

- For the four aminothiols there is a non negligible positive
charge on the sulfhydryl group, which is reduced in the case of I-102 S.

- For WR-2578 S and WR-1065, it is remarkable that the charge
of terminal NH_3 group is twice that on the NH_2 one.

- For cysteamine, WR-2578 S and WR-1065, methylenic groups are
equally positively charged when they are adjacent to an NH_3 or NH_2 ;
when a CH_2 group is adjacent to a sulfhydryl group, the charge is quasi
null.

- The charge distribution in I-102 S may be explained as a perturbation introduced by addition of glycyl residue to cysteamine through a pseudopeptide bond. A first result is clearly that the positive charge of ammonium is increased and the sulfhydryl charge is decreased to a value half of the cysteamine SH charge. On the other hand, it is noticeable that if we consider the global charge carried by the pseudopeptidic group : $(-0.139 + 0.318 - 0.281 + 0.219 = 0.117)$, the cationic rest charge $(+0.883)$ corresponds to the cysteamine portion in I-102 S : $(SH-CH_2-CH_2 \ldots NH_3)$. Then, it is remarkable to observe that the pseudopeptidic charge is quasi identical to the difference between the charge of methylenic group adjacent to ammonium in cysteamine $(+0.299)$ and the corresponding group in I-102 S $(+0.180)$.

INTERACTION BETWEEN DNA AND CATIONIC AMINOTHIOLS

Previous spectrophotometric studies have shown that interaction between DNA and cationic aminothiols is analogous to that of metallic cations. The melting point T_m of DNA is raised when the drug is added ; this cooperative phenomenon may be expressed by an empirical equation :

$$T_m = a \log D + b$$

where D is the concentration of added drug and a and b two adjustable parameters[21-23].

By using dielectric technique we have shown that the important dielectric increment of DNA solutions ($\Delta\epsilon' = \epsilon'_{DNA} - \epsilon'_{solvent}$) observed in the dielectric dispersion curve of permittivity ϵ' (in the frequency domain 0.5 - 10 MHz) is decreased when cysteamine is added to DNA[23]. It is known that the $\Delta\epsilon'$ increment of DNA in electrolytic solution is due to the polarizability created by the fluctuating atmosphere of counterions along the DNA polyanion. Thus, the observed dielectric decrement, when cysteamine interacts with DNA corresponds to a depolarization of the rod-like polyelectrolyte. This experimental fact confirms the existence of a strong electrostatic interaction between cationic sites of cysteamine and the anionic DNA phosphate sites ; this interaction is accompanied by an ejection of a fraction of "bounded" counterions out of DNA phosphate groups.

For illustrating this interaction between cysteamine and DNA, we have represented in figure 4 the result of a quantum chemical computation obtained by using a supermolecular modellisation[32]. In this theoretical simulation, DNA is figurated by a backbone containing two deoxyriboses and three phosphate sites, in the fixed B conformation of the double helix. Then, a deformable molecule of cysteamine approaches the central phosphate. Figure 4 represents the minimal energy of the

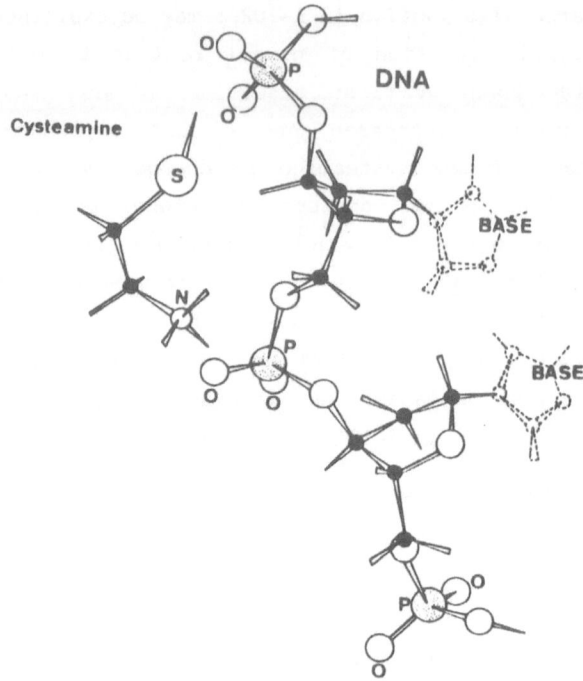

Fig. 4. Theoretical modellisation of the interaction between the monocation cysteamine
and DNA sugar-phosphate backbone, resulting from a quantum supermolecular computation (PCILO method).

supermolecule (DNA + cysteamine) obtained by PCILO computation ; the
figure shows clearly the bridge building made by ammonium and sulfhydryl
groups of cysteamine between two adjacent phosphate groups.

In the view to demonstrate this electrostatic interaction by
using two appropriate techniques (^{23}Na NMR and the Kerr effect),
we present before, some interesting properties of DNA - as a polyelectrolyte - which may be expressed with the help of few characteristic
parameters.

DNA as a polyelectrolyte (see figure 5)

When we dissolve the DNA polyanion in an 1.1 aqueous electrolyte
like NaCl, we can define for :

The electrolyte

If n is the number of monovalent cations (or anions) per cubic
meter, and introducing the $\mathfrak{L}$ constant[33] :

$$\mathfrak{L} = \frac{e^2}{4 \pi \varepsilon_0 k} \simeq 1.67 \times 10^{-5} \text{ mK}$$

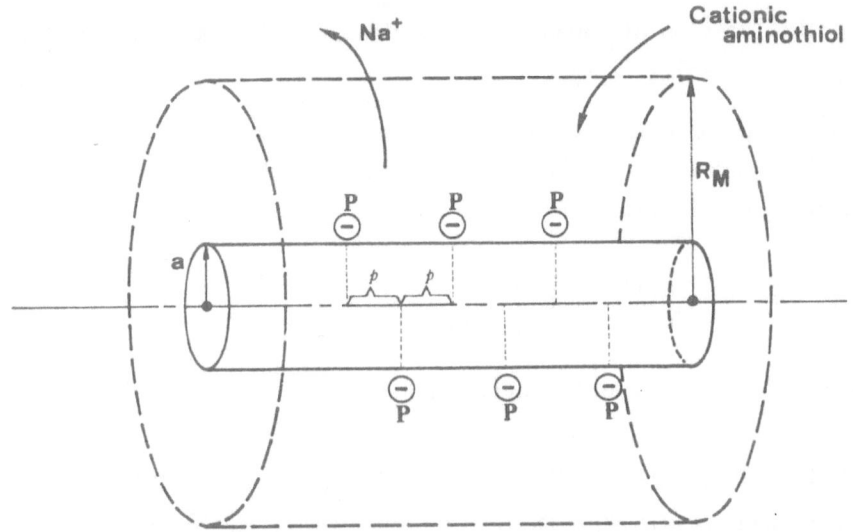

Fig. 5. The DNA polyelectrolyte in interaction with a cationic aminothiol. P are the centers of phosphate groups and p the distance between two consecutive phosphates projected onto the DNA double helix ; a is the radius of the DNA rod-like and R_M the radius of the cylindrical sheath containing the Manning's fraction η of condensed counterions Na^+.

(where : e is the proton charge, ε_0 the permittivity of free vaccuum and k the Boltzmann's constant), we can define two fundamental lengths :

* The Landau length

$$\ell = \frac{\mathcal{L}}{\varepsilon_r' T} = 7.15 \text{ Å} \quad \text{at } 25°C$$

with ε_r' = relative permittivity of water (= 78.33 at 25°C) ; T = absolute temperature.

* The Debye length

$$r_D = (8 \pi n \ell)^{-1/2}$$

The polyelectrolyte[34]

If P is the DNA phosphate site concentration (expressed in mol/liter) and p the distance between two consecutive phosphate groups projected onto the double-helix axis, we can define three fundamental parameters :

* The charge parameter of DNA : $\xi = \dfrac{\ell}{p}$

and ξ_h the ξ value for the DNA in ideal double-helix state : $\xi_h \simeq 4.2$, when $p_h = 1.7 \text{ Å}$.

* The fraction of bound counterions [according to Manning's model[35,36]]

$$\eta = 1 - \xi_h^{-1} = 0.76$$

and then, the concentration of bound counterions $= \eta P$.

* The radius of the cylinder containing the fraction of condensed counterions [following Le Bret and Zimm[37,38]]

$$R_M \simeq (2\,a\,r_D)^{1/2}\ \exp\left\{\frac{\xi_h - 2}{2(\xi_h - 1)}\right\}$$

where a is the radius of DNA cylinder.

^{23}Na NMR study of the interaction between DNA and cysteamine

The quadrupolar [23]Na nucleus is characterized in aqueous solution, by a single NMR peak with a relaxation rate :

$$R = \frac{1}{T_1} = \frac{1}{T_2} = \frac{2\pi^2}{5}\chi^2\,\tau_r = \pi\,\Delta\upsilon_{1/2}$$

where : χ is the quadrupole coupling constant and τ_r the correlation time of [23]Na ; $\Delta\upsilon_{1/2}$ is the linewidth of the peak (see figure 6). When DNA is dissolved in an NaCl solution, the measured relaxation rate R of sodium in the polyelectrolytic solution, may be defined - under rapid exchange conditions - by a two state model developped by Bleam et al.[39] :

$$R = R_F + \frac{(R_B - R_F)\,Na_B}{Na}$$

where : R_F and R_B are the relaxation rates of "free" and "bound" Na^+ ; Na_F and Na_B are the concentrations of "free" and "bound" Na^+ with regard to DNA, and the total sodium concentration is :

$$Na = Na_B + Na_F \quad \text{with } Na_B = \eta P$$

When cysteamine is added to DNA-NaCl solution, we can define the total concentration D of added drug :

$$D = D_B + D_F$$

with the same conventions adopted above. Figure 6 shows clearly the effect of the interaction between DNA and the monocation cysteamine ; we observe a decrease of the linewidth $\Delta\upsilon_{1/2}$ of [23]Na peak until a concentration of added drug $D \simeq 3/4\,P$, beyond this concentration there being a saturation plateau[40]. This behaviour may be interpretated by a decrease of Na_B. Then, introducing the new hypothesis :

$$\eta P = Na_B + D_B \quad \text{and} \quad R = R_F + \frac{(R_B - R_F)\,(Na_B + D_B)}{Na}$$

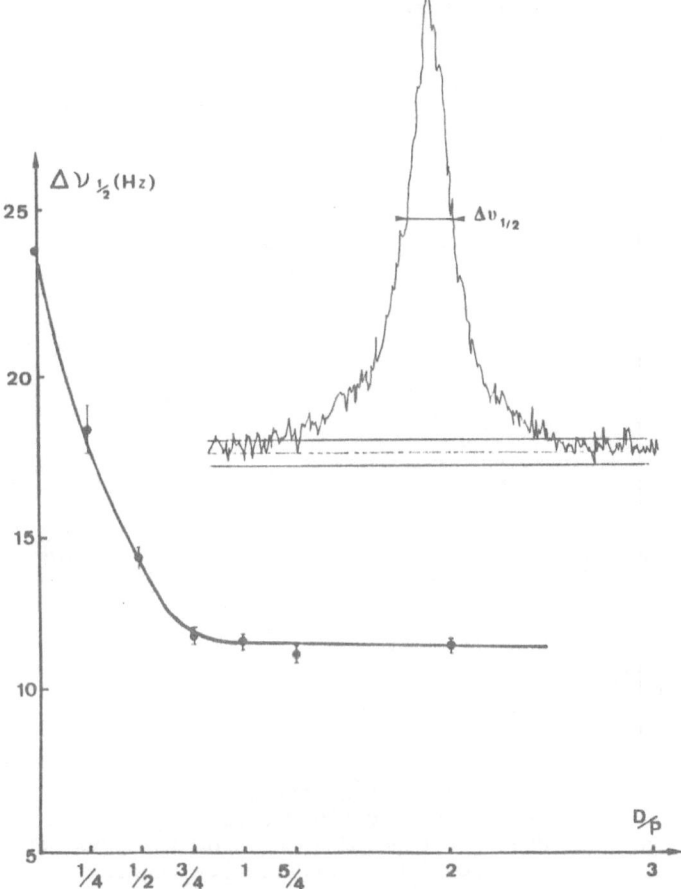

Fig. 6. Transform Fourier ^{23}Na peak of DNA in NaCl obtained at 23.81 MHz on a Brucker WH 90/DS NMR spectrometer. The $\Delta\upsilon_{1/2}$ linewidth is measured at the midway points of the peak. The curve corresponds to $\Delta\upsilon_{1/2}$ measurements of ^{23}Na$^+$ in a DNA-NaClsolution (10^{-2} mol/ liter) as a function of added cysteamine of D concentration. The concentration of phosphate sites P of DNA is 10^{-2} mol/liter.

A convenient graphic representation was introduced by means of the x parameter :

$$x = D_F - Na_B = D + \frac{(R_F - R)\,Na}{R_B - R_F}$$

Figure 7 shows the x variation versus the concentration of added drug. The region for $x < 0$ corresponds to $Na_B > D_F$ and the region for $x > 0$ corresponds to $Na_B < D_F$. When $x = 0$ (point A), we have $D_F = Na_B$ and $D/P \simeq \eta/2$. This representation shows clearly the cooperative effect of the electrostatic interaction of cysteamine cation with anionic

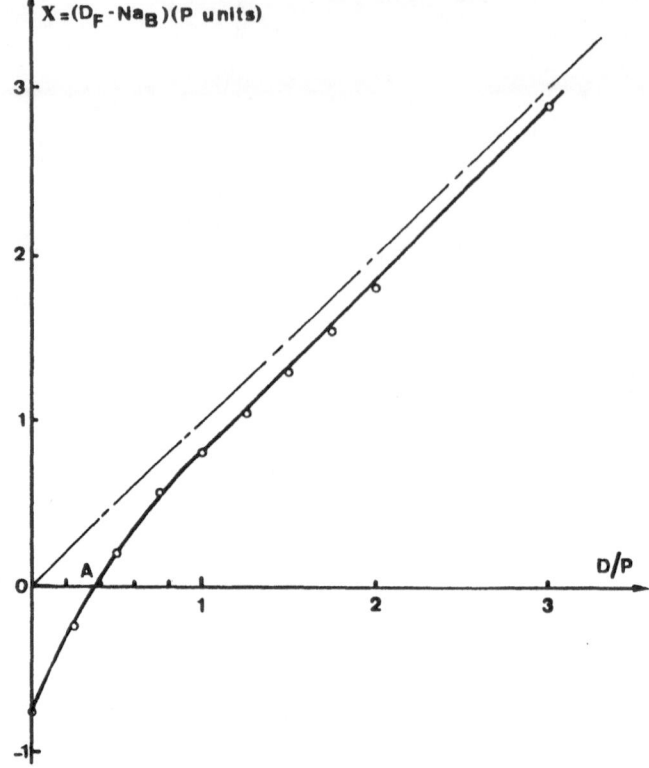

Fig. 7. Variation of the difference between free
cysteamine and bound sodium versus the
concentration of added drug to a DNA-NaCl
solution (10^{-2} mol/liter). D concentrations of added drug are expressed in DNA
phosphate sites.

phosphate sites and consequently the ejection of Na^+ counterions. The
curve is asymptotic to the bisector line corresponding to $Na_B = 0$ (and
then $x = D_F = D$). So, $Na_B \rightarrow 0$ when $x > 0$ and $D/P > \eta$.

Depolarization of DNA interacting with Aminothiols demonstrated by a Kerr Effect Experiment

When an electric field $\vec{E}$ is applied to a birefringent solution,
perpendicularly to a laser beam, the well known electrooptic Kerr effect
is induced as[41-46] :

$$\Delta n = n_{//} - n_\perp = B \lambda E^\beta$$

where : $n_{//}$ and $n_\perp$ are the refractive indexes of solution parallel
and perpendicular to $\vec{E}$, B is the Kerr constant, λ the light wavelength
and β a numerical parameter (see figure 8).

182

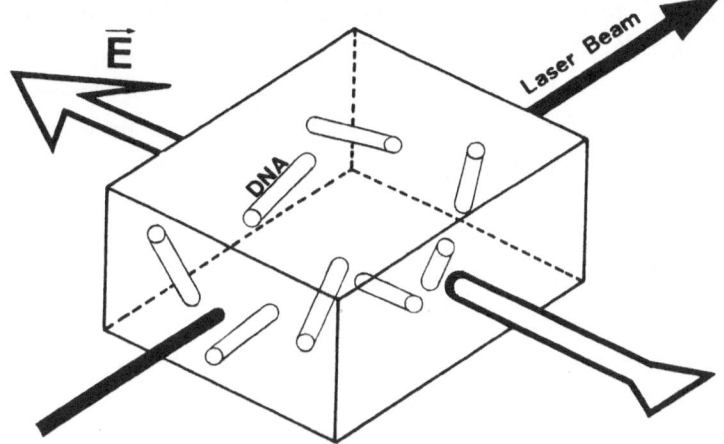

Fig. 8. The Kerr cell containing a DNA solution. A pulsed
electric field $\vec{E}$ is applied perpendicularly to
the laser beam. The electric birefringence is
defined by $\Delta n = n_{//} - n_{\perp}$, where $n_{//}$ and $n_{\perp}$ are
the refractive indexes of DNA solution parallel
and perpendicular to the direction of the applied
electric field $\vec{E}$.

Electric birefringence measurements of macromolecular solutions may
be conducted by using a Kerr effect apparatus based on square pulse
of electric field technique[47]. For rod-like molecules as DNA, the birefringence can be written[41] :

$$\Delta n = \frac{2 \pi \tilde{v} c}{n_s} \Delta g \, \Phi$$

where : $\tilde{v}$ is the solute partial specific volume, c is the concentration
and n_s is the optical index of solvent ; $\Delta g = g_1 - g_2$ is the difference
between the optical polarizabilities and Φ an orientation function.
Assuming a negligible permanent dipole moment for DNA, we can write :

$$\Phi = \frac{E^{\beta}}{15 \, k \, T} \Delta \alpha$$

with : $\Delta \alpha = \alpha_L - \alpha_T$ = induced electrical anisotropy of the macromolecule ; where α_L and α_T are the longitudinal and transverse electrical
polarizabilities of the DNA.

Taking into account that Δg may be considered as an invariant
and assuming a negligible α_T for the DNA rod-like, we are lead to :

$\Delta n \propto \alpha_L$ and then, the variations of DNA electric birefringence
may be directly correlated to modifications of electrical polarizability.

Figure 9 shows the Δn variations for a DNA solution when Na^+, cysteamine or WR-1065 are added. The observed decrease of Δn is thus correlated to an equivalent decrease in α_L. This effect was predicted, in the case of monovalent and divalent cations by Hornick and Weill[41], and according to the two phase model of cylindrical polyelectrolyte solution, it is possible to demonstrate that :

$$\Delta\alpha \simeq \alpha_L \propto \left\{ Ln \; \frac{r}{R_M} \right\}^{-1}$$

where : r is the radius of the cylindrical free volume offered to a single macromolecule and R_M is the radius of the cylinder containing the fraction of condensed counterions (see figure 5).

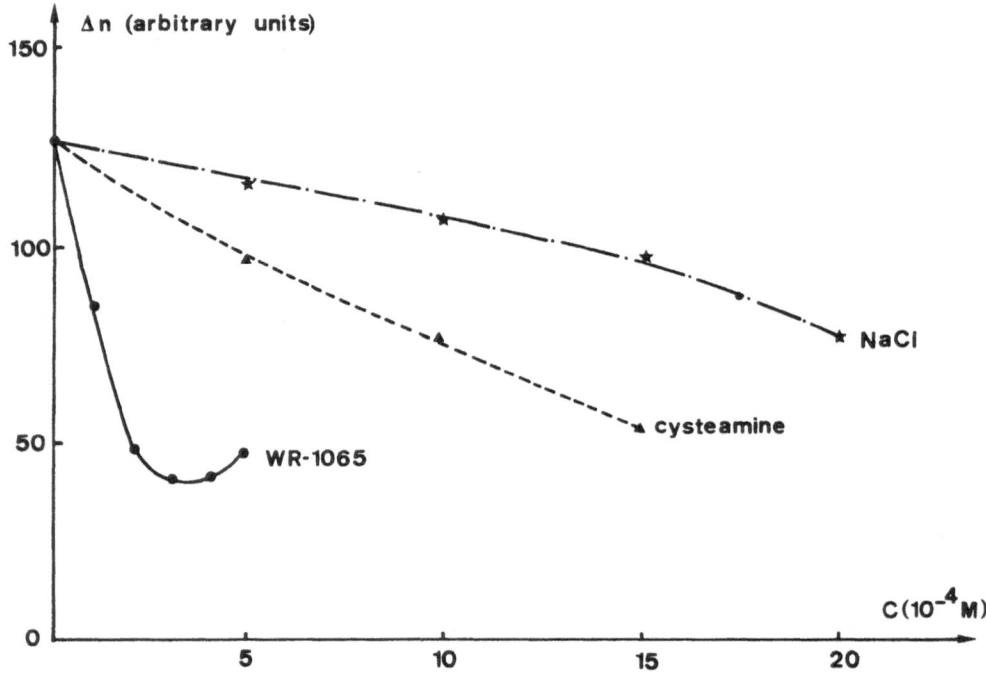

Fig. 9. Variation of the DNA solution (5×10^{-4} mol/liter P in 5×10^{-4} mol/liter NaCl) birefringence Δn versus the concentration C of added : Na^+, cysteamine monocation and WR-1065 bication. The applied electric pulse has a duration of 10 ms and an intensity $| E | = 400$ V x cm^{-1}.

Figure 10 shows the decrease of α_L in a simulation based on r = 800 Å and R_M calculated from Le Bret and Zimm formula.

So, our results reveal a very strong depolarization of DNA, when interacting with the two studied aminothiols. This electrostatic interaction is more important in the case of WR-1065 and, when we reach a

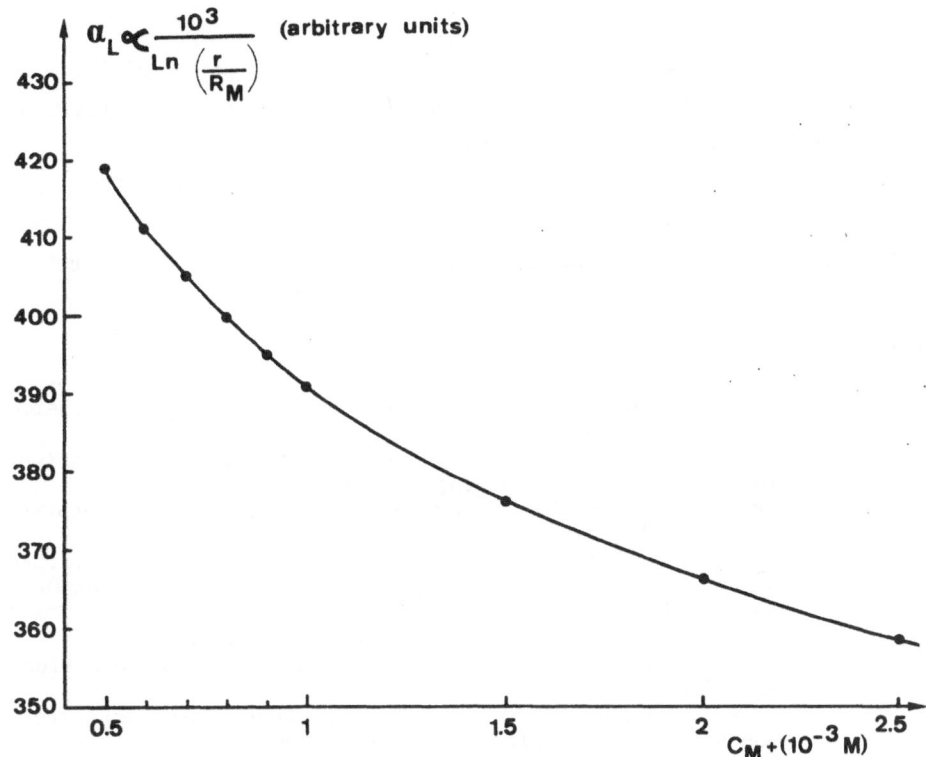

Fig. 10. Variation of the function $[Ln(r/R_M)]^{-1}$ (which is proportional to the DNA longitudinal electric polarizability α_l) versus the concentration of added counterions. This simulation corresponds to a cylindrical DNA of length $L \simeq 10^4$ Å at the concentration $P = 5 \times 10^{-4}$ mol/liter. R_M was calculated on the basis of $a = 10$ Å and $\xi_h = 4.2$.

concentration corresponding to one molecule of drug per one phosphate site, a precipitation occurs.

CONCLUSION

Quantum and topological similarities in aminothiol class of molecules suggest new lines of future research in the view of new perspectives in drug design, radioprotection or anticancer chemotherapy.

Experimental methods have been used to demonstrate that - at the biomolecular level - the electrostatic interaction between DNA phosphate sites and cationic aminothiols is one of the mechanisms implicated in radioprotection. The molecular sheath formed around the DNA may be involved in the reduction of injury induced by ionizing radiation : indirectly through the capture of induced free radicals in the electrolyte and directly by increasing the rigidity of DNA double helix.

ACKNOWLEDGMENTS

This research was sponsored by grant N° 86/114 from DRET and the author thanks Dr. H. Sentenac-Rommanou (Head of Chemical-Pharmacology Division, DRET, Paris) for stimulating discussions.

The contributions to this work by the following collaborators of the Biophysics Laboratory in Nice are also gratefully aknowledged : Drs H. Broch, H. Kranck, J. Lematre, G. Mallet, M.A. Rix-Montel and R. Viani.

REFERENCES

1. P. C. Jocelyn, "Biochemistry of the SH group", Academic Press, New York (1972).
2. W. O. Foye, Mechanisms of Radiation Protection by the Amino-thiols, Int. J. Sulfur Chem., 8:161 (1973).
3. Z. M. Bacq, "Sulfur Containing Radioprotective Agents", Pergamon, New York (1975).
4. J. M. Yuhas, On the Potential Application of Radioprotective drugs in solid tumor radiotherapy, in "Radiation-Drug Interactions in the Treatment of Cancer", G.H. Sokol and R.P. Maickel, eds., Wiley, New York (1980).
5. M. Fatome, La Radioprotection Chimique, Radioprotection, 16:113 (1981).
6. O. F. Nygaard and M.G. Simic, "Radioprotectors and Anticarci-nogens", Academic Press, New York (1983).
7. M. Vorlickova and E. Palecek, A study of changes in DNA Conformation caused by Ionizing and Ultra-violet Radiation by means of Pulse Polarography and Circular Dichroism, Int. J. Radiat. Biol., 26:363 (1974).
8. J. F. Ward, Molecular Mechanisms of Radiation-Induced Damage to Nucleic Acids, Adv. Rad. Biol., 5:181 (1975).
9. R. Frey and V. Hagen, Changes in DNA Secondary Structure after γ-Irradiation, Rad. Environ. Biophys., 12:111 (1975).
10. J. M. Sequaris, P. Valenta, H.W. Nurnberg and B. Malfoy, Voltametric Studies on the Bioelectrochemical Behaviour of Ultrasound Sonicated and γ-Irradiated Native DNA, Bioelectrochem. Bioenerg., 5:483 (1978).
11. J. W. Purdie, A Comparative Study of the Radioprotective Effects of Cysteamine, WR-2721 and WR-1065 in Cultured Human Cells, Radiat. Res., 77:303 (1979).
12. J. M. Yuhas and J.B. Storer, Differential Chemoprotection of Normal and Malignant Tissues, J. Natl. Cancer Inst., 42:331 (1969).
13. J. M. Yuhas, J.M. Spellman and F. Culo, The Role of WR-2721 in Radiotherapy and/or Chemotherapy, Cancer Clin. Trials, 3:211 (1980).
14. J. M. Yuhas and F. Culo, Selective Inhibition of the Nephroto-xicity of cis-Dichloroammineplatinium (II) by WR-2721 Without Alterning its Antitumor Properties, Cancer Treat. Rep., 64:57 (1980).
15. F. Lespinasse, J. Oiry, M. Fatome, P. Ardouin, J. Imbach, E.P. Malaise and M. Guichard, Radioprotection of EMT6 Tumor by a New Class of Radioprotectors Based on a Pseudo-Peptide Cysteamine Combination, Int. J. Rad. Oncol. Biol. Phys., 11:1035 (1985).

16.   J. Imbach, personnal communication.

17.   A. Meister, Selective Modification of Glutathione Metabo-
        lism, Science, 220:472 (1983).

18. J. M. Yuhas and T.L. Phillips, Pharmacokinetics and Mechanisms
        of action of WR-2721 and other protective agents, in:
        "Radioprotectors and Anticarcinogens", O.F. Nygaard
        and M.G. Simic, eds, Academic Press, New York (1983).

19.   D. Schulte-Frohlinde, Kinetics and Mechanisms of Polynucleo-
        tide and DNA strand break formation, in : "Radioprotectors
        and Anticarcinogens", O.F. Nygaard and M.G. Simic, eds,
        Academic Press, New York (1983).

20. J. F. Ward, Chemical Aspects of DNA radioprotection, in :
        "Radioprotectors and Anticarcinogens", O.F. Nygaard
        and M.G. Simic, eds, Academic Press, New York (1983).

21. M. A. Rix-Montel, D. Vasilescu and H. Sentenac, Dielectric ,
        Potentiometric and Spectrophotometric Measurements of
        the Interaction between DNA and Cysteamine, Stud. Biophys.,
        69:209 (1978).

22.   D. Vasilescu and M.A. Rix-Montel, Interaction of Sulfur-
        Containing Radioprotectors with DNA : A Spectrophotometric
        Study, Physiol. Chem. Phys., 12:51 (1980).

23.   D. Vasilescu, H. Broch and M.A. Rix-Montel, Mechanism
        of Aminothiol Radioprotectors Action at the Molecular
        Level, J. Mol. Structure (Theochem), 134:367 (1986).

24.   D. Vasilescu and R. Viani, Molecular Similarity in Amino-
        thiol Radioprotectors : A Randic Graph Approach, Int. J.
        Quantum Chem., Quantum Biology Symposium, 14:149 (1987).

25.   M. Randic and C.L. Wilkins, Graph Theoretical Study
        of Structural Similarity in Benzomorphans, Int. J. Quantum
        Chem., Quantum Biology Symposium, 6:55 (1979).

26.   M. Randic, Non-empirical Approach to Structure-Activity
        Studies, Int. J. Quantum Chem., Quantum Biology Symposium,
        11:137 (1984).

27.   M. Randic, Graph Theoretical Approach to Structure-
        Activity Studies : Search for Optimal Antitumors Compounds,
        in: "The Molecular Basis of Cancer", R. Rein, ed., A.R. Liss,
        New York (1985).

28. S. C. Grossman, B. Jerman Blazic Dzonova and M. Randic,
        A Graph Theoretical Approach to Quantitative Structure-
        Activity Relationship, Int. J. Quantum Chem., Quantum
        Biology Symposium, 12:123 (1986).

29. M. A. Rix-Montel, H. Kranck and D. Vasilescu, Electrochemical
        Behaviour of Aminothiol Radioprotectors, Bioelectrochem.
        Bioenergetics, 16:427 (1986).

30.   H. Broch, D. Cabrol and D. Vasilescu, Electrostatic
        Properties of Some Sulfur Containing Radioprotectors,
        Int. J. Quantum Chem., Quantum Biology Symposium, 9:111
        (1982).

31.   H. Broch and D. Vasilescu, Conformation and Electrostatic
        Properties Quantum Determination of the New Radioprotector
        and Anticancer Drug I-102, Int. J. Quantum Chem., Quantum
        Biology Symposium, 13:81 (1986).

32.   H. Broch, D. Cabrol and D. Vasilescu, Quantum Mechanical
        Simulation of the Interaction between the Radioprotector
        Cysteamine and DNA, Int. J. Quantum Chem., Quantum Biology
        Symposium, 7:283 (1980).

33.   D. Vasilescu, Sur une Notation Rationalisée des Paramètres
        Caractéristiques d'un Electrolyte 1.1 en Solution, J. Chim.
        Phys., 7-8:1131 (1974).

34. D. Vasilescu, H. Grassi and M.A. Rix-Montel, DNA as a Polyelectrolyte : Recent Investigations on the Na-DNA System, *in*: "Polyelectrolytes and their Applications", A. Rembaum and E. Selegny, eds, Reidel, Dordrecht (1975).

35. G. S. Manning, Limiting Laws and Counterion Condensation in Polyelectrolyte Solutions.I.Colligative Properties, J. Chem.Phys., 51:924 (1969).

36. G. S. Manning, On the Application of Polyelectrolyte "Limiting Laws" to the Helix-Coil Transition of DNA.I. Excess Univalent Cations, Biopolymers, 11:937 (1972).

37. M. Le Bret and B. Zimm, Monte Carlo Determination of the Distribution of Ions about a Cylindrical Polyelectrolyte, Biopolymers, 23:271 (1984).

38. M. Le Bret and B. Zimm, Distribution of Counterions around a Cylindrical Polyelectrolyte and Manning's Condensation Theory, Biopolymers, 23:287 (1984).

39. M. L. Bleam, C.F. Anderson and T. Record Jr., Relative Binding Affinities of Monovalent Cations for Double-Stranded DNA, Proc. Natl. Acad. USA, 77:3085 (1980).

40. D. Vasilescu and G. Mallet, Demonstration of the Interaction of Cysteamine with DNA using ^{23}Na NMR Technique, Biopolymers, 24:1845 (1985).

41. C. Hornick and G. Weill, Electrooptical Study of the Electric Polarizability of Rod-like Fragments of DNA, Biopolymers, 10:2345 (1971).

42. J. C. Bernengo, Doctoral Thesis, Lyon University (1970).

43. C. T. O'Konski and S. Krause, Electric Birefringence and Relaxation in Solutions of Rigid Macromolecules, *in* : "Molecular Electro-optics, part I", C.T. O'Konski, ed., M. Dekker, Inc., New York (1976).

44. C. Marion, B. Roux and M. Hanss, Orientational Interactions in Low-Concentration DNA Solutions, Biopolymers, 22:2353 (1983).

45. G. Weill and C. Hornick, Electric Polarizability of Rigid Polyelectrolytes, *in*: "Polyelectrolytes", E. Selegny, ed., D. Reidel, Dordrecht (1974).

46. N. Stellwagen, Electric Birefringence of Restriction Enzyme Fragments of DNA : Optical Factor and Electric Polarizability as a Function of Molecular Weight, Biopolymers, 20:399 (1981).

47. G. Mallet, J. Lematre, M. Leca and D. Vasilescu, Interaction of DNA with the Radioprotectors Cysteamine and WR-1065. A Kerr Effect Study, Applied Physics Comm., 7:57 (1987).

CYTOCHROME P-450, P-448 AND LIPID PEROXIDATION IN CELL DAMAGE

Oscar Torres-Alanís, Lourdes Garza-Ocañas, and
Alfredo Piñeyro-López

Depto. de Farmacología y Toxicología, Fac. de Medicina
Univ. Aut. de Nuevo León, Apartado Postal 146
Col. del Valle, N. L., México

CYTOCHROMES AND XENOBIOTICS

The human being is exposed to a great variety of extraneous substances (xenobiotics) which penetrate to the organism through the lungs, skin, or by ingestion either as compounds present in food or as medicaments.

Some xenobiotics are innocuous; however, others at the moment when they are metabolized or biotransformed by enzymatic or nonenzymatic processes cause reactions of a pharmacological or a toxic nature. The major role of biotransformation (detoxication) is to convert poorly excretable lipophilic compounds to more polar entities that can be readily excreted in the urine or in the bile. In the absence of metabolism such xenobiotics accumulate in the organism, increasing the potential for a toxic response. The xenobiotics that have high water/oil ratios (hydrophilic compounds) are rapidly excreted in urine.

The greater part of biotransformation reactions are carried out between the xenobiotic absorption and elimination (Fig. 1). Williams (1959) classified the pathways involved in phase I (non-synthetic) and phase II (synthetic).

Generally the reactions of phase I convert the original substance into a metabolite with more polarity, less liposolubility and is more excretable.

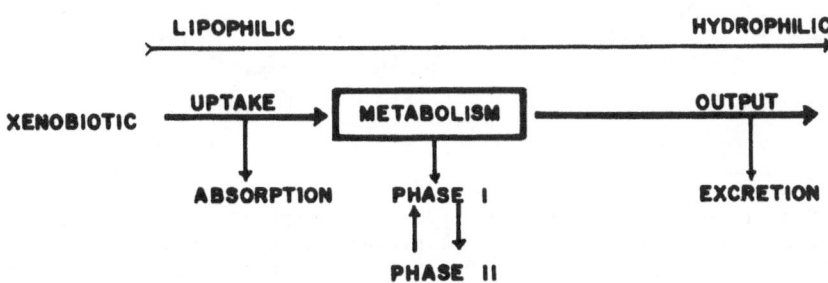

Fig. 1. Biotransformation of xenobiotics

Normally, one or more phase I reactions precede phase II metabolism. However, the phase II reactions can precede those of phase I.

The molecular oxidative changes in the organism are catalyzed by metabolizing enzymes which are located in the endoplasmic reticulum (ER), in cytoplasm and on rare occasions in mitochondria. ER consists of a membrane arranged as stacks of flattened cisternae, tubules, and vesicles. The membrane of an ER measures approximately 5-7 nm in thickness and part of its outer surface is studded with ribosomes which measure 15-20 nm in diameter. The portion of the ER with ribosomes is referred to as rough ER (RER), and the rest as smooth ER (SER). The cisternae of the ER are structurally continuous with the lumen of a nuclear envelope (Miyai 1984).

When ER is isolated by homogenization-differential centrifugation it is fragmented into small vesicles called __microsomes__. A major function of RER is protein synthesis; and those ribosomes bound to ER membranes synthesize the proteins secreted by the liver; whereas, free ribosomes synthesize those retained in the cell (Takagi and Ogata 1968; Hicks et al. 1969; Redman 1969; Rotschild et al. 1972; Schulze et al. 1975).

SER have many enzymes that are given the task of oxidative metabolism of xenobiotics in particular that contain the mixed-function oxidases or monooxygenases. Such an enzymatic system requires for its activity the presence of a reducing agent (NADPH) and molecular oxygen; in a characteristic reaction called monooxygenation an oxygen molecule is consumed by each substrate molecule, an oxygen atom appears in the product and the other in form of water.

In the oxidation-reduction process monooxygenase cytochrome P-450 intervenes. This system is so named because the reduced form of this pigment upon combination with CO has a soret maximum at about 450 nm. P-450 metabolizes several classes of endogenous compounds, including fatty acids, prostaglandins, steroids and vitamins. The P-450 system is membrane-bound and has two major components: cytochrome P-450, a hemoprotein; and NADPH cytochrome P-450 reductase, a flavoprotein that contains both FMN and FAD prosthetic groups. (for reviews, see Masters et al. 1980; Estabrook 1984; Levin et al. 1985).

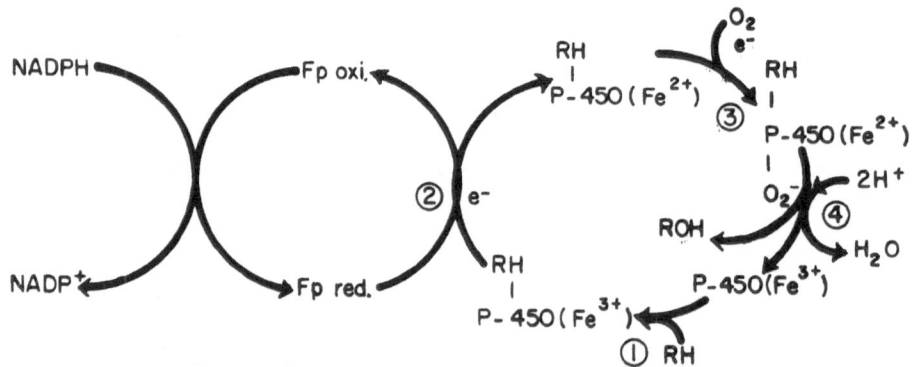

Fig. 2. Cytochrome P-450 system in the metabolism of drugs

P-450 contains one molecule of iron protoporphyrin IX as its prosthetic group. Normally, in microsomal preparations, this iron is in the ferric (Fe^{+3}) state. Oxydized P-450 first reacts with a molecule of substrate to form an enzyme-substrate complex (step 1) (Fig. 2). Next, an electron is donated to this complex from NADPH via NADPH-cytochrome P-450 reductase, which converts RH-P-450 (Fe^{+3}) to RH-P-450 (Fe^{+2}) (step 2). The P-450 (Fe^{+2})-substrate complex then reacts with molecular oxygen to form an oxycytochrome P-450 ternary complex (step 3). This complex may then accept one additional electron from NADPH via NADPH-P-450 reductase, or from NADH by cytochrome b_5 reductase to form the equivalent of a two-electron reduced complex of hemoprotein, oxygen, and substrate, which dissociates to yield an oxydized substrate, P-450 (Fe^{+3}) and water (step 4).

The majority of the substrates are highly liposoluble with low specificity. When they are administered in a repeated manner some of them have the capacity of inducing a cytochrome P-450 to elevate the velocity of synthesis or to reduce its degradation velocity. The induction process signals a relative increase in the rate of the novo synthesis or in the rate of activation of enzyme activity from pre-existing moieties, or in the rate of both, compared to the rate of breakdown. The induction produces a lowering of the inductor action transforming it into an inactive compound. but some xenobiotics are transformed by the metabolism of reactive intermediates, producing tissue toxicity.

Is the transformation to inactive compounds and reactive intermediates determined by P-450? or are there differences in induction, specificity and genetics?.

There are many forms of cytochrome P-450 that differ in substrate specificity and in the response to enzymatic inducers such as phenobarbital, 3-methylcholanthrene, β-naphthoflavone. In fact, an inducer does not increase the synthesis of only one isoenzyme; thus, polycyclic aromatic hydrocarbons induce by different extensions: at least two isoenzymes are induced in the rat and rabbit (Johnson and Müller-Eberhardt 1977), whereas phenobarbital is able to promote the synthesis of at least three isoenzymes. (Guengerich et al. 1982).

The two isoenzymes that have been the most carefully studied are cytochrome P-450 (cytochrome P-450$_b$ or LM$_2$, form II of the hepatic microsomes) which are induced by phenobarbital; and cytochrome P-448 (cytochrome P$_1$-450 or P-450$_c$ or LM$_4$) induced by 3-methylcholanthrene (polycyclic aromatic hydrocarbons).

Some differences between cytochrome P-450 and cytochrome P-448 are shown in Table 1; where SDS-gel electrophoresis reveals that all of the hemoprotein have a proteic component with a minimum molecular weight and a comparison of the substrate specificity shows that rat cytochrome P-448 has the highest activity for benzo(a)pyrene and zoxazolamine hydroxylation, and benzphetamine n-demethylation is elevated in rat cytochrome P-450 in comparison with other species. Other differences between cytochrome P-450 and P-448 have been reported (Thomas et al. 1984; Parke 1987): the cytochrome P-450 structure shows a sequence of aminoacids and nucleotides with high homology; but differ from those for P-448, the enzymatic induction by small molecule (phenobarbital) occurs in cytochrome P-450 while the carcinogenic-like 3-methylcholanthrene is formed by large flat molecules that induce enzymatically cytochrome P-448; the distribution of P-450 is mainly in liver and absent from placenta; however, cytochrome P-448 is extrahepatic, and is present in placenta. The substrate specificity is presumably determined by the kind of cytochrome that the oxidative site presents (Lu, A.Y.H. et al. 1971).

There are individual differences (age and sex) in enzymatic activity and differences in the susceptibility for induction that are genetically determined (Mbanefo et al. 1980; Nebert and Negishi 1984; Weber and Glawinski 1984). Gonzalez et al. (1988) have shown two distinct phenotypes of metabolizers in population studies of individuals given the antihypertensive drug debrisoquine, extensive metabolizers excretes 10-200 times more of the urinary metabolite 4-hydroxydebrisoquine than poor metabolizers, this form also is unable to metabolize many β-adrenergic blocking agents, antiarrythmics, antidepressants and dextromethorphan (Idle et al. 1979).

Table 1

Characteristics of cytochrome P-450 and P-448 in different animal species

PARAMETERS	RAT[a]		RABBIT[a]	MICE[a]	
	P-450	P-448	P-448	P-450A$_2$	P-450C$_2$
ELECTROPHORETIC PROFILE (m.w. daltons)	47000	53000	51000	50000	56000
SPECTRAL PROPERTIES specific content (nmoles/mg protein)	15-17	18-22	18-22	14-17	15-18
CO-DIFFERENCE SPECTRUM absorption maximum (nm)	450	447	447	451	450
SUBSTRATE SPECIFICITY[b]					
BENZPHETAMINE	52	2.5	1.0	5	8
BENZO(a)PYRENE	0.19	3.9	0.1	0.26	0.08
ETHOXY COUMARIN	4.1	56	N.D.[c]	3.7	0.3
ZOXAZOLAMINE	1.4	9.4	1.5	4.3	1.4

a– Cyt P-450 was purified from phenobarbital-pretreated rats and mice (B6D2F$_1$/1 mice) designated mouse Cyt P-450 A$_2$ and mouse Cyt P-450 C$_2$. Cyt P-448 was purified from 3-methylcholanthrene-pretreated rats and rabbits.
b– Enzyme activities expressed as nmoles/min/nmole hemoprotein.
c– N. D. = Not Determined.
   (Data from Levin et al. 1977).

Idle (1988), points out the importance of the individual differences which have varied as a natural consequence of evolution, giving to the organism a defense against xenobiotics, where several metabolic ways involve the super family of cytochrome P-450 which has eight known gene families, each consisting of 2-20 discrete cytochromes (Nebert and Gonzalez. 1987).

What is it that determines that the substrate is synthesized by cytochrome P-448? Can cytochrome P-450 under the influence of the inducer determine the formation of a specific hemoprotein and then start the synthesis?. Kahl et al. (1977), showed that synthesis can be due to a modification of the hemoprotein pre-existing in a post-translational step or that an alteration in the membrane environment of the cytochrome by the inducer might be considered. This is in accordance with Gielen et al., who concluded that the microsomal membrane and specifically the ligand environment of cytochrome P-450 is altered during polycyclic hydrocarbon treatment of mice genetically responsive to methylcholanthrene (Gielen et al. 1972).

The question of the different forms of induction and also the

activation or the production of metabolites were attempted to be clarified
in the excellent works of (Parke 1983; Ioannides et al. 1984), where the
cytochrome P-450 intervenes mainly in the oxidative detoxication (inactivation) while the cytochrome P-448, the flavoprotein monooxygenases, nonenzymic free radical hydroxylations and nonenzymic transoxygenations catalyzes
the activation of xenobiotics by biological oxidations. However, the form
in which the cytochromes activate and deactivate xenobiotics is uncertain
(Fig. 3), due to information about the behavior of molecules of substrates
not clearly specified as yet. Studies with the interaction of steroids and
liver microsomal P-450 have suggested that a unique spatial arrangement on
the plane of the heme of P-450 may exist whereby the site of a heme-bound
oxygen interaction with the substrate may be dictated by the immobilization
of the substrate molecule (Estabrook et al. 1975).

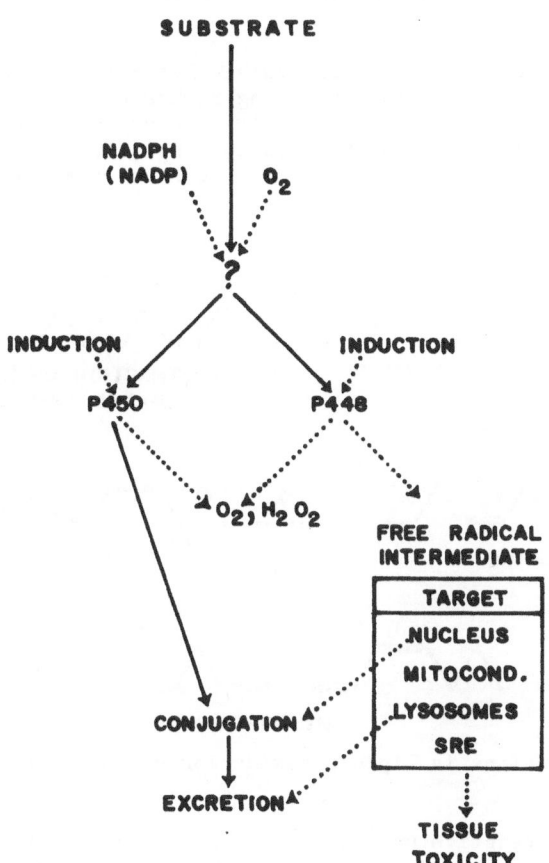

Fig. 3. Illustration of the important pathways of xenobiotics (substrate).

    Parke et al. (1987), have shown that P-448 has a preference for acti-
vating toxic chemical substances, and the reason for either an activation
or an oxidative detoxication is given because of the insertion of an oxygen
in the 'conformationally unhindered' position of the substrate molecule,
by cytochrome P-450, yielding metabolites, followed by rapid conjugation
or detoxication, by cytochrome P-448; the insertion of oxygen in a 'conform-

ationally hindered' position, results in the activation of the chemical to form reactive intermediates which react with intracellular macromolecules producing toxic damage, this process involves two different ways of oxygenation, catalyzed by two different families of cytochromes with different active sites.

LIPID PEROXIDATION

Not only cytochrome P-448 may be related to hydroxylation of benzo(a)-pyrene, metabolic activation and carcinogenesis; it is also related to the conversion of bromobenzene and CCl_4 and to toxic metabolites such as cytochrome P-450. Xenobiotics that undergo one-electron enzymatic redox cycling can activate oxygen-to-oxygen radicals, forming a superoxide-anion radical ($O_2^-\cdot$), hydrogen peroxide (H_2O_2), hydroxile radical ($OH\cdot$) and singlet oxygen ($^1O_2$). It is thought there is a relation between the activation of oxygen and its toxicity if the protective mechanisms of cells are overwhelmed.

The superoxide radicals are responsible for a limited number of toxic effects because a superoxide is a reducing rather than an oxidizing agent. $OH\cdot$ is an extremely potent oxidant, resulting from the reaction of $O_2^-\cdot$ with the two-electron reduction product of oxygen, H_2O_2, the reaction is catalyzed in biological systems in the presence of traces of metals (Fe, Cu).

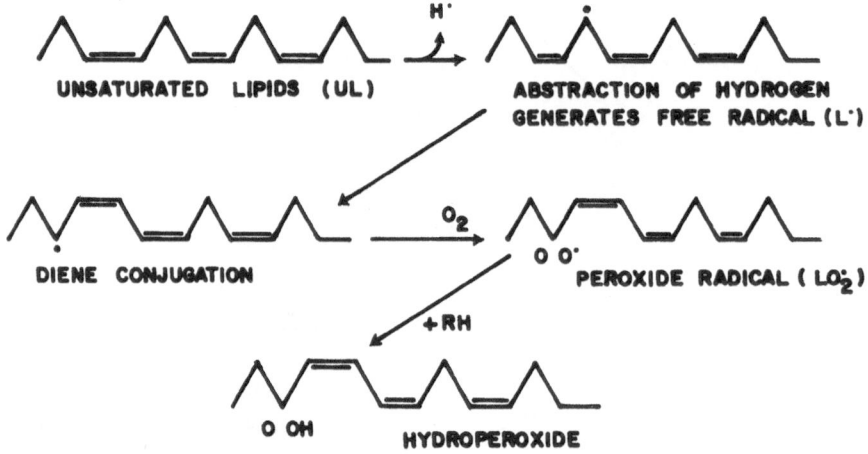

Fig. 4. Reactions in lipid peroxidation of unsaturated lipids

$OH\cdot$ is highly reactive and causes damage to lipids, proteins and acids. Protein damage by oxygen radicals leads to amino acid oxidation resulting in comformational changes and enzymic inactivation, the radical $OH\cdot$ damages DNA and leads to cytotoxicity, mutagenicity and carcinogenicity (Bus et al. 1984). $OH\cdot$ reacts with unsaturated lipids of the membrane (UL), forming radicals, initiating the process of lipid peroxidation (LP) (Fig. 4). The reaction is effected by steps: generation of lipid radicals are formed and termination reactions in which lipid free radicals ($L\cdot$) or a lipid peroxy radicals (LO_2^-) forming radical products. The production of lipid peroxidation results in the loss of functional integrity of the membrane, leading to disturbances in cell function and cell death.

Different quantities of polyunsaturated lipids distributed in phospholipids of the cellular membrane vary in tissue and different subcellular membranes in the same tissue due to the process of lipid peroxidation which will vary in each membrane and also will depend on the concentration of metal-catalytics: iron, etc., and on the presence of antioxidants such as superoxide dismutase, catalase, ascorbic acid, reduced glutathione (GSH) and lipid-soluble antioxidants (vitamin E) that protect from injury, preventing the formation of the potent oxidant, OH·. The balance between these factors (UL, metal catalyzers, antioxidants, and OH·) determine the rate of lipid peroxidation in vitro. If lipid peroxidation is considered as a result of pathological processes such as vitamin E deficiency, intoxication by CCl_4 or ethanol, damaging radiation, hyperbaric effects and aging (Barber et al. 1967), as well as many chemicals which are able to generate free radicals such as $O_2^-·$, OH·, and promote lipid peroxidation (see table 2).

Lipid peroxidation induced by adriamycin is used in tumor therapy because it is believed that the antitumor effects of anthracyclines are due to oxygen radical formation occuring during redox cycling in tumor cells (Mason et al. 1982). NADPH-cytochrome P-450 reductase is the principal enzyme that intervenes in the redox catalyzer cycle in drugs but not only the enzyme favors oxidative activation, but the quantity of oxygen available for oxidative reaction is important for this event. The procedures for the detection and quantification of lipid peroxidation in biological material are difficult. Thiobarbituric acid (TBA) reaction is the most widely used as an indirect method of the measurement of peroxidation in vitro (Plaa et al. 1976) and now the measurement of expired hydrocarbons such as ethane or pentane from beta-scission of lipid hydroperoxides is employed as an index of lipid peroxidation in vivo (Wendel et al. 1981; Clemens et al. 1985).

TABLE 2

Some chemicals that are able to generate intermediate free radicals ($O_2^-·$, OH·, CCl_3^- radicals, others) and may be acting in peroxidation reactions.

ANTHRACYCLINES	QUINONES
DAUNOMYCIN	MITOMYCIN C
DOXORUBICIN	AZQ
ADRIAMYCIN[1]	ETOPOSIDE (VP-16)
EPIRUBICIN	1,4-NAPHTOQUINONE
ACLACINOMYCIN A	BENZO(a)PYRENE-3,6-QUINONE
ANTHRACENEDIONES	β-LAPACHONE
MITOXANTRONE	STREPTONIGRIN
ALLOXAN	PARACETAMOL (ACETAMINOPHEN)
ADRIAMYCIN-Fe-COMPLEX	PARAQUAT
BLEOMYCIN-Fe-COMPLEX	AROMATIC NITRO COMPOUNDS
ISONIAZID	NITROFURANTOIN
IPRONIAZID	NIFURTIMOX
CARBON TETRACHLORIDE	BENZNIDAZOLE
CARBON DISULFIDE	METRONIDAZOLE
	NITRAZEPAM
	N-NITROSODIMETHYLAMINE

1- Commercial name of doxorubicin.

The truly known mechanisms of toxicity of free radicals in lipid per-oxidation are not well understood. If lipid peroxidation causes injury to the erythrocyte, liver or lungs we are shown only the result of peroxidation of unsaturated lipids. However nothing is known about the final event in the redox cycling process. Free oxygen radicals in this redox cycling are

dangerous but the intervention of the enzymatic system(s) are critical.
Toxicity is only an effect of a chemical agent, a dosis (concentration multiplied by time), the exposed biological species, the susceptibility for the
detoxication or activation genetically determined and the environment are
important factors in order to evaluate the toxicity or the toxicokinetics
of chemicals.

In the human being the state of health has an very important role; some
illnesses such as rheumatoid arthritis are able to potentialize the effects
of drugs utilized for the treatment of the disease state and produce adverse
effects (generation of free radicals) with alteration of metabolism (cytochrome P-450) (Parke and Parke 1987). Under normal conditions of health,
the existence of lipid peroxidation in living tissues suggests the possibility that the process plays a physiological role. The biosynthesis of
prostaglandins (Slater 1975) or the renewal of subcellular particles in
long-living cells are bases that may be considered for a possible physiological effect intervention of lipid peroxidation. If the effects of oxygen
radicals can be beneficial or toxic, it will depend on the above mentioned
factors and that protective mechanisms are overwhelmed.

Acknowledgements

The following are to be thanked for their contributions in this
investigation: A. Guajardo for her typescript, P. Bravo-Pérez for the
drawings and R.M. Chandler, for stylistic suggestions.

REFERENCES

Barber, A.L., and Bernheim, F., 1967, Lipid Peroxidation: Its measurement,
    occurrence, and significance in animal tissues, Gerontological Advan-
    ces., 2:355.
Bus, J.S., and Gibson, J.E., 1984, Role of activation oxygen in chemical
    toxicity, in: "Drug Metabolism and Drug Toxicity", J.R. Mitchell and
    M.G. Horning, eds., Raven Press, New York.
Clemens, M.R., Einsele, H., Remmer, H., Waller, H.D., 1985, Decreased sus-
    ceptibility of red blood cells to peroxidation in patients with
    alcoholic liver cirrhosis, Clin. Chim. Acta., 145:238.
Estabrook, R.W., Martinez-Zedillo, G., Young, S., Peterson, J.A., and Mc
    Carthy, J., 1975, The interaction of steroids with liver microsomal
    membrane P-450: A general hypothesis, J. Steroid. Biochem., 6:419.
Estabrook, R.W., 1984, Cytochrome P-450 and oxygenation reactions: a status
    report, in: "Drug metabolism and Drug Toxicity", J.R. Mitchell, and
    M.G. Horning, ed., Raven Press, New York.
Gielen, J.E., Goujon, F.M., and Nebert, D.W., 1972, Genetic regulation of
    arly hydrocarbon induction, II. Simple mendelian expression in mouse
    tissues in vivo, J. of Biol. Chem., 247:1125.
Gonzalez, F.J., Skoda, R.C., Kimura, S., Umeno, M., Zanger, U.M., Nebert,
    D.W., Gelboin, H.V., Hardwick, J.P., and Meyer, U.A., 1988, Charac-
    terization of the common genetic defect in humans deficient in debri-
    soquine metabolism, Nature., 331:441.
Guengerich, F.P., Dannan, G.A., Wright, S.T., Martin, M.V., and Kaminisky,
    L.S., 1982, Purification and characterization of rat liver microsomal
    cytochromes P-450: electrophoretic, spectral, catalytic, and immu-
    nological properties and inducibility of eight isozymes isolated
    from rats treated with phenobarbital of β-naphtoflavone, Biochemis-
    try., 21:6019.
Hicks, S., Drysdale, J.W., and Monro, H.N., 1969, Preferential synthesis of
    ferritin and albumin by different populations of liver polysomes,
    Science., 164:584.
Idle, J.R., and Smith, R.L., 1979, Drug Metab. Rev., 9:301.
Idle, J.R., 1988, Enigmatic variations, Nature., 331:391.

Ioannides, C., Lum, P.Y., Parke, D.V., 1984, Cytochrome P-448 and the activation of toxic chemicals and carcinogens, Xenobiotica., 14:119.

Johnson, E.F., and Müller-Eberhard, U., 1977, Resolution of two forms of cytochrome P-450 from liver microsomes of rabbits treated with 2,3, 7,8-tetrachloro dibenzo-p-dioxin, J. Biol. Chem., 252:2839.

Kahl, F.G., Zimmer, B., Galinsky, T., Jonen, H.G., and Kahl, R., 1977. Induction of cytochrome P-448 and 3-methylcholanthrene in the rat during inhibition of protein synthesis in vivo, in: "Microsomes and Drug Oxidation", Ullrich, I. Roots, A.G. Hildebrandt, R.W. Estabrook, A.H. Conney, eds., Pergamon Press, Oxford.

Levin, W., Ryan, D., Huvang, T.A., Kawalek, J., Thomas, P.E., West, S.B., and H. Lu. A.Y., 1977, in: "Microsomes and Drug Oxidation", V. Ullrich, I. Roots, A.G. Hildebrandt, R.W. Estabrook, A.H. Conney, ess., Pergamon Press, Oxford.

Levin, W., Thomas, P.E., Reik, L.Y., Ryan, D.E., Bendiera, S., Haniv, M., and Shively, J.E., 1985, Immunochemical and structural characterization of rat hepatic cytochrome P-450, in: "Microsomes and drug oxidation", A.R. Bobbis, J. Caldwell, F. De Matteis, and C.R. Elcombe, ed., Taylor and Francis, London.

Lewis, D.F.V., Ioannides, C., Parke, D.V., 1986, Molecular dimensions of the substrate binding site of cytochrome P-448, Biochem Pharmacol., 35:2179.

Lu, A.Y.H., Kuntzman, R., West, S., and Conney, A.H., 1971, Biochem. Res. Commun., 42:1200.

Lu. A.Y.H., Kuntzman, R., West, S., Jabobson, M., Conney, A.H., 1972, Reconstitued liver microsomal enzyme system that hydroxylates drugs, other foreign compounds and endogenous substrates. II Role of the cytochrome P-450 and P-448 fractions in drug and steroid hydroxylations, J. Biol. Chem., 247:1727.

Mason, R.P., 1982, in: "Free Radicals in Biology", Vol. V., W.A. Pryor, ed., Academic Press, New York.

Masters, B.S.S., 1980, The role of NADPH-cytochrome C (P-450) reductase in detoxication, in: "Enzymatic basis of detoxication", W.B. Jakoby, ed., Academic Press, New York.

Mbanefo, C., Bababunmi, E.A., Mahgoub, A., Sloan, T.P., Idle, J.R., and Smith, R.L., 1980, A study of the debrisoquine hydroxylation polymorphism in a Nigerian population, Xenobiotica, 10:811.

Miyai, K., 1984, Ultrastructural basis for toxic liver injury, in: "Toxic Injury of the Liver", E. Farber and M.M. Fischer, eds., Marcel Dekker, New York.

Nebert, D.W., and Negishi, M., 1984, Environmental and genetic factors influencing drug metabolism and toxicity, in: "Drug metabolism and drug toxicity", J.R. Mitchell and M.G. Horning, ed., Raven Press, New York.

Nebert, D.A., and Gonzalez, F.Y.A., 1987, Rev. Biochem., 56:945.

Parke, A.L., Parke. D.V., 1987, Genetic and environmental aspects of drug metabolism relevant to side-effects in arthritic drug metabolism relevant to side-effects in arthritic disease, in: Proceedings of the Second Symposium on the Adverse Effects of Non Steroidal anti-inflammatory Drugs", K. Rainsford, ed., CRC Press Inc.

Parke, D.V., 1982, Mechanisms of chemical toxicity: A unifying hypothesis, Reg Toxicol Pharmacol., 2:267.

Parke, D.V., 1983, Survey of drug-metabolizing enzymes, Biochem. Soc. Trans. II: 457.

Parke, D.V., 1985, The role of cytochrome P-450 in the metabolism of pollutants, Marine Environ Res., 17:97.

Parke, D.V., 1987, Activation mechanisms to chemical toxicity, Arch. Toxicol., 60:5.

Phillipson, C.E., Godden, P.M.M, Lum, P.Y., Ioannides, C., Parke, D.V., 1984, Determination of cytochrome P-448 activity in biological tissues, Biochem J., 221:81.

Plaa, L.G., Witschi, H., 1976, Chemicals drugs, and Lipid peroxidation, in: "Ann Review Pharm and Toxicol", H.W. Elliott, R. George and R. Okun, eds., Annual Review Inc, Palo Alto, California.

Redman, C.M., 1969, Biosynthesis of serum proteins and ferritin by free and attached ribosomes of rat liver, J. Biol. Chem., 244:4308.

Rothschild, M.A., Oratz, M., and Schreiber, S.S., 1972, Albumin metabolism, in: "Progress in Liver Diseases", Vol. 4, H. Popper and F. Shaffner, eds., Grune and Stratton, New York.

Schulze, H.U., and Staudinger, Hj., 1975, Struktur und funktion des endoplasmatischem retikulums, Naturwissenschaften., 62:331.

Slater, T.F., 1975, in: "International Meeting of Recent advances in Biochemical Pathology Toxic Liver Injury", p. 99 (Eds. M.V. Dianzini, G. Ugazio and L.M. Sena). Minerva Medica, Torino.

Takagi, M., and Ogata, K., 1968, Direct evidence for albumin biosynthesis by membrane bound polysomes in rat liver, Biochem. Biophys. Res. Comm., 33:55.

Thomas, P.E., Reidy, J., Reik, L.M., Ryan, D.E., Koop, D.R., Levin, W., 1984, Use of monoclonal antibody probes against rat hepatic cytochromes P-450 and P-450b to detect immunochemically related isozymes in liver microsomes from different species, Arch. Biochem. Biophys., 235:239.

Weber, W.W., and Glowiniski, I.B., 1980, Acetylation, in: "Enzymatic basis of detoxication", W.B. Jakoby , ed., Academic Press, New York.

Wendel, A., and Feverstein, S., 1981, Drug-induced lipid peroxidation in mice I, Biochemical Pharmacol., 30:2513.

Williams, R.T., 1959, "Detoxication mechanisms", 2nd ed, Chapman and Hall, Ltd., London.

OXYGEN IS A DANGEROUS FRIEND

Lester Packer

Department of Physiology/Anatomy
University of California
Berkeley, CA 94720

INTRODUCTION

We all realize that oxygen is a dangerous friend. Oxygen is something that we cannot live without, but it is also a substance that we cannot live with indefinately. This chapter will review two aspects of this subject. First, an overview will be given of some of the points relevant to the idea that oxygen is toxic by pointing out the underlying chemical and biological mechanisms that are important in the toxic effects of oxygen in biological systems. Secondly, some experimental approaches that have been undertaken in our laboratory, using the concept of studying the increased metabolism associated with exercise in order to follow the generation of oxygen free radicals (oxidants) and their consequences, in terms of changes in antioxidants will be described. Antioxidants are major cellular defense mechanisms that are important to preventing molecular, cellular and tissue damage due to free radicals and hence their understanding is pertinent to relating "Cell function and Disease."

Oxygen is toxic

This idea gained considerable credibility from work of physiologists more than four decades ago, who noted an inverse relationship between species lifespan and metabolic rate, and a direct relationship between body or brain weight and the mammalian species lifespan. This concept of the inverse relationship of metabolic rate to lifespan has been tested with cold-blooded animals such as fish or insects. For example, with insects, by modifying the environmental temperature, their metabolic rate can be set to operate at different levels. As the temperature is raised from 15 to about 30 degrees centigrade, the metabolic rate (oxygen consumption) progressively increases but the lifespan of the insects decreases proportionally. What is extremely interesting about such experiments is that the total amount of oxygen which the insects consume is unaffected regardless of the lifespan of the insects (or the metabolic rate). It is as if a certain amount of oxygen is allocated to individual insects during their lifespan and the individual can live its life either fast (by

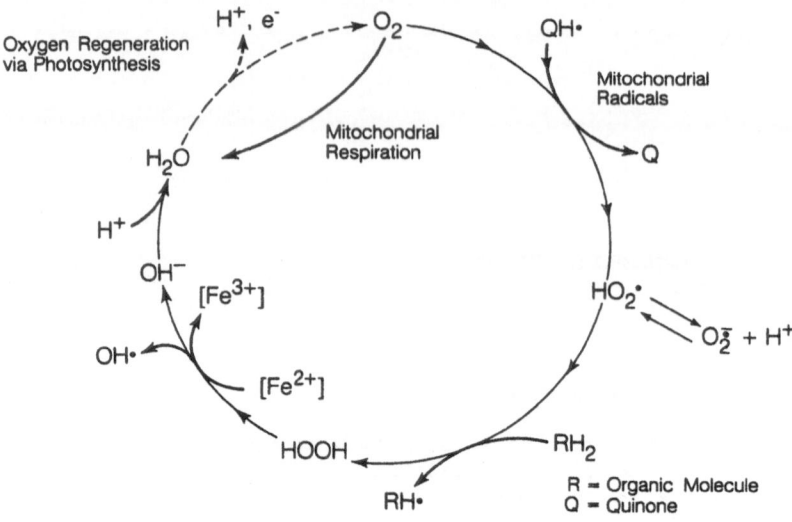

Fig. 1. The Oxygen Cycle. Reactions in the cell can lead to the
generation of active oxygen species. In this example,
mitochondrial ubiquinones of the electron transport chain in
the semiquinone state react with oxygen to form superoxide.
Superoxide radicals may exist either as the anion or as a weak
acid. Superoxide dismutase or water soluble reductants convert
superoxide to hydrogen peroxide, and the latter can be removed
by hydroperoxide removing enzymes such as glutathione
peroxidase. Alternatively, Fe^{+2} catalyzed Fenton reactions
can decompose hydroperoxides to form reactive oxygen species
such as the hydroxyl radicals. Completion of the reduction
cycle and formation of water, by the respiratory chain
involves a four electron transfer reaction by cytochrome
oxidase.

flying) or slow (by walking). The slowly metabolizing individuals can
benefit from a longer lifespan. Whether such experiments are important
for animals or human lifespan is still unknown.

How then could oxygen be toxic? We know that the consumption of
oxygen at the cellular level is due to electron transport reactions,
such as the respiratory electron transport system in the mitochondria,
where oxygen is reduced to water at the expense of the oxidation of
carbohydrate substrates (respiration) (cf. figure 1). One problem,
however, is that during the substrate oxidation, electrons can be
donated by the mitochondrial electron transport chain ubiquinone
components (ubiquinone semiquinone) to oxygen, generating a superoxide
radical anion. Since all aerobic cells contain the enzyme superoxide
dismutase or other substances such as ascorbic acid (Vitamin C) which
reduce superoxide to hydrogen peroxide, an accumulation of
hydroperoxides occurs. Although hydroperoxides (like hydrogen
peroxide) themselves are stable molecules, if transition metals are
present in their free reactive form in the cell, (e.g. Fe^{+2} or Cu^{+2}),
hydroperoxides can be decomposed to generate (by the Fenton reaction)
either hydroxyl radicals from hydrogen peroxide or alkoxyl radicals
from organic hydroperoxides. These radicals are readily reactive and
can damage biological macromolecules, including DNA, enzymes, and

structural components of cells. Such reactions can be causal to and accelerate aging and cancer.

Indeed, let us consider what are the effects of ionizing radiation. Ionizing radiation is known to accelerate cancer and aging. The main effects of ionizing radiation in biological systems are reactions with water. This is because water is the major component in intracellular and extracellular areas of the body to which ionizing radiation is exposed. Water can be decomposed by ionizing radiation to produce two main free radicals: a protonated electron which, in the presence of oxygen, will immediately form the superoxide radical anion, and a hydroxyl radical. So ionizing radiation works in water to produce the same two major free radical species that are also produced by the oxygen cycle during metabolism. One of my colleagues, Professor Bruce Ames, has often been quoted to say that "metabolism is like being irradiated all the time."

It is interesting to note that oxygen itself is also a free radical. In fact, it is a biradical. However, because two oxygen atoms are joined together in an O_2 molecule, and since the free electrons in O_2 spin in the same direction, oxygen is not a very reactive radical. However, the superoxide radical is somewhat more reactive and can initiate lipid peroxidation first forming the lipid peroxy radical. Oxygen can also initiate lipid peroxidation if a lipid first becomes a radical, after which it will more readily react with oxygen. Lipids may become radicals, since the unsaturated fatty acids which they contain are readily susceptible to electron addition or hydrogen extraction by electron transport reactions in the membranes in which they are located. Once lipid peroxy radicals have been formed, they can undergo chain reactions in membranes to form other radicals, transfering their radicals to proteins and nucleic acids to cause oxidative damage to these molecules and cross-linking reactions, reducing their biological activity. The removal of superoxide by superoxide dismutase would reduce the chances of this radical reacting with lipids to form lipid peroxy radicals. Indeed, Cutler's laboratory has demonstrated that species lifespan is proportional liver superoxide dismutase activity. This suggests that antioxidant enzymes are important in protecting against oxygen toxicity and aging.

It is well recognized that there is a correlation between inflammation and cancer. Here, the scenario is that normal tissues are damaged by the system in leukocytes which is specifically triggered to cause a respiratory burst of superoxide radicals by a system present in the plasma membranes. This system forms hydrogen peroxide which then can form other toxic products, either by metal catalyzed Fenton reactions or by the enzyme myeloperoxidase, which forms hypochlorus acid and subsequently more potent oxidizing species such as mono and di-substituted chloramines. This is a cellular mechanism normally used to remove foreign infectious agents such as bacteria or to target tumor cells, but in the process, normal cells may also be damaged. The high concentration of leukocytes in the lungs of people with asbestos fibers due to continuous recruitment of neutrophils, is an example where continuing inflammation exists, eventually leading to cancer. In the multi-step model system for cancer studied extensively in mouse skin, it is recognized that chemical carcinogens often involve the generation of reactive oxygen species which in turn may damage DNA and contribute to the irritation and inflammation that eventually lead to the progression of precancerous cells into rapidly growing cancer cells.

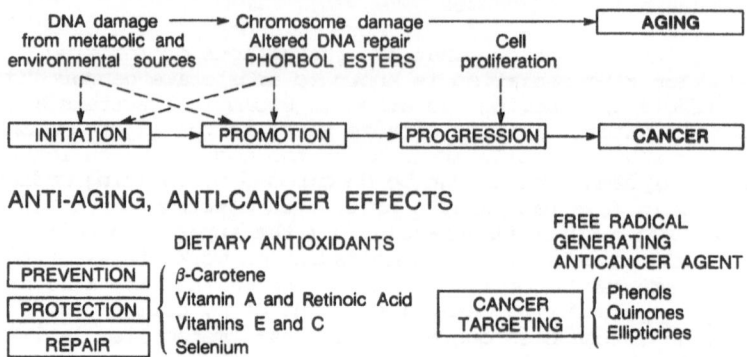

Fig. 2. Oxygen radicals a double-edged Sword. Active oxygen species have been implicated in the multi-step process of cancer and in the aging process, although active oxygen species may not be exclusive in this regard. Dietary antioxidants exhibit anti-cancer effects and effective anti-cancer agents are quinones which generate first superoxide, and then initiate a cascade of active oxygen species which serve to damage and target cancer.

These considerations immediately suggest that antioxidants, which protect against the propagation of free radical reactions, should have a beneficial effect on health and aging (cf. figure 2). Indeed, recent epidemiological studies have shown a correlation between lung cancer cases and number of cigarettes smoked, but when people who consume a higher concentration of yellow or green vegetables, or with people who have elevated levels of blood vitamin E and beta-carotene, there is a lower incidence of lung cancer.

So in the lipid and water soluble portions of cells, including both vitamin E and vitamin C respectively, anti-oxidants may have important anti-cancer and anti-aging effects. Vitamin E, for example, which reacts with the lipid peroxy-radical to form an organic peroxide, becomes in the process a radical itself. However, vitamin E (the alpha-tocopheroxyl radical) is not very dangerous because the free electron is delocalized around the aromatic ring structure of vitamin E. Vitamin E may decompose as a result of becoming a radical or it may be regenerated in biological systems, such as by vitamin C. When vitamin C reacts to quench free radicals, vitamin C also becomes a slowly reacting free radical (the semi-ascorbate radical) which readily reacts with itself to regenerate reduced and oxidized ascorbic acid molecules.

There are many anti-oxidants in the cell besides superoxide dismutase, vitamin E and vitamin C. Table I lists some of the common antioxidants in cells and extracellular fluids. Many of these substances or enzymes react to remove substances which may initiate free radical reactions (like hydroperoxides) rather than directly removing free radicals themselves as in the case of superoxide dismutase. Other substances are effective as antioxidants because they

Table I. Antioxidants

In Cellular Systems Membranes:	In Extracellular Fluids:
Vitamin E	Caeruloplasmin
Ubiquinones	Transferrin
Aqueous Compartments:	Lactoferrin
Ferritin	Albumin
Vitamin C	Haptoglobin
Glutathione	Superoxide dismutase (trace)
Superoxide dismutase	
Catalase	Uric acid
Glutathione peroxidase	Vitamin C
Glutathione-s-transferase	

can sequester (bind) and render metal ions unreactive for fenton
reactions that can lead to the formation of free radicals. Examples of
course are iron-binding substances such as transferrin ferritin,
lactoferrin. So we have many antioxidants, water soluble substances
and enzymes together with lipid soluble substances that should reduce
free radical damage in aging in pathology. Thus, there is evidence
that anti-cancer effects can be exhibited by dietary antioxidants. An
important unanswered question is to know what the right combination of
dietary antioxidants is to optimize anti-cancer or anti-aging effects.
Another strategy is to use the ideas that we learn about from these
reactions of oxidants and antioxidants to design drugs for example, to
specifically target the destruction of parasites or cancer cells.
Thus, we know that quinones are among most effective reagents that are
anti-parasitic or anti-cancer.

Oxidants and Antioxidants in Exercise

Most people believe that a regular program of exercise is good for
you and that regular exercise with proper training programs will improve
the performance of the cardiovascular system and the reduction of body fat
levels. These changes that have been widely documented as being significant
in the reduction of cardiovascular disease and in the reduction of age-
related diseases. However, a concept which has only recently been introduced
is that the increased metabolism associated with exercise may lead to
the production of oxidants having potentially harmful effects (Davies
et al. 1982, Packer, 1986 A, B, Packer, 1987). Some of our experiments
have shown that even mild exercise leads to dramatic changes in tissue

oxidative stress. I could paraphrase mild exercise as "the amount of exercise that it takes to stir a margarita or a martini cocktail." Together with my colleagues in the laboratory of Professor G.A. Brooks, we have performed experiments involving mild exercise in humans and studied the response of the glutathione antioxidant system. The glutathione system is centrally important in metabolism because much of the cells' reducing power is utilized to maintain large amounts of glutathione present in aerobic cells (between 1-10 millimolar) in its reduced condition. Metabolism channels its reducing power to keep glutathione more than 95% in its reduced form during normal metabolism. Reduced glutathione is important because it can react chemically or with the enzyme glutathione peroxidase to decompose organic or inorganic hydroperoxides, avoiding the possibility that they may become involved in generating free radicals that can cause tissue damage.

To study the effects of mild exercise, we recruited 8 human volunteers who were already well-trained individuals (Gohil et al. 1988). These subjects were set to perform at 65% of their VO_2 max on a bicycle ergometer. During the initial 15 minutes of exercise, their blood glutathione level dramatically changed, up to 60% of the reduced glutathione was oxidized. This more oxidized steady state level of glutathione was maintained for up to 90 minutes of exercise after which exercise was ceased. During the next few hours, the blood glutathione levels returned to normal. However, an interesting finding was that over a 3 day period, the total amount of glutathione in the blood underwent a net increase. This suggests that a single bout of exercise stimulates blood glutathione levels which may afford protection against a second or successive bouts of exercise. The oxidation of glutathione indicates that oxidants must have been produced in exercise, because the reaction of glutathione with hydryoperoxides causes its oxidation.

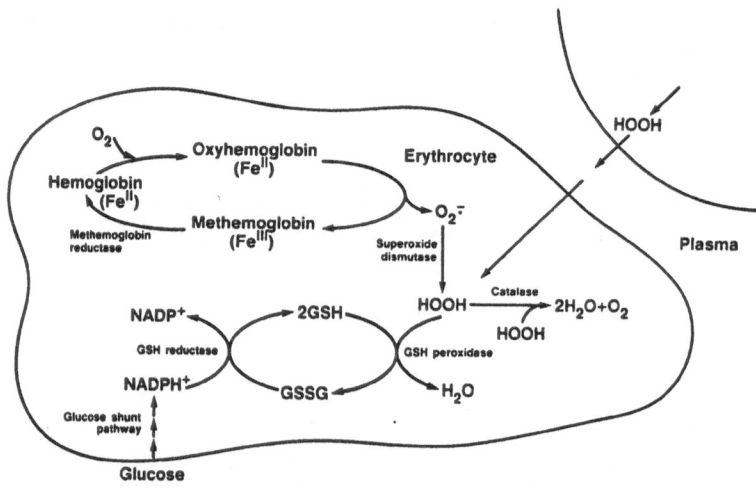

Fig. 3. The oxidative stress of exercise leads to large oxidation changes in the glutathione system of human blood. The diagram illustrates that the sources of active oxygen species, resulting in glutathione oxidation could arise endogenously in red bloods by the formation of methhemoglobin or alternatively, hydroperoxides produced in surrounding tissues may leak and enter red blood cells to also serve as the substrate for glutathione oxidation.

Since hydroperoxides arise from superoxide, it follows that electron transport reactions have generated superoxide, hydrogen peroxide, and oxidized glutathione, in that sequence. The steps involve the formation of oxidants (superoxide) and a change in the redox status of the major primary preventative cell antioxidant. The source of the superoxide and hydrogen peroxide have not yet, however, been definitely identified (cf. figure 3). The glutatione is certainly in the blood cell fraction and not in the plasma, but whether the reactions that form superoxide in the red blood cells (such as methemoglobin formation) are the main source of superoxide or whether hydrogen peroxide comes from surrounding tissues and enters into the circulating red blood cells is not yet known. Nevertheless, it is clear that a significant "oxidative stress" is generated even during mild exercise.

Vitamin E Effects

We have also been interested in the responses of vitamin E, the cells major chain breaking antioxidant, during exercise and exercise training (Quintanilha and Packer, 1982; Aikawa et al. 1984; Gohil et al. 1986; Packer, 1986). In a series of experiments, we have observed that if animals are endurance-trained, a large increase in the biogenesis of mitochondria (Davies et al. 1981; Kirkwook et al. 1987) in red skeletal muscles and a decrease in the tissue vitamin E content of animals is observed. These animals were given a dietary amount of vitamin E that is considered to be normal for rats (40 international units). In these animals, the increased amounts of mitochondria could be readily identified in the red muscles by specific increases in enzymes associated with the inner mitochondrial membrane, such as cytochrome-c reductase and by increased levels of ubiquinone and ubiquinols 9 and 10, specific quinones located in the inner mitochondrial membrane.

By comparing the changes in tissue vitamin E content to the ubiquinone content, very spectacular decreases in the vitamin E to ubiquinone ratio were calculated, following exercise endurance training (Gohil et al. 1987, Lang et al. 1987). This suggests that mitochondrial vitamin E is reduced by exercise training. This can be a sensitive test to use in other situations where oxidative stress may occur. The vitamin E status may change with animal performance (metabolism) or exposure to environmental toxic substances or dietary substances that result in the production of oxidants which deplete the vitamin E pool. The results suggest that perhaps higher dietary vitamin E may be required during regular exercise. Although it cannot be proven from these experiments, the circumstantial evidence is strong that oxidants and free radicals are produced and that free radical reactions are going on during the exercise period. In other studies with animals, it has also been found that glutatione is oxidized in red skeletal muscles accompanying exercise, particularly exercise to exhaustion, which causes large increases in oxidation of reduced glutathione not only in muscle tissues but in other tissues such as liver.

Since antioxidants are important in preventing certain diseases and aging, it is possible that increased physical activity could accelerate aging, at least of certain tissues that do not experience a positive adaptive increase as a result of exercise. Also, animals or people might be susceptible to toxic effects of substances that generate free radicals such as certain drugs that are metabolized by the liver. If glutathione in the liver become largely oxidized by exercise, the liver may not be able to remove active oxygen species that arise during drug detoxification, an idea that has been advanced by colleagues A. Quintanilha and M. Smith (personal communication).

Antioxidant Interactions

Water soluble and lipid soluble antioxidants are effective mainly by virtue of the fact that they are reducing agents which form relatively non-reactive radical species. They should in principle be able to interact with each other either directly or indirectly through metabolic processes involving the generation of reducing power. These inter-relationships involve a central role for superoxide dismutase, glutathione peroxidase (Gohil et al. 1988; Lang et al. 1987), and substances such as glutathione and vitamin C (ascorbic acid) and vitamin E play in these processes.

Vitamin E may be regenerated by some of these metabolic processes such as interaction with vitamin C or enzymes which have the characteristic of being able to reduce free radicals (cf Packer et al. 1986; Packer, 1987). This concept which has been designated "free radical reductase," could be important to investigate in the future to determine its potential significance for maintaining fully active antioxidant status. We have begun some investigations to establish whether free radical reductase activity exists. We seek to determine the importance of free radical reductase activity in the scavenging of persistent free radicals in cells and tissues with the aim of eventually identifying their importance in prevention of disease and the slowing of aging.

Concluding Remarks

Prevention is the most challenging field of medicine with which we are confronted today. In preventive medicine, we assess a certain degree of risk in adopting a certain lifestyle or selecting a particular dietary nutrient or antioxidant regime with an anticipated benefit for aging or age-related diseases. So it is a risk/benefit situation for which the answers will not be known for many years. Therefore, one of the important horizons for molecular medicine is to determine mechanisms of oxidant formation and then design antioxidant therapy to improve health and slow aging. In the future, these studies will be a central focus of the activities of many research laboratories seeking to advance our understanding of cell function and disease by developing molecular and cellular model systems for studying oxidants and antioxidants aiming to lend more precision to the rapidly developing field of prevention in medicine.

Acknowledgments

Research reported in this article carried out in our laboratory received support from the National Foundation for Cancer Research, Hoffmann-LaRoche Inc., and the National Institute of Health (CA47597).

References

Aikawa, K.M., Quintanilha, A.T., de Lumen, B., Brooks, G.A. and Packer, L., 1984, Exercise Endurance Training Alters Vitamin E Tissue Levels and Red Blood Cell Hemolysis in Rodents, Bioscience Reports, 4:253-257.

Davies, K.J.A., Packer, L. and Brooks, G.A., 1981, Biochemical Adaptation of Mitochondria, Muscle, and Whole-Animal Respiration to Endurance Training, Arch. Biochem. Biophys., 209:538-553.

Davies, K.J.A., Quintanilha, A.T., Brooks, G.A., and Packer, L., 1982 Free Radicals and Tissue Damage Produced by Exercise, Biochem. Biophys. Res. Commun., 107:1198-1205.

Quintanilha, A.T. and Packer, L., 1983, Vitamin E, Physical Exercise and Tissue Oxidative Damage, in: "Biology of Vitamin E, Proceedings of a Ciba Foundation Symposium, 7-10 March, 1983," R. Porter and J. Whelan, eds., London, Pitman Medical Ltd.

Gohil, K., Packer, L., De Lumen, B., Brooks, G.A., and Terblanche, S.E., 1986, Vitamin E Deficiency and Vitamin C Supplements: Exercise and Mitochondrial Oxidation, J. Applied Physiology, 60(6):1986-1991.

Gohil, K., Rothfuss, L., Lang, J., and Packer, L., 1987, Effect of Exercise Training on Tissue Vitamin E and Ubiquinone Content, J. Applied Physiol. 63(4):1638-1641.

Gohil, K., Viguie, C., Stanley, W., Brooks, G. and L. Packer, 1988, Blood Glutathione and Oxidation During Human Exercise, J. of Applied Physiology, 64(1):115-119.

Kirkwood, S.P., Packer, L. and Brooks, G.A., 1987, Effects of Endurance Training on a Mitochondrial Reticulum in Limb Skeletal Muscle, Arch. Biochem. Biophys. 255:80-88.

Lang, J., Gohil, K., Burk, R.F. and L. Packer, 1987, Selenium Deficiency, Endurance, Exercise Capacity and Antioxidant Status in Rats, J. Applied Physiol. 63(6): 2532-5.

Lang, J., Gohil, K., Rothfuss, L. and L. Packer, 1987, Exercise Training Effects on Mitochondrial Enzyme Activity, Ubiquinones, and Vitamin E, in: "Anticarcinogenesis and Radiation Protection," O.F. Nygaard, M. Simic, and P. Cerutti, eds., Plenum Press, New York.

Packer, L., Exercise, Aging, and Antioxidants, 1987, in: "Advances in Myochemistry, Proceedings of the 2nd ISM Congress (Rome, Italy, 8-10 October, 1987)," G. Benzi, ed., John Libbey & Co. Ltd., London.

Packer, L., 1986, Vitamin E, Physical Exercise and Tissue Damage in Animals, Med. Biol., 62:105-109.

Packer, L., 1986, Oxygen Radicals and Antioxidants in Endurance Exercise, in: Biochemical Aspects of Physical Exercise, G. Benzi, L. Packer, and N. Siliprandi, eds., Elsevier Science Publishers, Amsterdam.

Packer, L., Gohil, K., deLumen, B., and Terblanche, S.E., 1986, A Comparative Study of the Effects of Ascorbic Acid Deficiency and Supplementation on Endurance and Mitochondrial Oxidative Capacities in Various Tissues of the Guinea Pig, Comp. Biochem. Physiol., 83B:235-240.

FREE RADICAL MECHANISMS IN TISSUE INJURY

T. F. Slater

Dept. of Biology and Biochemistry
Brunel University
Uxbridge, Middx., United Kingdom

Free radicals are molecules or molecular fragments containing a single unpaired electron. In general, free radicals are reactive chemically, some (e.g. HO•) being extremely reactive. However, certain types of free radical, such as nitroxyl-radicals and free radicals stabilized by steric or delocalization features, are much less reactive and a few (e.g. diphenyl picryl hydrazyl) are stable enough to be crystallised and stored at temperatures above 0°. Table 1 gives the general structures of free radicals that will be discussed in this short review.

Table 1.   General structures of free radicals
that are discussed in the text

$^3O_2$	(ground-state oxygen)	RCO•	(alkoxyl)
$^1O_2$	(singlet oxygen)	RCOO•	(peroxyl)
$O_2^{•-}$	(superoxide anion radical)	RS•	(thiyl)
HO•	(hydroxyl)	RNO•	(nitroxyl)
$R_1\overset{•}{C}R_2$	(carbon-centred radical)		

Table 1 includes two substances that are not free radicals as defined above: dioxygen (O_2) and singlet oxygen ($^1O_2$). Dioxygen, the gas essential for life, has two unpaired electrons in its normal state; these two electrons are rather tightly coupled, and dioxygen in its ground state

is called a triplet. Singlet oxygen, on the other hand, contains no unpaired electrons but is generated from dioxygen by the absorption of
energy that raises oxygen to an excited higher state.

Free radicals can be formed in situ in our tissues and cells by (i)
the impact of radiation, which may be ionising, ultra-violet, visible or
thermal. Generally, a photosensitiser such as a porphyrin is required for
free radical production by visible light. (ii) redox-reactions catalysed
by transition metals such as iron or copper that can undergo unit changes
in valence state (e.g. $Fe^{3+} + e^- \longrightarrow Fe^{2+}$. (iii) enzymic-catalysis that
often involves flavoproteins or hemoproteins.

In general, there are very efficient protective mechanisms in our
cells and tissues that prevent any adventitious free radicals from causing
damage to structure and function. These protective mechanisms include
antioxidants and also enzymic processes that metabolise otherwise damaging
products of free radical reactions to less toxic products. If these protective mechanisms are overwhelmed, or are unusually ineffective, then
free radical production can result in serious tissue and cellular injury.
Figure 1 illustrates some of the major pathways by which reactive free
radical-mediated reactions can result in biologically disturbing consequences.

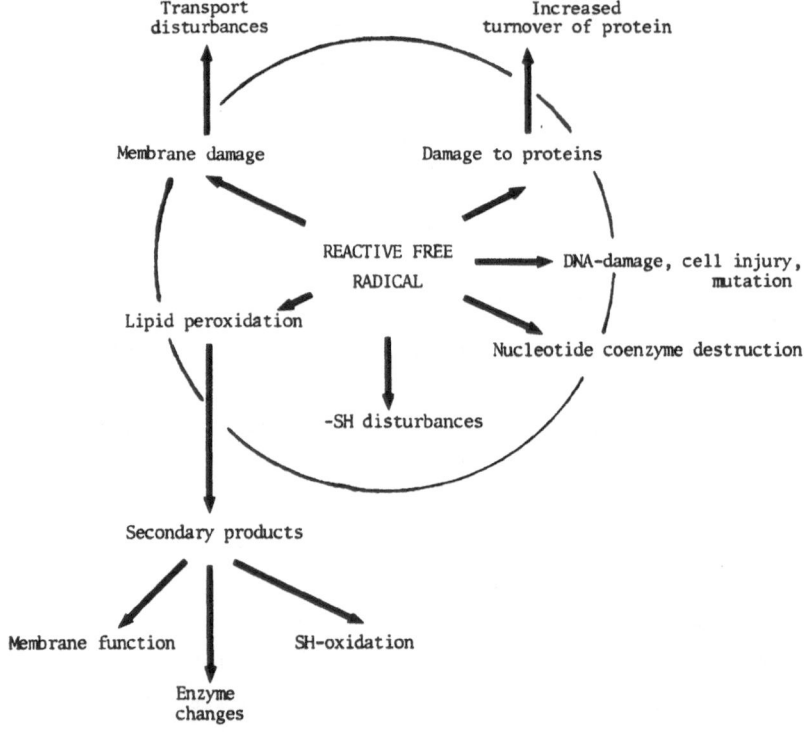

Fig. 1. Damaging reactions of free radicals (see 36,37)

This short review will illustrate free radical-mediated cell injury with a number of examples: (i) photosensitisation; (ii) iron-overload; (iii) metabolic activation; (iv) lipid peroxidation; and (v) liver tumours. The discussion will be necessarily brief; further details and full background documentation can be obtained from references 1-4.

1. PHOTOSENSITISATION

A number of clinically important diseases involve abnormal sensitivity of the skin to sunlight; for example, some porphyrias, Xeroderma pigmentosum, and drug-related disorders (5). Similar disturbances are known in domestic animals as with facial eczema and geel-dik-kop in sheep (6,7). The mechanisms for porphyrin-catalysed photosensitisation involves the absorption of a quantum of radiation (often in the Soret band region around 400nm) to give an excited singlet; transition to an excited triplet then occurs followed by reaction with dioxygen to yield singlet oxygen. The latter molecule is chemically reactive and can interact with a variety of biomolecules, including polyunsaturated fatty acids (PUFA's) to give damage. Reaction with PUFA's can initiate lipid peroxidation, a free radical-mediated process. The damaging effects of photosensitisation can be attenuated by quenching the singlet oxygen (e.g. with β-carotene) or by free radical scavengers. For an account of lysosomal disturbances in the epidermis in photosensitisation see (8). Visible light may also aggravate the damaging reactions of high oxygen levels on the premature retina; see (9). Although photosensitisation is usually regarded as a damaging reaction it is being used increasingly in the treatment of various accessible lesions including certain cancers; for an account of phototherapy see (10).

2. IRON OVERLOAD

The significance of transition metal ions in catalysing free radical reactions has been recognised for a long time. Ferrous or ferric ions can be of importance in this respect in a number of ways, for example through the Fenton reaction:

$$Fe^{2+} + H_2O_2 \longrightarrow Fe^{3+} + OH^- + HO^{\bullet}$$

In addition, these metal ions interact with lipid hydroperoxides (LOOH) to yield alkoxyl and peroxyl products :

$$LOOH + Fe^{2+} \longrightarrow LO^{\bullet} + Fe^{3+} + OH^-$$

$$LOOH + Fe^{3+} \longrightarrow LOO^{\bullet} + Fe^{2+} + H^+$$

The concentration of free Fe^{2+} and Fe^{3+} normally is controlled rigorously (see 11), but iron-catalysed reactions can become of significance if this control breaks down. This may occur as a result of cellular damage that is associated with disturbance of structure and of the compartmentation of iron within the cell (see 12), or when the intake of iron is excessive as in hemochromatosis (13). In addition, relatively small concentrations of iron can be of significance if mechanisms of protection are abnormally depressed. An example of the latter process is the muscle damage reported in piglets that were deficient in vitamin E and had been injected with an iron-dextrose preparation to control anaemia (14).

3. METABOLIC ACTIVATION

Many substances can be metabolised by a variety of enzyme systems in our bodies to intermediates that are chemically more reactive than the parent substance; this is called metabolic activation. The reactive intermediate can be an epoxide, carbonium-ion, aldehyde, unstable ester, or free radical. Table 2 gives a number of examples of compounds known to undergo metabolic activation to free radical products; it is important to note that some of these compounds (e.g. the polycyclic hydrocarbons) are also metabolically activated to other types of reactive intermediate that may be of much greater significance biologically (e.g. the dihydrodiol epoxides of polycyclic hydrocarbons) in the ensuing damage.

Table 2. Examples of compounds and classes of compounds that can undergo metabolic activation to free radical intermediates

Halogenoalkanes :	Nitro-compounds
CCl_4	Aromatic amines
$CHCl_3$	Azo-dyes
$CBrCl_3$	Nitrosoamines
Dibromoethane	Hydrazines
Halothane	Quinones
Halogenoalkenes	Polycyclic hydrocarbons

A very good example of a compound that undergoes metabolic activation to a reactive intermediate associated with tissue injury is carbon tetrachloride: a single dose of this substance can produce centrilobular necrosis and fatty degeneration of the liver when administered to rats and other species (see 15). The primary product of the metabolic activation of CCl_4 is the trichloromethyl radical, $^{\bullet}CCl_3$; this has been detected both in vitro and in vivo by the use of electron spin resonance (esr)-spin trapping (16,17). The $^{\bullet}CCl_3$ radical is rather unreactive in general, however, but does undergo a very rapid interaction with dioxygen to give the secondary product, $CCl_3OO^•$, the trichloromethyl peroxyl radical that is much more reactive (18). Aliphatic peroxyl radicals are normally not highly reactive but the chlorine atoms in the $CCl_3OO^•$ radical greatly increase its reactivity; a review of this subject with numerous rate constants is by Lal et al (19).

The primary and secondary free radicals produced from CCl_4 can cause metabolic perturbations and cell injury in various ways. For example, by chloromethylation (covalent binding) of $^•CCl_3$ to protein and lipid; change in redox status of SH/S-S; destruction of nucleotide coenzymes; and stimulation of lipid peroxidation. The liver injury caused by CCl_4 is dependent on a number of different damaging reactions; it is multicausal in nature (20).

4. LIPID PEROXIDATION

This is a free radical-mediated process that gives rise to many biologically active products (Table 3). Lipid peroxidation, in the context of cell injury, generally results from the attack of an oxidising free radical species (e.g. $CCl_3OO^•$) on PUFA's to yield PUFA carbon-centred radicals (PUFA$^•$) by hydrogen atom abstraction. The PUFA-free radical can undergo internal bond rearrangement to give a conjugated dienyl radical and this usually reacts quickly with O_2 to form the corresponding conjugated diene-peroxyl radical:

$$PUFA^• \ + \ O_2 \longrightarrow PUFAOO^•$$

However, if the concentration of dioxygen is very low and if the local concentration of hydrogen-donors (XH) is high, then conjugated dienes without a peroxyl entity can be formed :

$$PUFA^• \ + \ XH \longrightarrow PUFA \ (conjugated \ diene) \ + \ X^•$$

These non-peroxide dienes may be of significance in human disease (21).

Table 3. Products of Lipid Peroxidation

Leukotrienes	Alkanals
Lipid hydroperoxides	Alkenals
Hydroxy fatty acids	4-Hydroxy-Alkenals
Epoxy-fatty acids	Ketones
	Alkanes

As already mentioned, lipid peroxidation is associated with the formation of a variety of products (Table 3). Some of these have powerful biological effects even at very low concentrations. For example, PUFA-epoxides can affect hormone secretion at nM concentrations (22); lipid hydroperoxides can affect the prostaglandin cascade at μM concentrations (23); and the 4-hydroxy-alkenals have a number of important actions as summarised in Table 4.

Table 4.   Some Important Reactions of 4-hydroxy-alkenals
(see 24; 25; 26)

(a)	inhibition of DNA-synthesis
(b)	inhibition of glucose-6-phosphatase
(c)	inhibition of adenyl cyclase
(d)	reaction with thiols such as glutathione
(e)	increase in capillary permeability
(f)	inhibition of platelet aggregation by ADP
(g)	reaction with polyamines
(h)	stimulation of chemotaxis
(i)	effects on cell membrane receptors
(j)	effects on some oncogene expressions
(k)	effects on DNA-repair

A low level of lipid peroxidation can be detected even in apparently normal tissue. Although this may reflect the presence in the tissue sample of a relatively few injured cells, it is possible that lipid peroxidation may be physiologically important (27). The low concentrations of some products of lipid peroxidation required to produce important biological effects are consistent with this concept (28). In particular, the production of 4-hydroxy-alkenals may provide a coarse control of DNA-synthesis (29).

The formation of biologically reactive products of lipid peroxidation allows an understanding of how a spatially constrained free radical reaction can result in widespread systemic disturbances as a result of the diffusion of secondary, tertiary etc. products away from the initial locus of metabolic activation (30). This is illustrated in Fig. 2.

5. LIVER TUMOURS

Many types of liver tumour in experimental animals have a much lower rate of stimulated lipid peroxidation than corresponding samples of normal liver (for references see 31). This decreased rate of lipid peroxidation has been observed with intact tumour cells and with microsomal fractions prepared from tumour cells, and where the lipid peroxidation was stimulated by exposure to CCl$_4$ by the addition of iron-chelate or by exposure to γ-radiation (see 31). The reasons appear to be several: (a) a very low level of cytochrome P$_{450}$, which is involved normally in metabolic activation, and also in the initiation of new radical chains through the peroxidatic action of cytochrome P$_{450}$ on lipid hydroperoxides. (b) a small to moderate decrease in the content of PUFA's, which are the substrates for lipid peroxidation. (c) an increased level of lipid-soluble chain-breaking antioxidant.

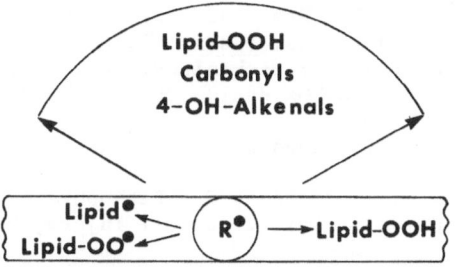

Fig. 2. Diffusion of products of a free radical reaction to result in action at a distance (from reference 30).

Antioxidants can be classified (32) as preventative or chain-breaking. Numerous studies on liver tumours have demonstrated that preventative antioxidants such as glutathione peroxidase catalase and superoxide dismutase are not increased in these cells; in addition, there is no consistent increase in glutathione, ascorbate or urate, which are chain-breaking antioxidants (see 33,34). However, α-tocopherol, a lipid-soluble chain-breaking antioxidant is increased in at least some liver tumours such as the Novikoff and Yoshida tumours (31,35). These liver tumours appear to have the capacity to accumulate α-tocopherol from the host.

The decreased rate of lipid peroxidation in these tumours may have two consequences that can be of considerable biological significance. Firstly, the increased antioxidant activity due to the increased content of α-tocopherol can be expected to decrease oxidative stress at a time when DNA is more susceptible to damage, i.e. during cell division. Secondly, the decrease in lipid peroxidation will be associated with a decreased production of 4-hydroxy-alkenals that normally may exert an inhibitory influence on DNA-synthesis (24). Under these conditions, damage to DNA may be decreased and cell division stimulated.

Acknowledgements

I am grateful to the Association for International Cancer Research Campaign for financial support.

REFERENCES

1. R.P.Mason and C.F.Chignell, Free Radicals in Pharmacology and Toxicology – Selected Topics, Pharmacol.Rev. 33:189 (1982).
2. H.Kappus, Lipid Peroxidation: Mechanisms, Analysis, Enzymology and Biological Relevance, in "Oxidative Stress", H.Sies, ed., Academic Press (1985).
3. T.F.Slater, Free Radical Mechanisms in Tissue Injury, Biochem.J. 220:1 (1984).
4. D.L.Tribble, T.Y.Aw and D.P.Jones, The Pathophysiological Significance of Lipid Peroxidation in Oxidative Cell Injury, Hepatology 7:377 (1987).
5. I.A.Magnus, "Dermatological Photobiology: Clinical and Experimental Aspects", Blackwells, Oxford (1976).
6. T.F.Slater, U.D.Sträuli and B.Sawyer, Sporidesmin Poisoning in the Rat. 1. Chemical Aspects, Res.Vet.Sci. 5:450 (1964).
7. N.T.Clare, "Photosensitisation in Diseases of Domestic Animals", Commonwealth Agricultural Bureaux, Farnham Royal, Bucks. (1952).
8. T.F.Slater and P.A.Riley, Photosensitisation and Lysosomal Damage, Nature, Lond. 209:151 (1966).
9. P.A.Riley and T.F.Slater, Pathogenesis of Retrolental Fibroplasia, Lancet (ii):265 (1969).
10. D.Kessel and T.J.Dougherty (eds.), "Porphyrin Photosensitisation", Plenum, New York (1981).
11. T.H.Bothwell, R.W.Charlton, J.D.Cook and C.A.Finch, "Iron Metabolism in Man", Blackwells, Oxford (1979).
12. R.L.Willson, Free Radicals and Tissue Damage: Mechanistic Evidence from Radiation Studies, in "Biochemical Mechanisms of Liver Injury", T.F.Slater, ed., Academic Press, London (1978).
13. B.R.Bacon, A.S.Tavill, G.M.Brittenham, C.H.Park and R.O.Recknagel, Hepatotoxicity of Chronic Iron Overload: Role of Lipid Peroxidation, in "Free Radicals in Liver Injury", G.Poli, K.H.Cheeseman, M.U. Dianzani and T.F.Slater, eds., IRL Press, Oxford (1985).

14. D.S.P.Patterson, W.M.Allen, S.Berrett, D.Sweasey and J.T.Done, The Toxicity of Parenteral Iron Preparations in the Rabbit and Pig with a Comparison of the Clinical and Biochemical Responses to Iron-Dextrose in 2-days old and 8-days old Piglets. Zbl.Vet.Med.A. 18:453 (1971).

15. T.F.Slater, "Free Radical Mechanisms in Tissue Injury", Pion, London (1972).

16. E.K.Lai, P.B.McCay, T.Noguchi and K-L.Fong, In vivo Spin-Trapping of Trichloromethyl Radicals formed from CCl₄, Biochem.Pharmacol. 28:2231 (1979).

17. E.Albano, K.A.K.Lott, T.F.Slater, A.Stier, M.C.R.Symons and A.Tomasi, Spin Trapping Studies on the Free Radical Products Formed by Metabolic Activation of Carbon Tetrachloride in Rat Liver Microsomal Fractions, Isolated Hepatocytes and in vivo in the Rat, Biochem.J. 204:593 (1982).

18. J.E.Packer, T.F.Slater and R.L.Willson, Reactions of the Carbon Tetrachloride-related Peroxy Free Radical (CCl₃O₂·) with Amino Acids: Pulse Radiolysis Evidence, Life Sci., 23:2617 (1978).

19. M.Lal, C.Schöneich, J.Mönig and K-D.Asmus, Rate Constants for the Reactions of Halogenated Organic Radicals, Int.J.Radiat.Biol., in press (1988).

20. T.F.Slater, Activation of Carbon Tetrachloride:Chemical Principles and Biological Significance, in "Free Radicals, Lipid Peroxidation and Cancer", D.C.H.McBrien and T.F.Slater (eds.) Academic Press, London (1982).

21. T.L.Dormandy, Free Radical Activity and Diene Conjugation in Man, in "Free Radicals in Liver Injury", G.Poli, K.H.Cheeseman, M.U.Dianzani and T.F.Slater (eds.), IRL Press, Oxford (1985).

22. J.Capdevila, N.Chacos, J.R.Falck, S.Manna, A.Negro-Vilar and S.R.Ojeda, Novel Hypothalamic Arachidonate Products Stimulate Somatostatin Release from the Median Eminence, Endocrinol, 113:421 (1983).

23. M.E.Hemler, H.W.Cook and W.E.M.Lands, Prostaglandin Synthesis can be Triggered by Lipid Peroxides, Archs.Biochem.Biophys., 193:340 (1979).

24. H.Esterbauer, Lipid Peroxidation Products:Formation, Chemical Properties and Biological Activities, in "Free Radicals in Liver Injury", G.Poli, K.H.Cheeseman, M.U.Dianzani and T.F.Slater (eds), IRL Press, Oxford (1985).

25. J.S.Hurst, T.F.Slater, J.Lang, G.Juergens, H.Zollner and H.Esterbauer, Effects of the Lipid Peroxidation Product 4-Hydroxy Nonenal on the Aggregation of Human Platelets,, Chem.Biol.Interactions 61;109 (1987).

26. H.Krokan, R.C.Graftstrom, K.Sundqvist, H.Esterbauer and C.C.Harris, Cytotoxicity, Thiol Depletion and Inhibition of O^6-Methyl Guanine-DNA Methyltransferase by Various Aldehydes in Cultural Human Bronchial Fibroblasts, Carcinogenesis 6:1755 (1985).

27. T.F.Slater, Free Radical Disturbances and Tissue Damage:Cause or Consequence ? in "Free Radicals: A Search for New Methodology", C.Rice-Evans and T.L.Dormandy (eds.), Richelieu Press, London (1988).

28. T.F.Slater, Lipid Peroxidation Products and Intercellular Messengers in Relation to Cell Injury, Agents and Actions, 22:333 (1987).

29. T.F.Slater, C.Benedetto, G.W.Burton, K.H.Cheeseman, K.U.Ingold and J.T.Nodes, Lipid Peroxidation in Animal Tumours: A Disturbance in the Control of Cell Division ? in "Icosanoids and Cancer", H. Thaler-Dao, A.Crastes de Paulet and R.Paoletti (eds.), Raven Press, New York (1984).

30. T.F.Slater, Biochemical Pathology in Microtime, in "Recent Advances in Biochemical Pathology: Toxic Liver Injury", M.U.Dianzani, G.Ugazio and L.M.Sena (eds.) Minerva Medica (1976).

31. K.H.Cheeseman, M.Collins, K.Proudfoot, T.F.Slater, G.W.Burton, A.C. Webb and K.U.Ingold, Studies on Lipid Peroxidation in Normal and Tumour Tissues: The Novikoff Rat Liver Tumour,Biochem.J., 235:507 (1986).

32. G.W.Burton, D.O.Foster, B.Perly, T.F.Slater, I.C.P.Smith and K.U. Ingold, Biological Antioxidants, Phil.Trans.R.Soc.Lond.B. 311:565 (1985).

33. R.Corrocher, M.Casaril, G.Bellisola, G.B.Gabrielli, N.Nicoli, G.C.Guidi and G.De Sandre, Severe Impairment of Antioxidant System in Human Hepatoma, Cancer 58:1658 (1986).

34. L.W.Oberley, Selected Abstracts on Modulation of Cancer by Superoxide Dismutase, in "Oncology Overview: International Cancer Research Data Bank Program", U.S.Dept. Health and Human Services, National Institutes of Health, Bethesda, U.S.A. (1983).

35. K.H.Cheeseman, S.Emery, S.P.Maddix, T.F.Slater, G.W.Burton and K.U. Ingold, Studies on Lipid Peroxidation in Normal and Tumour Tissues: The Yoshida Rat Liver Tumour, Biochem. J. 250:247 (1988).

36. T.F.Slater "Free Radical Mechanisms in Tissue Injury" 2nd Edition, in press.

37. T.F.Slater, Free radical-mediated tissue damage, Nutrition 87:46 (1987).

HUMAN URICASE LOSS: AN EVOLUTIONARY GAIN AGAINST DISEASE

Alfredo Varela-Echavarría, Luis Cañedo and Hugo A. Barrera-Saldaña

U.L.I.E.G., Departamento de Bioquímica, Facultad de Medicina, U.A.N.L., and The International Center for Molecular Medicine, Monterrey, Nuevo León, México

The peroxisomal enzyme uricase (urate: oxygen oxidoreductase E.C. 1.7.3.3.) is a tetramer of 32 kD subunits (1-4) that catalyzes the oxidation of uric acid to allantoin during purine metabolism. This enzyme has been found in liver homogenates of most vertebrates (fish, amphibian and mammalian species) but is absent in hominoids (gibbons, orangutans, chimpanzees, gorillas and man) (5). A common evolutionary origin for all uricases is suggested by the evidence depicted in Table 1, however previous attempts to determine the homology among uricases of evolutionarly distant species have been unsuccessful. Fujiwara et al (1) demonstrated immunorecognition among amphibian uricases using an antiserum prepared against native uricase from Rana catesbeiana but were unable to detect immunological cross reaction with fish or mammalian uricases. The failure to see cross reactivity among these proteins might reflect their degree of amino acid sequence divergence (6).

We used Western blot analysis (7) and antibodies against denatured rat uricase to investigate if the uricases from different species share antigen determinants as indicators of homology. In this work we were able to detect the enzymatic activity of uricase and immunological cross reactivity among the enzymes of evolutionarily distant species such as fish, amphibia and mammals but did not find uricase activity or immunological cross reactivity in those species previously reported to lack this enzyme (Reptilia, Aves and man) (Table 2) (8). Our results show that immunological cross reactivity correlates with the enzymatic activity found. This can be explained by the higher sensitivity of the Western blot techniques and also by the fact that by using a denatured unfolded rat uricase to produce the antiserum, a larger number of hidden antigenic sites corresponding to the most conserved inner part of the molecule (6) were recognized by the rabbit immune system, resulting in a more effective antiserum for comparative studies. Further support in regard to the absence of uricase in man was provided by the Southern blot studies of Lee et al. (9), who demonstrated in human genomic DNA, the presence of restriction fragments that are homologous to porcine uricase cDNA; however, a Northern blot analysis of human liver mRNA carried out by these investigators showed no hybridization signal. All these results support the hypothesis that a single mutational event could be responsible for the loss of uricase activity in man (5). The lack of this enzyme activity in

Table 1. Characteristics of liver uricases from evolutionarily distant vertebrates.

	Molecular weight(kD)[a]		Intracellular localization	Optimum pH	Ref.
	Tetramer	Monomer			
monkey		32			1
macaque		32			
pig	125	32			3
rat		32	Peroxisome	9	4,21
bull frog	120	30	Peroxisome		1
frog		32			
trout	130	32.5		8.8	2
mackerel	127	30		8.6	1,22
carp		32			
catfish		31			
bream		32			
tilapia		32			

[a]This information rests on data taken from references indicated. Data without reference indicate results obtained in our laboratory.

man, associated to the increase of uric acid reabsorption by the kidney (10) causes raised urate plasma levels. The high level of uric acid allows the body to limit the damage caused by harmful oxygen radicals.

The free radical species include singlet oxygen species (^{1}, O_2). The superoxide radical anion (O_2^-) peroxide radicals (O_2^2, $ROO°$) hydroxyl radicals ($OH°$) and alkoxyl radicals ($R°$, $RO°$). These radicals participate in an interacting chain of reactions that takes place in intra and extra cellular compartments, producing lipid peroxidation that disrupts the normal architecture of membranes and damages proteins and DNA (11). Free radicals are produced in phagocytosis, prostaglandin synthesis, the cytochrome P-450 and in catalyzed reactions by enzymes such as xanthine oxidase, aldehyde oxidase and flavin dehydrogenase (12); by non enzymatic means such as the reaction of oxygen with lipids and proteins and by environmental factors such as ionizing radiation, tobacco smoke, and smog (13). The magnitude of free radical production in man can be illustrated by citing two additional mechanisms. 1) The total amount of hemoglobin in man is about 750 gr. From this, about 3% (22 gr) undergoes autoxidation to methemoglobin, with the concomitant large production of superoxide radicals (10). 2) Under normal conditions, 97% of the oxygen in the blood is transported from the lungs to the tissues in chemical combination with hemoglobin. Since each gr of hemoglobin can bind 1.34 ml of oxygen, on the average 20 ml of O_2 are transported in each 100 ml of blood. From that, approximately 5 ml of oxygen are delivered to the tissues during circulation (14). When a person is at rest, the heart pumps close to five liters of blood per minute. That means that approximately 360 lt of oxygen at normal temperature and pressure are consumed each day and 98% of this oxygen is utilized in the mitochondria. Under normal conditions, oxidative phosphorylation is carried out step by step by the electronic carriers located in the inner mitochondrial membrane. These respiratory assemblies

Table 2. Homology among vertebrate uricases.

	Uricase activity of liver peroxisomes enriched fractions[a]. (U^b x 10^3/mg of protein).	Immunorecognition by anti-rat uricase antiserum. (by Western-blot).
bream	56.2	+
cat-fish	41.6	+
carp	55.7	+
tilapia	45.0	+
frog	3.4	+
turtle	-	-
chicken	-	-
rat	21.4	+
macaque	2.6	+
man	-	-

[a]Results shown are the average of three different samples with the exception of that of macaque in which only a sample was used.
[b]U is the enzyme activity that degrades 1 µmol of uric acid per minute in 50 mM borate buffer (pH 9) containing 100 mM uric acid.

transport the electrons from NADH or $FADH_2$ to O_2 to reduce it to H_2O. During this normal process some of the large amount of electrons can be donated to oxygen by the ubiquinone, semiquinone components of the respiratory transport chain generating superoxide radicals (15). In extrenous exercise the amount of oxygen consumption can be increased six or sevenfold with the concomitant production of free radicals. Processes like inflammation, hyperthermia, obesity and alcoholism also increases the production of free radicals in the body.

Defenses have evolved to limit the damage produced by the free radicals. In membranes: vitamin E, carotenoids, ubiquinones; in intracellular aqueous compartments: superoxide dismutase, catalase, glutathione peroxidase, glutathione s-transferase, storage proteins like ferritin and small molecules as vitamin C and glutathione; in extracellular fluids: ceruloplasmin, transferrin, lactoferrin, albumin, haptoglobin, vitamin C, bilirubin and uric acid (15-17).

Uric acid has been proposed as a powerful scavenger of singlet oxygen and hydroxyl radicals and as inhibitor of lipid peroxidation. The high plasma level of urate (about 300 µM) makes it one of the major antioxidants in humans. At physiological concentrations is as effective antioxidant as ascorbic acid, a molecule that humans can not synthesize due to the evolutionary loss of L-gulonolactone oxidase (18, 19). The absence of uricase in humans raises the serum levels of urate close to the solubility limit, predisposing human beings to suffer gout and other hyperuricemic diseases. Hyperuricemia (in excess of 7.0 mg/dl) is present in 2-18 percent of the general population. However the prevalence of gout in the western world is only 0.13 to 0.37% of the population. (20). By contrast cancer is one of the most important causes of death and radical reactions have been included as potential causative agents for these diseases (13). Other diseases included in the category of free radical diseases are: atherosclerosis, degenerative gout disease, senile dementia of the Alzheimer type, senile macular degeneration, Parkinson disease, cataracts, adult respiratory distress syndrome, Fanconi's anemia, Bloom syndrome, systemic lupus eritematosus, Bathen's disease, Insulin dependent diabetes mellitus, Laennec cirrhosis, preeclampsia, xeroderma pigmentosum and ataxia telangiectasia (13). The protective mechanisms against oxygen radicals in man have been associated to the evolutionary decrease in the age specific cancer rate and have been proposed as a major contributor in lengthening life of mammals (10). Therefore, the role of the antioxidant defenses to protect the human body from these diseases during the most part of its life has to be underlined. Being uric acid one of the major antioxidants in the body its action together with a similar action of other end products of degradative metabolic pathways such as bilirubin (10, 16, 17), may have been selected during evolution as beneficial mechanisms to reduce the incidence of disease and to extend the life span of the human beings.

REFERENCES

1.- S. Fujiwara, H. Ohashi and T. Noguchi, Comparision of intraperoxisomal localization form and properties of amphibian (Rana catesbeiana) uricase with those of other animal uricases, Comp. Biochem. Physiol. 86B (1): 23 (1987).
2.- J. E. Kinsella, B. German and J. Shetty, Uricase from fish liver: isolation and some properties, Comp. Biochem. Physiol. 82B: 621 (1985).
3.- O. M. Pitts, D. G. Priest and W. W. Fish, Uricase. Subunit composition and resistance to denaturants, Biochemistry 13: 888 (1974).
4.- T. Watanabe, T. Suga, and H. Hayashi, H. Studies on peroxisomes. VIII. Evidence for framework protein of the cores of rat liver peroxisomes, J. Biochem. 82: 607 (1977).
5.- T. B.Friedman, G. E. Polanco, J. Appold and J. E. Mayle, On the loss of uricolytic activity during primate evolution. I. Silencing of urate oxidase in a hominoid ancestor, Comp. Biochem. Physiol. 81B: 653 (1985).
6.- E. M. Prager, A. C. Wilson, The dependence of immunological cross-reactivity upon sequence resemblace among lysozymes. I. Microcomplement fixation studies, J. Biol. Chem. 246: 5978 (1971).
7.- H. Towbin, T. Staehelin, J. Gordon, Electrophoretic transfer of proteins from polyacrilamide gels to nitrocellulose sheets: procedure and some applications, Proc. Natl. Acad. Sci. U.S.A. 76: 4350 (1979).
8.- A. Varela-Echavarría, R. Montes de Oca-Luna and H.A. Barrera-Saldaña, Uricase protein sequences are conserved during vertebrate evolution but absent in man, FASEB Journal (in press).
9.- C. L. Lee, X. Wu, R. A. Gibbs, R. G. Cook, D. M. Muzny and C. T. Caskey, Generation of cDNA probes directed by amino acid sequence: cloning the urate oxidase, Science 239: 1288 (1988).
10.- B. N. Ames, R. Catheart, R.; Schwiers, E.; Hochstein, P. Uric acid provides an antioxidant defense in humans against oxidant and radical caused aging and cancer: a hypothesis, Proc. Natl. Acad. Sci. U.S.A. 78: 6858-6862 (1986).
11.- B. Halliwell and J. M. C. Gutteridge, "Free radicals in biology and medicine", Oxford University Press, New York (1986).
12.- B. E. Leibovitz and B. V. Siegel, Aspects of free radical reactions in biological systems: aging, Journal of Gerontology 35(1):45 (1980).
13.- D. Harman, Free radical theory of aging: Role of free radicals in origination and evolution of life, aging and disease processes. "Free radicals, aging and degenerative diseases", Alan R. Liss, Inc. (1986).

14.- A. C. Guyton, "Text book of medical phisiology", 5th ed., Philadelphia, W. B. Saunders Co. (1976).

15.- L. Packer, Oxygen is a dangerous friend,"Cell function and disease", Plenum Press, L. Cañedo, L. Todd, L. Packer and J. Jaz (eds.) (1988).

16.- R. Stoker, Y. Yamamoto, A. F. McDonagh, A. W. Glazer and B. N. Ames, Bilirubin is an antioxidant of possible physiologic importance, Science 235:1043 (1987).

17.- R. Stocker, A. N. Glazer and B. N. Ames, Antioxidant activity of albumin bound bilirubin, Proc. Natl. Acad. Sci. USA 84:5918 (1987).

18.- I. B. Chatterjee, Evolution and the biosynthesis of ascorbic acid, Science 182: 1271 (1973).

19.- P. Sato and S. Udenfriend, Scurvy-prone animals, including man, monkey, and guinea pig, do not express the gene for gulonolactone oxidase, Arch. Biochem. Biophys. 187(1): 158 (1978).

20.- E. Braunwald, K. J. Isselbacher, R. G. Petersdorf, J. P. Wilson, J. B. Martin, Fauci As eds. Harrison's. "Principles of internal medicine" , 11th ed. New York, McGraw Hill Co. (1985).

21.- B. Baum, G. Hubster and H. R. Mahler, Sudies on uricase. II. The enzyme-sustrate complex, Biochem. Biophys. Acta 22: 514 (1956).

22.- T. Noguchi, Y. Takada, S. Fujiwara, Degradation of uric acid to urea and glioxylate in peroxisomes, J. Biol. Chem. 254: 5272-5275 (1979).

RECOVERY OF FUNCTION AFTER TISSUE TRANSPLANTATION IN THE NIGROSTRIATAL
DOPAMINE SYSTEM

Jill B. Becker[1] and William J. Freed[2]

[1]The University of Michigan
Psychology Department and Neuroscience Program
Ann Arbor, MI and [2]NIMH Neurosciences Center at St. Elizabeths
Washington, D.C. 20032

Numerous reports spanning more than 70 years have established that
small fetal brain tissue grafts consistently survive and frequently
develop connections with the host brain. The basic methods for these
investigations were first presented in 1917 by Elizabeth Dunn. Dunn (1917)
reported the survival of immature rat cortex that had been transplanted
into the cortex of other animals and evidence of connections between the
grafts and the host brain. Shortly thereafter, Murphy and Sturm (1923)
described the "privileged" status of the brain, which permits small tissue
grafts to survive without rejection. Subsequently, Medawar (1948) discovered that graft rejection was an immunological phenomenon and that the
privileged status of the brain had an immunological basis. Numerous
investigators have confirmed that the brain is at least in part immunologically privileged (Albrink and Greene, 1953; Rosenstein and Brightman,
1978). Others have reported that not only does immature brain tissue
survive transplantation, but it frequently develops substantial fiber
connections with the host brain (Bjorklund et al., 1976; Le Gros Clark,
1940; Lund and Harvey,1981).

In spite of early success in transplanting tissue to brain, demonstration that brain grafts could have a functional impact on the behavior of
an animal has been reported only recently (Perlow et al., 1979; Freed et
al., 1980, 1981; Bjorklund et al., 1980a, 1980b). It was the application
of transplantation methods to the nigrostriatal dopamine (DA) system where
behavioral effects of brain grafts were first demonstrated. In the rat,
the substantia nigra, pars compacta (SN) contains approximately 3,500
neurons (Anden et al., 1966a; Dahlstrom and Fuxe, 1964). This small number
of neurons is the source of most of the DA in the brain. These DA-containing neurons project to a variety of forebrain regions, including the
striatum, nucleus accumbens, olfactory tubercle, and a number of allocort-
ical and neocortical regions (Anden et al., 1966b; Hokfelt and Ungerstedt,
1969; Moore and Bloom, 1978). It is estimated that each nigrostriatal DA
neuron makes about 500,000 synapses in the striatum (Anden et al., 1966a).
The bilateral destruction of these DA-secreting cells results in a
profound behavioral syndrome of adipsia, aphagia, bradykinesia, catalepsy,
and sensorimotor neglect (Dunnet and Iversen, 1982a; 1982b; Marshall et
al., 1974; Ungerstedt, 1971d). In humans, it is the loss of these DA
neurons that is thought to be responsible for a majority of the symptoms

of Parkinson's Disease (Horneykiewicz, 1966; Horneykiewicz and Kish, 1987; Rinne et al., 1980). The use of the rotational behavior model following unilateral nigrostriatal DA depletion in animals (Ungerstedt, 1971b; Arbuthnott and Ungerstedt, 1970) as a behavioral assay of nigrostriatal DA function has been useful in assessing the functional capacity of catechol-amine-rich tissue grafts (Perlow et al., 1979; Freed et al., 1981; Freed, 1983; Dunnett et al., 1982c, 1983).

Following a unilateral 6-hydroxydopamine (6-OHDA) lesion of the substantia nigra, rats or mice spontaneously turn in circles toward the damaged side (ipsiversive). Although spontaneous rotational behavior diminishes over time, vigorous turning behavior can be reinstated by the administration of a variety of drugs, particularly drugs that activate brain DA systems (Christie and Crow, 1971; Ungerstedt, 1971a, 1971b; Ungerstedt and Arbuthnott, 1970). For example, amphetamine typically produces vigorous ipsiversive turning, whereas apomorphine produces contraversive turning (Ungerstedt, 1971c). A large number of studies of this type have established that rats typically turn in the direction contralateral to the nigrostriatal DA system having the greatest activity. Thus, after a unilateral 6-OHDA lesion of the SN, the DA terminals in the striatum on the lesioned side degenerate, and amphetamine can release DA only from the intact side. Activity in the intact striatum induces turning towards the damaged side. DA receptors in both striata are intact follow-ing a 6-OHDA lesion. The fact that apomorphine induces contraversive turning is thought to indicate that the unilaterally denervated receptors have become hypersensitive, so that apomorphine induces greater activity on the lesioned side and the animal rotates toward the intact side (Ungerstedt, 1971c). A number of studies involving electrophysiological and receptor assay techniques have subsequently confirmed that unilatera-lly denervated striatal DA receptors do become supersensitive (Creese and Snyder, 1979; Kreuger et al., 1976; Schultz and Ungerstedt, 1978).

The rotational behavior model has been extremely useful in studying changes in the functional activity of the denervated striatum following transplantation of catecholamine containing tissues. Grafts of two types of tissue have been shown to be behaviorally effective: adrenal medulla and fetal substantia nigra (SN). When the adrenal medulla is isolated from the adrenal cortex, chromaffin cells exhibit an altered phenotype with neuron-like processes and an increased accumulation of DA (Unsicker et al., 1978; 1980; Olson et al., 1980). Grafts of adrenal medulla tissue have been shown to reduce apomorphine-induced rotational behavior (Freed et al., 1981; Freed, 1983) as long as adrenal tissue is obtained from young (2-4 month old) rats (Freed, 1983). These grafts have been shown to contain DA, and in surrounding host tissue there is a gradient of decreas-ing DA concentrations with increasing distance from the graft in host striatum (Freed et al, 1983; Becker and Freed, 1988a). The adrenal medulla grafts do not, however, actually innervate the host striatum. While data from both fluorescence histochemical studies and DA assays suggest that these grafts are secreting catecholamines, the mechanism through which functional activity is restored is still unknown.

Fetal SN tissue grafts adjacent to the unilaterally DA denervated striatum have also been shown to decrease apomorphine-induced rotational behavior (Freed et al., 1980; Perlow et al., 1979). This effect is specific to SN grafts and is not found after grafts of sciatic nerve, fetal cortex or fetal tectum (Freed, 1983). The rotational behavior induced by low doses of amphetamine is also altered by SN grafts. In fact, in some animals SN grafts not only decrease amphetamine-induced rotation, but actually cause the animal to start rotating in the opposite direction, as if more DA was being released on the side of the brain with the SN graft (Freed, 1983). Grafts of dissociated SN cells within the striatum also can reverse the direction of amphetamine-induced turning (Bjorklund et al., 1980b). These experiments demonstrate that the ability of catecholamine-rich brain grafts to ameliorate the behavioral effects of unilateral DA depletion is not a general property of brain grafts, but rather is specific to grafts that secrete catecholamines.

While it is assumed that grafts of adrenal medulla or fetal SN tissue influence neural activity in the host brain by releasing the neurotransmitter that they synthesize, whether this is actually occurring has not been demonstrated conclusively. The mechanism(s) through which these grafts alleviate the symptoms of DA depletion has become a question of significant clinical relevance with reports of clinical trials using autografts of adrenal medulla for the treatment of patients with Parkinson's Disease (Backlund et al, 1985; Lindvall et al., 1987; Madrazo et al., 1987). In only one of these studies, however, has significant long term improvement in the symposium of Parkinson's Disease been reported (Madrazo et al., 1987). It seems critical, therefore, to determine how adrenal medulla grafts produce behavioral recovery from the symptoms of striatal DA denervation. The rest of this chapter will describe the results of experiments designed to determine whether the release of catecholamines from intraventricular adrenal medulla grafts is associated with the behavioral efficacy.

In order to address questions about the release of catecholamines from intraventricular adrenal medulla grafts, we developed a method to measure the release of catecholamines in freely moving rats. We reasoned that grafts in the lateral ventricle would be likely to release catecholamines into the cerebrospinal fluid (CSF). Therefore, functional activity in the grafts could be assessed by sampling the CSF and measuring released monoamines and/or metabolites. The microdialysis system that was employed has been described in detail previously (Becker et al., 1988a). Concentrations of monoamine metabolites in CSF were determined in freely moving rats before and after intraventricular grafts of adrenal medulla or control tissue.. It had been anticipated that DA would be readily detectable in the CSF following adrenal medulla grafts (if not during basal conditions, at least after DA release was induced by treatment of the animals with amphetamine). This was not the case. Even with an assay sensitivity of 1-5 pg DA, no DA was detected either before or after 3 mg/kg amphetamine. Basal concentrations of DOPAC were, however, elevated post-graft (Becker et al., 1988b). In addition, during amphetamine-induced blockade of DA reuptake, the decrease in CSF concentrations of the DA metabolite dihydroxyphenylacetic acid (DOPAC) was greater in animals with adrenal medulla grafts than in the same animals prior to receiving the grafts. The amphetamine induced decrease in CSF DOPAC concentrations pre-graft was 0.051 ± 0.026 μm (mean±SEM). In contrast, after adrenal medulla grafts, the amphetamine-induced decrease in CSF concentrations of DOPAC was 0.139 ± 0.041 μm ($p < 0.05$, pre-graft vs. post-graft).

The increased neurochemical response to amphetamine in animals with adrenal medulla grafts suggests that even though the grafts are not releasing DA in quantities large enough to allow detection in the CSF, DA metabolism is increased. To test this hypothesis, we examined the effect of adrenal medulla vs. adrenal cortex grafts on DA turnover, using probenecid (200 mg/kg) to block the efflux of acidic metabolites. The rate of accumulation of DOPAC and homovanillic acid (HVA; free + conjugated metabolites) in the CSF following probenecid has been shown to be a good index of DA turnover in brain (Hutson et al., 1984) and we found that DA turnover was significantly higher in animals with adrenal medulla grafts compared to animals with adrenal cortex grafts or to pre-graft values (Becker and Freed, 1988a; 1988b).

In order to address the question of whether there is increased functional DA activity in the lesioned striatum adjacent to an adrenal medulla graft, DA release in the striatum was measured using microdyalisis in freely moving rats with adrenal medulla or control grafts (Becker and Freed, 1988b). Since grafts of adrenal medulla tissue in the lateral ventricle did not always result in a decrease in rotational behavior, animals were assigned to one of three groups based on the type of graft received and their behavior approximately two months post-graft. The groups tested were: 1) animals with adrenal medulla grafts that showed a decrease in rotational behavior; 2) animals with adrenal medulla grafts that did not show a decrease in rotational behavior; and 3) animals with adrenal cortex grafts. We found that the amphetamine-induced increase in striatal DA release was significantly greater for the animals that showed decreased rotational behavior than for either of the two control groups during the first hour after amphetamine. Therefore, behaviorally effective adrenal medulla grafts increase amphetamine-stimulated DA release in the striatum and increase CSF concentrations of DA metabolites, but do not increase CSF concentrations of DA or baseline concentrations of DA in striatal extracellular fluid.

How do adrenal medulla grafts induce an increase in striatal DA activity? There are at least two possibilities. One is that adrenal medulla grafts induce regrowth or rescue damaged striatal dopaminergic fibers, as has been suggested by Bohn and her colleagues for the MPTP lesioned mouse (Bohn et al., 1987). In this model, the regenerating or recovering DA fibers would mediate the increase in amphetamine-stimulated DA release found in animals with adrenal medulla grafts. If this is happening in the 6-OHDA rat, then one might expect to see enhanced catecholamine fluorescence in the striatum adjacent to adrenal medulla grafts, but this phenomena has not been observed (Freed et al., 1981; 1983; Stromberg et al. 1985; Becker and Freed, 1988a). It is nevertheless possible that a relatively small number of DA neurons have recovered or regenerated as a function of the adrenal medulla grafts, and that DA release and turnover are greatly increased in these neurons. In fact, trophic effects of cortical injury on grafted DA-containing neurites have also been reported (Freed et al., 1987; Freed and Cannon-Spoor, 1988).

It is also possible that DA secreted by the grafted adrenal medulla cells gains access to the striatum via the local circulatory system. Survival of adrenal medulla grafts depends on the development of a blood supply by anastomosis between blood vessels of the grafted tissue and blood vessels of the host (Rosenstein and Brightman, 1986). The anastomosis between grafted peripheral tissue and striatal blood vessels results in a loss of the blood-brain barrier at or near the junction and an apparent increase in the permeability of the associated host blood vessels to blood-borne proteins (Rosenstein, 1987). Therefore, the integrity of the blood-brain barrier is compromised not only within the graft but also in the host brain adjacent to the graft. This loss of the blood-brain barrier could allow DA released from adrenal chromaffin cells to enter the striatum. In order to test this hypothesis, we have measured serum DA concentrations in peripheral blood of animals with adrenal medulla grafts. Serum DA concentrations were found to be elevated compared to animals without adrenal medulla grafts (Becker and Freed, 1988b). More importantly, serum concentrations of DA were directly correlated (rho= -0.72; p<0.05) with the magnitude of the decrease in rotational behavior. In other words, animals with the highest serum DA concentrations showed the greatest decrease in rotational behavior (Becker and Freed, 1988b).

CONCLUSION

Adrenal medulla grafts increase DA turnover, amphetamine-stimulated striatal DA release, and DA concentrations in blood without producing an increase in CSF concentrations of DA or extracellular concentrations of DA in striatum. Moreover, the concentrations of DA in blood and the restoration of the striatal DA response to amphetamine are related to the behavioral efficacy of the grafts. Previous studies have reported that the blood-brain barrier is compromised adjacent to adrenal medulla grafts (Rosenstein, 1987) and a diffuse catecholamine fluorescence has been observed within the striatum adjacent to adrenal medulla grafts (Becker and Freed, 1988a). The present results, therefore, suggest that grafted adrenal chromaffin cells release DA into blood vessels. DA is thereby transported via the local circulatory system and gains access to striatal neurons through leaky blood vessels adjacent to the graft. If this is occurring, then in animals with adrenal medulla grafts the remaining (or rescued) DA terminals in striatum may take up DA that diffuses from the blood system and re-release it in response to amphetamine. Alternatively, the amphetamine-stimulated increase in striatal DA may reflect the release of DA from grafted chromaffin cells and its subsequent diffusion into the striatum via the circulatory system. Finally, while other beneficial effects of adrenal medulla grafts cannot be ruled out at this time, the results presented here suggest that the production of DA plays an important role in their behavioral efficacy.

The research reported here is a first step towards identifying the critical variables that may distinguish successful adrenal medulla grafts from unsuccessful ones. Whether the same mechanisms are involved in mediating the success of fetal substantia nigra grafts is not known at this time. The delineation of the mechanism(s) through which grafts of adrenal medulla tissue induce behavioral recovery of function following unilateral nigrostriatal-dopamine depletion may be very informative for development or evaluation of therapies used to treat patients with Parkinson's Disease. For example, if adrenal medulla grafts only provide the striatum with a source of dopamine that the host neurons take up and re-release, then with the continued degeneration of striatal dopamine neurons in Parkinson's Disease, eventually the grafts may lose their effectiveness in much the same way as L-DOPA loses its effectiveness. In contrast, if it is simply the presence of dopamine at dopamine receptors in striatum (to produce a decrease in dopamine receptor supersensitivity) that is responsible for the alleviation of symptoms, then an implantable source of dopamine (eg. Hargraves and Freed, 1987) may be superior to grafting procedures for the treatment of the majority of patients. On the other hand, an adrenal medulla graft may be an active, functional source of dopamine. While it is unlikely that they respond to changes in the ongoing behavior of the animal, adrenal medulla grafts may be responsive to pharmacological or hormonal agents. It so, appropriate medication could optimize dopamine release from the graft to suit the patient's changing needs. A fourth possibility (and the last two possibilities are not mutually exclusive) is that adrenal medulla grafts have other trophic effects on the host striatum (Bohn et al., 1987). If sprouting is occurring, it is possible that when grafting methods are optimized, adrenal medulla grafts may actually slow the progression of Parkinson's Disease. If these grafts also provide striatal dopaminergic neurons with dopamine, the prognosis for long-term relief of symptoms could be very good. Obviously, more research is required to address these questions.

ACKNOWLEDGMENTS

This research was supported by grant to JBB from the NIH (NS22157) and by grant from the American Parkinson's Disease Association. The authors would like to thank Dr. T.E. Robinson for his assistance in the development of the dialysis method and for comments on the manuscript.

REFERENCES

1. Albrink, W.S. and Greene, H.S.N., 1953, The transplantation of tissues between zoological classes, Cancer Research, 13:64-68.
2. Anden, N.E., Dahlstrom, A., Fuxe, K., Larsson, K., Olson, L. and Ungerstedt, U., 1966b, Ascending monoamine neurons to the telencephalon and diencephalon, Acta Physiol. Scand, 67:313-323.
3. Anden, N.E., Fuxe, K., Hamberger, B. and Hokfelt, T., 1966a, A quantitative study of the nigro-neostriatal dopamine neuron system in the rat, Acta Physiol. Scand, 67:306-312.
4. Arbuthnott, G.W. and Ungerstedt, U., 1975, Turning behavior induced by electrical stimulation of the nigro-neostriatal system of the rat, Exp. Neurol., 47:162-172.
5. Backlund, E-O., Granberg, P-O., Hamberger, B., Knutsson, E., Martens-on, A., Sedvall, G., Seiger, A. and Olson, L., 1985, Transplantation of adrenal medullary tissue to striatum in parkinsonism. First clinical trials, J. Neurosurg, 62:169-173.
6. Becker, J.B., Adams, F., and Robinson, T. E., 1988, Intraventricular microdialysis: A new method for determining the concentrations of monoamine metabolites in the CSF, J. Neuroscience Meth., (IN PRESS).
7. Becker, J.B. and Freed, W.J., 1988a, Neurochemical correlates of behavioral changes following intraventricular adrenal medulla grafts: in vivo microdialysis in freely moving rats, Progress in Brain Research, (IN PRESS).
8. Becker, J.B. and Freed, W.J., 1988b, Adrenal medulla grafts enhance functional activity of the striatal dopamine system following substantia nigra lesions, (SUBMITTED).
9. Bjorklund, A., Stenevi, U. and Svengaard, N.A., 1976, Growth of transplanted monoaminergic neurones into the adult hippocampus along the perforant path, Nature, 262:787-790.
10. Bjorklund, A., Dunnett,S.B., Stenevi, U., Lewis,M.E., and Iversen, S.D., 1980a, Reinnervation of the denervated striatum by substantia nigra transplants: functional consequences as revealed by pharmacological and sensorimotor testing, Brain Res., 199:307-333.
11. Bjorklund, A., Schmidt, R.H. and Stenevi, U., 1980b, Functional reinnervation of the neostriatum in the adult rat by use of intrapar-enchymal grafting of dissociated cell suspensions from the substantia nigra, Cell Tissue Res., 212:39-45.
12. Bohn, M.C., Cupit, L., Marciano,F. and Gash, D.M., 1987, Adrenal medulla grafts enhance recovery of striatal dopaminergic fibers, Science, 237:913-916.
13. Christie, J.E. and Crow, T.J., 1971, Turning behavior as an index of the action of amphetamines and ephedrines on central dopamine-containing neurons, Brit. J. Pharmacol., 43:658-667.
14. Creese, I., Prosser, T., and Snyder, S.H., 1978, Dopamine receptor binding: specificity, localization and regulation by ions and guanyl nucleotides, Life Sci., 23, 495-500.
15. Creese, I. and Snyder, S.H., 1979, Nigrostriatal lesions enhance striatal {³H} spiroperidol binding, Eur. J. Pharmacol, 56:227-281.
16. Dahlstrom, A., and Fuxe, K., 1964, Evidence for the existence of monoamine-containing neurons in the central nervous system..., Acta Physiol Scand, 62:1-55.

17. Dunn, E.H., 1917, Primary and secondary findings in a series of attempts to transplant cerebral cortex in the albino rat, J. Comp Neurol, 27:565-582.

18. Dunnett, S.B., Bjorklund, A., Stenevi, U. and Iversen, S.D., 1982c, in: "Chemical Transmission in the Brain", R.M. Buijs, P. Pevet and D.F. Swaab eds., Progress in Brain Research Vol. 55.

19. Dunnett, S.B., Bjorklund, A. and Stenevi, U., 1983, Transplant-induced recovery from brain lesions: a review of the nigrostriatal model, in: "Neural Tissue Transplantation Research", R.B. Wallace and G.D. Das eds., Springer-Verlag, New York, p. 191.

20. Dunnett, S.B. and Iversen, S.D., 1982b, Sensorimotor impairments following localized 6-hydroxydopamine and kainic acid-induced lesions of the neostriatum, Brain Res, 248:121-127.

21. Dunnett, S.B. and Iversen, S.D., 1982a, Regulatory impairments following selective 6-OHDA lesions of the neostriatum, Behav Res, 4:195-202.

22. Freed, W.J., 1983, Functional brain tissue transplantation: reversal of lesion-induced rotation by intraventricular substantia nigra and adrenal medulla grafts with a note on intracranial retinal grafts, Biol. Psychiat., 18:1205-1267.

23. Freed, W.J. and Cannon-Spoor, H.E., 1988, Cortical lesions increase reinnervation of the c striatum by substantia nigra grafts, Brain Res, 446:133-143.

24. Freed, W.J. Cannon-Sppor, H.E., de Beaurepaire, R., Greenberg, J.A. and Schwarz, S.S., 1987, Embryonic substantia nigra grafts: Factors controlling behavioral efficacy and reinnervation of the host striatum, in: "Cell and Tissue Transplantation into the Adult Brain", E. Azmitia and A. Bjorklund, Eds., Annals of the New York Academy of Sciences, 495:581-596.

25. Freed, W.J., Hoffer, B.J., Olson, L. and Wyatt, R.J., 1984, Transplantation of catecholamine-containing tissue to restore the functional capacity of the damaged nigrostriatal system, in: "Neural Transplants", J.R. Sladek and D.M. Gash, Eds., Plenum Publ. Co., New York, pp 373-406.

26. Freed, W.J., Karoum, F., Spoor, H.E. Olson, L., Morihisa, J. and Wyatt, R.J., 1983, Catecholamine content of intracerebral adrenal medulla grafts, Brain Res, 269:184-189.

27. Freed, W.J., Morihisa, J.M., Spoor, H.E., Hoffer, B.J., Olson, L., Seiger, A. and Wyatt, R.J., 1981, Transplanted adrenal chromaffin cells in rat brain to reduce lesion-induced rotational behavior, Nature, 292:351-352.

28. Freed, W.J., Perlow, M.J., Karoum, F., Seiger, A., Olson, L., Hoffer, B.J. and Wyatt, R.J., 1980, Restoration of dopaminergic function by grafting of fetal rat substantia nigra to the caudate nucleus: long-term behavioral, biochemical, and histochemical studies, Ann Neurol, 8:510-519.

29. Hargraves, R. and Freed, W.J., 1987, Chronic intrastriatal dopamine infusions in rats with unilateral lesions of the substantia nigra, Life Sciences, 40:959-966.

30. Hokfelt, T. and Ungerstedt, U., 1969, Electron and fluorescence microscopical studies on the nucleus caudatus putamen of the rat after unilateral lesions of the ascending nigro-neostriatal dopamine neurons, Acta Physiol Scanda, 76:415-426.

31. Hornykiewicz, O., 1966, Dopamine (3-hydroxytyramine) and brain function, Pharmacol. Rev., 1966:925-964.

32. Hornykiewicz, O. and Kish, S.J., 1987, Biochemical pathophysiology of Parkinson's Disease, Adv Neurol, 45:19-34.

33. Hutson, P.H., Sarna, G.S., and Curzon, G., 1984, Determination of the clearance of their acidic metabolites in conscious rats by repeated sampling of cerebrospinal fluid, J. Neurochem, 43:291-293.

34. Kreuger, B.K., Forn, J., Walters, J.R., Roth, R.N. and Greengard, P., 1976, Stimulation by dopamine of adenosine cyclic 3', 5'-monoaphosphate formation in rat caudate nucleaus: effect of lesions of the nigrostriatal pathway, Mol Pharmacol, 12:639-648.

35. Le Gros Clark, W.E., 1940, Neuronal differentiation in implanted foetal cortical tissues, J. Neurol Psychiat, 3:263-272.

36. Lindvall, O., Backlund, E-O, Farde, L., Sedvall, G., Freedman, R., Hoffer, B., Nobin, A., Seiger, A. and Olson, L., 1987, Transplantation in Parkinson's Disease: two cases of adrenal medullary grafts to the putamen, Ann Neurol., 22:457-468.

37. Lund, R.D. and Harvey, A.R., 1981, Transplantation of fetal tectal tissue in rats. I. Organization of transplants and pattern of distribution of host afferents within them, J Comp Neurol, 201:191-209.

38. Madrazo, I., Drucker-Colín, R., Díaz, V., Martínez-Mata, J., Torres, C. and Becerril, J.J., 1987, Open microsurgical autograft of adrenal medulla to the right caudate nucleus in two patients with intractable Parkinson's Disease, N. Engl. J. Med., 316:831-834.

39. Marshall, J.F., Richardson, J.S. and Teitelbaum, P., 1974, Nigrostriatal bundle damage and the lateral hypothalamic syndrome, J. Comp. Physiol Psychol. 87:808-830.

40. Medawar, P.B., 1948, Immunity to homologous grafted skin. II. The fate of skin homografts to the brain, to subcutaneous tissue, and to the anterior chamber of the eye, Brit. J. Exp. Pathol, 29:58-69.

41. Moore, R.Y. and Bloom, F.E., 1978, Central catecholamine neuron systems: anatomy and physiology of the dopamine systems, Ann Rev Neurosci, 1:129-169.

42. Murphy, J.E. and Sturm, E., 1923, Conditions determining the transplantability of tissue in the brain, J. Exp. Med, 38:183-197.

43. Olson, L., Seiger, A., Ebendal, T. and Hoffer, B.,1980, Advances in Biochemical Psychopharmacology, vol. 25, Raven Press, New York, pp 27-34.

44. Perlow, M.J., Freed, W.J., Hoffer, B.J. Seiger, A., Olson, L. and Wyatt, R.J., 1979, Brain grafts reduce motor abnormalities produced by destruction of the nigrostriatal dopamine system, Science, 204:643-647.

45. Rinne, U.K., Klinger, M.and Stamm, G., 1980, Parkinson's disease: Current Progress, Problems and Management, Elsevier Biomedical Press, Amsterdam.

46. Rosenstein, J.M., 1987, Adrenal medulla grafts produce blood-brain barrier dysfunction, Brain Res., 414:192-196.

47. Rosenstein, J.M. and Brightman, M.W., 1978, Intact cerebral ventricle as a site for tissue transplantation, Nature, 276:83-85.

48. Rosenstein, J.M. and Brightman, M.W., 1986, Alterations of the blood-brain barrier after transplantation of autonomic ganglia into the mammalian central nervous system, J. Comp. Neurol,250:339-351.

49. Schultz, W. and Ungerstedt, U., 1978, Striatal cell supersensitivity to apomorphine in dopamine-lesioned rats correlated to behavior, Neuropharmacology, 17:349-353.

50. Stromberg, I., Herrena-Marschitz, M., Ungerstedt, U., Ebendal, T., Olson, L., 1985, Chronic implants of chromaffin tissue into the dopamine-denervated striatum: Effects of NGF on grafts survival, fiber growth, and rotational behavior, Exp. Brain Res. 60:335-349.

51. Ungerstedt, U., 1971a, Striatal dopamine release after amphetamine or nerve degeneration revealed by rotational behavior, Acta Physiol. Scand., 82, Suppl. 367, 49-68.

52. Ungerstedt, U., 1971b, Postsynaptic supersensitivity after 6-hydroxy-dopamine induced degeneration of the nigrostriatal dopamine system, Acta Physiol Scand, 82:69-92.

53. Ungerstedt, U., 1971c, Postsynaptic supersensitivity after 6-hydroxy-dopamine induced degeneration of the nigro-striatal dopamine system, Acta Physiologica Scand. (Suppl 367) 82:69-92.

54. Ungerstedt, U., 1971d, Adipsia and aphasia after 6-hydroxydopamine induced degeneration of the nigrostriatal dopamine system, Acta Physiologica Scand. (Suppl 367) 82:95-122.

55. Ungerstedt, U. and Arbuthnott, G.W., 1970, Quantitative recording of rotational behavior in rats after 6-hydroxydopamine lesions of the nigrostriatal dopamine system, Brain Res., 24, 485-493.

56. Unsicker, K., Krisch, B., Otten, U. and Thoen, H., 1978, Nerve growth factor-induced fiber outgrowth from isolated rat adrenal chromaffin cells: impairment by glucocorticoids, Proc Natl Acad Sci, USA, 78:3498-3502.

57. Unsicker, K., Rieffert, B. and Ziegler, W., 1980, Advances in Biochemical Psychopharmacology, Vol. 25, Raven Press, New York, pp. 51-59.

NERVE GROWTH FACTOR IN VIVO ACTIONS ON CHOLINERGIC NEURONS

IN THE ADULT RAT CNS

Silvio Varon, Theo Hagg, H. Lee Vahlsing and Marston Manthorpe

Department of Biology, M-001, School of Medicine
University of California, San Diego
La Jolla, CA 92093 USA

INTRODUCTION

All living cells, including those in the nervous system, are subjected to extrinsic modulatory influences by agents in their microenvironment, be they humoral agents in the extracellular fluid or anchored agents in the extracellular matrix or on the surface of other cells. As long as such modulatory influences apply, cellular behaviors will reflect at any time the balance between stimulatory and inhibitory agents that reach the cell. For nerve cells, two classes of protein agents ("factors") have drawn increasing attention from the neurobiologist and, more recently, the exper- imental neuropathologist as well. NeuronoTrophic Factors (NTFs) are spe- cial proteins controlling survival, growth and functional capabilities of selected populations of neurons. Neurite-Promoting Factors (NPFs) specifi- cally stimulate the outgrowth of neuronal processes ("neurites") and may assist, but not substitute for, the required NTFs.

NTFs and NPFs draw conceptual strength and factual information from a variety of neurobiologic observations (Varon and Adler, 1981; Varon, 1985; Varon et al., 1988a,c). Most of them have been obtained from the study of neurons in the peripheral nervous system (PNS), during early development and, most often, by use of in vitro neuronal cultures. More recently, these concepts have been extended to the central nervous system (CNS) of adult mammals, leading to the articulation of the CNS neuronotrophic hypothesis illustrated in Table 1 (Varon et al., 1984, 1988a,b). If adult CNS neurons continue to depend on their endogenous NTFs for maintenance, function and repair, then a relative or actual deficit in endogenous NTF support should lead to neuronal dysfunction and degeneration and could underlie certain degenerative disorders as well as contribute to the known failure of adult CNS neurons to regenerate. If so, both such conditions may be alleviated by appropriate administration of the corresponding exogenous factors. Experi- mental evidence in support of this hypothesis has been obtained in the past two years with regard to at least one NTF (Nerve Growth Factor) and one adult CNS system (forebrain cholinergic neurons) (Hefti, 1986; Williams et al., 1986; Kromer, 1987; Fischer et al., 1987; Davis et al., 1987b).

Table 1. The CNS Neurotrophic Hypothesis

1. ADULT CNS Neurons Continue to Depend on their NTFs
   (for Maintenance, Function and Repair)
2. ENDOGENOUS NTFs are Supplied by their Innervation Territories
   (Post-synaptic Partners, Glia)
3. DEFICITS of Endogenous NTF Lead to Neuronal Damage
   (Dysfunction, Hypotrophy, Degeneration)
4. EXOGENOUS NTF ADMINISTRATION Prevents/Corrects Damage
   (Acute and Chronic Lesions)
5. NTF TREATMENTS May Help in Human Pathologies
   (Degenerative Diseases, Aging, Regeneration)

In this chapter, we shall review first some concepts and data concerning neuronotrophic and neurite promoting factors and then the recent information about their in vivo involvement in adult rat CNS repair. We shall conclude with a few brief comments and speculations regarding the possible use of such agents as therapeutic tools in human CNS pathologies.

NEURONOTROPHIC AND NEURITE-PROMOTING FACTORS

After a set of nerve cells has advanced from the proliferative (neuroblastic) to the postmitotic (neuronal) stage and has extended axons to the appropriate innervation territory, it undergoes a massive developmental neuronal death the extent of which depends on the size of the available innervation territory (Hamburger and Oppenheim, 1982; Cowan et al., 1984). It has been speculated that the neurons that survive this developmental death event are those that succeed in acquiring neuronotrophic factors from their innervation territory, binding them to specific receptors on the surface of their nerve endings, internalizing them by an endocytotic process, and bringing them by retrograde axonal transport to the neuronal somata where the NTF would exert its life-sustaining action (Varon and Adler, 1980).

In the late 1940's, Levi-Montalcini and collaborators discovered the first of these NTFs and called it Nerve Growth Factor, or NGF (Levi-Montalcini, 1966, 1982, 1987). The fortunate abundance of NGF in mouse submaxillary glands led to the subsequent purification and characterization of the mouse NGF protein, the generation of anti-NGF antibodies and, ultimately, the cloning of the human as well as the mouse NGF gene (Varon, 1975; Greene and Shooter, 1980; Ullrich et al., 1983). The mouse NGF is a basic, dimeric, 26 kda (kilodalton) protein. It traditionally addresses PNS neurons in dorsal root and sympathetic ganglia (as well as adreno-chromaffin cells and their tumoral derivatives, e.g. the PC12 clonal line) and promotes their survival, neuritic outgrowth and/or transmitter production depending on their developmental age. NGF target cells present two types of NGF receptors: i) the slow-dissociating and internalizing type I receptors, whose high affinity (10^{-11}M) matches the range of NGF activity, and ii) the faster-dissociating, lower affinity (10^{-9}M) type II receptors, for which a monoclonal antibody is now available and whose rat gene has been cloned. While initially thought to act only as a humoral agent, NGF was recently shown also to operate after anchorage to a substratum in both its neuronotrophic and its neurite-promoting capacities (Gundersen, 1985; Pettmann et al., 1988). A dramatic turn in the NGF story has been new evidence over the past five years that NGF also meets the criteria for an NTF addressing cholinergic neurons in the developing and adult rat CNS (cf. Varon et al., 1988a; Whittemore and Seiger, 1987; see also further on).

The NGF phenomenon has prompted the search for new NTFs. Several source materials have displayed trophic activities for various neuronal populations in vitro, but only a few new NTF proteins have been characterized thus far and none has yet reached the information level achieved with NGF. Brain Derived Neurotrophic Factor (BDNF) is a basic 24 kda protein obtained from pig brain, which acts on several PNS sensory ganglionic neurons (Barde et al., 1982). Neuroleukin is a monomeric, 56 kda protein which was recently purified from mouse submaxillary gland, is also present in muscle extracts, and addresses spinal motor neurons as well as dorsal root ganglion sensory neurons (Gurney et al., 1986). Ciliary Neuronotrophic Factor (CNTF) is a chick eye-derived acidic, monomeric, 24 kda protein addressing the cholinergic PNS neurons of ciliary ganglia and present in their intraocular innervation territory in spatial and temporal correlations with their developmental neuronal death event (Manthorpe and Varon, 1985). A similar CNTF is present in adult rat CNS tissue where it accumulates after mechanical or chemical lesions (Manthorpe et al., 1983b; Nieto-Sampedro et al., 1982, 1983), and it has been purified from adult rat peripheral nerve as an acidic, monomeric, 28 kda protein (Manthorpe et al., 1986b, 1988). Beside the ciliary ganglionic neurons, CNTF also addresses dorsal root and sympathetic ganglionic neurons (as well as adreno-chromaffin cells), thereby overlapping the neuronal spectrum of NGF action. CNTF can operate from a surface-anchored state as does NGF, but does not appear to stimulate neuritic outgrowth under such circumstances (Carnow et al., 1985; Rudge et al., 1987; Pettmann et al., 1988). Lastly, one should mention that neuronotrophic activities for CNS neurons have been recently reported for certain Growth Factors (i.e. proteins known to promote proliferation of various cells, including glial ones -- cf. Bradshaw and Prentis, 1987), specifically the basic Fibroblast Growth Factor FGF (Walicke et al., 1986; Morrison et al., 1986) and Epidermal Growth Factor EGF (Morrison et al., 1987).

Neuronal cell cultures have become an indispensible tool for the recognition and quantitation of NTF activities, their monitoring during NTF purification, and the investigation of NTF mechanisms of action (Varon et al., 1983). It is also the investigation of neural cultures which has led to the discovery of a second class of neuron-modulating proteins, the neurite-promoting factors. NPFs were first found in conditioned media from glial and other cell cultures. They anchor themselves particularly well to polycationic substrata (e.g., polyornithine- or polylysine-coated dishes) and, once anchored, strongly stimulate neuritic outgrowth from both PNS and CNS neurons (Varon and Adler, 1980, 1981). Characterization of a polycation-binding NPF from Schwannoma-conditioned medium led to the examination of several purified constituents of extracellular matrix and to the identification of laminin as a most potent NPF (Manthorpe et al., 1983a; Davis et al., 1985 a,b). Laminin is a large (900 kda) glycoprotein which is present in the basal lamina component of extracellular matrices and is particularly abundant in peripheral nerve, where it may play the crucial role of promoting and guiding peripheral axon regeneration. In the adult CNS, however, laminin is not usually detectable except in the basal lamina separating astroglial endfeet from blood vessels.

Another important contribution by in vitro neural cultures is the opportunity to evaluate identifiable cell populations as sources of NTFs and NPFs. With such an approach, NGF, CNTF and laminin-containing NPFs have all been shown to be produced and externalized in vitro by glial cell cultures of both peripheral (Schwann cells) and central (astroglia) origin (Varon and Manthorpe, 1982; Manthorpe et al., 1986a). NGF production by Schwann cells has subsequently been confirmed in adult rat peripheral nerve in vivo (e.g. Taniuchi et al., 1988). The recognition of glial cells as potential sources of neuronotrophic and neurite-promoting factors adds another dimension to the close partnership between neurons and glia already recognized in several other respects (cf. Varon and Somjen 1979).

The cholinergic system of the basal forebrain in adult rats provides an advantageous model for an in vivo evaluation of the CNS neuronotrophic hypothesis.

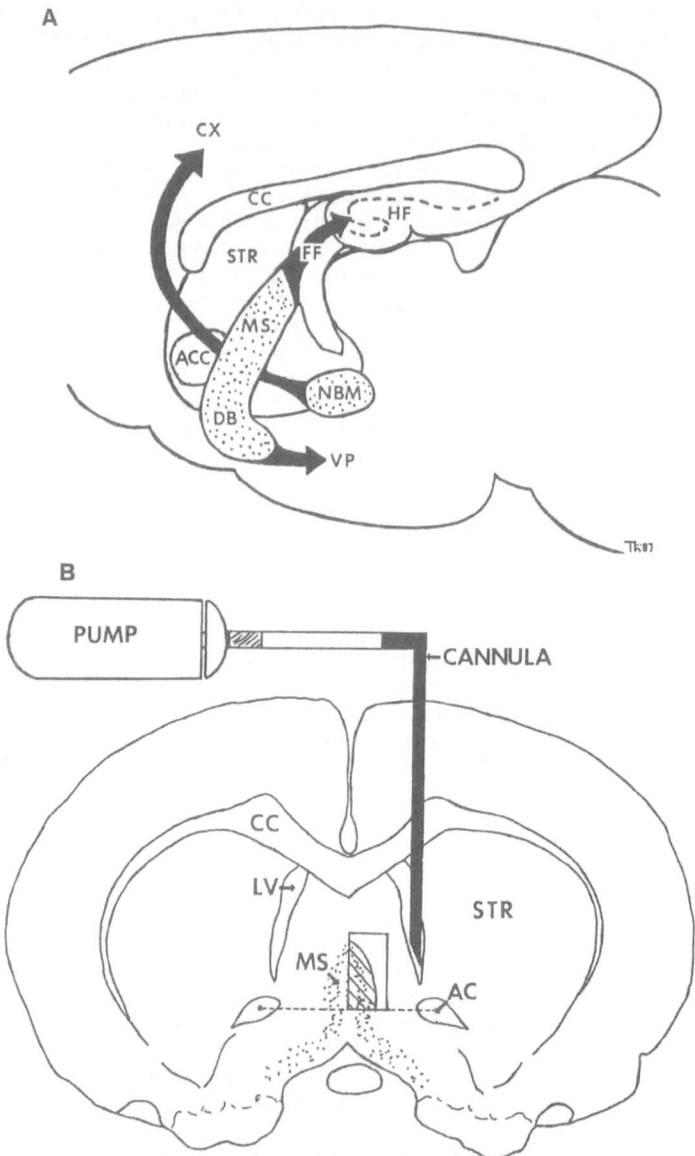

Fig. 1.   The septo-hippocampal model. A. Cholinergic neurons in the basal forebrain and their projections. MS, DB, NBM = medial septum, diagonal band and nucleus basalis magnocellularis. CX, CC, HF, STR, ACC (open areas) = neocortex, corpus callosum, hippocampal formation, striatum, nucleus accumbens. Arrows = projection pathways (FF = fimbria-fornix; VP = ventral pathway). B. Coronal section diagram illustrating the infusion device, location of the cannula tip in one lateral ventricle (LV), and the MS region to be counted for ChAT-positive MSC neurons.

As shown schematically in Fig. 1 A, cholinergic neurons in the medial septum (as well as in the vertical limb of the diagonal band) project their axons to the hippocampal formation largely through a well defined tract, the fimbria-fornix, on each side of the brain. Complete surgical transection of the fimbria-fornix can be carried out unilaterally without substantially affecting the corresponding contralateral structures. Such an acute lesion should interrupt the retrograde delivery of hippocampal endogenous NGF to the medial septum cholinergic (MSC) neurons and impose on them a neuronotrophic deficit. The resulting neuronal damage should, in turn, be minimized by a continuous administration of exogenous NGF, for example via a cannula implanted in the lateral ventricle on the same side of the lesion and connected with an osmotic minipump containing the NGF in an appropriate, stabilizing vehicle (Fig. 1 B). Both predictions have been unequivocally validated by three independent groups using different variations of this septo-hippocampal model (Hefti, 1986; Williams et al., 1986; Kromer, 1987).

Figure 2 provides a histologic overview of the effects on MSC neurons by the lesion and the NGF treatment. In cerebral coronal sections that display the medial septum on both sides, their subpopulation of cholinergic neurons can be distinguished from the other, more numerous nerve cells by histochemical staining for acetylcholinesterase (AChE) or immunochemical staining for choline acetyltransferase (ChAT), the two enzymes that control degradation and synthesis of the neurotransmitter acetylcholine. The unilateral fimbria-fornix transection causes within two weeks a massive disappearance of MSC neurons, providing a dramatic contrast between the contralateral (intact) and the ipsilateral (axotomized) sides (compare A and B). Continuous intraventricular infusion of NGF during the same two weeks prevents almost entirely this disappearance of MSC neurons (compare C and D; see also further on, Fig. 3).

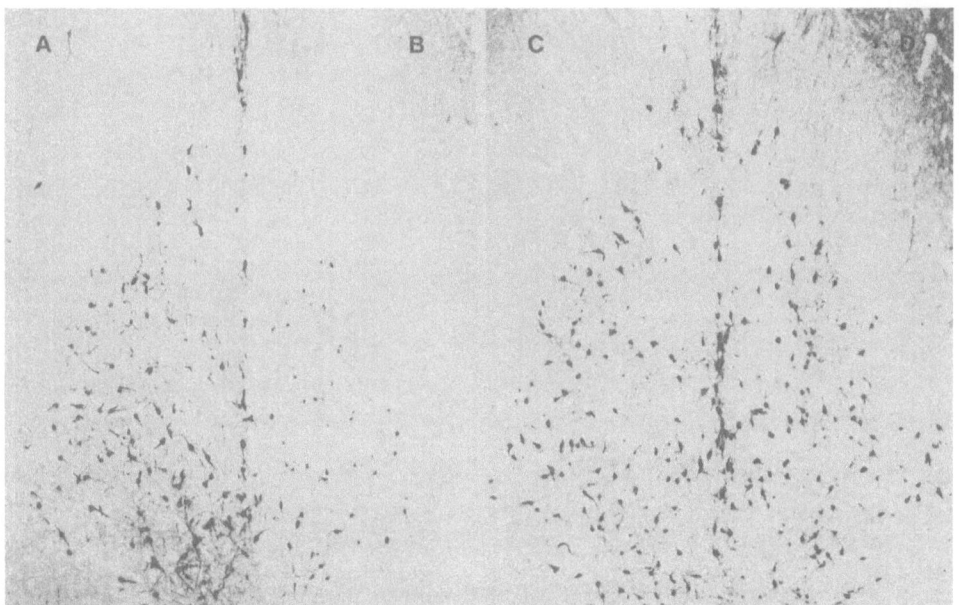

Fig. 2.   ChAT immunostaining in coronal sections of the medial septum. A,B = contralateral and ipsilateral sides 14 days after unilateral fimbria transection. C,D = same, in NGF-treated animals. Note the disappearance of MSC neurons in B and its prevention by NGF in D.

The axotomy-induced disappearance of MSC neurons and their protection by NGF were initially thought to reflect an actual death of these cells and their NGF-promoted survival, respectively — in line with the traditional role of NTFs on their target neurons. It is known, however, that in peripheral neurons disconnected from their innervation territory degeneration is increasingly delayed as the neurons advance toward their mature state. An alternate interpretation of the adult MSC neuronal disappearance, therefore, could be that they had lost their specialized functional markers (i.e. the cholinergic enzymes) by which they are identified, without necessarily undergoing irreversible damage over the time period investigated. That this was in fact the case has been demonstrated by more recent experiments (Hagg et al. 1988), illustrated in Figure 3.

In these experiments, adult rats with unilateral fimbria-fornix transection received 2-week long NGF infusions that started at different times after the main surgery. Undelayed NGF treatments (a) provide 100% protection against MSC disappearance. Delaying the NGF treatment by 7 days (b), at which time 50% of the MSC neurons have already disappeared, resulted in the reappearance of nearly all the "lost" cells. NGF treatments delayed by 14 (c) or 21 (d) days — at which times the maximal loss of 80% of the MSC neurons had already occurred — were also able to raise the number of ChAT-immunostainable neurons, although they failed to achieve a full recovery. The incompleteness of the latter recoveries may reflect a no longer adequate treatment (e.g. dose, duration) or may point to the progressive development of irreversible damages. In any case, these new results have several important implications, among them: (1) there may be a "grace" period between lesion and onset of treatment, which should greatly facilitate future clinical approaches, and (2) NGF can improve, among other cellular features, the functional capabilities of an adult, already damaged CNS neuron.

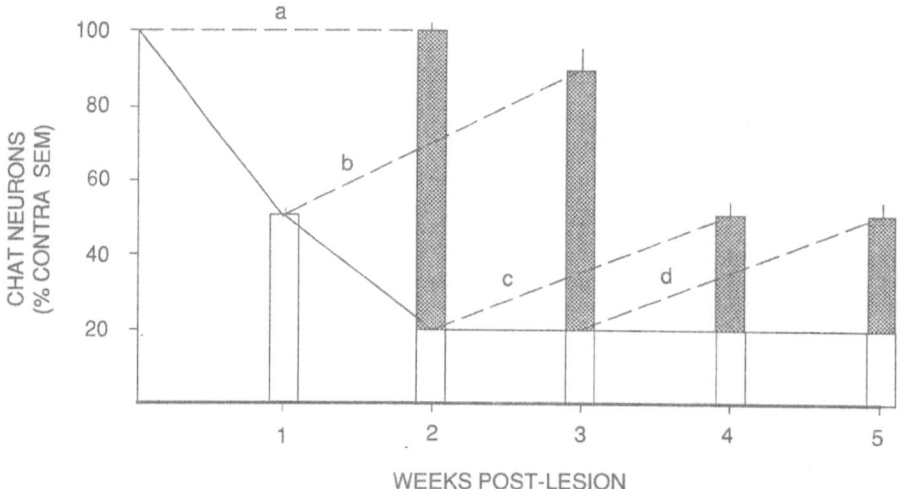

Fig. 3. Prevention and reversal of MSC neuron disappearance by 14-days treatments with NGF. Open bars = residual number of neurons without NGF treatments. Shaded bars = NGF-supported additional neurons. a: treatment started at lesion time. b,c,d = treatments started 1, 2 or 3 weeks after the lesion.

The cognitive activities of the brain (i.e. memory and learning processes) are carried out with the crucial participation of CNS cholinergic systems (cf. Fig. 1A), including the projections to the hippocampal formation by septal neurons and those to the neocortex by cholinergic neurons of the nucleus basalis. Accordingly, CNS cholinergic deficits accompany, and possibly underlie, the cognitive behavioral deficits which characterize brain aging as well as Alzheimer's disease (e.g. Bartus et al., 1982; Coyle et al., 1983; Hepler et al., 1985). If chronic cholinergic impairments, like those elicited by an acute experimental lesion, were to reflect an insufficient supply or utilization of endogenous NTFs (cf. Table 1) they may derive some benefits from exogenous NTF administration (Hefti, 1983). A recently published study on NGF treatments of aged rats strongly encourages such speculations (Fischer et al., 1987).

Two-year old rats were screened on a Morris watermaze task for their ability to identify and remember the location of a platform submerged (and thus not visible) in a swimming pool. A rat placed into the water facing the pool wall will swim at random until it bumps against the platform and climbs on it to escape the water. After a few trials, a normal rat will considerably shorten its swimming path and the time needed for a successful escape. In contrast, a cognitively impaired rat will improve its performance only after many more trials and possibly never reach the full efficiency of a normal one. Out of the 60 aged rats examined, 8 animals were selected for a normal (i.e. behaviorally unimpaired) performance and 24 others for a severely impaired behavior. Two months later, the impaired rats were fitted unilaterally with an intraventricular infusion device similar to that used in the acute lesion experiments (cf. Fig. 1B), and infused for 4 weeks with either vehicle only or vehicle + NGF. Both groups as well as the normally-behaving control set were re-tested at 2 and 4 weeks. The main results are summarized in Table 2.

Table 2. NGF Effects on Cognitive Behavior and Cholinergic CNS Cells in the Aged Rat

	Unimpaired	Impaired + vehicle	Impaired + NGF
ESCAPE LATENCY (sec $\pm$ SEM) (Number of animals)	(8)	(13)	(11)
Before treatment	8 ± 2	17 ± 5	18 ± 4
After 4-wk treatment	10 ± 5	15 ± 7	9 ± 4[a]
NUCLEUS BASALIS (Number of animals)	--	(4)	(5)
Neuronal size (μm^2) Left side (uninfused)	--	165 ± 4	156 ± 6
Right side (infused)	--	162 ± 4	185 ± 10[a]

[a] p < 0.01 compared to other value in the box

Although 2 weeks of NGF infusion did not appear to elicit any behavioral improvement, animals treated with NGF for 4 weeks revealed a learning and retention capability that was no longer distinguishable from that of the normal, unimpaired rats. Moreover, morphometric analyses of some of the initially impaired rat brains showed that the 4-week NGF treatment also increased significantly the size of nucleus basalis cholinergic cell bodies on the infusion side relative to the contralateral one (or the vehicle-treated animals) -- although these neurons remained considerably smaller than those of young rats (270 μm^2 $\pm$ 21).

These observations raise a number of important considerations for future research. First, the "atrophy" of CNS cholinergic neurons reported to accompany brain aging does not necessarily imply a permanent loss of these neurons, since NGF could increase their size. Secondly, NGF is able also to benefit chronically impaired CNS neurons, and these benefits extend beyond the neuronal cells to animal behaviors mediated by them. Thirdly, the effects of NGF on these "aged" neurons did not appear to reflect merely a protection from further damage or a stimulation of transmitter enzyme, since the 4 weeks of treatment needed are too short a period for the former and too long a period for the latter. Thus, one might be inclined to speculate about additional effects on reconstruction and/or reactivation of functional synapses in the cholinergic innervation territories.

AXONAL REGENERATION IN THE ADULT CNS

It is now well established that many adult rat CNS neurons retain the capability of growing new axons (or axonal branches) and will do so when provided with an appropriate terrain in which to grow them -- specifically, segments of peripheral nerves (Aguayo et al., 1982; Aguayo, 1985). Once again, therefore, the successful execution of a neuronal behavior depends in large part if not entirely on the nature of the neuronal microenvironment. Peripheral nerve promotes PNS axonal regeneration largely because of its Schwann cells. Schwann cells are not only arrayed in longitudinal chains (the Bungner bands) to provide oriented guidance, they are also known sources of both neuronotrophic factors such as NGF and CNTF and of neurite promoting factors such as the laminin of their basal laminae.

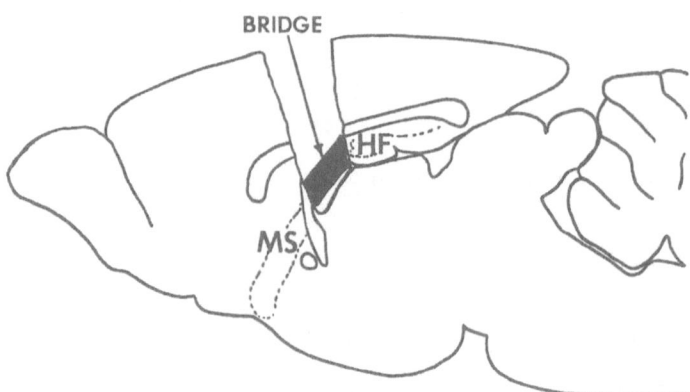

Fig. 4. Sagittal brain section diagram illustrating the location of the fimbria-fornix lesion and the emplacement of a regeneration bridge. MS = medial septum. HP = hippocampal formation.

The first requisite for CNS regeneration is to secure the maintenance of the axotomized neuron (cf. Varon, 1977). In the septo-hippocampal cholinergic model, this is now adequately achieved by administration of exogenous NGF to the MSC neurons. NGF is also known to stimulate neuritic growth, and indeed cholinergic fiber sprouting has been noted within the septum itself, ipsilateral to the NGF delivery (Williams et al., 1986). The next task, therefore, is to provide a "regeneration bridge" across the fimbria-fornix lesion, which would both promote and guide the NGF-supported cholinergic axons to traverse the gap and re-innervate their hippocampal territory — as schematically illustrated in Fig. 4. Three questions need to be addressed for an optimal solution: i) the nature of the bridge, ii) the entry problem (from the septum into the bridge) and iii) the exit problem (from the bridge into the hippocampal tissue). None of them have yet been addressed in conjunction with concurrent administration of trophic factors.

Earlier studies have explored grafts of fetal hippocampus as bridges for adult septal cholinergic axons (Kromer et al., 1981). AChE-positive fibers were seen to build a strong plexus within the graft and to extend beyond it into the hippocampus. The fetal graft fused well with septal and hippocampal host tissues at either side and may have presented appropriate endogenous NTFs and NPFs because of its fetal nature. Two major drawbacks to generalizing this approach, however, are i) the limits and ethical difficulties of securing fetal brain material and ii) the regional specificity that may underlie its effectiveness (e.g. hippocampal grafts for septal neurons). A more promising bridge material would be a segment of peripheral nerve, more readily available from the host itself and not apparently subject to a regional specificity. In one such attempt (Wendt et al., 1983), cholinergic fibers readily invaded the nerve graft and grew within it but did not succeed to any sizable extent in advancing across the bridge "exit" into the hippocampus. This apparent failure to abandon a nerve bridge and invade the host CNS tissue has been noted in several other nerve graft studies (Aguayo et al., 1982) with one recent, exciting exception dealing with optic nerve fibers (Vidal-Sanz et al., 1987).

The ideal regeneration bridge would be one using i) biological materials that are easily available and not region-specific, or ii) "surrogate" materials constructed to include the appropriate NTFs and NPFs. Two directions are being explored with this goal in mind. Naturally assembled extracellular matrix material can be prepared from human amnion membrane, readily available from usually discarded placentae. After removal of its single layer of epithelial cells, the amnion membrane provides a large sheet of cell-free extracellular matrix which comprises a laminin-containing "basement membrane" supported by a looser stroma of collagen fibers. This material can be rolled, folded or otherwise manipulated into various configurations. In vitro, CNS and PNS neurons will readily attach to the basement membrane side and grow on it extensive neurites which often display a parallel orientation (Davis et al., 1987a). An initial examination of this material implanted into the septo-hippocampal lesion gap has revealed cholinergic axons both along the bridge and in the adjacent hippocampal tissue, with no drastic alterations of the bridge itself or inflammatory reactions by the host (Davis et al., 1987b). A different approach involves the use of nitrocellulose paper bridges. Nitrocellulose can be supplied with electrophoretically transferred and still active NGF or other NTFs (Pettmann et al., 1988), or seeded with fetal astroglial (or other) cells presenting neurite-promoting surfaces (Silver and Ogawa, 1983). Eventually, biodegradable materials may be sought to replace the nitrocellulose substratum. Nitrocellulose bridges are being explored for both peripheral (Danielsen et al., 1988) and central (Smith et al., 1986) regenerations, but have not yet been investigated in the septo-hippocampal model.

Three major strategies are conceivable toward the alleviation of trau-
matic or pathologic damage of an identifiable set of neurons in the adult
CNS. It may be possible to replace irreversibly damaged neurons with a new
population of nerve cells of the same kind, provided that the new cells i)
are readily available, ii) can be made to survive within the host brain and
iii) will become integrated into the appropriate neural networks. The use
of homotypic fetal grafts illustrates this first approach. Alternatively,
one may attempt to compensate for the lost neurons by supplying the product
for which they are needed, i.e. the neurotransmitters or peptides normally
released by them. This second approach, exemplified by the use of L-DOPA
or of dopamine-producing grafts in Parkinsons' disease, rests on an impli-
cit assumption that the crucial role of the damaged neurons is not entirely
related to neural circuitry and that humorally-dispensed neuronal products
are equally successful in modulating target cell behaviors as do the neur-
ally dispensed ones. A third approach, underlined by the CNS neuronotrophic hypothesis (see Table 1), is to restore partially or totally the
functional competence of reversibly impaired neurons by providing an exogenous supplement of NTFs. Because NTFs are proteins, unlike most pharmacologic agents currently in use, we shall first address the interrelated
questions of NTF delivery systems and potential side effects.

Peripheral administration of NTF proteins faces the basic problem of
the blood brain barrier which normally prevents entry of macromolecules
into the CNS parenchyma. Even if treatments were devised to lower temporarily this barrier concurrently with multiple NTF injections, other
problems may be expected from the large amounts of peripheral NTF that
would be required to secure adequate intra-cerebral doses. Among them are
i) the NTF competence to elicit responses from PNS components (e.g. NGF
effects on sympathetic innervation), ii) potential toxicity on other
organs, unrelated or secondary to the PNS effects, and iii) immune reactions of the host to the NTF protein antigen. None of these problems are
likely to arise from local NTF administration within the CNS itself, where
the needed amounts of exogenous NTF are very small. Eventually, local supplies might be secured by the implantation of slow-release depots or of
suitably compatible living cells to act as long-term NTF sources (selected
tissue grafts, or NTF-superproducer cells engineered by genetic manipulations). Alternatively, one might capitalize on the known ability of astroglial cells to produce NTFs (Manthorpe et al., 1986a) and develop drugs
that can cross the blood brain barrier and stimulate local glia to increase
their NTF outputs.

For the time being, however, intraventricular infusion remains the
most direct and already available delivery system. There are obvious drawbacks. The inconvenience of this approach and the inherent risks of its
invasiveness make it suitable only to severe cases, where the potential
effectiveness of a treatment may itself be limited. For each agent, one
needs first to establish the stability of its biologic competence within an
appropriate vehicle as well as the unhindered continuity of the cannula
outflow. Most importantly, one needs to determine which other neurons
within the CNS, beside those to which the treatment is directed, i) may be
also responsive to the infused NTF and ii) can be reached by the NTF from
the infusion site. Given the generally assumed selectivity of these factors, only certain subsets of CNS neurons may be at risk, and even those
may not be affected by exogenous NTF if they are undamaged and already adequately supplied from endogenous sources. In any case, both of these risks
(responsiveness and accessibility) could be minimized by determining the
lowest effective dose of intraventricularly infused NTF and seeking further
dose reductions through concurrent treatments with synergistic agents --
several of which are now under investigation.

At the present state of the art, the best case for a neuronotrophic therapeutic approach in human disease can be made with regard to NGF and Alzheimer's disease, particularly on the strength of the reported NGF effects on brain age-related cholinergic and cognitive deficits (Will and Hefti, 1985; Fischer et al., 1988). Progress in this direction will require availability of recombinant human NGF, its testing in a primate model, and additional refinements of the infusion delivery system -- leading to the first clinical trials. With other CNS degenerative disorders (cf. Appel, 1982), the road will be much longer. It will require in vitro assay systems using dopaminergic neurons (for Parkinson's disease) and motor neurons (for motor neuron disease -- cf. Varon et al. 1982), the use of these assays to identify sources of suitable NTFs and monitor their purification, and the establishment of adequate animal models to test the purified NTFs in vivo. Even further down the road, but no longer a hopeless task, is the utilization of NTFs and NPFs for adult CNS regeneration. This will require combining the use of a specific NTF with an effective regeneration bridge, and then evaluating -- and, possibly, seeking promoting conditions for -- the resulting target reinnervation in terms of functional as well as morphologic criteria. We believe that the already available septal-hippocampal model may provide the necessary insight for future extensions to major human problems such as optic nerve and spinal tract lesions.

ACKNOWLEDGEMENTS: The work described here was partially supported by NINCDS grants NS-16349 (SV) and NS-25011 (MM) and by NSF grants BNS-85-01766 (SV) and BNS-86-17034 (MM).

REFERENCES

Aguayo, A. A., 1985, Axonal regeneration from injured neurons in the adult mammalian central nervous system, in: "Synaptic Plasticity," C. W. Cotman, ed., Guilford Press, New York, pp. 457-484.

Aguayo, A., Davis, S., Richardson, P., and Bray, G., 1982, Axonal elongation in peripheral and central nervous system transplants, Adv. Cell. Neurobiol. 3:215-234.

Appel, S. H., 1981, A unifying hypothesis for the cause of Amyotrophic Lateral Sclerosis, Parkinsonism, and Alzheimer disease, Ann. Neurol. 10:499-505.

Barde, Y-A., Edgar, D., and Thoenen, H., 1982, Purification of a new neurotrophic factor from mammalian brain, EMBO J. 1:549-553.

Bartus, R. T., Dean, R. L., Beer, B., and Lippa, A. S., 1982, The cholinergic hypothesis of geriatric memory dysfunction, Science 217:408-417.

Bradshaw, R. A. and Prentis, S. (eds), 1987, "Oncogenes and Growth Factors", Elsevier, New York.

Carnow, T. B., Manthorpe, M., Davis, G. E., and Varon, S., 1985, Localized survival of ciliary ganglionic neurons identifies neuronotrophic factor bands on nitrocellulose blots, J. Neurosci. 5:1965-1971.

Cowan, C. W., Fawcett, J. W., O'Leary, D. D. M., and Stanfield, B. B., 1984, Regressive events in neurogenesis, Science 225:1258-1265.

Coyle, J. T., Price, D. L., and DeLong, M. R., 1983, Alzheimer's disease: A disorder of cortical cholinergic innervation, Science 219:1184-1190.

Davis, G. E., Manthorpe, M., Engvall, E., and Varon, S., 1985a, Isolation and characterization of rat schwannoma neurite promoting factor: Evidence that the factor contains laminin, J. Neurosci. 5:2662-2671.

Davis, G. E., Varon, S., Engvall, E., and Manthorpe, M., 1985b, Substratum-binding neurite promoting factors: Relationships to laminin, Trends Neurosci. 8:528-532.

Davis, G. E., Engvall, E., Varon, S., and Manthorpe, M., 1987a, Human amnion membrane as a substratum for cultured peripheral and central nervous system neurons, Dev. Brain Res. 33:1-10.

Davis, G. E., Blaker, S. N., Engvall, E., Varon, S., Manthorpe, M., and Gage, F. H., 1987b, Human amnion membrane serves as a substratum for growing axons in vitro and in vivo, Science 236:1106-1109.

Danielsen, N., Pettmann, B., Vahlsing, H. L., Manthorpe, M., and Varon, S., 1988, Fibroblast Growth Factor effects on peripheral nerve regeneration in a silicone chamber model, J. Neurosci. Res., in press.

Fischer, W., Wictorin, K., Bjorklund, A., Williams, L. R., Varon, S., and Gage, F. H., 1987, Amelioration of cholinegic neuron atrophy and spatial memory impairment in aged rats by nerve growth factor, Nature 329:65-68.

Greene, L. A. and Shooter, E. M., 1980, The Nerve Growth Factor: Biochemistry, synthesis, and mechanism of action, Annu. Rev. Neurosci. 3:353-402.

Gundersen, R. W., 1985, Sensory neurite growth cone guidance by substrate adsorbed Nerve Growth Factor, J. Neurosci. Res. 13:199-212.

Gurney, M. E., Heinrich, S. P., Lee, M. R., and Yin, H-S., 1986, Molecular cloning and expression of neuroleukin, a neurotrophic factor for spinal and sensory neurons, Science 234:566-574.

Hagg, T., Manthorpe, M., Vahlsing, H. L., and Varon, S., 1988, Delayed treatment with nerve growth factor reverses the apparent loss of cholinergic neurons after acute brain damage, Exp. Neurol., in press.

Hamburger, V., and Oppenheim, R. N., 1982, Naturally occurring neuronal death in vertebrates, Neurosci. Comment. 1:39-55.

Hefti, F., 1983, Alzheimer's disease caused by a lack of nerve growth factor? Ann. Neurol. 13:109-110.

Hefti, F., 1986, Nerve Growth Factor promotes survival of septal cholinergic neurons after fimbrial transections, J. Neurosci. 6:2155-2162.

Hepler, D. J., Wenk, G. L., Cribbs, B. L., Olton, D. S., and Coyle, J. T., 1985, Memory impairments following basal forebrain lesions, Brain Res. 346:8-14.

Kromer, L. F., 1987, Nerve Growth Factor treatment after brain injury prevents neuronal death, Science 235:214-216.

Kromer, L. F., Bjorklund, A., and Stenevi, U., 1981, Regeneration of the septo-hippocampal pathways in adult rats is promoted by utilizing embryonic hippocampal implants as bridges, Brain Res. 210:173-200.

Levi-Montalcini, R., 1966, The Nerve Growth Factor: its mode of action on sensory and sympathetic neurons, Harvey Lect. 60:217-259.

Levi-Montalcini, R., 1982, Developmental neurobiology and the natural history of Nerve Growth Factor, Annu. Rev. Neurosci. 5:341-361.

Levi-Montalcini, R., 1987, The Nerve Growth Factor thirty-five years later, Science 237:1154-1162.

Manthorpe, M., and Varon, S., 1985, Regulation of neuronal survival and neuritic growth in the avian ciliary ganglion, in: "Growth and Maturation Factors," Vol. 3, G. Guroff, ed., J. Wiley & Sons, New York, pp. 77-117.

Manthorpe, M., Engvall, E., Ruoslahti, E., Longo, F. M., Davis, G. E., and Varon, S., 1983a, Laminin promotes neuritic regeneration from cultured peripheral and central neurons, J. Cell Biol. 97:1882-1890.

Manthorpe, M., Nieto-Sampedro, M., Skaper, S. D., Lewis, E. R., Barbin, G., Longo, F. M., Cotman, C. W., and Varon, S., 1983b, Neuronotrophic activity in brain wounds of the developing rat. Correlation with implant survival in the wound cavity. Brain Res. 267:47-56.

Manthorpe, M., Rudge, J., and Varon, S., 1986a, Astroglial cell contributions to neuronal survival and neuritic growth, in: "Astrocytes," Vol. 2, S. Fedoroff and A. Vernadakis, eds., Academic Press, New York, pp. 315-376.

Manthorpe, M., Skaper, S. D., Williams, L. R., and Varon, S., 1986b, Purification of adult rat sciatic nerve ciliary neuronotrophic factor, Brain Res. 367:282-286.

Manthorpe, M., Ray, J., Pettmann, B., and Varon, S., 1988, Ciliary Neurono-trophic Factors, in: "Nerve Growth Factors," R. Rush, ed., John Wiley & Sons Ltd., New York, in press.

Morrison, R. S., Sharma, A., DeVellis, J., and Bradshaw, R. A., 1986, Basic fibroblast growth factor supports the survival of cerebral cortical neurons in primary culture, Proc. Natl. Acad. Sci. USA 83:7537-7541.

Morrison, R. S., Kornblum, H. I., Leslie, F. M., and Bradshaw, R. A., 1987, Trophic stimulation of cultured neurons from neonatal rat brain by Epidermal Growth Factor, Science 238:72-75.

Nieto-Sampedro, M., Lewis, E. R., Cotman, C. W., Manthorpe, M., Skaper, S. D., Barbin, G., Longo, F. M., and Varon, S., 1982, Brain injury causes a time-dependent increase in neuronotrophic activity at the lesion site, Science 217:860-861.

Nieto-Sampedro, M., Manthorpe, M., Barbin, G., Varon, S., and Cotman, C. W., 1983, Injury-induced neuronotrophic activity in adult rat brain. Correlation with survival of delayed implants in a wound cavity, J. Neurosci. 3:2219-2229.

Pettmann, B., Powell, J., Manthorpe, M., and Varon, S., 1988, Biological activities of nerve growth factor bound to nitrocellulose paper by Western blotting, J. Neurosci., in press.

Rudge, J., Davis, G., Manthorpe, M., and Varon, S., 1987, An examination of ciliary neuronotrophic factors from avian and rodent tissue extracts using a blot and culture technique, Dev. Brain Res. 32:103-110.

Silver, J. and Ogawa, M. Y., 1983, Postnatally induced formation of the corpus callosum in acallosal mice on glia-coated cellulose bridges, Science 220:1067-1069.

Smith, G. M., Miller, R. H., and Silver, J., 1986, Changing role of forebrain astrocytes during development, regenerative failure, and induced regeneration upon transplantation, J. Comp. Neurol. 251:23-43.

Taniuchi, M., Clark, H. B., Schweitzer, J. B., and Johnson, E. M., 1988, Expression of Nerve Growth Factor receptors by Schwann cells of axotomized peripheral nerves: Ultrastructural location, suppression by axonal contact, and binding properties, J. Neurosci. 8:665-681.

Ullrich, A., Gray, A., Berman, C., and Dull, T.J., 1983, Human β-Nerve Growth Factor gene sequence highly homologous to that of mouse, Nature 303:821-825.

Varon, S., 1975, Nerve Growth Factor and its mode of action, Exp. Neurol. 48(3, part 2):75-92.

Varon, S., 1977, Neural growth and regeneration: A cellular perspective, Exp. Neurol. 54:1-6.

Varon, S., 1985, Factors Promoting the Growth of the Nervous System, Discussions in Neuroscience, Vol. II, No. 3, FESN (Foundation for the Study of the Nervous System), Geneva, Switzerland.

Varon, S., and Adler, R., 1980, Nerve growth factors and control of nerve growth, Curr. Topics Dev. Biol. 16:207-252.

Varon, S., and Adler, R., 1981, Trophic and specifying factors directed to neuronal cells, Adv. Cell. Neurobiol. 2:115-163.

Varon, S., and Manthorpe, M., 1982, Schwann cells: An in vitro perspective, Adv. Cell. Neurobiol. 3:35-95.

Varon, S., and Somjen, G., 1979, Neuron-Glia Interactions, Neurosci. Res. Prog. Bull. vol. 17, MIT Press, Cambridge, Mass.

Varon, S., Manthorpe, M., and Longo, F. M., 1982, Growth factors and motor neurons, in: "Human Motor Neuron Diseases," Adv. in Neurology, vol. 36, L. P. Rowland, ed., Raven Press, New York, pp. 453-472.

Varon, S., Adler, R., Manthorpe, M., and Skaper, S. D., 1983, Culture strategies for trophic and other factors directed to neurons, in: "Neuroscience Approached through Cell Culture," S. E. Pfeiffer, ed., CRC Press, Boca Raton, Florida, pp. 53-77.

Varon, S., Manthorpe, M., and Williams, L. R., 1984, Neuronotrophic and neurite promoting factors and their clinical potentials, Dev. Neuroscience 6:73-100.

Varon, S., Manthorpe, M., Davis, G. E., Williams, L. R., and Skaper, S. D., 1988a, Growth Factors, in: "Functional Recovery in Neurological Disease," Advances in Neurology, vol. 47, S. G. Waxman, ed., Raven Press, New York, pp. 493-521.

Varon, S., Hagg, T., and Manthorpe, M., 1988b, Neuronal growth factors, in: "Neural Regeneration Research for the Clinician," F. J. Seil, ed., Alan Liss, New York, in press.

Varon, S., Pettmann, B., and Manthorpe, M., 1988c, Humoral and surface-anchored factors in development and repair of the nervous system, in: "Cellular and Molecular Aspects of Neural Development and Regeneration," G. J. Boer, M. G. P. Feenstra, M. Mirmiran, D. F. Swab, and F. Van Haren, eds., Progress in Brain Res., Vol. 73, Elsevier, Amsterdam, in press.

Vidal-Sanz, M., Bray, G. M., Villegas-Perez, M. P., Thanos, S., and Aguayo, A. J., 1987, Axonal regeneration and synapse formation in the superior colliculus by retinal ganglion cells in the adult rat, J. Neurosci. 7:2894-2909.

Walicke, P., Cowan, W. M., Ueno, N., Baird, A., and Guillemin, R., 1986, Fibroblast growth factor promotes survival of dissociated hippocampal neurons and enhances neurite extension, Proc. Natl. Acad. Sci. USA 83:3012-3016.

Wendt, J. W., Fagg, G. E., and Cotman, C. W., 1983, Regeneration of rat hippocampal fimbria fibers after fimbria transection and peripheral nerve or fetal hippocampal implantation, Exp. Neurol. 79:452-461.

Whittemore, S. R. and Seiger, A., 1987, The expression, localization and functional significance of β-nerve growth factor in the central nervous system, Brain Res. Reviews 12:439-464.

Will, B. and Hefti, F., 1985, Behavioural and neurochemical effects of chronic intraventricular injections of Nerve Growth Factor in adult rats with fimbria lesions, Behavioural Brain Res. 17:17-24.

Williams, L. R., Varon, S., Peterson, G., Wictorin, K., Fischer, W., Bjorklund, A., and Gage, F. H., 1986, Continuous infusion of Nerve Growth Factor prevents basal forebrain neuronal death after fimbria-fornix transection, Proc. Natl. Acad. Sci. USA 83:9231-9235.

STRATEGIES FOR IMPROVING FETAL NEURONAL INTEGRATION WITH

ADULT BRAIN FOLLOWING BRAIN TRANSPLANTATION

Efrain C. Azmitia[1], Feng C. Zhou[2], and
Patricia M. Whitaker-Azmitia[3]

[1]Department of Biology, New York University
 Washington Square East, New York, NY 10003
[2]Department of Anatomy, Indiana University
 School of Medicine, Indianapolis, IN 46223
[3]Department of Psychiatry, State University
 of New York, Stony Brook, NY 11794

Key Words: Co-transplantation, laminin, astrocytes, serotonin,
 hippocampus, 5,7-dihydroxytryptamine, growth

Abstract

If transplanted fetal cells are to have maximal benefits in the
adult brain, the donor and host neurons must fully interact. In
this report, we have discussed procedures for the preparation of
the fetal cells to insure best survival and methods to identify
transplanted neurons whose neurotransmitter content is known.
Furthermore, we provide several strategies for reducing the glial
scar, enhancing fetal cell survival (co-transplantation),
inducing host brain trophic factors (prior lesions or preparation
of a soluble extract) and providing adhesive molecules to
encourage outgrowth. Many additional strategies need to be
developed to produce an optimal integration of fetal and adult
cells. Once achieved, brain transplantation can provide a
powerful and unparalleled tool to regulate cell function and
combat diseases.

A.   Introduction

"Once development was ended, the founts of growth and
regeneration of the axons and dendrites dried up irrevocably.
In adult centers the nerve paths are something fixed, ended
immutable. Everything may die, nothing may be regenerated"
(Ramon y Cajal, 1928). Since man first became aware of the
importance of the brain as a cognitive, emotional, and
regulatory center, the finality of death has been a harsh
reality to accept. Most neurons of the brain are present at
birth and survive for the life of the individual. However,

249

because of injury, stress, and disease, many neurons die prematurely. The procedure of neuronal (brain) transplantation provides a unique opportunity to replace the lost neurons with young healthy neurons. A great deal has already been written about the power of this transplantation paradigm to reverse loss of neuronal systems and restore normal function.

However, morphological examination of the adult brain following the transplantation of fetal neurons clearly shows that the integration is far from complete. In a typical case from our work, the donor fetal cell bodies remain isolated from the adult neuropile by a dense astrocytic border which forms hours after the trauma of implantation. Staining of the donor neurons shows that axons and dendrites are extremely dense within the transplant and only a relatively small number of fibers escape to invade the territory of the adult brain. This sparse reinnervation can be sufficiently robust to reverse some behavioral deficits, but clearly no one will argue that complete integration has occurred.

Our interest in this problem is long-standing and developed from our studies of the brain serotonergic system. These neurons develop early and extend their axons throughout the neural axis from the cerebral cortex to the spinal cord (see Azmitia, 1978, 1987). The fibers can be selectively destroyed by intracerebral microinjections of neurotoxin (Azmitia, 1986) including a variety of behavioral deficits which are tightly correlated with the loss of the 5-HT afferents. Given the ability of this one system to alter the physiology of its target area, and the opportunity to influence a variety of functions because of its expansive innervation pattern, the technique of transplantation of serotonergic neurons offers a unique situation for reengineering the mammalian brain.

Initial work in our laboratory showed that fetal serotonergic neurons could survive in and innervate the brain of adult and aged animals (Azmitia et al, 1981). However, these first studies also revealed the fact that the adult brain did not eagerly embrace the donor fetal neurons, they were kept largely isolated from the adult neuropile. This chapter describes some of the approaches we have devised over the year to promote a better integration between fetal and adult neurons. Much has been learned to encourage our efforts, but much remains to challenge our determination. For Cajal, the founts were closed, for us they have been open.

B.   Method of Injecting Fetal Cells and Tissue

An important factor for obtaining reliable results is to have a reproducible means of injecting small volumes (less than 1 ul) under visual inspection. We have developed a simple and inexpensive microsyringe (Fig. 1). A glass micropipette is pulled with a shank of at least 2 cm. The tip diameter is between 150 and 20 um for dissociated cells, and between 250 and 400 um for tissue suspensions. The tip is attached to a 19-gauge needle using 5-min. epoxy cement.

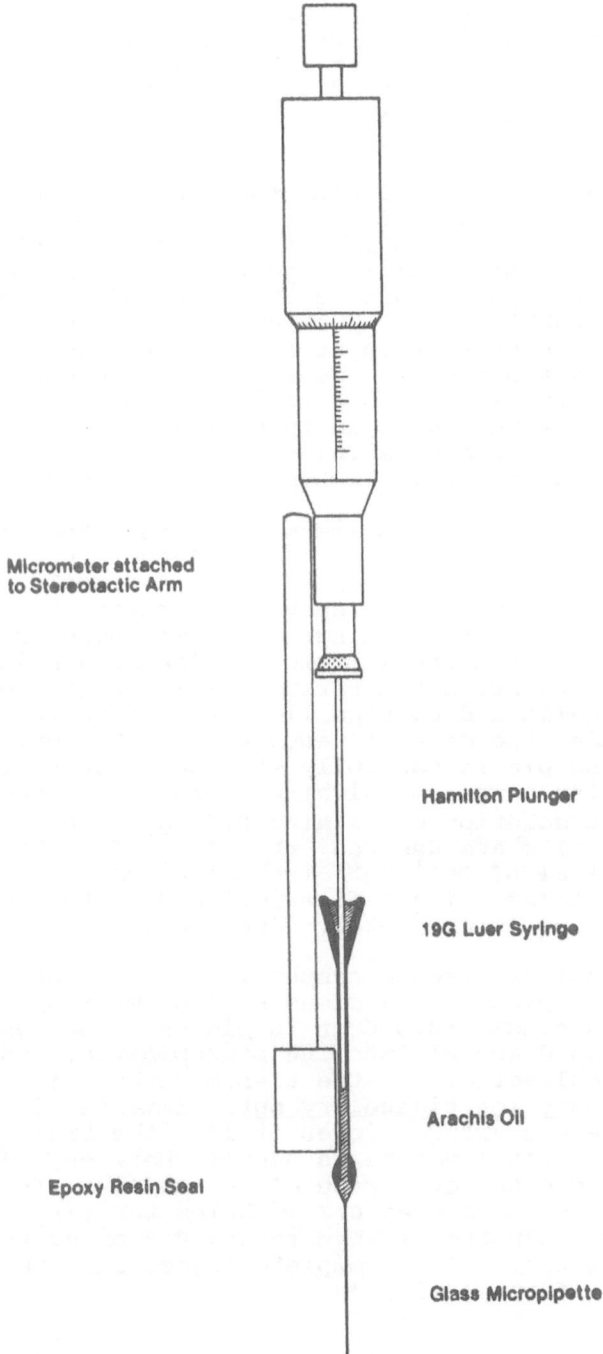

Micrometer attached
to Stereotactic Arm

Hamilton Plunger

19G Luer Syringe

Arachis Oil

Epoxy Resin Seal

Glass Micropipette

Fig. 1.   Diagram of micropipette system.

A plunger from a 10-ul Hamilton syringe (or similar-diameter wire) is attached to a micrometer (available from a good hardware store for about $20). The micrometer is secured to the vertical arm of a stereotaxic frame. The micropipette assembly is filled with a sterile light oil with no air pockets. The plunger is then coated with a heavy grease and slowly inserted into the needle. The turning of the micrometer will now displace 100 nl of oil for every 1 mm of vertical travel of the plunger.

The micropipette is filled with the donor tissue under visual inspection with the aid of a dissecting microscope. A small air bubble is drawn up to mark the top of the solution. The exact amount of tissue to be injected is drawn up into the micropipette by slowly turning the micrometer and observing that the air bubble moves up smoothly. The donor cells are ejected into the host brain at a slow rate (5-50 nl/min) under visual inspection to ensure smooth and complete delivery into the injection site. The pipette can be raised 0.5 mm if more than 500 nl is to be injected. At the conclusion of the pressure injection, the tip should be left in the brain for several minutes to allow for diffusion.

We normally inject minced-tissue or dissociated-cell suspensions into the brain. The rat fetuses are removed between 14 and 17 days of gestation by a sterile cesarian procedure. The brains are removed to sterile balanced salt solution (e.g., Hank's or Ringer's). To inject serotonergic neurons, a brainstem raphe strip is dissected using sterile procedures. The brainstem strip lies in the region between the mesencephalic and pontine flexures and extends vertically from just under the cerebral aqueduct to the ventral floor. The underlying pia is carefully removed. These strips (2 mm long, 1 mm wide, and 1 mm high) are placed in ice-cold balanced salt solution containing 0.5% glucose. If dissociated cells are desired, the strips from 20 fetuses are placed in 0.1 ml of Ca^{++}/Mg^{++} free solution, and the solution is gently agitated using a fire-polished Pasteur pipette until the tissue is completely dispersed.

When the animal receives a suspension transplant, a single raphe strip is placed on a clean surface with approximately 1 ul of MEM and minced into 6 or 10 pieces. The resultant slurry is then drawn up into the micropipette. Care should be taken to collect all of the tissue without introducing air bubbles. <u>Do not let tissue dry out</u>. Immediately after filling the microsyringe, lower it into the brain and begin to eject. The total volume is usually between 1.5 and 2.0 ul. This is ejected at a rate of 100 nl/min. In our work, the pipette tip is lowered 0.2 mm below the first tissue ejection. The pipette is then raised 0.2 mm every five turns of the micrometer. After complete injection, the pipette is left in place for two minutes.

C.   Descriptive Findings

In our initial studies, we used the technique of immunocytochemistry to detect fetal neurons when transplanted into a host brain. The ability to distinguish fetal-donor cells from the adult-host brain is essential to confirm the success of the transplantation procedure. Fetal mouse raphe and hippocampal tissue (from embryos with crown-rump length (CRL) 13-16 mm) was transplanted into adult and aged isogeneic mice to study the growth of serotonergic fibers between host and donor tissue after one month (Azmitia et al, 1981). An antibody raised against serotonin (5-HT) was used to immunocytochemically visualize 5-HT-containing cell bodies and fibers. Unilateral fetal transplants into the hippocampi of adult (4-6 mo.) or aged (24 mo.) mice matured and sent out processes which very densely innervated the transplant tissue itself and extended into the host hippocampus (Fig. 2). Within the transplanted tissue, an increased number contacts between 5-HT neurons were observed. In the host brain, the termination of fetal 5-HT fibers overlapped with the known 5-HT hippocampal lamination pattern in normal animals. Qualitative comparisons between control and transplanted areas indicated that the density of outgrowth into adult hippocampus was greater than into aged hippocampus. This observation was confirmed in later studies (Azmitia, 1987). In order to test if adult fibers could be induced to grow into a fetal target structure, fetal hippocampus was transplanted into an area of the brain (septum) which contained a dense innervation of 5-HT fibers. The results showed that adult 5-HT can grow deep within the donor fetal hippocampus. Therefore, immunocytochemical procedures can be used to monitor outgrowth from the fetal tissue to the host and ingrowth from the adult host to the fetal tissue. Furthermore, the apparent normal 5-HT lamination pattern produced by fetal raphe axons in adult hippocampus is consistent with reports that neuronal transplantation is effective in reversing the anatomical and behavioral deficits produced by homotypic denervation of a homotypic fiber system.

D.   Astrocytic Scar

It has long been known that any physical insult to the adult brain causes a pronounced astrocytic response which can lead to the formation of a glial scar (Cavanaugh, 1970; Latov et al, 1979). The formation of a glial scar can interfere with donor-host tissue interaction (especially in an aged brain where there is a large increase in astrocytic cells). In our work, raphe tissue (minced or dissociated) from fetal rat was microinjected into adult hippocampus or midbrain, and the response of astrocytes to the transplant was studied by immunocytochemical staining for the astrocyte specific marker, glial fibrillary acidic protein (GFA) (Azmitia and Whitaker, 1983). Astrocytes were observed to form a border around the transplanted cells. This border was visible as early as 24 hours post-transplant and was still present up to six months later (Whitaker-Azmitia et al, 1987). The transplanted serotonergic cell bodies, identified by immunocytochemical staining appeared at the edge of the border between fetal and adult tissue but not beyond. This border may be responsible for preventing the complete integration of the transplanted cells with the adult host hippocampal and midbrain tissue. A number of methods were tried to reduce the glial scar. This is not simple since the scar is formed by glial cell migration from both host and donor tissue by both migration and mitosis. Injections of

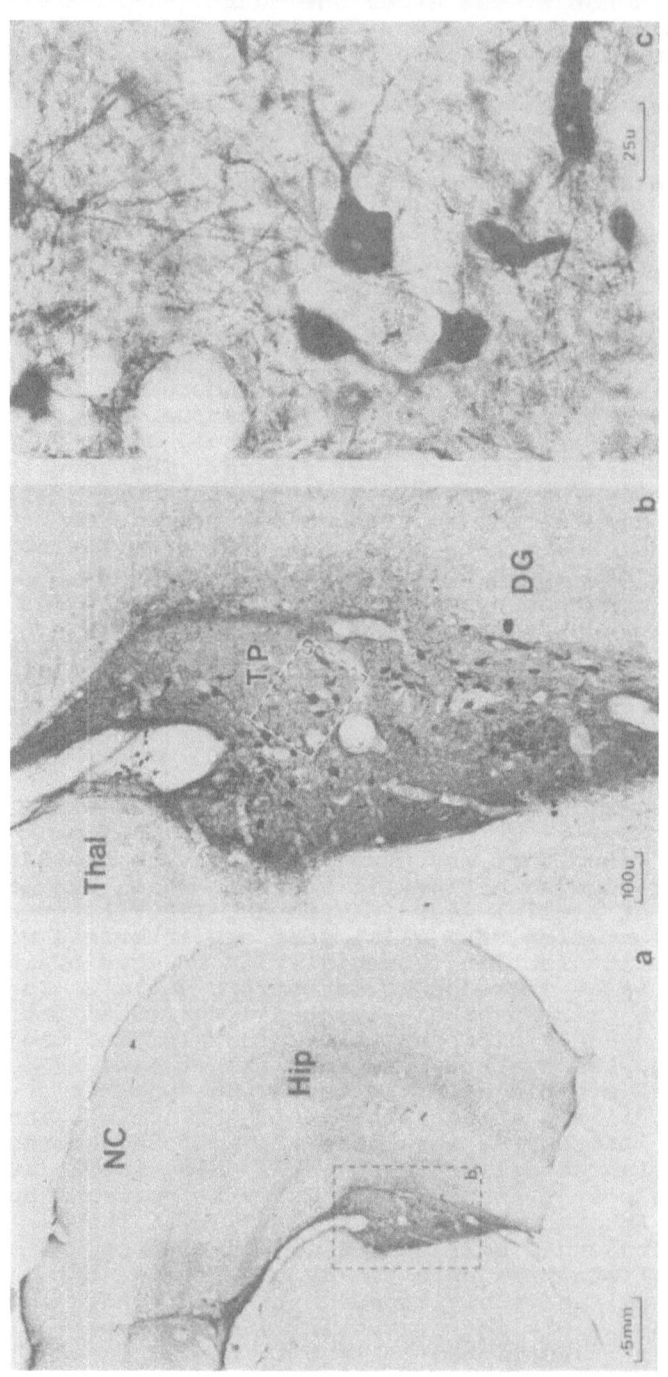

Fig. 2. Immunocytochemical localization of the serotonergic cells and fibers in fetal raphe 30 days after transplantation into the hippocampus of an aged mouse (24 mo.). a. A low magnification view of the transplant located adjacent to the lateral hippocampus. b. A higher magnification view showing the increased innervation of the transplanted raphe and the outgrowth of fibers into the host aged hippocampus (box). The degree of outgrowth into the host hippocampus is less than noted in the adult (Fig.2). c. Oil immersion photograph of a group of serotonergic neurons found within the transplant (box). The neurons are highly branched and are larger than observed in the fetal brain. (Reprinted from Azmitia et al., 1981)

254

antimitotic agents or into animals receiving high cortisol
treatments were not effective (unpublished reports). The
production of a decoy lesion made two weeks before the
transplantation injection did reduce the astrocytic reaction
(Shemer et al, 1985). Injection of a large number of
cultured glial primary cells or glial cell lines did reduce
the astrocytic reaction to the transplant (Whitaker-Azmitia
et al, 1987). In addition, phenytoin injections may also be
able to reduce the astrocytic response. However, it should
be mentioned that the role of astrocytes is not all bad,
since these cells make growth factors and can provide a
substrate (laminin) for axons to attach to during growth.
Much more work is needed in this field.

E.   Co-Transplants and Target Cells

In order to explore the potential benefits of target cells on
cell survival, adult neurons were transplanted (Azmitia,
1987). Neuronal transplantation of adult raphe tissue into
adult hippocampus showed that some 5-HT-IR adult neurons were
alive one week after transplantation. However, the size of
the transplant area itself was small, and the 5-HT-IR fibers
from the transplanted neurons formed dense swirls within the
transplant site. None of the donor 5-HT-IR fibers penetrated
into the adult hippocampus. One month after transplantation,
no 5-HT-IR neurons were visible in the host brain, and all
that remained of the transplanted tissue was a necrotic
region filled with macrophages.

However, better results were obtained when adult raphe cells
were co-transplanted with fetal hippocampal cells into the
dorsal hippocampus of an adult animal. After one month,
giant surviving 5-HT-IR neurons could be found within the co-
transplanted fetal hippocampus. These cells measured over
350 um^2 in area. These 5-HT-IR cells showed extensive
branching around the soma, and abnormal somatic spines on the
adult-donor neurons were apparent. This unusual morphology
was never observed in the 5-HT-IR cells, in the endogenous
raphe nuclei of the midbrain, or in the fetal neurons
transplanted into the adult animal. The feasibility of using
co-transplants to stimulate cell survival and process
outgrowth has not been adequately explored. A potential
problem is that a local circuit will be encouraged and
interaction with the target host brain may be actually
retarded.

F.   Host Trophic Factor

Partial removal of 5-HT fibers from the adult hippocampus by
injections of 5,7-dihydroxytryptamine (5,7-DHT) can induce
homotypic collateral sprouting of the undamaged 5-HT fiber
(Azmitia et al, 1978; Azmitia and Zhou, 1986; Zhou and
Azmitia, 1986). These results indicate the presence of an
endogenous trophic factor for 5-HT fiber which can induce
adult 5-HT sprouting. Can fetal 5-HT neurons be similarly
affected? The existence and specificity of the trophic
effects of specific denervation on the growth and survival of
fetal serotonergic (5-HT) or norepinephrinergic (NE) neurons
grafted into the hippocampus were assessed by means of HPLC

measurements and immunocytochemistry with 5-HT and tyrosine hydroxylase antibodies (Zhou et al, 1987). Fetal raphe cells (containing 5-HT neurons) were transplanted in normal hippocampus or two weeks after the specific removal of 5-HT afferents to the hippocampus with 5,7-DHT. A month after transplantation, the number of 5-HT immunoreactive neurons was not significantly different between the control and 5,7-DHT lesioned animals. However, transplanted raphe neurons had 400% more 5-HT synaptosomal high-affinity uptake and 380% higher content of 5-HT in the hippocampus with prior 5,7-DHT lesion than in control hippocampus. In morphometric studies after immunocytochemistry, the somatic area of the 5-HT neurons was 42% larger than that of control group, and there was an obvious increase in the occurrence of fiber processes with varicosities in the 5,7-DHT treated hippocampus. In addition, a larger increase in the 5-HT level could be achieved if transplanted fetal neurons in the control hippocampus were treated with the supernatant extracted from the hippocampus with 5,7-DHT lesion. This indicates the stimulatory factor was soluble factor and not dependent on the effects of intact neurons.

In contrast, the NE level of the implanted fetal locus coeruleus (containing NE neurons) was not significantly higher in the 5-HT denervated hippocampus than in control hippocampus a month after transplantation. These results suggest that 5-HT denervation in the hippocampus induces a trophic substance which promotes the maturation rather than survival of 5-HT neurons but not NE neurons.

G.   Laminin

One of the problems with fetal axonal growth into adult tissue may be the lack of a suitable substrate for the axons to attach when growing. Laminin has been shown to act as a surface adhesive molecule for neuronal process elongation in vitro (Zhou and Azmitia, 1988). To test whether laminin has a similar role in the adult brain, we sequentially injected laminin and transplanted fetal neurons into various brain regions to determine if the fetal neurons would preferentially grow along a laminin injection tract.

In the fetal brain, the raphe area of the rostral brainstem is rich in serotonergic (5-HT) neurons, the rostral ventral mesencephalon is rich in dopamine (DA) neurons, while the locus ceruleus is rich in norepinephrinergic (NE) neurons. These three populations were transplanted to the motor cortex, neostriatum or hippocampus of adult animals. The tract used for microinjection of cell suspension was then immediately filled with suspension medium laminin or laminin-collagen (type IV) mixture. In other animals, laminin or control solution was injected in a separate needle tract displacing 0.3-1 mm from the transplant injection tract.

Straight and thick 5-HT, DA or NE immunoreactive (IR) fibers (stained with anti-5-HT or anti-tyrosine hydroxylase antiserum) were predominant within the laminin-treated tracts, or were directed toward the laminin-treated parallel tracts when it was positioned less than 0.5 mm from the

transplant site. The density of 5-HT-, DA- and NE-IR fibers
in the injection tracts in all three brain areas was much
higher for laminin and laminin-collagen mixture than control
medium. Thin axonal fibers of fetal 5-HT and NE neurons were
observed surrounding the laminin-treated tracts, but not in
vehicle-treated tracts. Finally, laminin injection to the
hippocampus, motor cortex or neostriatum of the adult brain
did not stimulate sprouting of undamaged adult 5-HT or NE
fibers.

These results suggest that purified laminin can facilitate
and guide process outgrowth of 5-HT, DA and NE neurons during
early developmental stage, but does not induce sprouting on
these same fiber types in the adult brain.

H.   Conclusion

Nearly a century ago, Thompson first transplanted tissue into
an adult brain, but his and other early efforts were largely
ignored by the scientific community. The modern era began
with reports that fetal monoaminergic neurons could be
transplanted into the hippocampus of adult rats (Bjorklund et
al, 1976). This time not only was the scientific community
excited and appreciative, but the immense implication of this
procedure was recognized by the medical community. The first
successful report of a brain transplantation in humans was
reported in Parkinson's patients who received adrenal
medullary grafts (Madrazo et al, 1987). This rapid advance
from the laboratory bench to the operating table does not
mean that the basic research is complete. The data discussed
in our paper strongly indicates that the potential for
neuronal transplantation is only beginning to be realized.
The implication for replacing lost neurons is only one
application. Neuronal transplantation can be used to
potentiate the normal capacity of brain tissue; this avenue
for producing hyperfunction is still largely ignored.
Furthermore, neuronal transplantation can have a diagnostic
role to bioassay the neuronal environment of an aged or
diseased brain. Again, this approach has only been briefly
alluded to by a few workers. Finally, the power of this
technique will be useful in helping uncover many of the
unknown properties of normal brain cells during fetal
development, adult maintenance or age-related degeneration.

We have emphasized the question of integration. Our
hypothesis is that close interactions between donor fetal
neurons and host adult neurons is essential to achieve
maximal benefit from the new brain cells. Methods for
delivery of the cells and observing their fate in the adult
brain were discussed. The problem of donor tissue isolation
by formation of an astrocytic scar was emphasized. The
control of this scar formation is crucial for achieving
proper integration. Several methods were described for
advancing in this direction. Finally, the question of the
host tissue environment for enhancing survival and promoting
cell body migration and neurite extension were addressed.
The ability of prior selective damage for increasing the
availability of growth factors was covered. This approach
grew out of our experience of promoting homotypic collateral

sprouting of undamaged neurons in the adult brain by prior
5,7-DHT lesions (Azmitia et al, 1978; Azmitia and Zhou, 1986;
Zhou and Azmitia, 1986). In addition, growing axons require
a proper substrate as well as growth factors. Towards this
end, we presented exciting new evidence that direct
injections of exogenous laminin can promote cell body
migration and neurite extension into the adult brain.

The future of neuronal transplantation will be closely linked
to the future of growth regulatory factors. The more we
understand about the molecules (e.g., Ca^{++}, anti-oxidants,
gangliosides, neuropeptides and hormones) which can promote
cell survival and neurite extension, the more successful we
will become at maximizing donor fetal integration with host
adult neurons, and the greater the benefits can be extracted
from the donor tissue.

Acknowledgement: Thanks to Fernando Garcia Hernandez and
Tajrena Alexi for helpful comments on the preparation of the
manuscript, and to Luisa Fuentes for her secretarial assistance. Research supported by NSF BNS-86-07796.

REFERENCES

Azmitia, E.C., 1978. The serotonin producing neurons in the
midbrain median and dorsal raphe nuclei. In: The Handbook
of Psychopharmacology, 9:233-314. Eds. Iversen, L.L.,
Iversen, S.D. and Snyder, S. Plenum Press, N.Y., 1978.

Azmitia, E.C., Buchan, A.M., and Williams, J.H., 1978.
Structural and functional restoration by collateral
sprouting of hippocampal 5-HT axons. Nature, 274:374-377.

Azmitia, E.C., Perlow, M.J., Brennan, M.J., and Lauder, J.M.,
1981. Fetal raphe and hippocampal transplants in adult and
aged C57BL/6N mice; an immunohistochemical study. Brain
Res. Bull., 7:703-710.

Azmitia, E.C. and Whitaker, P.M., 1983. Formation of a glial
scar following microinjection of fetal raphe neurons into
the dorsal hippocampus or midbrain of the adult rat: An
immunocytochemical study. Neurosci. Lett. 38:145-150.

Azmitia, E.C., 1986. Re-engineering the brain serotonin
system: localized application of neurotoxins and fetal
neurons. In Myoclonus, Eds. Fahn, S., Marsden, C.D. and Van
Woert, M., Adv. Neurol., 43:493-507.

Azmitia, E.C. and Zhou, F.C., 1986. Chemically induced
homotypic collateral sprouting of hippocampal serotonergic
afferents. In Process of Recovery from Neural Trauma,
edited by Gilad, G.M., Gorio, A. and Kreutzberg, G.W.
Experimental Brain Res. (Suppl. 13), pp. 129-141.

Azmitia, E.C., 1987. Transplantation of fetal and adult
raphe cells into hippocampus of adult and aged mice. New
York Academy of Science, Vol. in Cell and Tisssue
Transplantation into the Adult Brain. (Edited by Azmitia,
E.C. and Bjorklund, A.)

Azmitia, E.C., 1987. The primate serotonergic system; progression towards a collaborative organization. Chapt. 7 in <u>Psychopharmacology, Generation of Progress</u>, Plenum Press, Ed. H. Meltzer, New York.

Azmitia, E.C., Bartus, R. and Whitaker-Azmitia, P.M., 1988. Tissue culture models of aging and dementia; search for neuronotrophic and neuronotoxins. Proceedings of the "Models of Aging and Dementia", NIA sponsored meeting in Washington, D.C., in press in <u>Neurobiology of Aging</u>.

Bjorklund, A., Stenevi, U., Svendgaard, N.A., 1976. Growth of transplanted monoaminergic neurons into the adult hippocampus along the perforant path. <u>Nature</u>, 262:787-790.

Cavanaugh, J.B., 1970. The proliferation of astrocytes around a needle wound in the rat brain. <u>J. Anat.</u> 106:471-487.

Madrazo, I., Drucker-Colin, R., Diaz, V. Martiniez-Mata, J., Torres, C., and Becerril, J., 1987. Open microsurgical autograft of adrenal medulla to the right caudate nucleus in two patients with intractable Parkinson's disease. <u>New England Journal of Medicine</u>, 316:831-4.

Ramon y Cajal, S., 1928. <u>Degeneration and Regeneration in the Nervous System</u>, Oxford University Press, London.

Whitaker-Azmitia, P.M., Doering, L., Gannon, P., Ramirez, A., and Azmitia, E.C., 1987. The onset and duration of astrocytic response to fetal cells. New York Academy of Science, Vol. in <u>Cell and Tissue Transplantation into the Adult Brain</u>. (Edited by Azmitia, E.C. and Bjorklund, A.)

Zhou, F.C. and Azmitia, E.C., 1985. Induced homotypic sprouting of serotonergic fibers in hippocampus: II, an immunocytochemical study. <u>Brain Research</u>, 337:337-348.

Zhou, F.C. and Azmitia, E.C., 1988. Denervation of serotonergic fibers in the hippocampus induced a trophic factor which enhances the maturation of transplanted serotonergic but not noradrenergic neurons. <u>J. Neurosci. Res.</u>, 17:235-248.

Zhou, F.C. and Azmitia, E.C., 1988. Role of laminin in directing fibers from transplanted serotonergic, noradrenergic and dopaminergic neurons in adult brain. <u>J. Chem. Neuroanat.</u> In Press.

FETAL BRAIN TRANSPLANTS INDUCE RECOVERY OF MORPHOLOGICAL AND LEARNING DEFICITS OF CORTICAL LESIONED RATS

Federico Bermúdez-Rattoni, Juan Fernández and
Martha Lilia Escobar

Instituto de Fisiología Celular, Universidad
Nacional Autónoma de México, México, D.F.

INTRODUCTION

The recovery from brain injuring has recently been obtained using the fetal brain transplant technique in adult mammal brains. Thus, it has been established that transplanted neurons differentiate and make connections with the host brain (Frotscher and Zimmer, 1986). Moreover, there are studies that have been able to show biochemical and functional changes due to such transplants (Bjorklund and Stenevi, 1977; Drucker-Colín et al, 1984). Up until recently, some studies have shown cognitive function recuperation (Bjorklund and Stenevi, 1985; Dunnett et al, 1985).

In 1983 Labbe and coworkers, reported that rats with cortical transplants (E22) in the frontal neocortex were able to learn a spatial alternation task in fewer trials than lesioned control rats or rats with cerebellar implants. The recovery effects were seen just one week after transplantation. Moreover, it is noteworthy that cerebellar tissue transplants where atrophied while the frontal implants survived and were healthy and well integrated with the host tissue (Labbe et al., 1983; Stein et al., 1985). However, in 1987 Dunnett et al., tried to replicate the anterior report and found that neocortical grafts (E21) produced a short-lasting improvement in the t-maze alternation performance, therefore they concluded that "... the short-lasting recovery in delayed alternation performance is attributable to diffuse influences of the embryonic tissue on the lesioned host brain rather to a reconnection of the damaged circuites" (Dunnet et al., 1987).

Other studies have shown that the transplantation of either adult or embryonic frontal cortex accelerated the recovery of frontal cortex lesioned rats on a reinforced alternation task (Kesslak et al., 1986). The authors found that the rate of behavioral recovery correlates with the size of the surviving transplants. The recovery showed by animals with cortical grafts could be partially answered by the findings of Sharp and González (1986). In this study, they reported that there was an increase of survival thalamic neurons by frontal cortical grafts as compared with those cortical lesioned animals. Moreover, they showed the existence of reciprocal connections between the thalamus and the graft by HRP-WGA injections.

The reasons for reconnection between host-graft are not well understood. Nevertheless, Chang and coworkers (1986) propose that while the factors determining the cortical arrange are intrinsic to the graft, the factors that determine the innervation between graft and host depends on the cellular environment which surrounds the grafted tissue. The first proposal has been demonstrated with grafts taken from frontal or occipital cerebral cortex that were placed into the occipito-parietal region of newborn rats. Results showed that the grafts developed normal pattern of lamination, with its original orientation, i.e.; the grafts had inverted orientation if they were placed upside down (Chang et al., 1986). The demonstration of the second comes from the same study in which the authors showed that regardless where the cortical area of the grafted tissue was taken, the transplants consistently received projections from those thalamic nuclei that normally innervated the adjacent host cortex. These results suggest that while immature cortical tissue may has an intrinsic, and perhaps autonomous, ability to develop lamination, the afferent and efferent cortical projections are most likely specified by extrinsic factors. However, for other authors the development of specific cell types and connectivity of the homotopic grafts, were mediated by intrinsic factors, as well as by the presence of some enzymes fundamental for the neurotransmitters synthesis (Smith et al., 1985). The authors indicated that the tissue taken at later stages of embryonic development (after cell migration and cortical plate formation) contains neurons that will express the synthetic enzyme for glutamic acid decarboxylase (GAD). When this happens, the GAD-labeled neurons in the surrounding host brain do not sprout into the transplants. On the other hand, neocortex taken at an early stage of development, in which the cell division and migration are just beginning, fails to express GAD and presumably contains no GABAergic neurons. Under these conditions the host GAD-positive neurons sprout profusely into the transplant. Therefore, the expression of some enzymes fundamental for the neurotransmitter synthesis, is very important for the innervation of host-graft (Smith et al., 1985).

In this regard, other studies had been made to investigate the innervation of grafts by host tissue. In 1986 Ebner and Erzurumlu, demonstrated that neocortical tissue transplants were innervated by thalamocortical axons of different ages hosts. With this purpose, the authors analized groups of newborn, and 30 days old rats, which received neocortical grafts. The most profuse reinnervation was observed in the group of newborn rats, while the subjects with 30 days showed fewer projections. One year before Ross and Ebner (1985) identified the differential capacity of several thalamic nuclei (ventrobasal complex; VB and posteromedial nucleus; POm) to innervate transplants localized in the somatosensorial cortex S_1. The grafts received afferents from POm but not from VB, which showed the differential capacity of both thalamic nuclei to innervate neocortical grafts. Finally, Hamasaki et al. (1987) transplanted lateral geniculate nucleus (LGN) from fetal rats into the visual cortex (VC) of neonatal rats, their results indicated that synaptic connections were established reciprocally between the transplanted LGN and the host VC. The presence of connections were observed through electro-physiological methods. All these researches, suggest the potential plasticity of the neocortical grafts, as well as their capacity to reestablish reciprocal connections with the host tissue.

CONDITIONED TASTE AVERSIONS

A wide varieties of animals can associate flavor with toxic effects apparently as a result of the coevolution of protective mechanisms on the host species (See Garcia et al., 1985; Garcia et al., 1977). These flavor-illness associations have been demonstrated in many laboratories and with different species (Garcia et al., 1977). Thus, taste is readily associated with illness producing the conditioned taste aversions (CTA) after a single taste-illness experience. Unlike most other demonstrations of classical conditioning the delay between the taste (conditioned stimulus; CS) and the illness (unconditioned stimulus; US) could be an hour or more, and it is possible to have a strong taste aversion. The audio-visual signals are poor CS for a toxic US, they acquire little or no aversive properties following a single toxic US. In contrast, the audio-visual signal can be readily associated with the footshocks US, whereas tastes are poor CSs in shock avoidance conditioning (García et al., 1982; 1985). This difference in conditioning has been termed cue-consequences specificity (García and Koelling, 1966; Domjan, 1985).

One of the advantages to use this paradigm in the study of neural recovery by brain transplants, rests in the knowledge of the neural pathways involved in the CTA conditioning. Therefore, it is possible to know if fetal brain transplants could recover the previous damaged CTA pathways, and this could be correlated with functional recovery.

The neural mediation of conditioned taste aversion has been established with the use of anatomical, electrophysiological and behavioral methods (See Fig. 1). Thus, it has been established that the nucleus solitarius (NTS; the first gustatory relay) receives heavy visceral input from the hepatic branch of the vagus (sensitive to stomach irritating toxins) as well as inputs from the area postrema (sensitive to blood-irritating toxins) and the vestibular system (sensitive to nausea-causing motion). The NTS also receives primary taste afferents from the entire tongue via nerves VII and IX and from the larynx and parynx via X (Travers et al., 1987). Neurons responded to both gustatory and visceral stimuli are found in the pontine taste area of the parabrachial complex (second gustatory relay).

There are two main projections from the pontine taste area. One major projection of fibers passes to ventral forebrain structures, such as the amygdala, lateral hypothalamus, and the substantia innominata (Norgren, 1974). The second projection ascends ipsilaterally in the central tegmental bundle to the posterior ventromedial and ventromedial nucleus of the thalamus, a zone identified as a relay site for gustatory and lingual afferents. (Norgren and Leonad, 1973; Kiefer, 1985). The thalamic taste area projects to the gustatory neocortex (GN), a small band located in the anterolateral part of the cortex, 1 mm wide by 3 mm long along the rhinal sulcus in rat (Norgren and Wolf, 1975). Recently, Lasiter described a direct projection from GN to the amygdaloid complex via the internal capsule. The trajectory of these projections were established by application of horseradish peroxidase (HRP) in the GN which produced retrograde cellular labeling within the ipsilateral and basolateral amygdaloid nucleus (Lasiter, 1982).

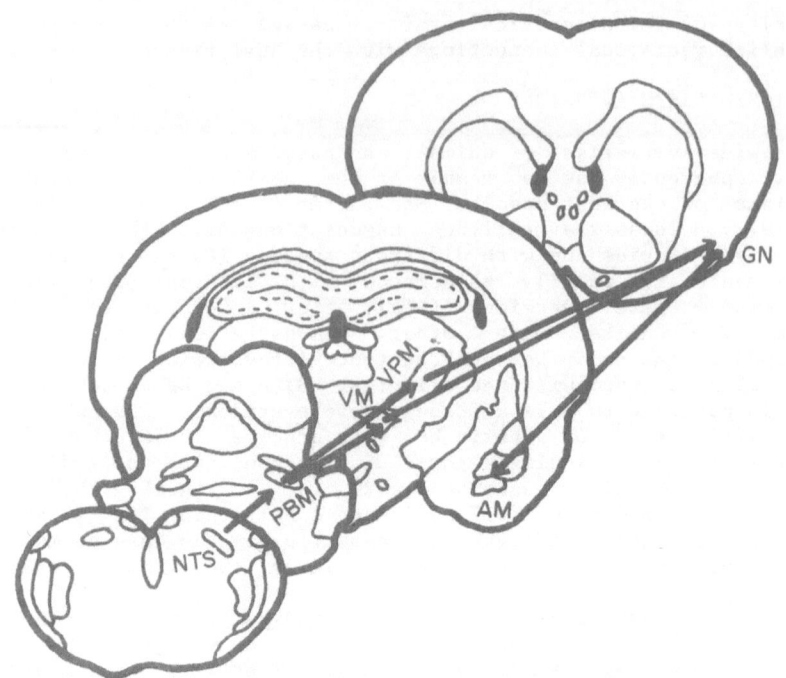

Fig. 1. Schematic drawing outlining the gustatory and visceral
pathways in the rat. Abbreviations: (NTS) solitary
nucleus, (PBM) medial parabrachial nucleus of the
pons, (VM, VPM) ventromedial and posterior
ventromedial thalamic nuclei, (AM) amygdala, (GN)
gustatory neocortex.

Several studies have shown that the GN area is involved in the
associative aspects of taste responding, but not in the hedonic
responses to taste (Kiefer, 1985). Thus, rats lacking GN have a
disrupted acquisition of taste aversions compared with the proper
controls (Kiefer, 1985; Bermúdez-Rattoni et al., 1987 and Lasiter and
Glanzman, 1985). Moreover, rats that were trained to avoid specific
taste and followed by ablation of the GN, in the postoperative tests
the lesioned animals showed no evidence of retention (Braun et al.,
1981). On the other hand, the hedonic response of a lesioned GN rats
appear to be normal. Since, it has been demonstrated that GN lesioned
animals consume above water base-lines sucrose as well as low
concentrations of sodium chloride. In addition, GN rats are able to
reject quinine and acid solutions as normal rats do (Kiefer, 1985;
Grill, 1985). These results indicated that GN integrity is not
necessary for normal taste responsiveness. Moreover, it has been
demonstrated that taste responsiveness remained intact even in
decerebrate rats (Grill and Norgren, 1978).

In a first series of experiments, we showed that the recovery of the lost ability to acquire taste aversions due to GN lesions is possible with homotopic cortical fetal brain transplants. Briefly, male Wistar rats, were randomly assigned to one of two groups. Large bilateral electrolytic lesions were made in one group to encompass the gustatory neocortex (Krieg's areas 13 & 14; See Fig. 1), the other group remained as unoperated control. Following post operative recovery the animals were deprived for 24 hrs., and trained to drink water daily during 5-minute trials for 10 days (See Fig. 3). The consumption volume was taken every day. On the acquisition trial, 0.1% saccharin was presented as a CS and followed 30 minutes later by intragastric LiCl (190 mg/kg) as US. An extinction trial was given after two water intake baseline measures; during extinction the CS was presented again, and the test volume scores were taken, this sequence was repeated once more.

Two days after the second extinction trial, the lesioned animals were divided randomly in four groups: One group received cortical homotopic grafts (GGN); other group heterotopic cortical occipital (GON) grafts; other group received heterotopic tectal (GT) graft; the last group remained without transplant as a lesioned control (LxGN). Seventeen-day old fetuses were removed from the abdominal cavity of pregnant rats. The fetal brains were taken (See Fig. 2), and the temporo-parietal area (above the rhinal sulcus) for the GGN group, occipital area for the GT group, and the tectal area for the GT group (See Seiger, 1985), were dissected under a microscope. The blocks of tissue were all then stereotaxically placed into the GN area with the same stereotaxic coordinates used to make the previous lesion (Bermúdez-Rattoni et al., 1987). After eight weeks of recovery, the four groups of rats were retrained in the same behavioral procedure described above. Results indicated that lesioned animals tested before transplantation showed the expected disrupted taste aversions when compared with the unoperated controls (Fig. 4). The postgraft results revealed that the animals with homotopic and heterotopic occipital grafts recovered the ability to acquire the taste illness association and were not significantly different from the control group. On the other hand, the groups which received the heterotopic tectal transplants or remained without transplant did not show any behavioral recovery (See Fig. 4; Bermúdez-Rattoni et al., 1987).

FETAL GN GRAFTS PRODUCED RECONNECTIVITY WITH THE HOST TISSUE

In other series of experiments we demonstrated with HRP histochemistry, that the transplants were able to reestablish connectivity with those areas that have been described as having normal connections with the GN. We followed the horseradish peroxidase protocol according to Mesulam (1982) technique and counterstained with thionine. The slices were examined and photographed under bright and dark field microscopy for the presence and location of retrogradely labeled neurons. Briefly, some control animals received a unilateral injection (0.5 ul) of HRP in the amygdala. The GN, occipital and tectal grafted animals received the same solution in the amygdala, always ipsilateral to the graft (Escobar et al., submitted).

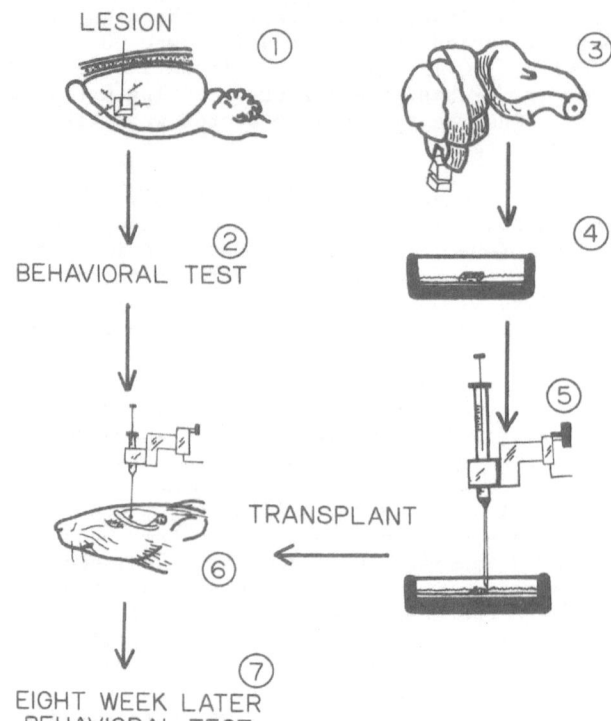

LESION ①

③

② BEHAVIORAL TEST

④

⑤

TRANSPLANT

⑥

⑦

EIGHT WEEK LATER
BEHAVIORAL TEST

Fig. 2. Schematic representation of the methodological sequence. 1. Electrolytic lesions in the GN. 2. Behavioral test (see Fig. 3). 3, 4 and 5 transplant procedure; obtention of solid tissue from embryonic rats donnors; solid tissue block was put into a petry dish and aspirated by 100 ul Hamilton syringe. 6. Transplants were about 5 ul of embryonic tissue implanted with stereotaxic methods. 7. Eight weeks post-graft the behavioral training was given once more.

Results from HRP revealed that sections from control animals which received the enzyme in the amygdala, showed reaction product boundaries extended 1 mm in diameter surrounding the area of the tracer application (Escobar et al., submitted). HRP labelled cells were always found in the ipsilateral gustatory neocortex (Krieg's area 13 and 14), and in the ventromedial nucleus of the thalamus (See Fig. 5).

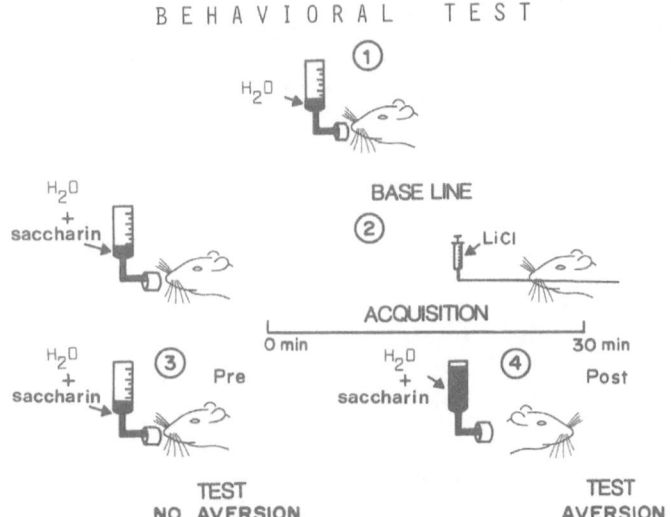

Fig. 3. Schematic representation of behavioral training sequence. 1. Training for 10 days to drink water for 5 minutes. 2. Acquisition day; presentation of saccharin disolved in the water (.1%) followed 30 min later with an infusion of LiCl (IG; 190 mg/kg). 3. Test day with saccharin in the water; Lesioned animals did not showed taste aversion, consuming similar quantities of the solution as baseline day. 4. Normal animals reject the solution, showing taste aversion.

The HRP histochemistry results support previous observations, that there is normal connectivity between amygdala and gustatory neocortex (Lasiter, 1982), since applications of HRP in the amygdala clearly produced labelled neurons in the ipsilateral gustatory neocortex of normal rats. Moreover, retrograde cellular labeling was found in the VPM of the thalamus, resulting from HRP applications in the amygdala. These results are in close agreement with those found by Lasiter (1982) and Kretek and Price (1974).

In general, the GN and occipital brain transplants appeared to be
healthy and placed in the appropriate target area of the host brain.
In both GGN and GON brain transplants we found scarce HRP labeled cells,
although we found a good ammount of HRP labeled cells in the VPM and
VM nucleus of thalamus in the same animals. The low density in the
grafts have been previously described in studies that used HRP as a
tracer for marking projections between host-graft tissues. Thus, a
few labeled cells have been found when fetal brain transplants have
been made in the hippocampus, occipital and somatosensory cortex,
(Bjorklund and Stenevi, 1979; Jaeger and Lund, 1981; Kromer et al.,
1980; Ross and Ebner, 1985). The reason of the low density of labeled
cells has not yet been established (See, Jones, 1975). One hypothesis
is that the fetal brain transplant is still under development and
therefore their neurons are just starting to make connections with the
host tissue. However, it has been shown that developing neurons are
more efficient for incorporating peroxidase (Kristensson and
Sjostrand, 1972). It is possible therefore, that the fetal brain
transplant makes connections with its host, although not in a complete
and normal fashion.

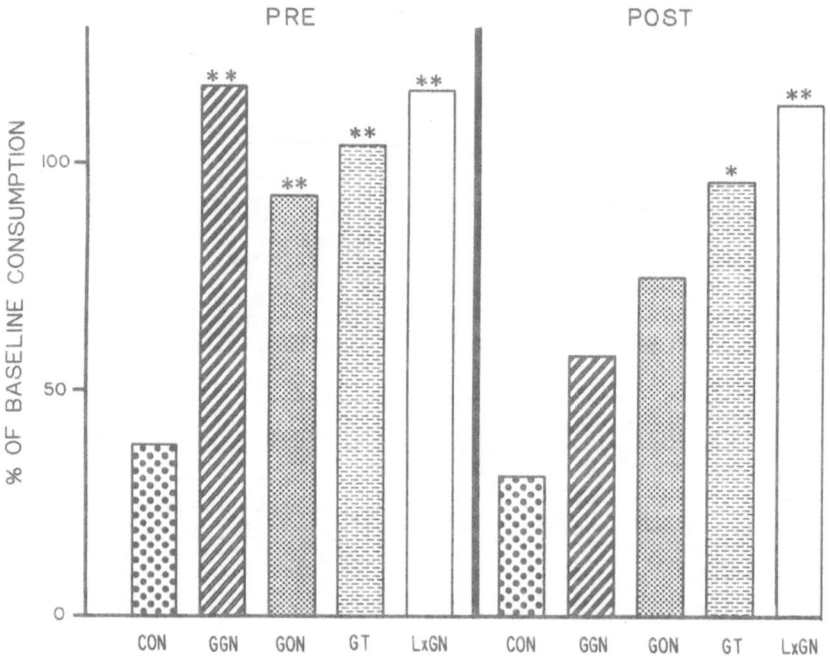

Fig. 4. The bars illustrate the amount of saccharin
consumption by control, grafted and lesioned groups.
Consumption is expressed as the percentage of each
group's previous day water baseline, left side shows
the results from one taste test trial prior to
transplant. Right side shows the results of one taste
test trial eight weeks after transplant; * p < 0.05,
** p < 0.01 (Dunnet test). (For description of groups
see text).

On the other hand, the heterotopic tectal transplants did not integrate well with its host. There was a heavy glial invasion, necrosis, abundant vacuoles and very scant vascularization. Therefore, a complete lack of HRP labeled cells were found in the grafted tissue. Nevertheless, there were HRP labeled cells in the VM and VPM nucleus of the thalamus of the same rats (Escobar et al., submitted).

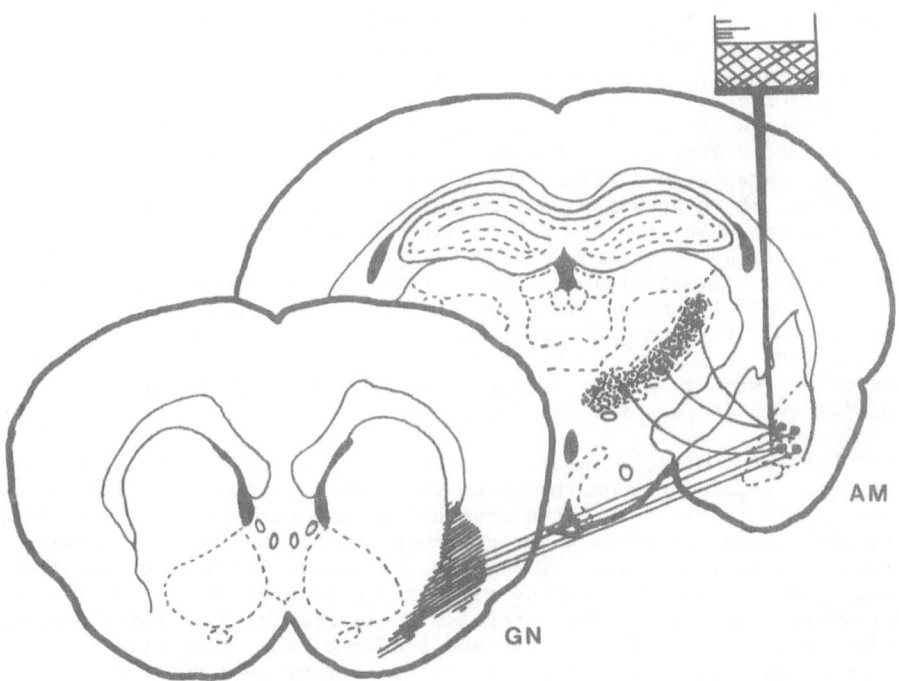

Fig. 5. Schematic representation of HRP injection in the amygdala (AM) and retrogradely labeled neurons within the gustatory neocortex (GN) and ventromedial nucleus of thalamus (VPM).

DISCUSSION

It is clear that gustatory neocortex lesions in rats produced disruption in the ability to associate the taste with its visceral consequences, these results have been reported several times (Braun et al., 1982; Kiefer, 1985). Moreover, the fetal brain transplants produced a significant recovery in the ability of lesioned rats to associate the taste with the visceral consequences (Bermúdez-Rattoni et al., 1987). The possibility of spontaneous recovery was excluded, because the animals with GN lesions that did not receive any transplant, were unable to acquire the CTA after 8 weeks even with two acquisition trials (Yirmiya et al., 1987; Bermúdez-Rattoni et al., 1987). In contrast, in a previous report it was found that animals with lesions in amygdala showed spontaneous recovery after eight weeks post operation, when tested in taste aversion conditions (Bermúdez-Rattoni et al., 1987). Similar spontaneous recovery has been reported with large ablations of frontal cortex in an alternation task after six weeks postoperation (Dunnett et al., 1987).

Elsewhere we attempted to explain these functional differences between amygdala and GN. One possible explanation is that amygdala lesion produced reorganization of other elements in the neuronal network (Prado-Alcalá et al., 1978; Prado-Alcalá and Cobos Zapiaín, 1979). This idea has been demonstrated with functional alteration of the neostriatum. Thus, chemical alterations (i.e., microinjection of anticholinergic drugs) of the neostratum produces severe disruption of learned tasks (Prado-Alcalá et al., 1978). However, if the animals are overtrained, similar functional alterations do not produce disruption of the same learned tasks (Prado-Alcalá and Cobos Zapiaín, 1979). Therefore, it is possible to conclude that after the overtraining, the encoding necessary for the performance could be transferred to another neuroanatomical or neurochemical system. In our experimental conditions, overtraining could have been produced by repeated acquisition of taste aversion trials. Therefore, a plausible explanation for the differential effects between the cortical and amygdaloid lesions, is that for taste aversion learning the GN is a permanent memory store, whereas the amygdala only intervenes as an initial step storage for CTA (Bermúdez-Rattoni et al., 1987).

Our preliminary results showed that there is some recovery gradient regarding upon the place where graft tissue was taken. In Fig. 4 it is clear that the best behavioral recovery are from those animals which received homotopic cortical tissue (GGN), followed by those of occipital tissue (GON). Those animals which received tectal heterotopic tissue or remained without any transplant did not show any behavioral recovery. These results indicated that for taste aversion only cortical fetal brain tissue produced the recovery. Several authors employing other areas and different behavior tasks have reported that some tissue specificity is needed for anatomical and functional recovery. Stein and coworkers in 1985 made heterotopic cerebellar transplants into the frontal cortex and did not find any functional recuperation in a maze learning task. Moreover, they found a lack of integration of the cerebellar grafts with the host tissue as compared with the frontal graft integration. Similar results have been found when retina grafts are transplanted to a non-visual system location such as the cerebellum; the grafts do not form projections into the host brain, and the ganglion cells within the transplant degenerated (McLoon et al., 1985). In our results, the heterotopic tectal transplants did not integrate with its host, and there were heavy glial invasion, necrosis and very scant vascularization. Moreover, there were a lack of HRP labeled cells in the grafted tissue. The animals with tectum grafts did not recover the ability to associate taste with its visceral consequences (Bermúdez-Rattoni et al., 1987). These results give further support to the idea that some tissue specificity is needed for behavioral recovery.

ACKNOWLEDGEMENTS

This research was supported by Grant from CONACyT:PCEXCNA-050290. We thank Mrs. Maria Teresa Torres for preparing the manuscript.

REFERENCES

Bermúdez-Rattoni, F., Fernández, J., Sánchez, M.A., Aguilar-Roblero, R. and Drucker-Colín, R., 1987, Fetal brain transplants induce recuperation of taste aversion learning., Brain Res., 416:147-152.

Bjorklund, A. and Stenevi, U., 1977, Reformation of the severed septo-hipocampal cholinergic pathway in the adult rat by transplanted septal neurons, Cell. Tissue Res., 185:285-302.

Bjorklund, A. and Stenevi, U., 1979, Reconstruction of the nigrostriatal dopamine pathway intracerebral nigral transplants, Brain Res., 177:555-560.

Bjorklund, A. and Stenevi, U., 1985, "Neural Grafting in the Mammalians CNS", Elsevier, Amsterdam.

Braun, J.J., Kiefer, S.W., and Ovellet, J.V., 1981, Psychic agensia in rats lacking gustatory neocortex, Exp. Neurol., 72:711-716.

Brown, J.J., Lasiter, P.J. and Kiefer, S.W., 1982, The gustatory neocortex of the rat, Physiol. Psychol., 10:13-45.

Chang, F.L.F., Steedman, J.G. and Lund, R.D., 1986, The lamination and connectivity of embryonic cerebral cortex transplanted into newborn rat cortex. J. Comp. Neurol., 244:401-411.

Domjan, M., 1985, Cue-consequence specificity and long-delay learning revisited. Ann. N.Y. Acad. Sci. 443:54-66.

Drucker-Colín, R., Aguilar-Roblero, R., García-Hernández, F., Fernández-Cancino, F. and Bermúdez-Rattoni, F., 1984, Fetal suprachiasmatic nucleus transplants: diurnal rhythm recovery of lesioned rats, Brain Res., 311:353-357.

Dunnett, S.B., Ryan, C.N., Levin, P.D., Reynolds, M. and Bunch, S.T., 1987, Functional consequences of embryonic neocortex transplanted to rats with prefrontal cortex lesions. Behav. Neurosci. 101:489-503.

Dunnett, S.B., Toniolo, G., Fine, A., Ryan, C.N., Bjorklund, A. and Iversen, S.B., 1985, Transplantation of embryonic ventral forebrain neurons to the neocortex of rats with lesions of nucleus basalis magnocelularis II. Sensorimotor and learning impairments, Neuroscience, 16: 787-797.

Ebner, F.F. and Erzurumlu, R.S., 1986, Innervation of embryonic neocortical cell suspensions by thalamocortical axons of different aged hosts, Soc. Neurosci. Abstr., 12:973.

Escobar, M., Fernández, J., Guevara-Aguilar, R. and Bermúdez-Rattoni, F., 1988, Fetal brain grafts induce recovery of learning deficits and connectivity in rats with gustatory neocortex lesion. Submitted

Frotscher, M. and Zimmer, J., 1986, Intracerebral transplant of the rat fascia dentata: a Golgi/electron microscope study of dentate granule cells, J. Comp. Neurol., 246:181-190.

Garcia, J. and Koelling, R.A., 1966, Relation of cue to consequence in avoidance learning, Psychonom. Sci., 4:123-124.

Garcia, J., Rusiniak, K.W. and Brett, L.P., 1977, Conditioned food-illness aversion in wild animals: Caveant canunici., in: "Operant-Pavlovian interaction", H. Davis and H. Hurwitz, ed., Lawrence Erlbaum Associates, New Jersey.

Garcia, J., Lasiter, P.S., Bermúdez-Rattoni, F. and Deems, D., 1985, A general theory of aversion learning, Ann. N.Y. Acad. Sci., 443:8-20.

Garcia, J., Rusiniak, K.W., Kiefer, J.W. and Bermúdez-Rattoni, F., 1982, The neural integration of feeding and drinking habits, in: "Conditioning", C. Woody, ed., Plenum Press, N.Y.

Grill, H.J., 1985, Physiological mechanisms in conditioned taste aversions, Ann. N.Y. Acad. Sci., 443-67-88.

Grill, H.J. and Norgren, R., 1978, The taste reactivity test. II. Mimetic responses to gustatory stimuli in chronic thalamic and chronic decerebrate rats, Brain Res., 143:281-297.

Hamasaki, T. et al., 1987, Electrophysiological study of synaptic connections between a transplanted lateral geniculate nucleus and the visual cortex of the host rat, Brain Res., 422:172-177.

Jaeger, C.B. and Lund, R.D., 1980, Transplantation of embryonic occipital cortex to the brain of newborn rats, Exp. Brain Res., 40:265-272.

Jaeger, C.B. and Lund, R.D., 1981, Transplantation of embryonic occipital cortex to the brain of newborn rats: A Golgi study of mature and developing transplants, J. Comp. Neurol., 200:213-230.

Jones, E.G., 1975, Possible determinants of the degree of retrograde neuronal labeling with horseradish peroxidase, Brain Res., 85: 249-253.

Kesslak, J.P., Brown, L., Steichen, C. and Cotman, C.W., 1986, Adult and embryonic frontal cortex transplants after frontal cortex ablation enhance recovery on a reinforced alternation task, Exp. Neurol., 94:615-626.

Kiefer, S.W., 1985, Neural mediation of conditioned food aversions. J. Ann. N.Y. Acad. Sci., 443:100-109.

Krettek, J.E., and Price, J.L., 1974, A direct input from the amygdala to the thalamus and the cerebral cortex, Brain Res., 67:169-174.

Kristensson, K. y Sjostrand, J., 1972, Retrograde transport of protein tracer in the rabbit hypoglossal nerve during regeneration, Brain Res., 45: 175-181.

Kromer, L.F., Bjorklund, A. and Stenevi, U., 1980, Innervation of embryonic hippocampal implants by regenerating axons of cholinergic septal neurons in the adult rat, Brain Res., 210:153-171.

Labbe, R., Firl, A. Jr., Mufson, E.J. and Stein, D.G., 1983, Fetal brain transplants: reduction of cognitive deficits in rats with frontal cortex lesions, Science, 221:470-472.

Lasiter, P.S., 1982, Cortical substrates of taste aversion learning. Direct amygdalocortical projections to the gustatory neocortex do not mediate conditioned taste aversion learning, Physiol. Psychol., 10: 377-383.

Lasiter, P.S. and Glanzman, D.L., 1985, Cortical substrates of taste aversion learning: Involvement of the dorsolateral amygdaloid nuclei and temporal neocortex in taste aversion learning. Behav. Neurosci., 99:257-276.

McLoon, L.K., McLoon, S.C., Chang, F.L.F., Steedman, J.G. and Lund, R.D., 1985, Visual system transplanted to the brain of rats, in: "Neural grafting in the mammalians CNS", Bjorklund, A. and Stenevi V. eds., Elsevier Science Publishers, B.V.

Mesulam, M.M., 1982, Tracing neural connections with horseradish peroxidase, John Wiley and Sons ed., N.Y.

Norgren, R., 1974, Gustatory afferents to ventral forebrain. Brain Res., 81:285-295.

Norgren, R. and Leonard, C.M., 1973, Ascending central gustatory pathways, J. Comp. Neurol., 150:217-238.

Norgren, R. and Wolf, G., 1975, Projections of thalamic gustatory and lengual areas in the rat, Brain Res., 92:123-129.

Prado-Alcalá, R.A., Bermúdez-Rattoni, F., Velázquez-Martínez, A. and Bacha, G.M., 1978, Cholinergic blockade of the caudate nucleus and spatial alternation performance in rats: overtraining-induced protection against behavioral deficit, Life Sci., 23:889-896.

Prado-Alcalá, R.A. and Cobos-Zapiaín, G., 1979, Interference with caudate nucleus activity by potassium chloride. Evidence for a moving engram, Brain Res., 172: 577-583.

Ross, D.T. and Ebner, F.F., 1985, Neocortical transplants in the S1 cortex of adult mice receive thalamic afferents from the posteromedial (POm) but not the ventrobasal (VB) nucleus, Soc. Neurosci. Abstr., 11:974.

Seiger, A., 1985, Preparation of immature central nervous system regions for transplantation, in: "Neural grafting in the mammalian CNS", Bjorklund, A. and Stenevi, U., eds., Elsevier Science Publishers, B.V.

Sharp, F.R. and González, M.F., 1986, Fetal cortex transplants attenuate thalamic atrophy caused by neonatal frontal cortex lesions: WGA-HRP and amino acid studies of connections between motor cortex and thalamus of neonatally lesioned rats with and without transplants. Soc. Neurosci. Abstr., 12:1473.

Smith, L.M., Hohmann, Ch. and Ebner, F.F., 1985, The development of specific cell types and connectivity in neocortical transplantation, in: "Neural grafting in the mammalian CNS", A. Bjorklund and U. Stenevi, ed., Elsevier Science Publishers, B.V.

Stein, D.G., Labbe, R., Firl, A. and Mufson, E.J., 1985, Behavioral recovery following implantation of fetal brain tissue into mature rats with bilateral, cortical lesions, in: "Neural grafting in the mammalian CNS", A. Bjorklund and U. Stenevi, eds., Elsevier Science Publishers, B.V.

Travers, J.B., Travers, S.P. and Norgren, R., 1987, Gustatory neural processing in the hindbrain, Ann. Rev. Neurosci., 10:595-632.

Yirmiya, R., Zhou, F.C., Holder, M.D., Deems, D.A. and Garcia, J., 1987, Ammelioration of lesion induced deficit in taste-illness integration by gustatory neocortex transplantation. Schmitt Neurological Sciences Symposium, Univ. Rochester, N.Y.

ORAL VACCINATION OF HUMAN VOLUNTEERS WITH galE MUTANTS

OF ENTEROPATHOGENIC Escherichia coli

Alejandro Cravioto, Rosa E. Reyes, Francisca Trujillo, Angel
Tello and Virginia Vázquez

Instituto Nacional de Ciencias y Tecnología-DIF
Mexico City, Mexico

INTRODUCTION

Diarrhea constitutes the single most important cause of death in infants and children born in developing areas of the world (Puffer and Serrano, 1973). The widespread use of oral rehydration in recent years has lead to a decrease in the mortality rate due to this disease. Treatment albeit does not prevent moderate to severe malnutrition, the main complication of acute infectious diarrhea. The long term effects of this dyad on the physical growth and mental development of affected individuals have been well documented (Cravioto and Arrieta, 1986). For these reasons, the aim of several research groups during the past ten years has been the development of specific vaccines to be used in prevention programs against diarrhea. This approach has been specially apparent in relation with cholera and typhoid fever, where parental vaccines developed earlier have been substituted by oral presentations prepared with dead organisms, live attenuated strains or combinations of both (Levine and Tramont, 1986; Levine et al., 1987).

Epidemiological evidence obtained in countries where cholera is not present have shown that enteropathogenic strains of Escherichia coli (EPEC) constitute one of the main etiologic agents of diarrhea in small children who live in these areas (Guerrant et al., 1983; Cravioto et al., 1988). Ability to attach and efface the microvillous surface of intestinal epithelial cells seems to be the main pathogenicity trait of these strains (Levine et al., 1985; Knutton et al., 1987). Human volunteer studies by Levine et al. (1985) have shown that the adhesive ability of EPEC is related with the presence of a plasmid-coded outer membrane protein (OMP) of 94 kilodaltons (kDa). Loss of this plasmid reduces the capacity of an EPEC strain to cause diarrhea in adult volunteers.

Over the past three years we have obtained mutants of an EPEC strain of serotype O111:H2 capable of expressing this 94 kDa OMP, but lacking the ability to survive in the intestine of mammals due modifications in the galE region of its chromosome. After extensive experimental testing we decided to carry-out a safety and immunogenicity trial using two of these galE mutants in adult human volunteers. The preliminary results of this investigation are the subject of the present communication.

MATERIALS AND METHODS

VACCINE STRAINS

Two galE mutants of an O111:H2 EPEC strain were used in the vaccine trial. One of these strains was a spontaneous mutant obtained under selective pressure using indicator nutrient agar (Difco) plates with 5.5 mM D-galactose and 0.005% (wt/vol) neutral red. On these plates gal⁻ strains yielded yellow colonies with a surrounding clear yellow zone. A single gal⁻ strain (DIF 47) was selected and tested for sensitivity to galactose-induced lysis at 0.2 mM galactose. Strain DIF 47 lacks both the enzyme uridine 5'-diphosphate-galactose 4-epimerase and the ability to incorporate exogenous galactose into its lipopolysaccharide (Elbein and Heath, 1965). Full experimental details to support these findings will be published elsewhere.

The second galE mutant (strain DIF 67) used in this trial was obtained by transposon insertion mutagenesis and minicell analysis following the method described by Hone et al. (1987).

Both galE mutants DIF 47 and DIF 67 harbour a 60 mDal plasmid coding for the production of the 94 kDa OMP related with adherence to intestinal cells and production of diarrhea in human volunteers (Levine et al., 1985). The presence of this plasmid is also related with the ability of these strains to adhere to HEp-2 tissue culture cells (Cravioto et al., 1979) with a localized pattern (Scaletzky et al., 1984). Strain HS a non-typable E. coli isolated from the feces of a healthy adult was used as control strain in the trial.

VACCINE TRIAL

Twenty-two healthy adults of both sexes were invited to participate in the trial after it was approved by the Ethics and Human Experimentation Committee of the Institute. Written consent was obtained after each volunteer after being informed of the aims, design and possible complications of the trial, following the guidelines of the Center for Vaccine Development of the University of Maryland in Baltimore, kindly obtained from Dr. Myron M. Levine.

A double blind study design was used. The 22 volunteers were divided at random into three groups and inoculated orally with the vaccine strains grown overnight in nutrient broth (Difco), washed twice with phosphate buffered saline (PBS) pH 7.2 and diluted to the appropriate concentration of 6×10^{10} bacteria/ml in commercial apple juice. To supress gastric acidity, each volunteer was given 2g of sodium bicarbonate by mouth 15 minutes before being inoculated with the bacterial strain.

Fecal samples were obtained on the day of the begining of the trial and then daily for 10 days. Samples were inoculated on MacConkey, Tergitol 7, and Xylose-lysine desoxycholate (XLD) agars (Merck) and enriched with selenite and tetrathionate broths. Growth on these plates was identified by biochemical tests after overnight incubation at 37ºC. Enrichment broths were inoculated on MacConkey, XLD, Brilliant-green and Salmonella-Shigella agar (Merck) and incubated overnight at 37ºC. Growth on these plates was identified as above.

Serial hundred-fold dilutions of 1g of feces were inoculated in triplicate on MacConkey plates and incubated overnight at 37ºC. Bacterial growth on the plate with the highest number of single colonies was transferred to Whatman 541 filter paper (Whatman) and lysed by the method described by Maas (1983). Filters were then hybridized with a specific DNA probe (Nataro et al., 1985) constructed from a 1 kb fragment of the EPEC adherence factor (EAF) plasmid of strain E2348 (E. coli O127:H6). This plasmid codes for the 94 kDa OMP previously mentioned. The probe was labelled with a commercial random primer extension system (DuPont/NEN) using $[\alpha\text{-}^{32}P]$ dCTP at 54ºC, a temperature adjusted for the altitude of Mexico City.

The filters were mounted between plastic sheets and placed on X-OMAT film (Kodak) in cassettes with Cronex-lighting plus intensifying screens (DuPont) and exposed for 24-48 hours at -70ºC. The films were developed and analysed according to appropriate controls hybridized simultaneously with each group of filters.

Venous blood was obtained from each volunteer after overnight fasting on the day of the oral inoculation and then 10, 30 and 45 days after. Aliquots of serum obtained from these samples were kept at -70ºC until use, after incubation at 50ºC for 30 min to destroy complement and absorption of heterologous antibodies.

Serum was examined for the presence of specific antibodies to O111 LPS of E. coli and a whole-cell preparation of mutant strain DIF 47, with a direct hemagglutination assay (Lemieux et al., 1974). In brief, washed sheep red blood cells (RBC) were coated with LPS obtained by boiling for 60 min whole cell preparations after overnight growth in nutrient agar (Difco) plates and standardized by nephelometry at 640 nm. End point titrations of two-fold dilutions of the problem sera were detected after incubation for 30 min at 37ºC with coated sheep RBC and centrifugation at 2,000 x g for 3 min at 15ºC. Appropriate control rabbit antiserum against these same antigens was used in each assay.

RESULTS AND DISCUSSION

All volunteers involved in the study had been healthy during the 30 days previous to the trial, and none of them developed diarrhea after inoculation with the galE mutants or the control strains. Daily physical examination did not reveal the presence of clinical signs or symptoms related with the oral inoculation of live organisms. These results indicate that even at very high dosis both galE mutants were safe when used in humans, supporting previous findings of oral inoculation trials with galE mutants of Salmonella typhi or Salmonella typhimurium (Wahdan et al., 1982; Robertsson et al., 1983).

Analysis of the fecal samples obtained from the volunteers showed that both galE mutants were able to colonize the upper intestine of these individuals within 24-48 hours of the oral inoculation. By the third day post-inoculation gal$^+$ strains harbouring a plasmid coding for the 94 kDa OMP represented between 45-75% of the fecal aerobic microbiota. These strains were not detected in any of the volunteers before the inoculation or in those vaccinated with the control strain.

As seen in Fig. 1, colonization curves for both galE mutants were significantly different. Colonization in this case was defined as the presence of a predominant aerobic bacteria in the fecal samples grown at 37ºC. Differences in colonization by the two galE strains became apparent after the forth day post inoculation when strain DIF 47, the spontaneous mutant, rose sharply to become almost the exclusive aerobic bacteria in the feces, while strain DIF 67, the cloned galE mutant, remained below 75%. After the fifth day post-inoculation presence of the latter strain decreased sharply and disappeared from the feces completely by day 10. In contrast, strain DIF 47 remained as the predominant aerobic bacteria in the feces from days 5 to 8, and then decreased rapidly over the next two days, but was still present in the fecal samples by day 10 post-inoculation.

These findings correlated with the serological response to both O111 LPS and the whole cell preparation of each galE mutant in the post-inoculation sera of vaccinated individuals (Figs. 2-3). In both cases a four-fold serum antibody rise was seen to both O111 LPS and the whole cell preparation. Response in volunteers inoculated with strain DIF 47 was significantly higher to that of individuals vaccinated with strain DIF 67, the cloned mutant. The reasons for these differences are the subject of current research, as well as the specific

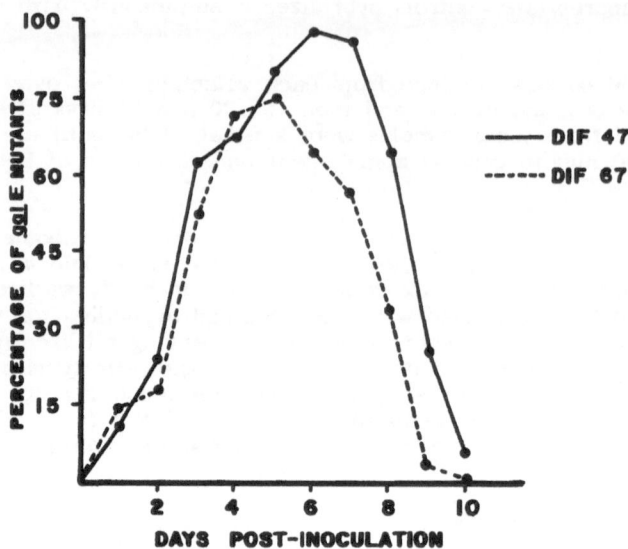

Fig. 1. Percentage of galE mutants harbouring an EAF plasmid coding for the production of a 94 kDa OMP in the feces of vaccinated volunteers.

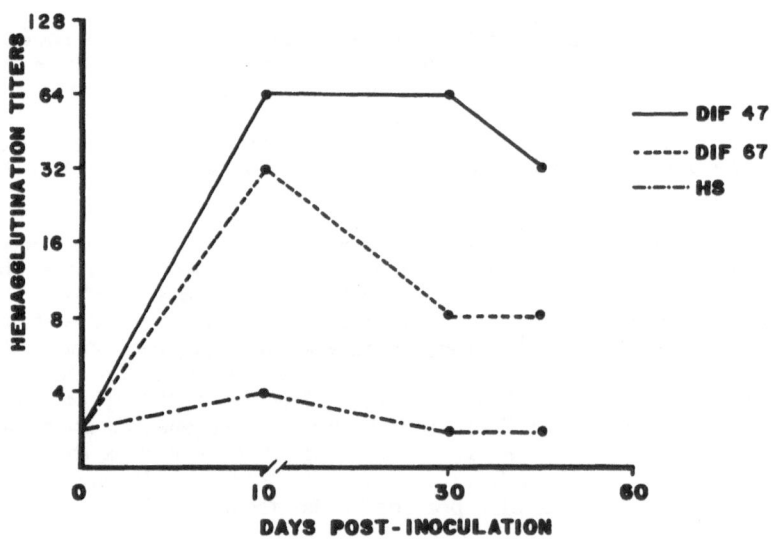

Fig. 2. Serological response to O111 LPS antigen in the sera of vaccinated volunteers.

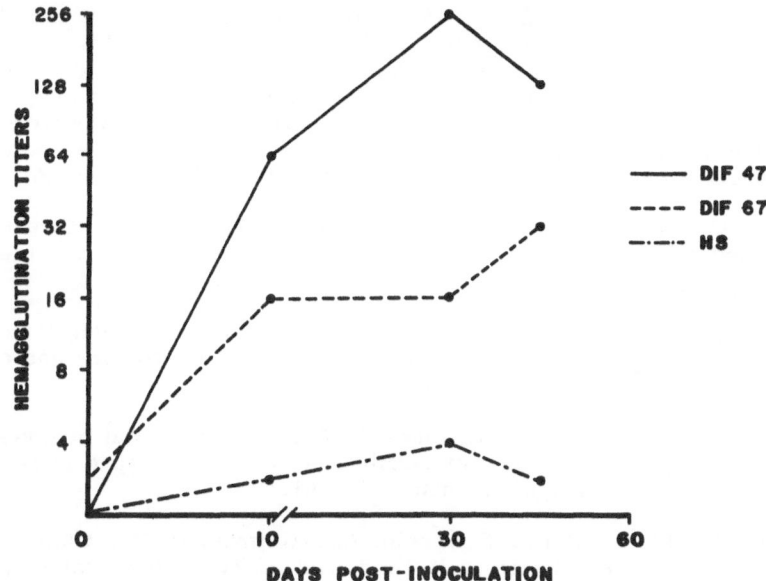

Fig. 3. Serological response to a whole-cell preparation of a galE mutant
of Escherichia coli in the sera of vaccinated volunteers.

serum response of the volunteers to the plasmid-coded 94 kDa OMP after it has
been purified to homogeneity.

It is evident from the results that the spontaneous galE mutant (DIF 47)
is a better colonizer and gives a better immune response than the cloned
mutant DIF 67. The reasons for this differences are probably related with
modifications caused by the cloning experiments in parts of the outer membrane
or the lipopolysaccharide core of this latter mutant. These apparent
disadvantages, however, should be weighed against the inability of this cloned
mutant to revert to a gal+ strain, a constant possibility in any spontaneous
mutant such as strain DIF 47, although no evidence for this was found in the
course of the present or previous investigations using this latter strain.

The preliminary results presented show that both galE mutants used in
this trial were safe in humans and capable of producing an immunological
response after oral inoculation. Their protective capacity remains to be
determined in future experiments using fully pathogenic EPEC strains to
challenge these same vaccinated individuals.

REFERENCES

Cravioto, A., Gross, R.J., Scotland, S.M., and Rowe, B., 1979, An adhesive
    factor found in strains of Escherichia coli belonging to the traditional
    enteropathogenic serotypes. Curr. Microbiol., 3:95.

Cravioto, A., Reyes, R.E., Ortega, R., Fernández, G., Hernández, R.,
    and López, D., 1988, Prospective study of diarrhoeal disease in a
    cohort of rural Mexican children: incidence and isolated pathogens
    during the first two years of life. Epidem. Inf., 100: in press.

Cravioto, J., and Arrieta, R., 1986, Nutrition, mental development, and
    learning, in: "Human Growth", F. Faulker, and J.M. Tanner, ed.,
    Plenum Publishing Corporation, New York.

Elbein, A.D., and Heath, E.C., 1965, The biosynthesis of cell wall lipopoly-saccharide in Escherichia coli. J. Biol. Chem., 240:1919.

Guerrant, R.L., Kirchhoff, L.V., Shields, D.S., Nations, M.K., Leslie, J., de Sousa, M.A., Araujo, J.G., Correia, L.L., Sauer, K.T., McClelland, K.E., Trowbridge, F.L., and Hughes, J.M., 1983, Prospective study of diarrheal illness in Northeastern Brazil: patterns of disease, nutritional impact, etiologies, and risk factors. J. Infect. Dis., 148:986.

Hone, D., Morona, R., Attridge, S., and Hackett, J., 1987, Construction of defined galE mutants of Salmonella for use as vaccines. J. Infect. Dis., 156:167.

Knutton, S., Lloyd, D.R., and McNeish, A.S., 1987, Adhesion of entero-pathogenic Escherichia coli to human intestinal enterocytes and cultured human intestinal mucosa. Infect. Immun., 55:69.

Lemieux, S., Avrameas, S., and Bussard, A.E., 1974, Local hemolysis plaque assay using a new method of coupling antigens on sheep erythrocytes by glutaraldehyde. Immunochemistry, 11:261.

Levine, M.M., Black, R.E., Ferreccio, C., Germanier, R., Chilean Typhoid Committee, 1987, Large-scale field trial of Ty21a live oral typhoid vaccine in enteric coated capsule formation. Lancet, i:1049.

Levine, M.M., Nataro, J.P., Karch, H., Baldini, M.M., Kaper, J.B., Black, R.E., Clements, M.L., O'Brien, A.D., 1985, The diarrheal response of humans to some classic serotypes of enteropathogenic Escherichia coli is dependant on a plasmid encoding an enteroadhesive factor. J. Infect. Dis., 152:550.

Levine, M.M., and Tramont, E.C., 1986, Vaccines against enteric infections, in: "Frontiers in Gastroenterology", P. Rozen, ed., S. Karger, Basel.

Maas, R., 1983, An improved colony hybridization method with significantly increased sensitivity for detection of single genes. Plasmid, 10:296.

Nataro, J.P., Baldini, M.M., Kaper, J.B., Black, R.E., Bravo, N., and Levine, M.M., 1985, Detection of and adherence factor of entero-pathogenic Escherichia coli with a DNA probe. J. Infect. Dis., 152:560.

Puffer, R.R., and Serrano, C.V., 1973, Patterns of mortality in childhood. PAHO Sci. Publ. 262, Panamerican Health Organization, Washington, D.C.

Robertsson, J.A., Lindberg, A.A., Hoiseth, S., and Stocker, B.A.D., 1983, Salmonella typhimurium infection in calves: protection and survival of virulent challenge bacteria after immunization with live or inactivated vaccines. Infect. Immun., 41:751.

Scaletzky, I.C.A., Silva, M.L.M., and Trabulsi, L.R., 1984, Distinctive patterns of adherence of enteropathogenic Escherichia coli to HeLa cells. Infect. Immun., 45:534.

Wahdan, M.H., Série, C., Cerisier, Y., Sallam, S., and Germanier, R., 1982, A controlled field trial of live Salmonella typhi strain Ty21a oral vaccine against typhoid: three year results. J. Infect. Dis., 145:292.

THE ROLE OF OUTER MEMBRANE PROTEINS FROM GRAM-NEGATIVE BACTERIA AS VACCINES WITH SPECIAL EMPHASIS IN TYPHOID FEVER: Monoclonal antibodies against *S. typhi* porins

Armando Isibasi[1], Vianney Ortiz[2], José Moreno[1], Jorge Paniagua[1], Martín Vargas[2], César González[1], and Jesús Kumate[3]

[1]Laboratory of Immunochemistry. Unidad de Investigación Biomédica, Instituto Mexicano del Seguro Social. P.O. Box 73-032, México, D.F., México, [2]Laboratory of Experimental Amebiasis, Instituto Nacional de Higiene, [3]Subsecretary of the Secretaría de Salud, México, D.F., México

INTRODUCTION

Gram negative bacteria contain in their outer membrane a group of proteins that have been classified into major and minor according to their level of expression (1). The major members of the outer membrane proteins (OMP) include the Braun's lipoprotein, the porins (*ompC, ompF, ompD, phoE*, etc.) and the heat-modifiable protein (Omp A) (2). The porins are a group of proteins with molecular weights (Mr) ranging from 36 to 38 kd per monomeric unit (3). In the membrane, porins are expressed as non-covalently linked homotrimers. These proteins function in the transmembrane transport of nutrients and ions (4). Moreover, some members of the porin family (*lamB*) function as receptors for phagi (5). Biochemical and molecular genetic studies have indicated that there is extensive homology among the porins of Gram negative bacteria, although some species-specific sequences have been demonstrated (6,7).

Previous studies have used OMPs from gram negative bacteria as immunogens to study the role of the specific immune response against these proteins in the protection against the infection by the bacteria (8-13). Thus, immunization with OMPs from *Neisseria meningitidis* and *N. gonorrhoeae* result in protection against the infection caused by the relevant bacterium (8,9). In subsequent studies it was shown that immunization with OMPs from other Gram negative bacteria such as *Haemophilus influenzae* (10), *Shigella flexneri* (11), and *Pseudomonas aeruginosa* (12) also are capable of inducing a protective status in experimental animals. Recently, a vaccine elaborated with OMPs from *N. gonhorreae* has become available for human use (13).

Recent studies have demonstrated that OMPs from a rough strain of *S. typhimurium* can protect mice against the infection with the bacterium as long as these OMP are linked to a polysaccharide which functions as an adjuvant (14,15). Similarly, rabbit anti-OMP serum is protective against *S. typhimurium* infection in mice (16). Protein antigens might have an advantage over the polysaccharidic ones because the former have the capacity of inducing antibodies of higher affinity (17) besides being capable of eliciting cell-mediated immunity (18) which should result in longer-lived immunity against the immunogen.

Currently available vaccines against typhoid fever containing either acetone or heat-killed *Salmonella typhi*, are of limited value because of both short-lived protection and serious side effects caused mainly by the presence of endotoxin (19). Some attempts to obtain a better vaccine have yielded the orally administered strain *S. typhi* Ty21a, developed by Germanier et al (20) which is deficient in the enzyme UDP-galactose-4 epimerase. This vaccination is devoid of side effects and proved very effective in a field trial carried out in Egypt, showing protection of 92% (21). In latter studies performed in Chile (22), however, the results were not as impressive and protection ranged from only 19 to 67%, which is not acceptable for a general use vaccine.

The nature of the protective immunogens derived from *S. typhi* remains controversial. Antibodies against the somatic antigen lypopolysaccharide "O" correlate with the history of infection but not with protection against *Salmonella* both in humans with typhoid (23) and in mice infected with *S. typhimurium* (24). Similarly, antibodies against the flagellar antigen "H" are not protective (23). Moreover, despite studies indicating the ability of the capsular polysaccharidic antigen Vi to induce high titers of serum antibodies; its role in the induction of protective immunity against *S. typhi* infection (25) is controversial. Although one study (26), with a 17 month follow up would suggest that immunization with Vi confers a protective status against typhoid in human volunteers. For these reasons, the recently acquired attention toward the role of the OMPs of Gram negative bacteria in the induction of specific immunity seems justified.

One problem in the study of typhoid fever is the fact that its causative agent *Salmonella typhi* is virulent only for man and chimpanzees (27), which precludes easy and extensive experimental murine manipulations. Mice infected with the natural mouse pathogen *Salmonella typhimurium* are the most widely accepted model for the study of the pathogenesis and the immune response in typhoid (28). This model, however, has the disadvantage of using a different strain of *Salmonella,* what makes it of spurious value as a test for vaccines and brings up the need to confirm the validity of the observations obtained with such a model in animals, and eventually humans, infected with *S. typhi*, the etiologic agent of typhoid fever. To solve this problem, mice injected intraperitoneally with *S. typhi* in mucin (29) has been used. With the use of this model, we have recently examined the role of the OMPs of *S. typhi* in the active induction of protection against *S. typhi* in mice (30). Moreover, the role of an anti-OMP rabbit antiserum in the passive transfer of immunity was also assessed. We found that OMPs may function as immunogens in the induction of protection against *S. typhi*. In addition, partial cross protection against at least two species of *Salmonella* was obtained (31).

The development of a vaccine elaborated with protein components of any Gram-negative bacterial strain has as a major difficulty the obtention of high amounts of lipopolysaccharide (LPS) free purified protein. To solve this problem, we have attempted different approaches including the obtention of porins directly from the bacteria by the method described by Nikaido (32). In addition, we have performed electroelution of the porins directly from preparative polyacrilamide gels of OMPs isolated as described by Schnaitman (33). These techniques, despite resulting in porins contaminated with <1% LPS, have the disadvantage of being time-consuming and that the yields of porins are somewhat variable from time to time both quantitatively and qualitatively.

The development of monoclonal antibodies (mAb) against the porins by the hybridoma technique described by Köhler and Milstein (34) might offer the advantage of elaborating an immunoabsorbent for the purification of these proteins, which could be useful to obtain high yields of homogeneous batches of purified LPS-free porins. Besides its role as immunoabsorbents, and more importantly; mAbs might help to characterize the immunodominant epitopes of *S. typhi* porins, which could help in the design of a potential synthetic vaccine for the prevention of typhoid fever.

MATERIALS AND METHODS

Bacterial strains

The virulent strain *Salmonella typhi* 9,12,d,Vi; was originally isolated from a patient with typhoid fever and has been maintained in culture in our laboratory since 1979.

Mice

Balb/c mice were originally obtained from the Jackson laboratories and have been bred in our animal facilities.

Isolation of outer-membrane proteins

OMPs were obtained as described by Nikaido (32). Briefly, cultures of *S. typhi* were grown in minimum salts medium (medium A) with 5% yeast extract and 12.5% glucose. Cultures were harvested and resuspended in 0.01 M HEPES buffer. Bacteria were disrupted by sonication and centrifuged at 7,000 g for 10 min. The supernatant was centrifuged at 100,000 g for 45 min, then the pellet containing the cell walls was resuspended in 2% sodium dodecyl sulfate (SDS), 10 mM Tris-HCl, pH 7.7. This suspension was kept at 32°C for 30 min and centrifuged at 100,000 g for 30 min at 20°C; this procedure was repeated once. In order to free the porins from the peptidoglycan, the pellet was resuspended in a buffer containing 50 mM Tris-HCl (pH 7.7), 0.4 M NaCl, 1% SDS, 5mM EDTA, 0.05% 2 mercaptoethanol (2ME) heretofore referred to as NaCl buffer. The suspension was kept at 37°C for 2 hours and then centrifuged at 100,000 g at 25°C for 20 min, after which the proteins were recovered in the supernatant fraction. This supernatant was passed through a Sephacryl S200 column (Pharmacia Fine Chemicals, Upsala, Sweden) in NaCl buffer at a rate of 6 ml/hour. Proteins were detected in the chromatographic fractions by OD at 280 nm and their concentration was measured as described by Lowry et al (35). LPS contamination of the protein samples was determined by measuring the concentration of β-hydroxy myristic acid by reverse phase high-preasure liquid chromatography (HPLC, Beckman Instruments, Stanford, CA).

Development of anti-porin monoclonal antibodies

Balb/c mice were immunized intraperitoneally (i.p.) with 70 μg of *S. typhi* porins emulsified in complete Freund's adjuvant (CFA). Immunizations were repeated every 30 days during the following three months with 50 μg of porins without CFA. Three days after the last immunization, mice were bled and the serum concentration of anti-porin antibodies was measured by an ELISA technique (36). Mice with high anti-porin antibody titers were killed by cervical dislocation, the spleens were sterilely removed and splenocytes were obtained by passing the tissue through a nylon mesh. Cells were fused at a 2:1 ratio with the HGPRT-deficient mouse plasmocytoma SP2/0 Ag14 (A gift of Dr. Peter E. Lipsky, University of Texas Health Science Center, Dallas, TX) with 50% polyethylene glycol 2000 (Sigma Chemical Company, St. Louis, MO), as described (34). Cells were cultured in hypoxantine aminopterin thymidine (HAT) containing Dulbecco's modified Eagle's medium (Sigma) supplemented with 20% fetal bovine serum (Sigma); penicillin, 500 U/ml; gentamicin, 0.04 mg/ml, and 2ME, 5×10^{-5} M. Three weeks later, supernatants from growing wells were tested for anti-porin antibody activity by ELISA and those giving a positive reaction were cloned by limiting dilution at 1 cell per well and were expanded and frozen until further characterization. The immunoglobulin isotypes of the mAbs were directly determined in the culture supernatants by ELISA with a commercial kit purchased from Bio-Rad Laboratories (Richmond, CA).

Sodium dodecyl sulfate polyacrilamide (SDS-PAGE) gel electrophoresis and Western blotting

SDS-PAGE was performed under reducing conditions by using the discontinuous buffer systems of Laemmli (37) in a vertical slab gel electrophoresis unit (LKB Instruments). The separating gel contained 11.2% acrylamide, 2.5% bis-acrylamide, 0.19% SDS in 0.35 M Tris-HCl buffer pH 8.8. The stacking gel had 5% acrylamide, 0.1% SDS in 0.12 M Tris-HCl buffer pH 6.8. The running buffer was 0.025 M Tris-HCl, 0.192 glycine, pH 8.3 and 0.1% SDS. Electrophoresis was performed at 30 mA. These porins were compared with OMPs isolated as described by Schnaitman, utilized in our previous studies (31).

Electrophoretic transfer of OMPs from polyacrylamide gels to nitrocellulose paper (NCP) was accomplished in a LKB Transphor electroblotting unit by means of the transfer buffer described by Towbin et al (38). The proteins were electrophoresed at 100 mA for 18 h. NCPs were immersed in PBS (0.15 M NaCl 0.01 M phosphate buffer, pH 7.2) with 6 mM EDTA, 0.25% gelatin and 0.1% Tween 20. After washing in 0.1% Tween 20 (PBS-T), the NCPs were incubated for 3 h with the hybridoma supernatant in 0.25% gelatin-PBS. After washing in PBS-T, the NCPs were incubated for 1.5 h at 25ºC with optimal concentrations of horseradish peroxidase conjugated goat anti-mouse immunoglobulins (Cooper Biomedical, Malvern, PA). After washing the NCPs twice in PBS-T and PBS, were immersed in PBS with 2 mM 4-chloro-2-naphtol and 0.08% H_2O_2. After a wash in tap water the NCPs were air dried in the dark and photographed.

Evaluation of passive protection

For the passive protection studies groups of 10 mice each (18→20 g) were given 200 µl of either SP2 (control) or hybridoma supernatant intravenously and challenged one hour later with 50 or 100 LD_{50} of *S. typhi* in Tris-EDTA buffer with 5% mucin. i.p.. Protection was defined as the percent survival during the 10 days following the challenge. Control groups consisted of 10 mice given 200 µl of Tris-EDTA buffer with 5% mucin i.p., and either SP2 or hybridoma supernatant was given. The LD_{50} for *S. typhi* was calculated as described by Reed and Muench (39).

RESULTS

Porin preparation

Figure 1 shows the SDS-PAGE electrophoretic pattern of *S. typhi* porins eluted from the Sephacryl S200 column, in which two dominant bands with a Mr of about 38 kd are seen. Contamination with other proteins was insignificant compared to the OMPs in which besides porins there are additional bands with Mr from 17 to 70 kd. LPS contamination was 0.047% as determined by measuring β-hydroxy myristic acid by HPLC (not shown).

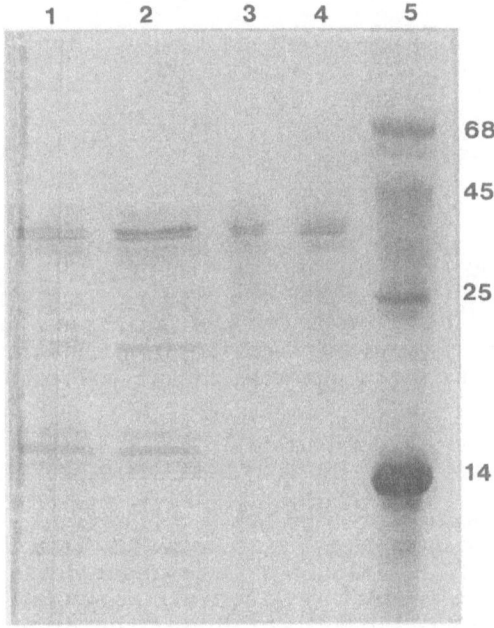

Figure 1. SDS-PAGE of OMPs (lanes 1 and 2) and porins (lanes 3 and 4). Lane 5 has the molecular weight markers.

Anti-porin hybridomas

After cloning the fused porin-immune splenocytes with the SP2 plasmocytoma, 46 hybridomas were obtained. Eight of these have been characterized and are described herein. The rest was frozen awaiting further characterization.

Of the eight hybridomas characterized, six showed reactivity against porins and two were directed against the polysaccharidic fraction of the LPS. Absorption studies (not shown) confirmed the single specificity of the mAb produced by these hybridomas.

Anti-porin antibody secreting hybridomas were subcloned by limiting dilution at 0.3 cells per well, and the specificity was again determined. Two subclones of the porin-reactive hybridoma 5F2 hereafter designed as P1 and P2 were obtained (Figure 2); these were reactive only against porins but not LPS. On the other hand, cloning of the mainly LPS-reactive hybridoma 5E9 gave rise to the hybridomas L1 and L2 which are reactive only against LPS (Figure 2). In additional experiments (not shown) it was found that all the mAb obtained belonged to the IgM isotype.

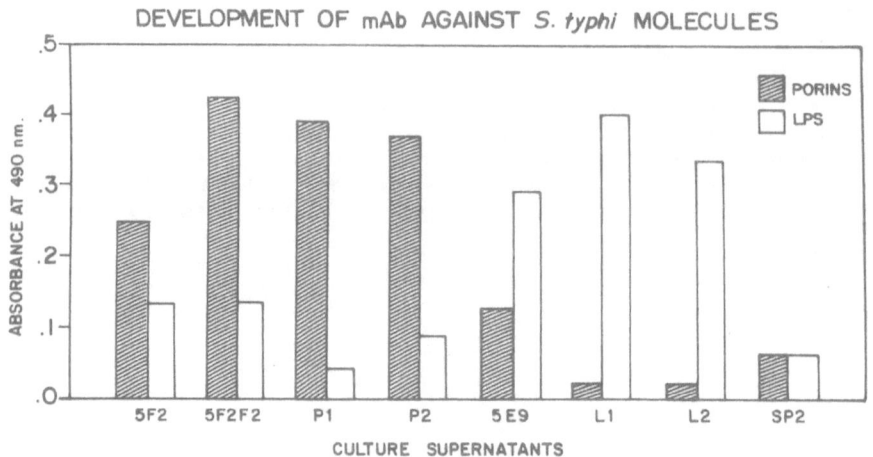

Figure 2. Reactivity of mAb against porins and LPS. The specificity was tested by an ELISA technique with peroxidase-labeled goat anti-mouse immunoglobulin. 5F2 is the original anti-porin mAb and P1 and P2 are subclones of it. 5E9 is the original anti-LPS hybridoma and their subclones are L1 and L2. Results are expressed as absorbance at 490 nm.

Western blot analysis of porin-reactive hybridomas

Figure 3 shows the Western blots obtained after reacting supernatants from hybridomas P1 and P2 with NCPs transferred with *S. typhi* OMPs isolated as described by Schnaitman (which as shown in figure 1 contain several proteins beside the porins). As seen, both hybridomas react against the bands migrating in the Mr around 38 kD, same as the *S. typhi* porins purified as described above.

TABLE I
PASSIVE PROTECTION INDUCED BY ANTI-OMPs mAb

LD$_{50}$[1]	Supernatant				
	P1	P2	L1	L2	SP2
	% mice survival after 10 days				
20	60	40	20	20	30
100	30	40	20	20	30

[1] Lethal dose for 50% of animals = 1.5×10^5 *Salmonella typhi*.

Pasive transfer of protection with mAb-containing supernatants

Finally, in order to determine the role of anti-porin mAbs in the protection of mice against *S. typhi* infection, passive immunization studies were performed by injecting hybridoma supernatants to naive mice one hour before the challenge with the live bacterium in mucin. The results shown in table 1 suggest that anti-porin antibodies might confer slight protection against a challenge with 20 LD$_{50}$ of *S. typhi*. This results are preliminary, however, since the concentration of mAb in the hybridoma supernatant is not available yet. Comparing these results with our previous studies using a rabbit anti-OMP heteroantiserum, indicate that mAb are not as effective as a polyclonal antiserum in the induction of a protective status. As seen also, LPS-specific mAb did not confer any protection to mice challenged in the same manner.

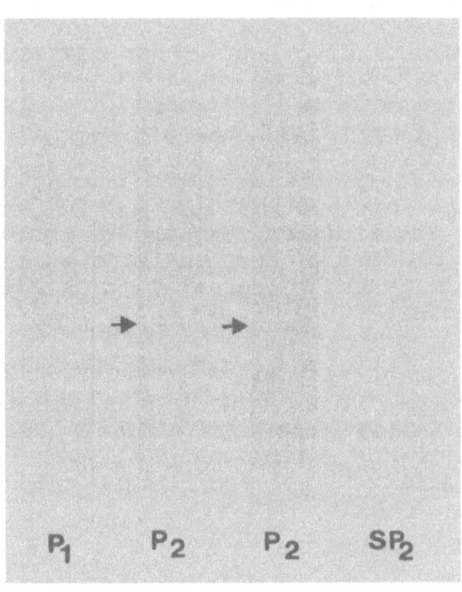

Figure 3. Western Blot of OMPs reacted with anti-porin mAb P1 and P2. The transferred NCPs were developed with goat anti-mouse IgM.

DISCUSSION

The current studies were undertaken in order to obtain mAbs against the porin components of the *S. typhi* OMPs. Balb/c mice were immunized with a highly purified porin preparation obtained as described by Nikaido which was shown to contain less than 0.05% LPS. Despite this low contamination, from the eight hybridomas so far characterized, two are directed against LPS whereas six are specific for the porins. Preliminary results suggest that one of these mAbs might have some protective activity against *S. typhi* in mice, albeit it is low.

Monoclonal antibodies are a highly useful tool for the study and isolation of antigens (40). The availability of mAbs directed against the porins of *S. typhi* provides a reagent to obtain affinity purified porins directly from bacterial lysates. In addition, these mAbs might be useful to analyze the immunologic cross reactivities among the Gram negative porins. The mAb may not be as useful, however for the induction of protection, since one of the important features of a protective antibody response is the polyclonality. This has been shown for tetanus toxin and other soluble antigens (41). However, some studies have suggested that mAb against porins might be of value (albeit somewhat reduced) against the bacterial infection (42). The results presented herein are in agreement with these findings but are still too preliminary to draw any conclusion.

It is interesting the fact that despite the LPS concentration in our porin preparation was 0.047 µg/ml (therefore each mice received <0.012µg per injection), we were still able to obtain a 25% of LPS-specific mAbs. This suggests either that our method for determining LPS is not as sensitive as initially thought or that the immunodominance of LPS is such that a protein antigen used for vaccination must be completely devoid of LPS to obtain an immune response against the desired determinants (although in some cases it might be acting as an adjuvant). From the current results, we cannot explain the reasons for this preferential response to LPS.

Our current efforts are directed toward the definition of the immunodominant epitopes of *S. typhi* porins with the use of these mAb. In addition, we are preparing antiidiotypic antibodies hoping to obtain the antiidiotypic "internal image" of the antigen and test its value as a protective immunogen against typhoid fever.

Acknowledgments. This work was supported by a project grant No. PSSABNA-023449 from the Consejo Nacional de Ciencia y Tecnología, México. The authors thank Dr. Peter E. Lipsky for his generous gift of the SP2/0 Ag14 cell line.

REFERENCES

1. M.N. Hall, and T.J. Silhavy, 1981. Genetic analysis of the major outer membrane proteins of *Escherichia coli*. Annu Rev Genet 5:91

2. M.J. Osborn, and H.C. Wu, 1980. Proteins of the outer membrane of Gram-negative bacteria. Annu Rev Microbiol 34:369

3. B. Lugtenberg, and L. Van Alphen, 1983. Molecular architecture and fuctioning of the outer membrane of *Escherichia coli* and other Gram-negative bacteria. Biochem Biophys Acta 737:51

4. H. Nikaido, and M. Vaara, 1985. Molecular basis of bacterial outer membrane permeability. Microbiol Rev 49:1

5. M. Schwartz, 1983. Phage receptor (Lam B protein) in *Escherichia coli.* Methods Enzymol 97:100

6. T. Mizuno, M.Y. Chou, and M. Inouye, 1983. A comparative study on the genes for three porins of the *Escherichia coli* outer membrane DNA sequence of osmoregulation omp C gene. J Biol Chem 258:6932

7. Y. Fujio, 1987. DNA and amino acid sequences of outer membrane proteins and lipoproteins. In: Bacterial outer membranes as model system, M. Inouye ed., John Wiley and Sons p 419

8. T.M. Buchanan, W.A. Pearce, G.K. Schoolnick, and R.J. Arko, 1977. Protection against infections with *Neisseria gonorrhoeae* by immunization with outer membrane protein complex and purified pili. J Infect Dis 136(suppl):132:137

9. L.I. Wang, and C.E. Frash, 1984. Development of a *Neisseria meningitidis* Group B serotype 2b protein vaccine and evaluation in a mouse model. Infect Immun 46:408

10. P.A. Guling, G.H. McCracken Jr, C.F. Frich, K.H. Johnston, and E.J. Hansen, 1982. Antibody response of infants to cell surface-exposed outer membrane proteins of *Haemophilus influenzae* type b after systemic *Haemophilus* disease. Infect Immun 37:82

11. G. Adamus, M. Mulczka, D. Witkowska, and E. Romanowska, 1980. Protection against keratoconjunctivitis shigellosa induced by immunization with outer membrane proteins of *Shigella sp.* Infect Immun 30:321

12. H.E. Gilleland, M.G. Parker, J.W. Matthews, and R.D. Berg, 1984. Use of purifed outer membrane protein F (porin) preparation of *Pseudomonas aerouginosa* as a protective vaccine in mice. Infect Immun 44:49

13. M.S. Blake, and E.C. Gotschlin, 1983. Gonococcal membrane proteins: speculation on their role in pathogenesis. Prog Allergy 33:298

14. N. Kussi, M. Nurminen, H. Saxén, M. Valtonen, and P.H. Makela, 1979. Immunization with outer major membrane proteins in experimental salmonellosis of mice. Infect Immun 25:857

15. S.B. Svenson, M. Nurminen, and A.A. Lindberg, 1979. Artificial *Salmonella* vaccines: O-antigenic oligosaccharide-protein conjugates induce protection against infection with *Salmonella typhimurium.* Infect Immun 25:863

16. N. Kussi, M. Nurminen, H. Saxén, and P.H. Mäkela, 1981. Immunization with major outer membrane protein (porin) Preparations in experimental murine salomonellosis: Effect of lipopolysaccharide. Infect Immun 34:328

17. H.N. Eisen, and G.W. Siskind, 1969. Variations in affinities of antibodies during the immune response. Biochemistry 3:996

18. E.R. Unanue, 1984. Antigen-presenting function of the macrophage. Annu Rev Immunol 2:395

19. I. Joó, 1970. International conference on the application of vaccines against viral, rickettsial and bacterial disease of man. Washington., D.C. PAHO/WHO Scientific Publication 226:339

20. R. Germanier, and E. Furer, 1975. Isolation and characterization of Salmonella typhi gal E mutant Ty21a: a candidate strain fcr a live typhoid vaccine. J Infect Dis 131:553

21. M.H. Wandan, C. Serie, Y. Cerisier, S. Sallam, and G. Germanier, 1982. Controlled field trial of live Salmonella typhi strain Ty21a oral vaccine against typhoid: three years results. J Infect Dis 145:292

22. R. Germanier, 1986. The live oral typhoid vaccine Ty21a: recent field trial results. In: Sclavo International Conference on bacterial vaccines and local immunity. p.10

23. R.B. Hornick, S.E. Greisman, T.E. Woodward, H.L. Dupont, A.T. Dawkins, and M.J. Snyder, 1970. Typhoid fever, pathogenesis and immunological control. Parts I and II. N Engl J Med 283:686, 739

24. B.A.D. Stocker, and P.H. Mäkela, 1986. Genetic determination of bacterial virulence, with special reference to Salmonella. Curr Top Microbiol Immunol 124:149

25. J.D. Robbins, and J.B. Robbins, 1984. Re-examination of the immunopathogenic role of the capsular polysaccharide (Vi antigen) of Salmonella typhi. J Infect Dis 47:436

26. I.L. Acharya, C.L. Lowe, R. Thapa, V.L. Gurubacharya, M.B. Shresta, D. Bact, M. Cadoz, D. Schulz, J. Armand, D.A. Bryla, B. Trollfors, T. Cramton, R. Schneerson, and J.B. Robbins, 1987. Prevention of typhoid fever in Nepal with the Vi capsular polisaccharide of Salmonella typhi. A preliminary report. N Engl J Med 317:1101

27. S. Gaines, A. Sprinz, and J. Tully, 1968. Studies on infection and immunity in experimental typhoid fever. VII. The distribution of Salmonella typhi in chimpanzee tissue following oral challenge and the relationship between the numbers of bacilli and morphological lesions. J Infect Dis 118:293

28. A.D. O'Brien, 1986. Influence of host genes on resistance of inbred mice to lethal infection with Salmonella typhimurium. Curr Top Microbiol Immunol 124:37

29. J. Spaun, 1964. Studies on the influence of the immunization in the active mouse protection test with intraperito-
neal challenge for potency assay to typhoid vaccines. Bull
WHO 31:793

30. A. Isibasi, E. Calva, V. Ortiz, M. Fernández, A. Hernández, y J. Kumate, 1986. Vacunas contra la fiebre tifoidea
a partir de antígenos de membrana externa. In: Avances en
el uso de vacunas. p 109. Juan Garza, ed., Gerencia General de Biológicos y Reactivos. México

31. A. Isibasi, V. Ortiz, M. Vargas, J. Paniagua, C. González,
J. Moreno, and J. Kumate, 1988. Protection against *Salmonella typhi* infection in mice immunization with outer-
membrane proteins isolated from *Salmonella typhi* 9,12,
d,Vi. Submitted for publication.

32. H. Nikaido, 1983. Proteins forming large channels from
bacterial and mitochondrial outer membranes: porins and
phage lambda receptor protein. Methods Enzymol 97:85

33. C.A. Schnaitman, 1971. Effect of ethylendiamine tetracetic
acid, Triton X-100 and lisozyme on the morphology and
chemical composition of isolated cell walls of *Escherichia coli*. J Bacteriol 108:553

34. G. Köhler, and C. Milstein, 1975. Continuous culture of
fused cells secreting antibody of predefined specificity.
Nature 256:495

35. O.H. Lowry, N.J. Rosebrough, A.L. Farr, and R.J. Randall,
1951. Protein measurement with the folin-phenol-reagent. J
Biol Chem 193:265

36. E. Engvall, and P. Perlman, 1971. Enzyme-linked immunoabsorbent assay (ELISA). Immunochem 8:874

37. U.K. Laemmli, 1970. Cleavage of structural proteins during
the assembly of the head of bacteriophage T4. Nature
227:680

38. H.T. Towbin, T. Staehlin, and J. Gordon, 1979. Electrophoretic transfer of proteins from polyacrylamide gels to
nitrocellulose sheets: procedure and some applications.
Proc Natl Acad Sci USA 76:4350-4354

39. L.J. Reed and H.A. Muench, 1938. A simple method of estimating fifty percent end points. Am J Hyg 27:493

40. C. Milstein, M.R. Clark, G. Galfre, and A.C. Cuello, 1980.
Monoclonal antibodies from hybrid myelomas. In: Immunology. M. Fourgerau, ed., p17, Academic Press. New York

41. E. Habermann, and K. Goretzki, 1985. Monoclonal antibodies
against tetanus toxin and tetanus toxoid. In: Monoclonal
antibodies against bacteria. Macario, A.J.L. and Conway de
Macario E. eds. p191 Academic Press. New York

42. S. Sawada, M. Suzuki, T. Kawamura, S. Fujinaga, Y. Masuho,
and K. Tomibe, 1984. Protection against infection with
Pseudomonas aeruginosa by passive transfer of monoclonal

antibodies to lipopolysaccharides and outer membrane proteins. J Infect Dis 150:570

STRUCTURAL AND FUNCTIONAL MODIFICATIONS

OF PLASMODIUM-INFECTED ERYTHROCYTES

M.H. Rodríguez

Centro de Investigaciones
y de Estudios Avanzados
IPN, México City, and
Centro de Investigación
de Paludismo
Tapachula, Chis. México

INTRODUCTION

Human malaria is a disease produced from infection with any of four
species of Plasmodium parasites: P. falciparum, P. vivax, P. malariae and
P. ovale. The most common of these is the infection with P. vivax, the
most severe is produced by P. falciparum. The classical clinical manifestations of malaria, recurring chills and fever, are caused by the
synchronous rupture of erythrocytes during production of broods of young
forms of parasites. The more severe form of the disease, cerebral malaria,
is the indirect consequence of modifications produced by the parasite on
the host cell.

Induced changes of the infected erythrocyte are related to the
parasite's need to adapt its environment for its own metabolic requirement. Basic modifications include primarily structural changes on the
parasitized cell's plasma membrane. These changes increase the cell's
permeability to essential metabolites and, in at least one instance,
induce active transport of key metabolic elements.

Besides the use of a host cell's metabolic machinery, another
advantage of intracellular parasitism is avoidance of recognition and
destruction by elements of the host's immune system. On the other hand,
adaptation of the infected cell by means of plasma membrane modifications
has the inexpediency of exposing foreign proteins on the cell surface.
Malaria parasites avoid immune destruction by several mechanisms, including
antigenic variation and, in some species, a further erythrocyte plasma
membrane modification. This modification induces capillary retention where
metabolic conditions are suitable and where they may evade destructive
reticuloendothelial cells.

An understanding of the mechanisms involved in host cell adaptation
of malaria parasites, could provide specific methods for a metabolic
blockade using pharmacological agents, and possibly suggest parasite
protein candidates for vaccination. This paper addresses several known
erythrocyte alterations induced by malaria parasite infection. A brief
description of the parasite life cycle is presented in order to emphasize

the erythrocytic stage in the context of the human infection, and its
relation to disease symptomology. Next, the main metabolic pathways of
intraerythrocytic parasites are reviewed only to indicate those cases where
the parasite is dependant on the host-cell metabolic machinery. Following a
description of erythrocyte structural modifications, and a review of the
present evidence for host cell remodelation mechanisms, a model for molecular
exchange among the multiple infected erythrocyte membranes is discussed.

LIFE CYCLE OF MALARIA PARASITES

The general life cycle of all species of Plasmodium is similar.
However, some variations are responsible for very important differences in
the overall effect that the infection produces in the host. In humans,
Plasmodium parasites are injected as sporozoites by infected mosquitoes.
These thread-like forms are transported in the blood for a very short period
of time and eventually reach liver parenchymal cells where they grow and
multiply by asexual schizogony producing large quantities of invasive forms
called merozoites.

No clinical manifestations are associated with this period of parasite
development. Variations in the schizogony rate are probably responsible for
the differences observed in the prepatent period between the four human
malaria species. A very important difference between P. vivax and the other
three species is that in the former, some of the sporozoites that invade
liver cells will develop as already described, but others will remain
unicellular for various periods of time. Reactivation of these dormant
forms called hypnozoites[2] are responsible for disease "relapses" commonly
found in this type of malaria.

Hepatic merozoites leaving the heptocytes are liberated into the
bloodstream where they invade erythrocytes. In these cells the parasite
will grow and develop new merozoites that will continue the cycle by
reinvasion of yet other erythrocytes. After a variable number of intraerythrocytic multiplication cycles, some of the merozoites, after cell
invasion, will transform and mature into sexual forms called micro and
macrogametocytes. These forms remain intracellular until ingested by a
susceptible mosquito. In the mosquito, gamete fertilization is followed by
sporogonic multiplication and mature sporozoites migrate in the salivary
glands ready to be injected into the next human host.

Clinical manifestations are associated with the intraerythrocytic
development of the parasite. In all infections, fever and chills coincide
with the rupture of the schizont infected erythrocytes. In the case of P.
falciparum, several severe conditions may develop. In this infection, only
young parasites are observed in peripheral blood. Schizogony occurrs in
postcapillary vessels where infected erythrocytes may block the flow of
blood producing isquemia of vital organs. This may result in coma or death
by stroke (cerebral malaria), heart failure, or colera-like gastrointestinal symptoms.

As a rule, malaria parasites have a certain degree of host selectivity.
In nature, this selectivity has involved to a high level of specialization
of parasites for specific host. In the human malaria, the parasites have a
selective preference for a host-cell type. P. vivax has a preference for
reticulocytes and young erythrocytes, while P. falciparum invades erythrocytes of any age. This different selectivity may be due to specific needs.
In the case of P. vivax. This need may only be fullfilled by younger cells
that have not lost all their metabolic machinery. The method of host-cell
selection is not yet clear. P. falciparum seems to recognize specific
glycoproteins[3-6] on the erythrocyte's surface while Duffy blood antigens are

probably the receptors for P. vivax.[7,8] Several parasite proteins have been identified in malaria parasites as counterpart candidates in the erythrocyte recognition phenomenon.[9,10]

After initial atachment of the merozoite to the erythrocyte, which could ocurr on any part of the merozoite surface, the parasite reorients itself in order to counterpoint its apical end to the mammalian cell surface[11]. A junction is formed[12] between the parasite and the erythrocyte and the contents of specialized parasite organelles (rhoptries and meronemes) are discharged. The end result is the invagination of the erythrocyte membrane, with the parasite attached at its apical end, and the junction formed by condensation of submembranous material in both cells. This junction which first appears at the point of contact between the apical end of the parasite and the erythrocyte, remains as a ring at the initiation of the invaginated vacuole. Following invasion, the parasite inside the host cell is surrounded by a parasitophorous vacuole and the site of entrance is sealed off.

METABOLISM OF INTRAERYTHROCYTIC MALARIA PARASITES

Studies on the metabolism of malaria parasites have been carried out for many decades. For a review see Sherman[13,14] The principal drawback of most of these studies is the difficulty in obtaining pure parasite fractions. Other deficiencies include the inherent limitations of in vitro manipulation, and in many cases the absence of proper controls. Metabolic studies have been performed using diverse plasmodial species (avian, rodent and primate) and some important differences have been found. Nevertheless, in certain cases the presence of some metabolic pathway is only infered from one specie to the other. Bearing in mind the above limitations, enough data have been accumulated to have a good understanding of the main metabolic requirements of these parasites. Only aspects relevant to erythrocyte modification are discussed.

Lipid Metabolism

Intraerytrocytic malaria parasites grow and multiply inside the parasitophorous vacuole initialy formed by invagination of the cell's plasma membrane[15-18]. This parasitophorous vacuole lacks the proteins found on the original erythrocyte membrane. Shortly after invasion, the lipid composition of this membrane is altered to that of parasite origin.

No lipid reserves have been found in plasmodia, but a comparison indicates that the lipid content in parasitized cells is up to five times that of non-infected cells.[19-21] These lipids are associated with a great number of membranous organelles located in the parasite cytosol. As will be discussed later, other membranous organelles are found in the erythrocyte cytosol and more lipid molecules have to be incorporated in the parasitophorous vacuole to allow parasite growth.

The lipid composition of Plasmodium membranes is typically richer in phospholipis, di- and triacylglycerols diacylphosphatidylethanolmine, phosphatidylinositol and saturated fatty acids than the erythrocyte membrane [20,22-25]. Besides a decrease in the amount of cholesterol in the erythrocyte plasma membrane, and an increased ratio phospholipid to cholesterol in the parasites membranes[26], the toal cholesterol of the infected cells increases. Increased cholesterol levels within the erythrocyte plasma membrane could be responsible for the increase in its permiability and osmotic fragility observed as the parasite develops and reaches mature schizogony[18].

Plasmodium parasites do not synthesize fatty acids de novo[13], as there is no evidence for cholesterol synthesis. To maintain membrane production, the parasite uses phospholipids and fatty acids and probably cholesterol from circulatory plasma. These molecules are modified through desaturation and chain elongation by parasite enzymes. The mechanisms for the exchange of lipid material from the surface of the erythrocyte to the parasite could be provide by the movement of vesicles in both directions as will be discussed later.

Protein Metabolism

Malaria parasites have a ready source of amino acids from the erythrocyte's cytosol proteins. The most abundant of these is haemoglobin. Although micropinocytosis occurs, a specialized organelle named the cytostome[27] is in charge of phagocytosis through the parasitophorous vacuole membrane. Microvesicles containing the endocytic material are derived from the cytostome. Proteases, cathepsin D and aminopeptidase synthesized from parasite cytoplasm ribosomes, migrate to the vesicles where digestion occurs.[28] The final residue of haemoglobin digestion is the so-called malaria pigment, which contains insoluble heme (haemozoin).

Amino acids are additionally taken from the plasma, and incorporation of cystine, hystidine, methionine, leucine, and isoleucine has been documented when supplied in culture medium.[29,30] Mammalian reticulocytes possess specific and concentrating amino acid transport systems, although the mature erythrocyte has lost most of this capacity. The mechanisms of amino acid transport in infected erythrocytes are in need of clarification. Some possible mechanisms include: increase in diffusion due to modifications in the lipid composition of the plasma membrane, and depletion of intra-cellular ATP,[31] although modifications of the amino acid carriers by the parasite can not be ruled out.

Protein synthesis in Plasmodium is of the eukaryotic type and the parasite utilizes its own assembly machinery for protein production.[32,33] Protein synthesis appears to occur in ribosomes located in the cytosol and along side of the extensive membranous organelles in the metabolically active parasite. No organelles for protein modification and sorting (Golgi compartment) have been identified. How the parasite tags its proteins for the different compartments of the infected cell and even transports to the exterior milieu is still unknown.

Glucose, Oxygen and other Requirements

Malaria parasites of mammals utilize glucose as the principal source of energy.[13] The lack of stored glucose[34] in the parasite indicates that this sugar has to be supplied from the plasma. Glucose is metabolized to lactic acid by anaerobic glycolysis.[35] Although O_2 consumption by the parasite increases in the presence of glucose, except for the presence of cytochrome oxidase,[36] none of the oxidative pathway enzymes have been found in malaria parasites. Growth in culture of some species (e.q. P. falciparum) is optimal optimal only in microaerobic enviroments.

Living in an aerobic organism, and particularly inside erythrocytes, Plasmodium has developed ways to avoid destruction by activated oxygen. Most aerobic organisms synthesize superoxide dismutase (SOD), an enzyme that converts superoxide into hydrogen and peroxidase, that in turn is converted into water by catalases. In the case of Plasmodium, no synthesis of SOD occurs since the parasites incorporated readily available SOD from the erythrocytic cytoplasm.[37,38] The enzyme is probably taken into phagocytic vacuoles and due to its intrinsic resistance to most proteases, is not digested. The mechanisms as to how the enzyme leaves the digestive vacuole and remains in the parasite plasmalemma are still unknown.

Several parasite-proteins have been identified in the infected erythrocyte membrane, several are secreted into the plasma or culture medium. Although many of these proteins from monkey, human and rodent malarias[39-43] have been identified, their function remains unknown. For the purposes of this paper, only protein examples from P. falciparum whose functions are better studied, will be referred to.

P. falciparum Proteins Exported to the Erythrocyte Surface

Most of the studies to identify parasite proteins from infected cells have been performed by metabolically labelling the cells with radioactive amino acids or carbohydrates, followed by separation in gel electrophoresis and visualization using fluorography. Since there is no protein synthesis machinery in the mature erythrocyte, it is assumed that all labelled material results from parasite synthesis. Protein localization in situ have been achieved by surface labelling with I^{125} and lactoperoxidase, or by removing surface proteins with enzymes such as trypsin or pronase. In some cases, localization has been carried out using immunoelectron-microscopy.

In experiments using ^{35} S-methionine (^{3}H) mannose and (^{3}H) glucosamine Perkins[44] identified at least eleven surface proteins in P. falciparum infected erythrocytes. Three of them with molecular weight (Mr) of 40 kilodaltons (Kd) were cleaved with pronase. Eight proteins bands (Mr between 150 and 40 Kd) were labelled with the radioactive glycosamine and mannose residues. Three of these glycoproteins were removed by trypsin and all of them were modified by pronase treatment. This indicated that their localization on the surface occurs in different forms. None of these surface proteins were unaffected by neuraminidase, indicating the lack of sialic acid residues. No function has been attributed to these proteins.

The lack of sialic acid is a characteristic that differentiates these glycoproteins from those of mammalian origen in the erythrocyte surface. Recently, a neuraminidase-like activity that removes sialic acid residues from glycoprotein A of infected erythrocytes[45], has been identified in P. falciparum. A possible consequence of asialic proteins on the surface of infected erythrocytes, and modified glycophorin, is prevention of further invasion by other merozoites and enhancement of attachment of the infected cell to endothelial cells in capillaries. How the neuraminidase-like activity is exerted at the erythrocyte surface level still needs investigation.

Other proteins produced during parasite development were initially recognized from the sera of malarious patients. These proteins are recognized by sera from patients of endemic areas and have been classified on the basis of their heat lability[46-48] Many of these proteins are released into the plasma at the moment of schizont rupture. Others are secreted and appear in the medium during younger parasite stages. The best studied of these is a protein containing 35 % histidine (Pf HRP II, histidine rich protein), 40 % alanine and 12 % aspartate[49]. The functions of these secreted proteins are unknown and are only mentioned here to indicate the fact that parasite proteins become not only constituent elements of the erythrocyte membrane, but are also exported from the cell.

The absence of circulating mature parasites during P. falciparum malaria, is associated with the neoformation of structures called knobs on the surface of the infected cell.[50-51] These structures visualized by transmission electronmicroscopy, have a diameter of 100 nm and are formed by cup-shaped electron-dense material situated immediately under the erythrocyte membrane, forming a protrusion[52].

These knob protrusions also appear during in vitro parasite maturation from ring to trophozoites. It has been demonstrated that cytoadherence of these infected cells to human umbilical vein endothelial cells or a line of malignant melanoma cells, occurs at the point of the knobs.[53,54] P. falciparum strains that have lost their capability in culture to produce knobs, have also lost cytoadherence. However, it is also true that some knob-producing strains do not adhere to endothelial cells. This indicates that knobs are required, but not sufficient for cytoadherence.

Further evidence that others components located at the knob formation are responsible for endothelial cell attachment have been produced using antibodies directed to these protrusions. Initial experiments using immune sera against different parasite isolates indicate that the immune response that blocks cytoadherence is strain specific.[55-57] In further experiments, it was found that strain specific immune serum added to cultures inhibited the adherence of knob-producing infected cells to melanoma cells and produced detachment of bound cells. Using electron microscopy, it was found that antibody was deposited on the extracellular side of knob protrusions.[58]

At least three proteins of parasite origin have been identified in association with the knobs. Another histidine rich protein (PfHRP I)[59,60] whose molecular weight varies with the parasite strain has been located in the electron-dense material of the knobs.[61,62] The fact that this protein can not be extracted with low 1% Triton x 100 and remains within the insoluble material extractable with SDS, along with several of the erythrocyte cytoskeletal proteins, indicates that this protein is associated to the host cell cytoskeleton[63]. The association of the protein and the knobs is based on electron microscopic analysis of this insoluble material present as electron-dense cups. PfHRP I does not directly mediate cytoadherence.

Other proteins, PfEMP I[57] and II[49] (erythrocyte membrane protein) of Mr 300 Kd, has been identified on the knobs using monoclonal antibodies. Pf MP II has a similar distribution as PfHRP I.[64] The finding that antibodies to PfEMP I, located on the exterior surface of the knobing plasma membrane[57,58] produce inhibition of cytoadherence in a strain specific manner, indicates that this protein is directly involved in the cytoadherence phenomenon.

The exact structure of the knob, and the assembly of the participating parasite proteins is still speculative. It is posible that Pf EMP II and Pf HRP I function as anchorage points to the cytoskeleton; the interaction with the lipid bilayer and/or erythrocyte membrane proteins could be mediated by portions or these proteins. PfEMP I associated with this structure would be exposed on the surface of the cell.

Structural Modification in Plasmodium-Infected Erythrocytes

Structural erythrocyte modifications produced by malaria infections have been known since the early days of malariology, and have been the basis for the differential microscopic diagnosis in Giemsa-stained blood smears. Typically, in P. vivax and P. ovale, minute red granules are seen in the infected erythrocyte and have been called Schuffner's dots. While no mature forms of P. falciparum are seen in peripheral blood, in trophozoite-infected cells, the granules are fewer and deeper stained and are called Maurer's clefts. Using electron microscopy, Schuffner's dots in P. vivax and P. ovale correspond to caveola located on the erythrocyte plasma membrane.[67] The caveola are surrounded by small vesicles containing electron-dense material. This material is antigenically of parasite origin. Other membrane formations in the cytoplasm of the infected cells correspond to the Maurer's clefts[65],[66] These are found free in the cytosol and in some cases continuity of the clefts with the parasitophorous vacuole has been documented.[67]

The nature of these formations remained speculative until recent years, when some evidence has been gathered for the inference that they may function as transport organelles to and from the parasite and the infected cell membrane.[67] In knob positive infected cells electron-dense material (EDM) similar to that observed in the knobs could be seen accumalated in the space between the parasite and the parasitophorous vacuole. This material was also associated with the cytoplasm surface of the membranes of the clefts, and continuity of the EDM associated membranes with a knob formation was demonstrated.

Membrane organelles are not the only means for parasite proteins movements to the host cell surface. In the case of PfHRP II[49], identified by immunogold techniques, clusters of gold particles were observed in the cytoplasm of the parasite, and on the external surface of host cells. In the erythrocyte cytoplasm, gold particles were mainly seen forming clusters and seldom associated with membrane vesicles. In contrast, PfEMP 2, a contituent knob protein is transported via membrane vesicles.[64]

Secretion into the parasitophorous vacuole could explain the mechanism of protein transport in Plasmodium-infected cells. The movement of lipid material of parasite origin requires other mechanisms which is probably mediated through membrane exchange. In experiments using a rodent model, P. chabaudi we have documented the formation of vesicles and clefts from the parasitophorous vacuole as well as the accumulation of electron-dense material in the infected cell cytoplasm (figs, 1 and 2). In parasite preparations using immune lysis with anti-erythrocyte antisera, the parasite remained sheathed by the parasitophorous vacuole. Short-term culture of these preparations produced swelling of intraparasitic vesicles and the appearance of membrane vesicles in the parasitophorous space (fig. 3). Although these formations could be interpreted as artifacts due to manipulation, the evidence of excreted membrane vesicles in the parasitophorous vacuole may indicate the initial route for the movement of lipids to the infected erythrocyte membranes.

Movement of macromolecules from the surface of the infected cell to the parasite has been documented only in case of transferrin. Several lines of evidence indicate that although an available source of iron is haemoglobin, the parasite can not obtain iron from this source and has to obtain iron from the plasma. In P. falciparum cultures it was demonstrated that Dexferroxamine, an iron chelator agent, completely inhibited parasite growth at lower concentrations that are needed to chelate iron from haemoglobin[68] and that ^{59}Fe bound to transferrin was incorporated by the parasite.[69]

In metabolically active mammalian cells[70], iron is obtained from a carrier protein in the plasma, ferrotransferrin. Receptors on the cell surface bind iron-loaded transferrin and the receptor-transferrin complex is internalized in endocytic vesicles. Following iron dissociation, the complex is recycled to the cell surface where apotransferrin is liberated and the receptor binds more ferrotransferrin.

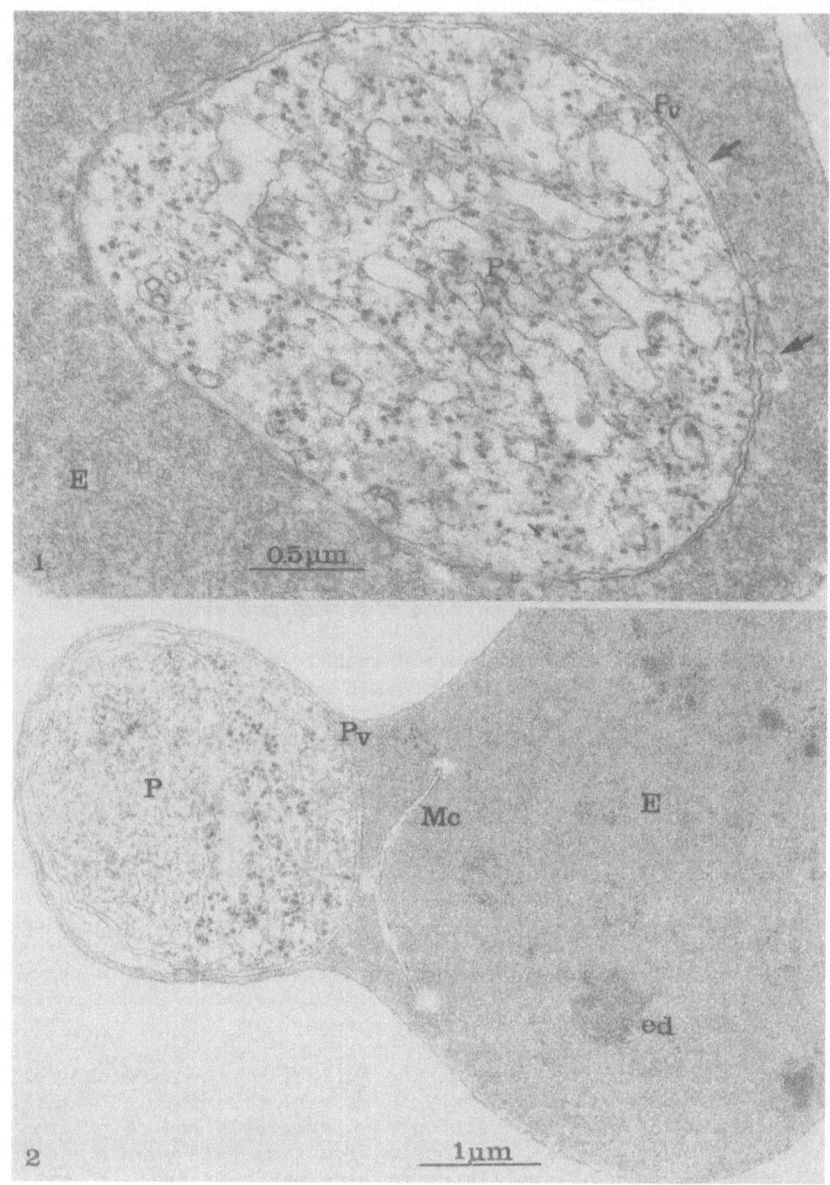

Figs. 1 and 2. Electron microscopy of <u>Plasmodium chabaudi</u> in
in mouse erythrocytes. The parasite (P) is surrounded by the
parasitophorous vacuole (Pv). Membrane projections (arrows)
deriving from the parasitophorous vacuole and membrane formations
corresponding to Maurer's clefts (Mc) as well as electron-
dense material (ed) aggregates are seen in the erythrocyte
cytoplasm (E).

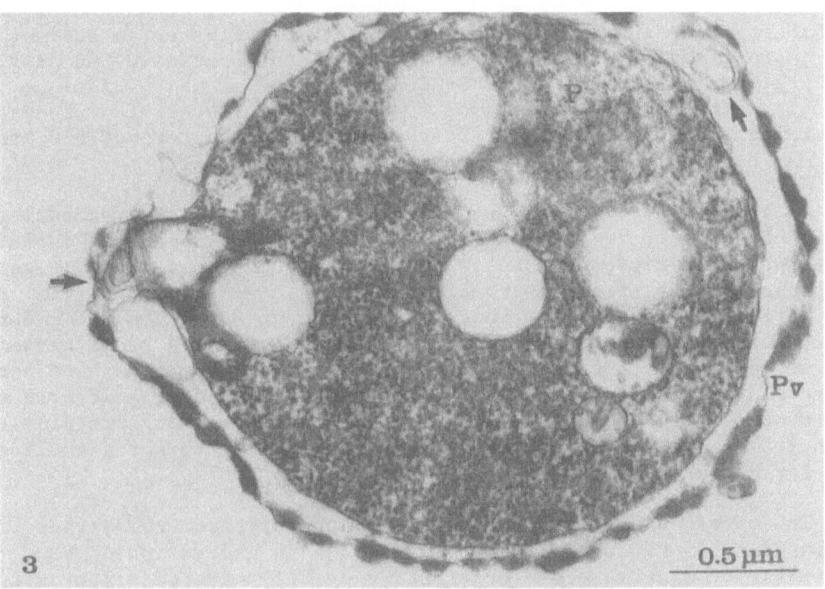

Fig. 3. Electron microscopy of a P. chabaudi trofozoite
obtained by immune lysis using rabbit anti-mouse erytrocyte
serum. The parasite (P) remained sheathed by the
parasitophorous vacuole (Pv). Membrane formation are seen
in the space between the parasite plasma membrane and the
parasitophorous vacuole (arrows).

In experiments in vitro with the Palo Alto, Uganda strain of P.
falciparum,[71] it was shown that fluorescein labelled-transferrin added to
the culture medium, first appeared as vesicles in the periphery of the
infected cell, and within 30 minutes concentrated around the parasitophorous vacuole. The presence of transferrin in the vesicles was further
demonstrated by immunoperoxidase staining using anti-transferrin antibodies. In controls using uninfected cells, no fluorescent vesicles were
seen in the cytoplasm. In experiments using infected cells with an excess
of unlabelled transferrin to FITC-transferrin, the uptake of the labelled
protein greatly decreased. Although transferrin receptors are abundant in
reticulocytes, they are totally lost after erythrocyte maturation. Since,
P. falciparum was cultured in mature erythrocytes in these experiments,
the internalization of transferrin indicated the possibility that the
parasite had produced its own receptors and located them on the
erythrocyte surface.

The presence of parasite derived proteins capable of binding
ferrotransferrin was documented by affinity chromatography of preparations
of ^{35}S-methionine labelled P. falciparum protein on anti-transferrin
sepharose or ferrotransferrin-sepharose columns. In both experiments, a
major protein of Mr 93 Kd was obtained. In experiments using ^{125}I ferrotransferrin, followed by protein extraction, and electrophoresis,
incorporation of the label was seen in a protein of 200 Kd, which may
represent the receptor-ligand complex.

Parasite proteins capable of binding ferrotransferrin have been
documented by others using the Gambian clone FRC-3/A_2. In these
experiments[72] a parasite protein isolated from membranes of infected

erythrocytes (Mr 102 Kd) was seen to bind ferrotransferrin but not apotransferrin. The difference on the Mr of this putative receptor with that previously reported could be indicative of strain variations. Interestingly, the 102 Kd protein was found to be acylated via 1,2-diacyl-sn-glycerol probably linked by a phosphodiester to the rest of the protein. This may be required for its membrane association.

Some difficulties found by other investigators[7,3] in saturating the uptake of transferrin in P. falciparum infected cells and in isolating the transferrin-receptor complex, have introduced controversy regarding the production of specific transferrin receptors by the parasite. Despite this, transferrin uptake by infected erythrocytes has been documented, and the possibility of pinocytosis of non-specifically bound transferring has been proposed. Although studies on receptor saturation have not been reported by others, enough data on the specificity of transferrin vs apotransferrin in the uptake phenomenon, and on parasite proteins as well as their localization in the erythrocyte membrane, supports the idea of a receptor mediated mechanism.

The movement of macromolecules from the erythrocyte surface towards the parasite, introduces a new complication in understanding the mechanisms of macromolecule trafficking in the infected erythrocyte. A previous model proposed that vesicles of the Maurer's clefts could be the route for transport to the host cell membrane[74] Parasite components could be secreted into the parasitophorous vacuole and vesicles budding from the vacuole or temporary membrane channels would be in charge of transport to the host cell surface. The secretion of two different proteins, PfEMP 2 a knob constituent protein, and PfHRP II that is secreted outside the infected cell, indicate that there exists at least two routes for protein export. One in which PfEMP 2[66] is associated with membrane vesicles, and the other moves independent of the vesicles.[49] A route for other membrane proteins transport not related to knob formation (like a transferrin receptor) has not been identified.

The route of proteins from the host cell surface, through the cytosol towards the parasite has only been visualized by light microscopy in the FITC- Fe. transferrin uptake experiments. The presence of vesicles has only been infered by the shape of the fluorescent bodies, and a supposed analogy with the internalization of transferrin-receptor complexes in mammalian cells in clathrin covered endocytic vesicles[75] To date no clathrin formations have been identified in the membrane vesicle complex of erythrocytes infected with plasmodia. But clearly, the parasite should have developed mechanisms for traffic sorting to distinguish import and export proteins and from these, those that will be secreted to the environment and those that will be assembled into specialized erythrocyte neoformations. More information on parasite proteins synthesis especially in protein precursor metabolism could indicate whether lead protein sequences and/or specific carrier proteins in the membrane organelles are involved in this process.

In summary, malaria parasites inside erythrocytes have developed specialized machinery for metabolic adaptation to their host cells. The principal changes inflicted on the erythrocyte are: an increased permeability to some essential metabolites (i.g. glucose and amino acids); the induction of neoformations (knobs) and neoantigens (neuraminidase modification of erythrocyte glycophorins) for cytoadherence in endothelial cells of capillaries where metabolic conditions are propitious and provide relative safety from destruction by elements of the immune system; and the insertion of proteins into the erythrocyte surface that function as receptors for essential metabolites (transferrin receptor). At least some of these modifications are mediated by membrane organelles of parasite origin situated in the cytoplasm of the host cell. The host cells adaptation ensures the survival and multiplication of the parasite, but some of its consequences results in the severe pathology seen principally in P. falciparum malaria. A better understanding of the host adaptation mechanisms may provide new targets for the development of pharmacological agents directed against the disruption of the parasites metabolic requirements.

Acknowledgements

The author is grateful to Dr. V. Tsutsumi for assistance in electron microscopy and to Dr. J. Ramsey for helpful suggestions on this manuscript. Part of the work reported here was supported by a grant from the MacArthur Foundation.

REFERENCES

1. K.N. Brown, Antigenic variation and immunity to malaria. In Parasites in the Immunized Host: mechanisms of survival. Ciba Foundation Symposium 25. Elsiever. Excerpta Medica. North-Holland. Associated Scientific Publishers. Amsterdam, London, New York. 1974.
2. W.A. Krotoski, Discovery of the hypnozoite and a new theory of malarial relapse. Trans. Roy. Soc. Hyg. Trop. Med. 79:1-11, 1985.
3. L.H. Miller, J.D. Haynes, F.M. McAuliffe, T. Shirosi, J.R. Durocher and M.H. McGinnis. Evidence for differences in erythrocyte surface receptors for malarial parasites. Plasmodium falciparum and Plasmodium knowlesi. J. Exp. Med. 146:277-281, 1977.
4. M. Jungery, G. Pasvol, C.I. Newbold and D.J. Weatherall. A lectin-like receptor is involved in invasion of erythrocytes by Plasmodium falciparum. Proc. Natl. Acad. Sci. USA, 80:10b8-1022, 1983.
5. M. Jungery. Invasion of human red cells by Plasmodium falciparum requires two steps involving different portions of red cell sialoglycoprotein molecules. In Red Cell Membrane Glycoconjugates and Related Genetic Markers. J.P. Cartron, P. Rouger and C. Salmon eds. Librairie Arnett, Paris. 1983
6. G. Pasvol, M. Jungery, D.J. Weatherall, S.F. Parsons, D.J. Anstee and M.J.A. Tanner. Glycophorin as possible receptor for Plasmodium falciparum. Lancet. 2:279-950, 1982.
7. L.H. Miller, S.J. Mason, J.A. Dvorak, M.H. McGuinnis, and I.K. Rothman. Erythrocyte receptors for (Plasmodium knowlesi) malaria: Duffy blood group determinats. Science. 189:561-562, 1975.
8. L.H. Miller, F.M. McAuliffe and S.J. Mason. Erythrocyte receptors for malaria merozoites. Amer. J. Trop. Med. Hyg. 26:204-208, 1977.
9. M. Jungery, D. Boyle, T. Patel, G. Pasvol and D.J. Weatherall. Lectin-like polypeptides of Plasmodium falciparum bind to red cell sialoglycoproteins. Nature. (London) 301:704-705, 1983.
10. R. Udomsangpetch, K. Lundgren, K. Berzins, B. Wahlin, H. Perlmann, M. Trye-Blomberg, J. Carlsson, M. Wahlgreen, P. Perlmann and A.Kjorkman. Human monoclonal antibodies to Pf 155, a major antigen of malaria parasite Plasmodium falciparum. Science. 231:57-59, 1986
11. J.A. Dvorak, L.H. Miller, W.C. Whitehouse and T. Shiroisi. Invasion of erythrocytes by malaria merozoites. Science. 187:748-749, 1975.
12. M. Aikawa, L.H. Miller, J. Johnson and J. Rabbege. Erythrocyte entry by malarial parasites: a moving junction between erythrocyte and parasite. J. Cell. Biol. 77:72-81, 1978
13. I.W. Sherman. Biochemistry of Plasmodium (malarial parasites). Microbiol. Rev. 43:453-495.
14. I.W. Sherman, Metabolism and surface transport of parasirized erythrocytes in malaria. In Malaria and the Red Cell. Ciba Foundation Symposium 94. Pitman. London. 1983.

15. T.M. Seed and J.P. Kreier. Surface properties of extracellular malaria parasites: electrophoretic and lectin-binding characteristics. Infect. Immun. 14:1339-1347, 1976.

16. D.J. McLaren, L.H. Bannister, P.I. Trigg and G.A. Butcher. A freeze-fracture study on the parasite-erythrocyte interrelationship in Plasmodium knowlesi infections. Bull. W.H.O. 55:199-204, 1977.

17. D.J. McLaren, L.H. Bannister, P.I. Trigg and G.A. Butcher. Freeze fracture studies on the interaction between the malaria parasite and host erythrocyte in Plasmodium knowlesi infections. Parasitology. 79:125-139.

18. S. Langreth. Electron microscope cytochemistry of host-parasite membrane interactions in malaria. Bull. W.H.O. 55:171-178, 1977.

19. M.G.N. Angus, K.A. Fletcher and B.G. Maegraith. Studies on the lipids of Plasmodium knowlesi-infected rhesus monkeys (Macaca mulatta). IV Changes in erythrocyte lipids. Ann. Trop. Med. Parasitol. 65:429-439 1972.

20. G.G. Holz Jr. Lipids and the malarial parasite. Bull. W.H.O. 55:237-248, 1977.

21. C.W. Lawrence and R.J. Cenedella. Lipid content of Plasmodium berghei infected rat red blood cell. Exp. Parasitol. 26:181-186, 1969.

22. D.H. Beach, I.W. Sherman and G.G. Holz Jr. Lipids of Plasmodium lophurae, and of erythrocytes and plasma of normal and P. lophurae-infected Pekin ducklings. J. Parasitol. 63:62-75, 1977.

23. R.A. De Zeeuw, J. Wijsbeek, R.C. Rock and G.J. McCormick. Composition of phospholipids in Plasmodium knowlesi membranes. Proc. Helmintol. Soc. Wash. 39:412-418, 1972.

24. K.N. Rao, D. Subrahmanyam and S. Prakash. Plasmodium bergehei: lipids of rat red blood cells. Exp. Parasitol. 27:22-27, 1970

25. R.C. Rock, J.C. Standefer, R.T. Cook, W. Little and H. Sprinz. Lipid composition of Plasmodium knowlesi membranes: comparison of parasites and microsomal subfractions with host rhesus erythrocyte membranes. Comp. Biochem. Physiol. 388:425-437, 1971

26. T.M. Seed and J.P. Kreier. Plasmodium gallinaceum: erythrocitic membrane alterations and associated plasma changes induced by experimental infections. Proc. Helminthol. Soc. Wash. 39:387-411, 1972.

27. C. Slomianny, G. Prensier and E. Vivier. Ultrastructural study of the feeding process of erythrocytic Plasmodium chabaudi trophozoite. Mol. Biochem. Parasitol. Suppl. 695, 1982.

28. C. Slomianny, P. Charet and G. Prensier. Ultrastructural localization of enzymes involved in the feeding process in Plasmodium chabaudi and Babesia hylomiysci. J. Protzool. 30:376-382, 1983.

29. H. Polet and M.E. Conrad. Malaria: extracellular amino acid requirements for in vitro growth of erythrocytic forms of Plasmodium knowlesi. Proc. Soc. Exp. Biol. Med. 127:251-253, 1968.

30. I.W. Sherman. Transport of amino acids and nucleic precursors in malarial parasites. Bull. W.H.O. 55:265-276, 1977.

31. I.W. Sherman and L. Tanigoshi. Incorporation of 14C-amino acids by malarial plasmodia (Plasmodium lophurae). VI. Changes in the kinetic constants of amino acid transport during infection. Exp. Parasitol. 35:369-373, 1974.

32. I.W. Sherman, R.A. Cox, B. Higginson, D.J. McLaren and J. Williamson. The ribosomes of the simian malaria Plasmodium knowlesi. I. Isolation and characterization. J. Protozool. 22:568-572, 1975

33. I.W. Sherman. The ribosomes of the simian malaria Plasmodium knowlesi II A cell free protein synthesizing system. Comp. Biochem. Physiol. 538:447-450, 1976.

34. B. Dasgupta. Polysaccharides in the different stages of the life cycles of certain sporozoa. Parasitology. 50:509-514, 1960.

35. L.W. Scheibel and W.K. Pflaum. Carbohydrate metabolism in Plasmodium knowlesi. Comp. Biochem. Physiol. 37:543,553. 1970

36. L.W. Scheibel and J. Miller. Glycolytic and cytochrome oxidase activity in plasmodia. Mil. Med. 134:1074-1080, 1969.
37. A.S. Fairfield, S.R. Meshnick and J.W. Eaton, Malaria Parasites adopt host cell superoxide dismutase. Science. 221:764-765, 1983
38. A.S. Fairfield and J.W. Eaton. Host superoxide dismutase incorporation by intraerythrocytic plasmodia.in Malaria and the Red Cell. Progress in clinical and biological research 155. J.W. Eaton and G.J. Brewer eds. Alan R. Liss Inc. New York, 1984.
39. H.M. Vincent and R.J. Wilson. Malaria antigens on infected erythrocytes. Trnas. Roy. Soc. Trop. Med. Hyg. 74:452-455, 1980
40. R. Schmidt-Ullrich, J. Lightholder and M.T.M. Monroe. Protective Plasmodium knowlesi Mr 74,000 antigen in membrane of schizont-infected rhesus erythrocytes. J. Exp. Med. 158:146-158, 1983.
41. R. Schmidt-Ullrich, D.F.H. Wallach and J. Lightholder. Two Plasmodium knowlesi-specific antigens on the surface of schizont-infected rhesus monkey erythrocytes induce antibody production in immune host. J. Exp. Med. 150:86-89, 1979.
42. S.B. Aley, J.W. Barnwell, W. Daniel and R.J. Howard. Identification of parasite proteins in membrane preparation enriched for the surface membrane of erythrocytes infected with Plasmodium knowlesi. Mol. Biochem. Parasitol. 12:69-84, 1984.
43. P.M. Knopf, G.V. Brown, R.J. Howard and G.F. Mitchell. Immuno-precipitation of biosynthetically-labelled products in the identification of antigens of murine red cells infected with the protozoan parasite, Plasmodium berghei. AJEBAK. 57:603-605, 1979.
44. M. Perkins. Surface proteins on schizont-infected erythrocytes and and merozoites of Plasmodium falciparum. Mol. Biochem. Parasitol. 5:55-64, 1982
45. M.T. Makler. P. falciparum invasion on human red cells and cytoadherence to endothelial cells is dependent upon a parasite produced glycosidase. Biochem. Biophys. Res. Comm. 143:461-466,1987
46. R.J.M. Wilson. The production of antigens by Plasmodium falciparum in vitro. Int. J. Parasitol. 4:537-547, 1974.
47. R.J.M. Wilson and R.K. Bartholomew. The release of antigens by Plasmodium falciparum. Parasitology. 71:183-192, 1975.
48. L. Rodriguez da Silva, M. Loche, R. Duval and L.H. Perrin. Plasmodium falciparum polypeptides released during in vitro cultivation. Bull. W.H.O. 61:105-112, 1983.
49. R.J. Howard, S. Uni, M. Aikawa, S.B. Aley, J.H. Leech, A.M. Lew, T.E. Wellems, J. Rener and D.W. Taylor. Secretion of a malaria histidine-rich protein (PfHRP II) from Plasmodium falciparum-infected erythrocytes. J. Cell. Biol. 103:1269-1277, 1986.
50. S.A. Luse and L.H. Miller. Plasmodium falciparum malaria: ultra-structure of parasitized erythrocytes in cardiac vessels. Am. J. Trop. Med. Hyg. 20:655-660, 1971.
51. M. Aikawa, J.R. Rabbege and B.T. Wellde. Junctional apparatus in erythrocytes infected with malarial parasites. Z. Zellforsch. Mikrosk. Anat. 124:72-75, 1972
52. M. Aikawa, J.R. Rabbege, I. Udeinya and L.H. Miller. Electron microscopic of knobs in P. falciparum-infected erythrocytes. J. Parasitol. 62:435-437, 1983.
53. I.J. Udeinya, J.A. Schmidt, M. Aikawa, L.H. Miller and I. Green. Falciparum malaria-infected erythrocytes specifically bind to cultured human endothelial cells. Science. 213:555-557, 1981
54. I.J. Udeinya, P.M. Graves, R. Carter, M. Aikawa and L.H. Miller. Plasmodium falciparum: the effect of time in continuous culture on binding to human endothelial cells and melanomic melanoma cells. Exp. Parasitol. 56:207-214, 1983.
55. S.B. Aley, J.A. Sherwood and R.J. Howard. Knob-positive and knob-negative Plasmodium falciparum differ in expression of a strain specific malarial antigen on the surface of infected erythrocytes. J. Exp. Med. 160:1585-1590, 1984.

56. T.J. Hadley, J.H. Leech, T.J. Green, W.A. Daniel, M. Wahlgren, L.H. Miller and R.J. Howard. Plasmodium falciparum: a comparison of knobby (K^+) and knobless (K^-) parasites of two strains. Mol. Biochem. Parasitol. 9:271–278, 1983

57. J.H. Leech, J.W. Barnwell, L.H. Miller and R.J. Howard. Identification of a strain-specific malarial antigen exposed on the surface of Plasmodium falciparum-infected erythrocytes. J. Exp. Med. 159:1567–1575, 1984.

58. I.J. Udeinya, L.H. Miller, I.A. McGregor and J.B. Jensen. Plasmodium falciparum strain-specific antibodies blocks binding of infected erythrocytes to amelanomic melanoma cells. Nature (Lond) 303:429–431, 1983.

59. A. Kilejian. The biosynthesis of the knob protein and a 65,000 Dalton histidine-rich polypeptide of Plasmodium falciparum. Mol. Biochem. Parasitol. 12:185–194, 1984.

60. J.V. Ravetch, R. Feder, A. Pavlovec and G. Blobel. Primary structure and genomic organization of the histidine-rich protein of the malarial parasite Plasmodium lophurae. Nature (Lond) 312:616–620, 1984.

61. A. Kilejian. Characterization of a protein correlated with the production of knob-like protrusions on membranes of erythrocytes infected with Plasmodium falciparum. Proc. Natl. Acad. Sci. USA. 76:4650–4653, 1979.

62. J.H. Leech, J.W. Barnwell, M. Aikawa, L.H. Miller and R.J. Howard. Plasmodim falciparum malaria: association of knobs on the surface of infected erythrocytes with a histidine-rich protein and the erythrocyte skeleton. J. Cell. Biol. 98:1256–1264, 1984

63. J.H. Leech, S.B. Aley, L.H. Miller and R.J. Howard. Plasmodium falciparum malaria: cytoadherence of infected erythrocytes to endothelial cells and associated changes in the erythrocyte membrane. In Malaria and the Red Cell. Prog. Clin. Biol. Res.J.W. Eaton and G.J. Brewer eds. Alan R. Liss Inc. New York, 1984.

64. R.J. Howard, J.A. Lyon, S. Uni, A.J. Saul, S.B. Aley, F. Klotz, L.J. Panton, J.A. Sherwood, K. Marsh, M. Aikawa and E.P. Rock. Transport of an Mr 300,00 Plasmodium falciparum protein (PfEMP 2) from the intraerythrocytic asexual parasite to the cytoplasmic face of the host cell membrane. J. Cell. Biol. 104:1269–1280, 1987.

65. M. Aikawa, L.H. Miller and J. Rabbege. Caveola-vesicle complexes: the plasmalemma of erythrocytes infected by Plasmodium vivax and Plasmodium cynomolgi. Am. J. Pathol. 79:285–300, 1975.

66. M. Rudzinska and W. Trager. The fine structure of throphozoites and gametocytes in Plasmodium coatneyi. J. Protozool. 15:73–88, 1968.

67. M. Aikawa, Y. Uni, A.T. Andrutis and R.J. Howard. Membrane-associated electron-dense material of the asexual stages of Plasmodium falciparum: Evidence for movement from the intracellular parasite to the erythrocyte membrane. Am. J. Trop. Med. Hyg. 35:30–36, 1986.

68. C. Raventos-Suarez, S. Pollack and R.L. Nagel. Plasmodium falciparum inhibition of in vitro growth by desferrioxamine. Am. J. Trop. Med. Hyg. 31:919–922, 1982.

69. S. Pollack and J. Flemming. Plasmodium falciparum takes up iron from transferrin. Br. J. Haematol. 58:289–293, 1984.

70. J.W. Larrick and P. Cresswell. Modulation of cell surface iron transferrin receptor by cellular density and state of activation. J. Supramol. Struct. 11:579–586, 1979.

71. M.H. Rodriguez and M. Jungery. A protein on Plasmodium falciparum-infected erythrocytes functions as a transferrin receptor. Nature. 324:388–391, 1986.

72. K. Haldar, C.L. Herderson and G.A.M. Cross. Identification of the parasite transferrin receptor of Plasmodium falciparum-infected erythrocytes and its acylation via 1,2-diacyl-sn-glycerol. Proc. Natl. Acad. Sci. USA. 83:8565–8569, 1986.

73. S. Pollack and V. Schenelle. No transferrin recéptors on P. falciparum The 36th Annual Meeting of the Am. Soc. Trop. Med. Hyg. Abstract 363, 1987.
74. R.J. Howard. Alterations in the surface membrane of red blood cells during malaria. Immunol. Rev. 61:67-107, 1982.
75. J.L. Golstein, R.G.W. Anderson and M.S. Brown. Coated pits, coated vesicles and receptor-mediated endocytosis. Nature. 279:679-685, 1979.

INFLAMMATORY RESPONSE IN AMEBIC LIVER ABSCESS

L. E. Muñoz

Unidad de Hígado, Facultad de Medicina y Hospital Universitario "Dr. José E. González", Universidad Autónoma de Nuevo León, Monterrey, N.L. (México)

INTRODUCTION

In the inflammatory response humoral and cellular effector systems participate during acute as well as chronic inflammation (1,3). The main effector systems that are involved in the acute inflammatory process are: a) the complement system in particular C3a, C3e, C5a, C5b, C567 and factor B (2); b) coagulation and fibrinolytic systems, and c) acute phase reactants (2). Some cells are attracted to the site of inflammation by humoral processes. These cells are polymorphonuclear (PMN); basophils; eosinophils; monocytes; lymphocytes and platelets mainly. Fluid mediators derived from these cells participate in inflammatory regulation (2,3). Plasma mediators of acute inflammation inhibit or amplify these mechanisms.

Some of the histopathological description from autopsies in patients that died of amebic liver abscess (ALA) suggest that tissue inflammation is rare (4,5). However, other studies, have shown intense inflammation in intestinal amebiasis (5). PMN cell infiltration may be seen in human ALA as well as granulomas (5).

Studies done in ALA show that there is an intense, acute inflammatory reaction. In the following sections studies on complement system, serum immunoglobulins, phagocytic cells, serum chemoattractant activity and inhibitors of chemotaxis in patients with ALA will be discussed. Finally a possible role of the acute inflammatory response in the pathogenesis of ALA is put forward.

THE COMPLEMENT SYSTEM

Entamoeba histolytica trophozoites activate the complement system via classical (6,7) and alternative (6,8,9) pathways. When ameba extracts are studied the complement system activation proceeds via both pathways. However, activation of the classical pathway is more evident (6). In

vitro studies have shown that although pathogenic strains of
E. histolytica are capable of activating the complement
system they are resistant to lysis (7). Complement component
concentrations of the classical and alternative pathways in
patients with ALA are suggestive of an increased complement
utilization (10). Patients that recover under medical
treatment (Group 1) show increased serum concentrations of C3
and factor H as compared to the control group. On the other
hand, patients who needed abscess drainage in addition to
medical treatment (Group 2) had significantly diminished
levels of C1q, C3, factor B and factor H (10) (table 1).
These data suggest that an increased utilization of the
complement system rather than a decreased synthesis takes
place. The relationship of complement components suggest that
complement activation takes place mainly through the
classical pathway in patients that recover under medical
treatment, whereas in patients that require surgical
treatment activation proceeds via both pathways (10). It is
interesting to note that patients that needed abscess
drainage had been ill for longer periods of time (10)
suggesting that the complement activation increases with the
natural course of the disease. Three out of five patients
that exhibited low plasma C3d levels showed an increase
during the convalescence period. Furthermore, there was a
significant direct correlation between plasma levels of C3d
and serum albumin levels; and SGOT, suggesting that
complement activation is secondary to the presence of the
ameba (10). Whether or not the complement system participates
in the elimination of the parasite in vivo remains to be
elucidated. If the ameba is resistant to complement lysis in
vivo, the biological activities generated upon complement
activation may have deleterious effects on the host possibly
contributing to the initial damage. Some of these biological
activities are opsonization by C3b; induction of leukocytosis
by C3e; generation of chemotactic factors such as C5a
(11,12), and antibody production by factor H (13).

SERUM IMMUNOGLOBULINS

        Serum anti-ameba antibodies have been found to be mainly
of the IgG2 subclass (14). The suppression of the
hemagglutination reaction by the treatment of sera with
mercaptoethanol suggest that IgM is also involved (14).
Studies done in our laboratory suggest that IgA and IgE may
also be involved (see below).

        Reports on serum concentrations of IgM in patients with
ALA have been controversial. Some studies have reported
increased values (15,16), whereas normal serum concentrations
have been found by other investigtators (17,18,19). When an
early IgM antibody response by the ELISA method was sought
for in patients with ALA (20), very poor activity was
detected in the IgM fraction, similar to that found in the
controls. It has been suggested that specific anti-ameba IgM
shows minor elevations and only for short periods of time
(19,20). Studies in our laboratory have shown significantly
diminished IgM serum concentrations (171.8 ± 69.4 mg/dl; mean
± 1 SD) in patients with ALA, as compared to the control
group (279 ± 95 mg/dl; p<0.001) which is also evident when
patients are grouped according to whether they had been ill

310

Table 1. Serum or plasma levels of components of classical and alternative complement pathways, the regulatory protein factor H and C3d in patients with ALA. From Munoz & Salazar, 1987 (10).

Serum or Plasma Levels	Group 1 (n=13)	Group 2 (n=16)	Total Group (n=29)	Controls (n=22)	P <
C1q (relative conc.)	39.0 ± 7.0	32.4 ± 10.9	35.5 ± 10	38.8 ± 5.0	0.05 *
C4 (% pool)	129.5 ± 31.4	88.6 ± 44.3	107.0 ± 43	111.7 ± 27.0	0.01 **
C3 (% pool)	145.0 ± 25.6	92.3 ± 31.9	116.0 ± 39	110.8 ± 17.6	0.05 * 0.001 <
Factor B (mg/dl)	20.6 ± 5.5	15.5 ± 6.4	18.0 ± 6	21.6 ± 5.3	0.05 ** 0.01 *
Factor H	49.7 ± 7	35.7 ± 10.9	41.5 ± 11	42.7 ± 8.0	0.05 ^^ 0.01 ^^
C3d (mg/l)	10.0 ± 3.7	7.8 ± 6.6	8.5 ± 5	9.0 ± 2.0	0.001 ** N. S.

Values are given as mean + 1 S D

* Group 2 vs. control group

** Group 1 vs. group 2

^ Group 1 vs. control group and vs. group 2

^^ Group 1 vs. control group

for 1, 2 or 3 weeks (21). A decrease in helper T cells and an increase in suppressor T cells have been reported in patients with ALA (22). This finding could explain a decrease in IgM synthesis in this sort of patient.

IgG serum concentrations have been found to be increased in ALA (15,18,19,23). However, some authors have reported normal IgG serum concentrations (16,17). When patients were grouped according to the time they had been ill we found that IgG serum concentrations were not significantly elevated in those patients who had been ill for 1 to 7 days (1,903 ± 913.6 mg/dl; controls 1,469 ± 418 mg/dl). However, IgG was significantly elevated in patients who had either been ill for 8 to 14 days (2,552.5 ± 1,106.7 mg/dl; p<0.01) or 15 to 21 days (2,167.5 ± 910 mg/dl; p<0.002) (21).

IgA serum levels have been found to be normal (15,17-19) or increased (16) in ALA patients. IgE serum levels have also been found to be increased (17). We have found both IgA and IgE serum levels to be increased in these patients (IgA:ALA = 421.5 ± 220.5 mg/dl, controls 221.6 ± 77 mg/dl, p<0.001; IgE:ALA = 783.7 ± 662 mg/dl, controls 381 ± 521 mg/dl p<0.002). It is interesting to note that both immunoglobulins were more significantly elevated in those patients that had been ill for up to two weeks (IgA 404 ± 104 mg/dl, p<0.0005; IgE 1,065 ± 798.5 mg/dl, p<0.005). When patients had had a 3-week clinical picture IgE serum concentrations were within normal limits (21). Both immunoglobulins are present in the gastrointestinal tract, IgA being the most abundant (24,25). Studies on the distribution of IgE immunocytes in the body indicate that these are found mainly close to the mucosa (24,26). This data suggest that there may be a continuous challenge of immune cells with E. histolytica in the gastrointestinal tract during the early course of ALA.

Studies on serum immunoglobulins differ according to author. Studied patients may differ with regard to the clinical course, immune response, geographical distribution. On the other hand, the strain of ameba, amount of innoculum may also be responsable for differing results.

Another evidence that suggests that serum antibodies other than IgG are important in the response to ameba infection is that when the number of serum antigen-antibody systems identified by immunoelectrophoresis (27,28) is studied in relation to IgM, IgG, IgA and IgE serum concentrations a direct significant correlation with IgG (r=+ 0.69; p<0.01; n=8); IgA (r=+0.80;p<0.001; n=8), and IgE (r=+0.70; p<0.01; n=8) are found. However, no relationship with IgM can be demonstrated (r=+0.38; p NS; n=8) (21). The number of antigen-antibody systems found in the aforementioned study varied from 2 to 8. The most significant correlation found was with IgA. It has been suggested that polymeric IgA antibodies transport the amebic antigen from the systemic circulation to the bile (29), which again may contribute to the elevated serum concentration of this immunoglobulin.

PHAGOCYTIC CELLS

Only a few studies have been reported on mononuclear cell functions. The capacity of the mononuclear phagocytic system of the hamster to eliminate Candida albicans after being infected with E. histolytica was found to be impaired (30). Peripheral monocytes from normal volunteers have been found to exhibit a decreased chemotaxis to axenic ameba (31). Phagocytosis of Staphylococcs aureus (32,33); latex, and sheep red blood cells (33) have been found not to differ between patients with ALA and a control population. Bactericidal capacity of mononuclear cells has been found to be either normal (32) or decresead (33) in these sorts of patients.

Phagocytosis and chemotaxis of PMN cells have been reported; S. aureus phagocytosis by PMN having been found to be significantly diminished in patients with ALA as compared to intestinal amebiasis and the control group (32).

Studies done by cinemicrography showed that PMN cells exhibit chemotaxis towards pathogenic and non-pathogenic strains of E. histolytica (34). Upon contact pathogenic strains produce degranulation and death of PMN cells. In addition, the ameba phagocytize the PMN cell (34). On the other hand, PMN cells can phagocytize non-pathogenic strains (34). Preliminary studies done in our laboratory show that trophozoite extracts are powerful chemoattractants for PMN cells, 1.4 µg/ml of total extract produces a comparable chemotaxis to that produced with 1 mg/ml of casein (unpublished observations).

When peripheral PMN cell chemotaxis from patients with ALA was studied a normal mean value was found as compared to the control group (35). However, a significantly inverse correlation could be demonstrated between PMN cell chemotaxis and the number of days patients had been ill ($r=-0.45$; $p<0.01$; $n=32$); serum antibody titer ($r=-0.46$; $p<0.01$; $n=30$) in the total group of patients. In addition, a significantly inverse correlation between PMN cell chemotaxis and number of peripheral leukocytes was seen in patients that required abscess drainage in addition to medical treatment ($r=-0.70$; $p<0.01$; $n=13$) (35). In this report 16 patients were subsequently studied in the convalescence period (table 2): 4 patients who initially showed an increased chemotactic value (normal range 65 to 80.6 µm; mean + 2 SD) had a second one within normal limits. Four patients with an initial decreased PMN cell chemotaxis showed an improvement in a subsequent determination, three of whom were within normal limits. Eight patients exhibited a normal chemotactic value in the acute illness and during convalescence or the follow-up period (35).

SERUM CHEMOATTRACTANT ACTIVITY OF PATIENTS WITH ALA

In this section unpublished results from work done in our laboratory on serum chemoattractant activity of patients with ALA will be discussed. In view of the fact that abnormalities in patients' PMN cell chemotaxis were observed, and that these abnormalities subsided during the

Table 2. Polymorphonuclear cell chemotaxis was performed in patients either during the convalescence period or during 25 to 27 months in the follow-up period. Time was recorded since the day patients started with symptoms. From Munoz & Salazar, 1986 (35).

Pat No.	Acute Illness		Convalescence		Follow-up	
	Chemotaxis μm	Day No.	Chemotaxis μm	Day No.	Chemotaxis μm	(months)
1	76	5	69.5	36	73	27
2	87	3	66	23	71	25
3	51	20	77	50	-	-
4	68.5	5	69	19	-	-
5	56.5	7	65	21	-	-
6	80	8	71.5	241	-	-
7	58.5	31	69.5	61	-	-
8	84	12	69.5	25	-	-
9	75.5	14	64.5	28	-	-
10	81.5	7	66.5	18	-	-
11	77.5	10	68	33	-	-
12	44	60	61	74	-	-
13	73.5	5	-	-	72	25
14	72.5	14	-	-	70	27.5
15	66.5	12	-	-	72.5	27
16	85	2	-	-	74	26

Table 3. Non-activated and LPS-activated serum chemoattractant activity from patients with ALA and controls on control PMN cell.

Patient Number	10 % ALA Patients' Serum				5 % ALA Pat.S. 5 % Control S.	10% Control Serum		Casein	Gey
	LPS-Act.*	Non-Act.*	LPS-Act.^	Non-Act.^	LPS-Activated	LPS-Act.	Non-Act.		
2	62	56	66.5	-	-	69	-	71.5	32
3	40.5	49.5	44	44.5	52.5	71	-	71.5	32
5	47	40.5	48.5	39.5	44	71	-	71.5	30.5
10	51	56.5	52.5	62	47.5	81.5	50.5	72.5	30.5
13	48	50	60.5	43.5	50.5	81.5	50.5	72.5	30.5
14	65.3	66	62.5	49	72.5	81.5	50.5	72.5	30.5

* :Acute disease; ^ : Convalescence; results are expressed in μm; S.: serum
Number of patients corresponds to patients from table 2

convalescence period (see above), the capacity of patients' serum to induce chemotaxis of normal PMN cells was determined. In addition, normal serum was mixed with patients' serum in equal proportions and chemotaxis of normal cells studied looking for a serum inhibitor. Sera of 6 patients were included in these experiments which were done using the Boyden chamber modified by Wilkinson (35,36). The lower chamber was prepared with 10% non-activated and E. coli lipopolysaccharide (LPS)-activated serum (30 μg/ml) (Difco, U.K.), after incubation at 37°C for 30 minutes. Control sera were treated in the same manner. In experiments where patients' sera were mixed with normal sera and used as chemoattractant 5% of each in Gey's culture medium was used. One mg/ml of casein (Merck, A.G.) was used as positive control. In the upper chamber 200 μl of 2 x 10-6 of normal PMN cells in Gey's culture medium were placed. Migration of cells was assessed after 1 hour incubation at 37°C by the `leading front' technique.

Sera from 6 patients of those included in table 2 were studied. Results are shown in table 3. Five out of six LPS-activated serum from patients with ALA exhibited a defective chemoattractant activity (normal range 62.9 to 83.7 μm). This defect was seen in 5 out of 6 patients during the convalescence period. It is interesting to note that the defect was found even in those patients whose cells had shown a normal or increased chemotactic response (table 2). The presence of a serum inhibitor was suggested by the experiments in which patients' sera (during acute illness) were mixed with normal sera where a defective chemoattractant activity persisted in 4 out of 5 patients. In patients no. 3, 10, 13 and 14 non-activated serum promoted a chemotaxis similar to LPS-activated serum or even higher, as opposed to control serum where the former gives lower values. This may be explained by means of chemotactic factor inactivator production as a consequence of complement activation in vitro with LPS (see below).

INHIBITORS OF CHEMOTAXIS

Regulation of the inflammatory process is mediated by inhibitors of chemotaxis (37). These inhibitors may be cell- (38,39) or chemotactic factor- (40,41) directed. Cell-directed chemotactic inhibitors have been identified as IgA and to a lesser extent IgG (38) acting through its Fc components (39). Normal human serum contains a chemotactic factor inactivator (CFI) (40,42). CFI regulates complement-mediated inflammatory processes. Its action is enzymatic in nature and there are two components of this inactivator. One inactivates complement-derived chemotactic factor, C5a. The other inactivates chemotactic factors of bacterial origin (37,40,42). Abnormal quantities of CFI are present in those diseases in which anergy is manifest such as Hodgkin's disease, lepromatous leprosy and sarcoidosis (43). Cell- (44) and chemotactic factor- (45) directed inhibitors have been described in patients with malignant tumors.

CFI has been described to be present in the gamma and non-gamma fraction of plasma (42). Preliminary studies in our laboratory show a similar inhibition of chemotaxis by CFI

Table 4. Effect of increasing amounts of partially purified CFI present in the gamma and non-gamma fractions of normal human plasma on control PMN cell chemotaxis.

LPS - Activated NHP Plus								10% LPS-Activated NHP	10% Non-Activated NHP	1mg/ml Casein	Gey's Medium
CFI in Gamma Fraction (µl)				CFI in non-Gamma Fraction (µl)							
70	180	360	720	70	180	360	720				
54.5	57.5	64.5	58	54	57.5	64	60	70	45	71.5	25
57.5	56.5	51	48.5	55	51.5	52.5	47.5	71.5	42.5	73.5	30
54.5	57.5	58	46.5	51	57.5	56.5	51.5	71.5	42.5	73.5	30

Results are expressed in µm
NHP: normal human plasma

317

Table 5. Effect of CFI from patients with ALA and normal
controls on normal PMN cell chemotaxis.

Source of CFI	LPS-Act NHP + CFI	LPS-Act. NHP	Non-Act. NHP	Casein 1 mg/ml	Gey's medium
Patient No.					
1	31	68	40.5	71	30
2	41	68	40.5	71	30
3	47	68	40.5	71	30
4	29	68	40.5	71	30
5	50	79	43.5	72.5	30
6	47	78	42	71	30
Control No.					
1	51	79	43.5	72.5	30
2	63.5	79	43.5	72.5	30
3	65	79	43.5	72.5	30
4	52.5	79	43.5	72.5	30

Results are expressed in μm

present in the gamma and non-gamma fractions of normal human
plasma (NHP) upon 45% saturation with ammonium sulphate using
the Berenberg & Ward technique (40) for the non-gamma
fraction, and for the gamma fraction Robins, et al.,
technique (41). Table 4 shows the effect of varying amounts
of partially purified CFI present in the gamma and non-gamma
fractions of 3 samples of NHP on control PMN cell chemotaxis
(unpublished observations). CFI was added to LPS-activated
NHP in the lower chamber. Chemotaxis stimulated by 1 mg/ml of
casein; 10% non-activated and LPS-activated NHP were used as
controls (see previous section). A similar chemotactic
inhibition was observed when the gamma and non-gamma
fractions were tested. Till and Ward (42) found that when
saturation with ammonium sulphate over 60% is used the
activity of CFI decreases substantially in the non-gamma
fraction. Therefore, we have subsequently studied the gamma
fraction for CFI activity.

In another series of experiments 360 μl of the gamma
fraction (CFI) of plasma from patients with ALA and controls
were added to LPS-activated NHP in the lower chamber and
tested for chemoattractant activity with PMN cells from
normal controls. Table 5 shows that CFI from patients with
ALA inhibit chemotaxis from normal PMN cells to a greater
extent than CFI from controls. The effect of CFI was seen
regardless of the time patients had been ill. Patients of
whom CFI activity was studied had had a clinical course from

one day to two weeks. When CFI activity of 7 controls was compared with the 6 patients studied a significant difference was found (p<0.01).

No relationship of PMN cell chemotaxis and major serum immunoglobulin concentrations could be demonstrated (IgM: r=-0.22, p NS, n=28; IgG: r=-0.02, p NS, n=28, and IgA: r=-0.04, p NS, n=28)(unpublished observations). Therefore, it is unlikely that IgA participates in the chemotactic inhibition in spite of these patients having increased serum IgA levels.

INFLAMMATORY RESPONSE IN THE PATHOGENESIS OF ALA

It is evident that multiple factors play an important role in the pathogenesis of ALA. The mechanisms that the parasite utilizes to survive and proliferate play key roles in the pathogenesis of amebiasis (14,46). On the other hand, host factors in response to the presence of the parasite may also be important in the development of ALA. From the data presented here it can be concluded that immune host mechanisms that participate in the acute inflammatory process may be important in the elimination of the parasite. However, overstimulation of some of these mechanisms may participate in inducing liver damage. Activation of the complement system upon antigen invasion is one of the initial inflammatory mechanisms that amplifies the immune response due to biological activities interacting with humoral and cellular mechanisms (11,12). If E. histolytica trophozoites are resistant to complement lysis in vivo , this could be a key factor promoting an intense inflammatory response due to constant stimulus by trophozoites. Hyperactivation of the complement system may enhance the immunoregulatory effects of its proteins and fragments (47). In addition, circulating immune complexes have been described in patients with amebiasis (48,49) which may contribute to complement activation. Circulating ameba free antigen has also been described (50), which together with immune complexes and complete trophozoites could account for complement activation. Patients with ALA have been shown to have circulating antibodies directed against intracellular liver cell components (51). These antibodies are directly related to the titer of serum antibodies to E. histolytica. Thus, participation of complement in antibody-mediated cytotoxic reactions during the development of ALA could be one of the mechanisms in the pathogenesis of ALA.

Another ongoing fact with complement activation is the chemotaxis of PMN cells which is increased early in the course of the disease and diminishes with its natural course. C5a is the most powerful chemoattractant in vivo. As a result of complement activation increasing amounts of plasma CFI may inhibit chemotaxis in those patients who have been ill for longer periods of time. Activated neutrophils can be cytotoxic to hepatocytes,(52,53) and these cells are present in the liver surrounding trophozoites in the early stages of experimental ALA (54). Therefore, neutrophils may be present in the liver as a result of chemotactic stimuli generated by trophozoites, complement-derived cytotaxins and injured tissue debris, which in turn activate the neutrophils. Activated neutrophils produce reactive oxygen intermediates

which are highly toxic, and can cause tissue damage when they
overwhelm or circumvent the cellular protective mechanisms
(53). PMN cells and macrophages release reactive oxygen
intermediates into the phagolysosomes or the extra-cellular
space, this is an important physiological mechanism by which
phagocytes kill bacteria but also has been implicated as a
mechanism of tissue injury (55). Although activated
macrophages can also produce liver injury by the above-
mentioned mechanism it is unlikely that they participate to a
great extent in the initial damage in ALA because in early
experimental ALA they are rare (54). Furthermore, ameba
produces cellular immunosuppression (14). Monocytic functions
of chemotaxis and bactericidal activities may be diminished
(30,33) in patients with ALA. While axenic ameba produces a
decreased chemotaxis of normal mononuclear cells (31) it is a
powerful chemoattractant for normal PMN cells (personal
observation).

Experiments were an inhibitor of serum chemoattractant
activity was suggested in patients with ALA as well as those
where the partial isolation and effect of CFI are shown
suggest that the inflammatory response is present regardless
of the period of time patients had been ill. The effect of
CFI may be seen as early as one day after initiation of
symptoms.

With regard to serum immunoglobulins it is worth
commenting that although most available studies indicate
total serum concentrations, and few analyze specific-anti-
ameba antibodies the fact that there are important
abnormalities in serum concentrations which are influenced by
the clinical stage may highlight the various roles that they
play. IgG and IgM are known to activate the classical pathway
of complement (56). IgG has an important function as an
opsonin. IgA activates the alternative pathway of this system
(56). Therefore, participation of immunoglobulins in the
acute inflammatory process is important. Among the various
immunoregulatory functions of the complement system
involvement of immunoglobulins is also important, either
enhancing by C3 degradation products like C3d or inhibiting
by C5a its synthesis (47). An interesting point on the
relationship between serum immunoglobulin concentrations and
serum antigen-antibody systems is whether or not a continuous
invasion of E. histolytica in the intestine accompanies the
natural course of ALA.

It is evident from these data that an intense
inflammatory response accompanies human amebic liver abscess.
An adequate balance between the inflammatory process and its
regulatory systems will prevent tissue damage. Consequences
of the inflammatory response such as biological activities
and immunoregulation by the complement system, and
interaction with PMN cells may cause deleterious effects when
this balance is altered.

Acknowledgements

To Oliva G. Salazar for her excellent technical
assisstance in serum inhibitors experiments. To Professor
Robert M; Chandler for stylistic suggestions.

REFERENCES

1. D. H. Wright, Immunological aspects of diagnostic histopathology, in: "Clinical Aspects of Immunology," P.G.H. Gell, R.R.A. Coombs and P. J. Lachmann, eds., Blackwell Scientific Publications, London, 783p (1975).

2. P. M. Henson, Antibody and immune-complex-mediated allergic and inflammatory reactions, in: "Clinical Aspects of Immunology," P. J. Lachmann and D. K. Peters, eds., Blackwell Scientific Publications, London, 687p (1982).

3. R. N. Pinckard, The "new" chemical mediators of inflammation, in: "Current Topics in Inflammation and Infection," Williams and Wilkins, London, 38p (1982).

4. H. Brandt and R. Perez-Tamayo, "Amibiasis," 1a. ed., La Prensa Médica Mexicana, México (1970).

5. D. Trissl, Immunology of Entamoeba histolytica in human and animal host, Rev. Infect. Dis. 4:1154 (1982).

6. S. Meri, G. Richaud and E. Linder, Complement activation by antigenic fractions of Entamoeba histolytica, Parasite Immunol. 7:153 (1985).

7. S. L. Reeds, J. G. Curd, I. Gigli, F. D. Gillin and A. I. Braude, Activation of complement by pathogenic and nonpathogenic Entamoeba histolytica, J. Immunol. 136:2265 (1986).

8. L. Ortiz-Ortiz, R. Capin, N. R. Capin, B. Sepulveda and G. Zamacona, Activation of the alternative pathway of complement by Entamoeba histolytica, Clin. Exp. Immunol. 34:10 (1978).

9. G. Huldt, P. Davies, A. C. Allison, H. U. Schorlommer, Interactions between Entamoeba histolytica and complement, Nature 277:214 (1979).

10. L. E. Muñoz and O. G. Salazar, Complement activation in patients with amebic liver abscess, J. Hepatol. 5:30 (1987).

11. H. J. Müller-Eberhard and R. D. Schreiber, Molecular biology and chemistry of the alternative pathway of complement, Adv. Immunol. 29:1 (1980).

12. N. K. Pangburn and H. J. Müller-Eberhard, The alternative pathway of complement, Spinger Semin. Immunopathol. 7:163 (1984)

13. G. C. Tsokos, G. Inghirami, C. D. Tsoukas, J. E. Balow and J. D. Lambris, Regulation of immunoglobulin secretion by factor H of serum human complement, Immunol. 55:419 (1985).

14. B. Sepulveda and A. Martinez-Palomo, Immunology of amebiasis by Entamoeba histolytica, in: "Immunology of Parasitic Infections," S. Cohen and K. Warren, eds., Blackwell Scientific Publications, Great Britain, 170p (1982).

15. A. A. Abioye, E. A. Lewis and H. Mc Farlane, Clinical evaluation of serum immunoglobulins in amoebiasis, Immunol. 23:937 (1972).

16. V. V. Ravi, S. Mithal, A. N. Malaviya and B. N. Tandom, Immunologic studies in amebic liver abscess, Indian J. Med. Res. 63:1732 (1975).

17. A. Dasgupta, Immunoglobulins in health and disease. III. Immunoglobulins in the sera of patients with amoebiasis, Clin. Exp. Immunol. 86:163 (1974).

18. J. O. S. Osisanya and D. C. Warhurst, Specific antiamebic

immunoglobulins and the cellulose acetate precipitin test in Entamoeba histolytica infection, Trans. Roy. Soc. Trop. Med. Hyg. 74:605 (1980).

19. N. K. Ganguly, R. C. Mahajan, D. V. Datta, S. Sharma, P. N. Chuttani,and A. K. Gupta, Immunoglobulin and complement levels in cases of invasive amoebiasis, Indian J. Med. Res. 67:221 (1978).

20. O. Muñoz, R. Hernandez-Velarde, E. Cruz-Mejia and M. C. Martinez, Es posible distinguir entre infeccion hepatica antigua y reciente mediante el analisis inmunoenzimatico. Arch. Invest. Med. (Mex.) 17(Suppl.):327 (1986).

21. M. J. Ibarra, "Aspectos de la respuesta inmune humoral en pacientes con absceso hepatico amibiano," Tesis de licenciatura para Q.F.B. , U.A.N.L. Facultad de Ciencias Quimicas (inedita), Monterrey, N.L. Mex. Septiembre. 1987.

22. R. A. Salata, A. Martinez-Palomo, H. W. Murray, L. Canales, N. revino, E. Segovia, C. F. Murphy and J. I. Ravdin, Patients treated for amebic liver abscess develop cell- mediated immune responses effective in vitro against Entamoeba histolytica, J. Immunol. 136(7):2633 (1986).

23. A. Perches, R. Kretschmer, E. Lee, B. Sepulveda, Determinacion de inmunoglobulinas en suero de pacientes con amibiasis invasora, Arch. Invest. Med. (Mex.) 1(Suppl.):s97 (1970).

24. H. Bazin, The secretory antibody system, in "Immunological Aspects of the Liver and Gastrointestinal Tract," A. Ferguson, R. N. M. Mac Sween eds., M. T. P. Press Ltd., Great Britain, 33p (1976).

25. J. R. David, Host-Parasite interface: Immunology, in: "Tropical and Geographical Medicine," K. S. Warren and A. F. Mahmoud, eds., Mac Graw-Hill, New York, 125p (1985).

26. K. Ishizaka and T. Ishizaka, Biological functions of gamma IgE antibodies and mechanisms of reaginic hypersensitivity, Clin. Exp. Immunol. 6:25 (1970).

27. I. M. Krupp and J. Powell, Comparative study of the antibody response in amebiasis, Am. J. Trop. Med. Hyg. 20:421 (1971).

28. I. M. Krupp, Definition of the antigenic pattern of Entamoeba histolytica, and immunoelectrophoretic analysis of the variation of patient response to amebic disease, Am. J. Trop. Med. Hyg. 26:387 (1977).

29. R. Campos-Rodriguez, O. Diaz-Guerra, C. Barranco-Tovar, A. Isibasi-Araujo, J. Kumate-Rodriguez, Papel de la IgA en la eliminacion de antigenos amibianos, Arch. Invest. Med. (Mex.) 17:353 (1986).

30. R. Capin, A. Gonzalez-Mendoza, L. Ortiz-Ortiz,Disminucion de la actividad del sistema fagocitico mononuclear en hamsters infectados con Entamoeba histolytica, Arch. Invest. Med. (Mex.) 11(Suppl. 1) :235 (1980).

31. R. Kretschmer, M. C. Salinas-Carmona, M. Lopez-Osuna and M. A. Avila, Efecto de Entamoeba histolytica sobre la quimiotaxis de monocitos humanos, Arch. Invest. Med. (Mex.) 11(Suppl 1):147 (1980).

32. T. N. Gosh and F. C. Sen, Phagocytic function in amebiasis, Indian J. Med. Res. 71:207 (1980).

33. N. J. Gill, N. K. Ganguly, R.C. Mahajan, S.R. Bhusnurmath

and J.B. Dilawari, Monocyte functions in human amebiasis, Indian J. Med. Res. 76:674 (1982).

34. R. L. Guerrant, J. Brush, J. I. Ravdin, J. A. Sullivan and G. L. Mandell, Interactions between Entamoeba histolytica and human polymorphonuclear neutrophilis, J. Infect. Dis. 143:83 (1981).

35. L. E. Muñoz-Espinosa and O. G. Salazar-Flores, Systemic inflammation in patients with amebic liver abscess: Chemotaxis of polymorphonuclear cells, Arch. Invest. Med. (Mex.) 17(Suppl.): 313 (1986).

36. P. C. Wilkinson, "Chemotaxis and Inflammation, "Churchill Livingstone, Great Britain, 168p (1974).

37. E. L. Becker and P. A. Ward, Chemotaxis, in: "Clinical Immunology," C. W. Parker, ed., Saunders Co., Philadelphia 272p (1983).

38. D. E. Van Epps and J. R. Jr Williams, Suppression of leukocyte chemotaxis by human IgA components, J. Exp. Med. 144:1227 (1976).

39. D. E. Van Epps, K. Reed and J. R. C. Jr. Williams, Suppression of human PMN bactericidal activity by human IgA paraproteins. Cell. Immunol. 36:363 (1978).

40. J. A. Berenberg and P. A. Ward, Chemotactic factor inactivator in normal human serum, J. Clin. Invest. 52:120 (1973).

41. R. A. Robins, R. K. Zetterman, T. J. Kendall, G. L. Gossman, H. P. Monsour and R. I. Rennard, Elevation of chemotactic factor inactivator in alcoholic liver disease, Hepatol. 7(5):872 (1987).

42. G. Till and P. A. Ward, Two distinct chemotactic factor inactivators in human serum, J. Immunol. 114(2):843 (1975).

43. R. A. Clark, Disorders of granulocyte chemotaxis, in: "Leukocyte Chemotaxis: Methods, Physiology and Clinical Implications," J. I. Gallin and P. G. Quie, eds., Raven Press, New York, 329p (1978).

44. E. G. Maderazo, T. F. Antoni and P. A. Ward, Serum associated inhibitor of leukotaxis in humans with cancer, Clin. Immunol. Immunopathol. 9:166 (1978).

45. J. P. Brozna and P. A. Ward, Antileukotactic properties of tumor cells. J. Clin. Invest. 56:616 (1975).

46. S. Said-Fernandez, J. Vargas-Villarreal and J. Castro-Garza, Mecanismo multifactorial de la actividad citolitica de Entamoeba histolytica, Arch. Invest. Med. (Mex.) 17(Suppl):173 (1986).

47. J. M. Weiler, Complement and the immune response, in: "Complement in Health and Disease," K. Whaley, ed., M. T. P. Press Ltd., Norwell, MA. U. S. A. 289p (1987).

48. S. Pillai, A. Mohimen, A solid phase sandwich radioimmunoassay for E. histolytica antigen and the detection of circulating antigen in amebiasis, Gastroenterol. 83:1210 (1982).

49. T. K. Maitra, N. K. Jalan, A. Mohimen, Detection of E. histolytica immune complexes in human tissues by a solid phase sandwich assay, Memorias del X Seminario sobre Amibiasis, 45p, Octubre, Mexico. City, (1986).

50. B. Sepulveda, Amibiasis: Host-pathogen biology, Rev. Infect. Dis. 4:836 (1982).

51. G. M. Faubert, E. Meerovitch, J. Mc. Kaughlin, the presence of liver auto-antibodies induced by Entamoeba histolytica in the sera from naturally infected humans

and immunized rabbits, Am. J. Trop. Med. Hyg. 27:892 (1978).

52. P. Mavier, A. M. Preaux, J. Rosenbaum, B. Guigui and D. Dhumeaux, Toxicity of PMN cells against hepatocytes. Evidence for an oxygen-dependent and independt mechanism, J. Hepatol. Suppl. 2:S282 (1985).

53. M. J. P. Arthur, Reactive oxygen intermediates and liver injury, J. Hepatol. 6:125 (1988).

54. V. Tsutsumi, R. Mena-Lopez, F. Anaya-Velazquez and A. Martinez-Palomo, Cellular bases of experimental amebic liver abscess formation, Am. J. Pathol. 117:81 (1984).

55. J. C. Fantone, P. A. Ward, Role of oxygen-derived free radicals and metabolites in leukocyte-dependent inflammatory reactions, Am. J. Pathol. 107:397 (1982).

56. P. J. Lachmann and D. K. Peters, Complement, in: "Clinical Aspects of Immunology," P. J. Lachmann and D. K. Peters, eds., 4th. Ed., Blackwell Scientific Publications, London, 18p (1982).

BIOLOGICAL DETERMINANTS OF HOST-PARASITE RELATIONSHIP IN MOUSE CYSTICER-
COSIS CAUSED BY Taenia crassiceps: INFLUENCE OF SEX, MAJOR HISTOCOMPATIBI-
LITY COMPLEX AND VACCINATION*

Larralde, C.+, Sciutto, E.+, Grun, J.°., Díaz, M.L.+, Govezensky, T.+, and Montoya, R.M.+

+Instituto de Investigaciones Biomédicas, Universidad Nacional Autónoma de México, AP 70-228, México, D. F. 04510,
°Dept. Biochemistry, Th. Jefferson U., Philadelphia, PA. USA

The metacestode of Taenia solium is frequent cause of serious
neurological illness of humans in Mexico and other countries of Latin America, Asia and Africa. Man is the only carrier of the adult worm and sole
responsible of transmission to pigs and other humans via inadequate
disposal of faeces. Recent years have see advancement in therapy (1,2)
diagnosis (3-5) and pathology of human and porcine disease (6,7). Copious
accounts of recent developments in cysticercosis are collected in (8), (9)
is an exhaustive review of all literature on taeniasis/cysticercosis,
while (10) is the most authorative and comprehensive account of human
cysticercosis.

Factors determining the risk of humans contracting cysticercosis are
thought to be mainly related to magnitude and frequency of exposure to eggs
of T. solium. Evidence implicating biological factors in susceptibility is
tenuous. No impressive association of human cysticercosis with
histocompatibility antigens was found in a doubtfully representative study
performed recently (11), nor do the few terminal cases studied make a strong
case for immunosupression determining human disease (12). However,
inklings of biological factors being involved are present in a recent
report of women showing more frequently than males severe inflammation
in neurocysticercosis (13). Also suspicious of biological mediation is the
lack of correlation between positive serology and social factors
conventionally associated to high risk of infectious disease, such as low
income and scholarity and defective personal hygiene (14). Further, the
very heterogeneous clinical pictures and forms of evolution of the disease,
some curing spontaneously while others progress relentlessy to fatal
outcomes or live on essentially asymptomatic (10), together with the
parasites'sensitivity to drugs, some resistant to praziquantel and others
to albendazol (15), all argue for a complicated network of factors and
events belonging to parasite (16), host and environment, concurring in the
pathogenesis of cysticercosis.

*This work was supported in part by Consejo Nacional de Ciencia y
Tecnología de Mexico and Química Hoechst de México.

Systematic exploration of the role of biological factors in susceptibility to cysticercosis is hardly possible studying man, and most laborious and costly in pigs. However, there are other tapeworms – T. crassiceps and T. taeniformis – whose metacestodes affect mice, that are most suitable for experimentation and have already provided with some evidence for the genetic background of the host influencing the outcome of infection with T. taeniaformis (17) and of immunity affecting installation of T. crassiceps (18). Mice harbor the cysticerci of T.crassiceps in their peritoneal cavities as a chronic infection causing some inflammation in serose intestinal surfaces at late

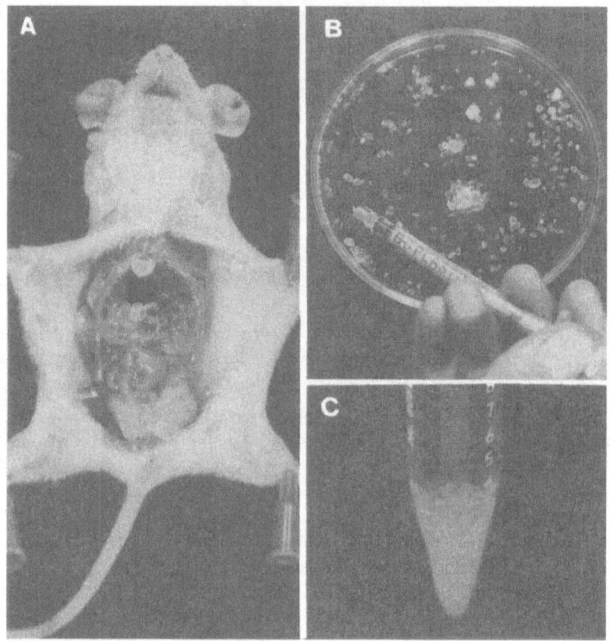

Figure 1. T. crassiceps cysticercosis in the peritoneal cavity of mouse infected six months before with five live cysticerci. The disease's accesibility to experimentation and measurement of parasite load is immediately apparent.

stages of disease quite like that caused by T. solium in basal meninges of man. Furthermore, anatomically, the murine parasite resembles that of T. solium, although somewhat smaller, and shares the seemingly convivial behavior of the human parasite, not causing major damage to neighboring structures of the host, other than space occupation and scant inflammation, as evidenced by conventional light microscopy and NMR imaging (19). In nature mice contract the disease ingesting eggs present in the environment contaminated with faeces from small carnivores –like foxes, cats, and perhaps others (20)– bearing the adult worm in their intestine.

Experimental infection is simply attained by injecting metacestodes in the peritoneal cavity of mice, where they reproduce by budding, presumably asexually. In normal conditions one may harvest them by the hundreds a couple of months after infection (21). If truly asexual, T. crassiceps capability to multiply by budding, allows for control and uniformity of the parasite's genetic characteristics. Most conveniently, cysticerci of T. crassiceps fare well for weeks in conventional tissue culture conditions and for days in minimal media.

In here we present preliminary observations in mice infected with T. crassiceps indicating that biological factors are indeed involved in host susceptibility to cysticerci. Results point to the significant participation of immunological, endrocrinological and genetical determinants on the rate of parasite growth and replication inside the host, opening very exciting possibilities of studying the interaction of these three prominent organic systems upon the host-parasite relationship, a unique biological phenomenom seemingly heedless of simple rules of thumb concerning immunology, histocompatibility and causation of disease (21).

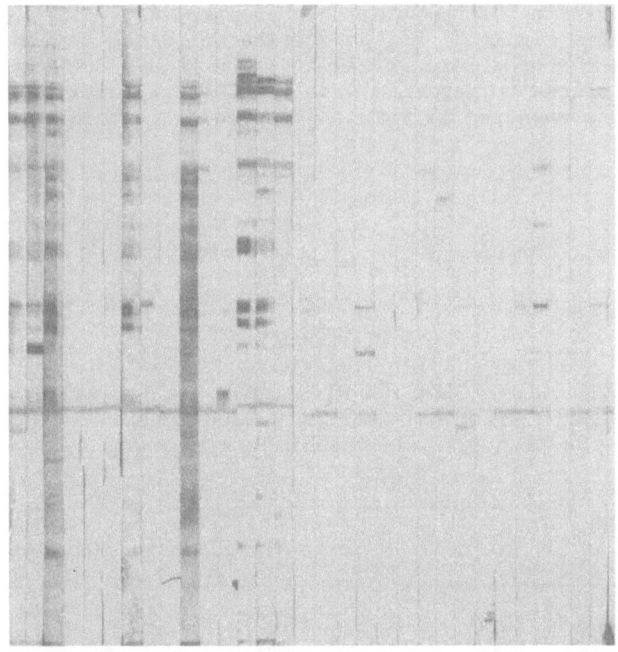

Figure 2. Western blots of protein antigens from the vesicular fluid of T. crassiceps cysticerci reacting with sera of neurocysticercotic patients (left) and of healthy donors (right).

1) Immunological Determinants
Antigenic crossreactivity between the human and the mouse parasite was definitively established by Western Blots (Figure 2), for a number of proteins differing in molecular weight. Furthermore, Tables I and II show that vaccination of susceptible mice with antigen extracts of T. solium cysticerci reduces the parasite load developed from a challange with T. crassiceps.

TABLE I

EFFECTS OF VACCINATION WITH Taenia solium ANTIGENS
UPON PARASITE LOAD OF MICE CHALLENGED WITH Taenia crassiceps

Parasite load ul.	Relative dose of vaccine			
	0.00	0.01	0.10	1.00
Individual Mouse	2000	1000	1500	0
	1500	400	0	600
	2000	400	200	0
	1000	2000	2000	50
	1000	2000	2500	1000
Mean	1500	1160	850	300

Decrement of T. crassiceps parasite load with increasing vaccine dose in female (SJL X Balb/c) mice, challanged five months after vaccination with five live cysticerci implanted intraperitoneally. Vaccine was a sterile 50% ammonium sulphate protein extract (3X precipitated) of a single T. solium cysticercus dissected from the skeletal muscle of a parasitized pig; ten fold dilutions of the extract were prepared to generate widely different vaccination protocols. Parasite load was measured as volume of harvested parasites one month after challenge.

TABLE II

EFFECTS OF VACCINATION WITH Taenia solium ANTIGENS
FROM DIFFERENT SOURCES UPON PARASITE LOAD
OF MICE CHALLENGED WITH Taenia crassiceps

	Saline	Antigens From Vesicular Fluid	Antigens From Complete Parasite
Parasite loads per animal, in ul	100	0	200
	1000	600	600
	1000	1000	400
	1000	900	500
	2000	600	0
		800	10
		300	0

Decrement of T. crassiceps parasite load due to vaccination in female Balb/c mice. Vaccine (1 mg total protein of T. solium cysticerci/mouse) was administered two months before challange with five cysticerci of T. crassiceps implanted intraperitoneally. Parasite load was measured one month after challenge.

These results establish the sharing of antigens between the murine and the human parasites, as well as the effective influence of vaccination in reducing parasite load. These findings are of interest in that T. crassiceps in mice closely resembles T. solium in antigenic constitution, validating it's use as a disease model and as an alternative source for antigens to be used in immunodiagnosis of neurocysticercosis Likewise, crossimmunity from T. solium to T.crassiceps allows some optimism about the inverse also holding, at least in pigs. Reducing the parasite load in pigs, via an effective vaccine, could lower the endemia of adult T. solium tapeworm infection and the number of eggs in the environment, as consequence.

TABLE III

SEXUAL AND HISTOCOMPATIBILITY DIFFERENCES IN SUSCEPTIBILITY
OF MICE TO Taenia crassiceps AS MEASURED BY TOTAL PARASITE COUNT

Strain	H Haplotype 2	Males	Females	Fem/Male
Balb/c	d	200	640	3.2
C57Bl	b	116	750	6.5
C3A	k	130	860	6.6
SWR	q	290		3.1
DBA	q		600	2.1
SJL	s	200	540	2.7

Greater female susceptibility to T. crassiceps is here shown in five strains of mice differing in H2haplotypes. Parasite loads are given as the mean number of parasites collected from each of ten animals, three weeks after challange with five cysticerci in the peritoneal cavity. Statistical analysis (Two Way ANOVA) indicated to significant variation between sexes and among strains.

2) Sexual histocompatibility Determinants

Mouse females' greater susceptibility to T. crassiceps is demostrated in Table III: in all strains of mice tried, the ratios of parasite loads in ten challanged females, as measured thirty days after injecting five cysticerci in the peritoneal cavity of each animal, was always greater than twice that of the respective males. Statistical analysis of results in Table III indicated to significant variation due to strains, thus pointing to the importance of genetic background, presumably the Major Histocompatibility Complex, in susceptibility of mice to T. crassiceps. Experimental design does not allow definitive conclusions on the role of histocompatibility because the strains used, although differing in their H2 haplotypes, also differ, and to unknown extent, in their genetic complement. Careful selection of mouse strains is required to elucidate this point. However, there is no statistical doubt that the susceptibility associated to sex varied in the different strains used. This we consider an important finding. Although the literature is not lacking in reports of sex associated differences in susceptibility to experimental infections, rarely do they point to females being more susceptible or to histocompatibility association (21,28-33) and the consequent implications for immune reactivity (33). Thus, we have great hopes in T. crassiceps cysticercosis as a potent model for studing the significance of "immunoneuroendocrine interactions" in a macroscopic event such as infection, instead of microscopic phenomena in simplified systems such as isolated cells, subcellular particles and the like (reviewed in 22-27).

In closing, it would be adventurous to claim these results are already relevant to our understanding and dealings with human disease in terms other than the most general: parasitic disease installs and progresses as the net result of a complicated and delicate network of interactions among the organic constitutents of both host and parasite. That sex – presumably hormones – make such a great difference in host-parasite relationship involving helminths was indeed surprising. Dilucidation of the mechanisms involved in the sexual differences in susceptibility is yet to be done; direct hormonal influence on the parasite or through immunological mediation are the most prominent – non-disjunctive – possibilities.

References

1. Sotelo, J., Torres, B., Rubio, D.F., Escobedo, F. and Rodríguez-Carvajal, J. (1985). Praziquantel in the treatment of neurocysticercosis. Long term follow up. Neurology 35: 752-755.

2. Escobedo, F., Penago, P., Rodríguez-Carvajal, J. and Sotelo, J. (1987). Albendazole therapy for neurocysticercosis. Arch. Intern. Med. Vol. 147, 738-741.

3. Flisser, A., and Larralde, C. (1986). Cysticercosis. In: Schantz, P. et al. Immunodiagnosis of Parasitic Infections 1: 109-161.

4. Schantz, P.M. (1987). Improvements in the serodiagnosis of helminthic zoonoses. Veterinary Parasitology 25: 95-120.

5. Larralde, C., Laclette, J.P., Owen, Ch. S., Madrazo, I., Sandoval, M., Bojalil, R., Sciutto, E., Contreras, L., Arzate, J., Díaz, M.L., Govezensky, T., Montoya, R.M. and Goodsaid, F. (1986). Reliable serology of Taenia solium cysticercosis with antigens from cyst vesicular fluid. ELISA and hemagglutination tests. Am. J. Trop. Med. 35(5): 965-973.

6. González, D., Rodríguez-Carbajal, J., Aluja, A., and Flisser, A. (1987): Cerebral cysticercosis in pigs studied by Computed tomography and Necropsy. Vet. Parasitol. 26: 55-69.

7. Rabiela, M.T. (1985) Pruebas morfológicas de que C. cellulosae y C. racemosus son de T. solium. Arch. de Inv. Med. (Mex) 16: 83-95.

8. Flisser, A., Willims, K., Laclette, J.P., Larralde, C., Ridaura, C. and Beltrán, F. (Eds) (1982) Cysticercosis: Present State of Knowledge and Perspectives. Academic Press. (New York).

9. Gemmell, M., Matyas, Z., Pawlowski, Z., Soulsby, E., Larralde, C., Nelson, G.S., Rosicky, B. (Eds) (1985) Guidelines for surveillance prevention and control of taeniasis/cysticercosis. VPH/83.49 World Health Organization Geneva, Switzerland.

10. Aluja, A., Escobar, A., Escobedo, F., Flisser, A., Laclette, J.P., Larralde, C., Madrazo, I., Velázquez, V., Willms, K. (1987). Cisticercosis. Una recopilación actualizada de los conocimientos básicos para el manejo y control de la cisticercosis causada por Taenia solium. Biblioteca de la Salud. Fondo de Cultura Económico. Inst. Nal. de Salud Pública. México

11. Correa, D., Gorodezky, C., Castro, L., Rabiela, Ma. T. and Flisser, A. (1986) Detection of MHC products on the surface of Taenia solium cysticerci from humans. Rev. Lat-Amer. Microbiol. 28: 373-379.

12. Flisser, A., González, D., Plancarte, A., Tovar, A., Correa, D., Rodríguez del Rosal, E. y Aluja, A. (1987) Modificación de la respuesta inmune y la reacción inflamatoria en cerdos con cisticercosis después del tratamiento con praziquantel. En Memorias del VII Congreso Nac. de Inmunología. Soc. Mex. Inmunol. pp. 26.

13. Del Brutto, O., García, E., Talamas, O., Sotelo, J. (1988) Sex-related severity of inflamation in parenquimal brain cysticercosis. Archives of Internal Medicine 48: 544-546.

14. Woodhouse, E., Flisser, A. and Larralde, C., "Seroepidemiology of human cysticercosis in Mexico", (1982) In: Flisser, A., Willms, K., Laclette, J.P., Larralde, C., Ridaura, C. and Beltrán, F. (Eds). Cysticercosis: Present State of Knowledge and Perspectives, 11-23 Academic Press, New York.

15. Sotelo, J., Escobedo, F. and Penago, P. (1988). Albendazol versus praziquantel for therapy in neurocysticercosis a controlled trials. Arch. of Neurology, in press.

16. Yakoleff-Greenhouse, V., Flisser, A., Sierra, A., and Larralde, C. (1982) Analysis of antigen variation in cysticerci of Taenia solium. J. of Parasitology, 68: 39-47.

17. Mitchell, G.F., Genetic Variations In Resistance of Mice to Taenia taeniaeformis: Analyses of Host-protective Immunity and Immune Evasion. In "Cysticercosis. Present State of Knowledge and Perspectives", pp. 575-584 (Flisser, A., Willms, K., Laclette, J.P., Larralde, C., Ridaura, C., and Beltrán, F. (Eds.) Academic Press, New York, 1982.

18. Good, A.H., Siebert, A.E. Jr. Robbins, P. and Zaun, S. (1982) Modulation of the Host Immune Response by Larvae of Taenia crassiceps In. Cysticercosis. Present State of Knowledge and Perspectives (Flisser, A., Willms, K., Laclette, J.P., Larralde, C., Ridaura, C., and Beltrán, F. (Eds) pp. 593-609 Academic Press, New York.

19. Haselgrove, J., Grun, J., Owen, Ch. and Larralde, C. (1987) Magnetic resonance imaging of parasitic tapeworm larvae Taenia crassiceps cysticerci in the peritoneal cavity of mice. Magnetic Resonance in Medicine 4: 517-525.

20. Freeman, R.S. (1962) Studies on the biology of Taenia crassiceps (Zeder, 1800) Rudolphi, 1810 (Cestoda) Can. J. Zool. 40: 969-990.

21. Smith, K.J., Esch, G.W. and Kuhn, R.E.) (1972). Growth and development of larval Taenia crassiceps (Cestoda) I. Aneuploidy in the anomalous ORF strain. Int. J. Parasit. 2:261-263.

22. Besedovsky, H.O., Del Rey, A.E., and Sorkin, E. Immuneneuroendocrine Interactions. J. Immunol (Suppl.) 135(2): 750s-754s. 1985.

23. Plaut, M. Lymphocyte Hormone Receptors. Ann Rev. Immunol. 5: 621-629, 1987.

24. Smith, E.M., Harbour-McMenamin, D., and Blalock, J.E. Lymphocyte Production of Endorphins and Endorphin-Mediated Immunoregulatory Activity. J. Immunol. (Suppl.) 135(2):779s-782s , 1985.

25. Besedovsky, H.O., Del Rey, A., Sorkin. E., Da Prada, M., and Keller, H.H. Immunoregulation Mediated by the Sympathetic Nervous System. Cell Immunol 48: 346-355, 1979.

26. Ahmed, S.A. Penhale, W.J. and Talal, N. Sex Hormones, Immune Responses, and Autoimmune Diseases. Mechanisms of Sex Hormone Action. Am. J. Pathol. 121(3): 531-551, 1985.

27. Blalock, J.E., Harbour-McMenamin, D., and Smith, E.M. Peptide Hormones Shared by the Neuroendocrine and Immunologic System. J. Immunol. (Suppl) 135(2): 858-861 1985.

28. Reddington, J.J., Stewart, G.L., Kramar, G.W., and Kramas, M.A.: The effects of host sex and hormones on Trichinella spiralis in the mouse. J. Parasitol. 67(4): 548-555, 1981.

29. Giannini, M.S.H. (1986): Sex-influenced response in the pathogenesis of cutaneous leishmaniasis in mice. Parasite Immunology 8: 31-37.

30. Greenblatt, H.C., and Rosenstreich, D.L. (1984): Trypanosoma rhodesiense infection in mice: sex dependence of resistance. Infection and Immunity 43(1): 337-340.

31. Charniga, L., Stewart, G.L., Gramar, G.W., and Stanfield, J.A. (1981): The effects of host sex on enteric response to infection with Trichinella spiralis. J. Parasitol. 67(6): 017-922.

32. Mitchell, G.F., Rajasekariah, G.R., Rickard, M.D. (1980): A mechanism to account for mouse strain variation in resistance to the larval cestode, Taenia taeiniformis. Immunology 39: 481-489.

33. Evans, W.S., Novak, M., and Basilevsky, A. (1985): Effects of enviromental temperature, sex, and infection with Hymenolepis microstoma on the liver bile duct weights of mice. J. Parasitol. 71(1): 106-109.

34. Terres, G., Morrison, S.L., and Habicht, G.S. (1968): A quantitative difference in the immune response between male and female mice. Proc. Soc. Exp. Biol. & Med. 127: 664-667.

DETECTION OF THE HUMAN PAPILLOMAVIRUS GENOME IN CERVICAL CANCER

Manuel L. González-Garay[1], Juan F. González-Guerrero[2] and Hugo A. Barrera-Saldaña[1,3]

[1]Unidad de Laboratorios de Ingeniería y Expresión Genéticas. Departamento de Bioquímica, Facultad de Medicina de la U.A.N.L. Monterrey, N.L. México

[2]Servicio de Oncología, Hospital Universitario "Dr. José Eleuterio González" Monterrey, N.L. México

[3]International Center for Molecular Medicine. Monterrey, N. L. México

INTRODUCTION

Cervical cancer is one of the primary neoplasias responsible for deaths of women in the world. In 1975, the world incidence of this sickness was approximately five hundred thousand cases, a value comparable to the number of women in which breast cancer was detected in the same year (1).

The frequency of this type of cancer varies according to geographic regions, Latin America being one of the areas with the highest incidence (2). In Mexican women cervical cancer is the most frequent neoplasia (3).

The relationship between cervical cancer and sexual behavior has been known for over one hundred and fifty years (2, 4). Recent evidence has accumulated indicating that certain types of human papillomavirus, such as types 16 (5), 18 (6), 31 (7) and 35 (8) are closely associated with gynecological cancer.

As shown in table 1 papillomavirus type 16 has been consistently found in samples of cervical cancer from women of different regions in percentages between twenty five and sixty seven (7, 9-24). Furthermore, this type of papillomavirus has also been found in precancerous lesions (10, 25, 11, 26, 27, 24, 7, 28, 12), in other gynecological lesions (29) and in some cervical tissues obtained from healthy women (11, 27, 28, 29, 30).

For these reasons, we considered it important to detect the genome of this type of virus in our population and begin a search for carriers of

Table 1. Frecuency of HPV-16 in cervical cancer
of women from different countries.

Country (Reference)	Percentage	Cases
Germany (10)	61.1	18
Germany (16)	67	6
Italy (23)	46.15	13
England (13)	45	11
England (11)	33	3
England (15)	87.5	8
U.S.A. (17)	33	27
U.S.A. (14)	25	12
U.S.A. (7)	36	39
U.S.A. (9)	40	5
Japan (12)	50	6
Japan (18)	36	50
Japan (19)	40	5
Japan (24)	55	9
Mexico (21)	31	16
Panama (22)	60	20
Kenya and		
Brasil (10)	34.8	23
Brasil (20)	42	19

this infectious agent. This in turn could be beneficial for the prevention
of gynecological cancer.

RESULTS AND DISCUSSION

Twenty samples were collected in our University Hospital "Dr. José
Eleuterio González", from women with cervical invasive epidermoid carcinomas, by either biopsy or surgical remotion. The samples were selected on
the basis of clinical and histological criteria to include specimens from
the four stages of severity of the disease (37), going from the less
severe (stage I) to the more advanced state (stage IV). These consisted of
four samples from the first stage, nine of the second, three from the
third and four from the fourth stage.

We also collected one specimen of a vaginal carcinoma, one cervical
intraepithelial neoplasia (CIN III), one cervical sample showing only
inflammatory changes and one vulvar carcinoma. We not only decided to
obtain tissue samples but, in addition as a means of control, we obtained
blood samples from each patient. As a negative control six tissue specimens from normal patients without cytological signals of abnormalities
were employed. The characteristics of the samples included in this study
are shown in table 2.

To detect human papilloma virus DNA in our samples we followed the
experimental strategy described in figure 1.

DNA was purified from each sample by the procedure described by
Gariglio et al (21). We obtained approximately 1.6 μg of DNA per mg of
cervical samples and twenty ug of DNA per ml of peripheral blood.
Comparable DNA yield values have been reported (38, 39, 40, 41, 42).

Table 2. Characteristics of the samples analyzed.

No.	CODE	PATIENT	AGE	LESSION
1.-	T1	MCRS	59	CC IIB
2.-	T2	IEB	34	CC R
3.-	T3	LMVJ	60	CC IIB
4.-	T4	GCM	61	CC IB
5.-	T5	MBB	65	CC IA
6.-	T6	MAGA	45	CC IIIB
7.-	T7	MIPR	43	CC IIB
8.-	T8	MLTH	30	CC IIB
9.-	T9	JBP	53	CC IIIA
10.-	T10	JARL	35	CC IV
11.-	T11	APP	48	CC IA
12.-	T12	MEMR	33	CC IB-IIA
13.-	T15	PLA	49	CC II-IIA
14.-	T16	MSM	58	CC II
15.-	T17	PML	54	CC IV
16.-	T18	JHE	47	CC IIB
17.-	T19	SFR	35	CIN III
18.-	T20	BBA	54	CC IIB
19.-	T21	EMM	36	CC IIB
20.-	TB	AMM	52	CC IIIB-IV
21.-	TC	CRG	73	CC IV
22.-	TF	MYMM	45	VC
23.-	NT4	CRM	48	CC IIB
24.-	A1	MSP	53	None
25.-	A2	VGG	38	None
26.-	A3	OER	44	None
27.-	A4	MRM	39	None
28.-	A5	CLM	55	None
29.-	A6	MTS	43	None
30.-	R1	MLZL	45	IC

[a] CC Cervical carcinoma
[b] CCR Relapse of a Cervical carcinoma
[c] CIN Cervical intraepithelial Neoplasia
[d] VC Vulvar Carcinoma
[e] IC Uterine cervix with inflammatory changes

The dot blot hybridization technique was used as the primary screening method for identification of positive samples, five μg aliquots of cellular DNA were denatured by incubation in 1.5 M NaCl and 0.5 M NaOH for at least one h at 37°C and filtered onto a nylon membrane in a 96-well manifold (34). The filter was baked for 2 h under vacuum and then prehybridized at 42 °C for 16 h in a solution containing: 50 mM Tris-HCl (pH 7.9), 1 M NaCl, 5 mM EDTA, 20 % formamide, 0.1 % SDS, 5 X Denhardt's solution (35) and 250 ug/ ml of salmon sperm DNA. Hybridization was performed for at least 24 h at 42 °C, essentially in the same prehybridization buffer as above, but containing in addition 10 % dextran sulphate and 1×10^6 cpm/ml of the probe (specific activity > 2×10^8 cpm/ug) The probe consisted of a plasmid (pHPV-16) provided by Dr. P. Gariglio, that contains the HPV-16 DNA cloned in the unique Bam HI site of pUC 8 (31). It was used to transform Escherichia coli HB101 and after chloramphenicol amplification its DNA was purified by lysis with triton x-100 followed by isopycnic centrifugation in cesium chloride-ethidium bromide gradients

(32). Finally, 250 ng of purified plasmid DNA were radioactively labeled by the random oligonucleotides labeling technique (33), resulting in having the radiolabeled probe for the hybridization studies.

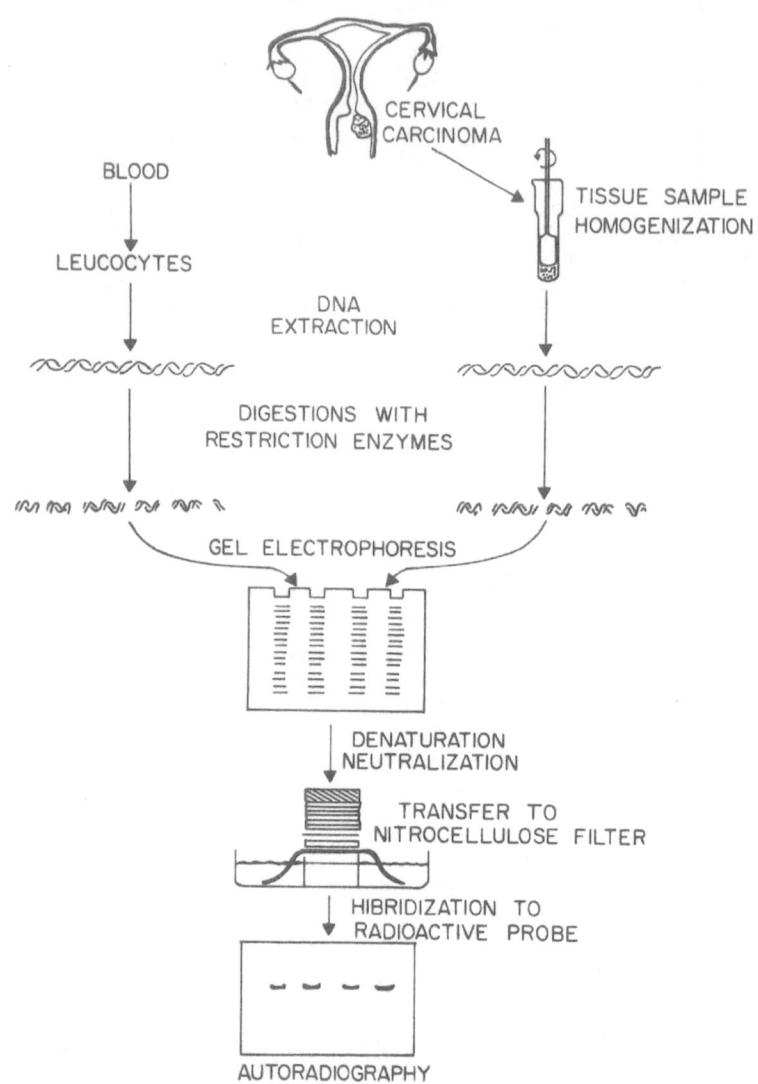

Fig. 1. Experimental strategy used to detect human papilloma virus DNA in cervical carcino-mas. The leucocytic DNA was used for con-trol purposes. The probe consisted of radioactively labeled HPV-16 cloned DNA.

In order to discriminate hybridization against HPV-16 versus other non-HPV-16 papilloma viruses, the hybridized filter was washed under two differents conditions. For moderate stringency conditions the washes were in 2X SSC and 0.5 % SDS at 65 °C, and for high stringency conditions (for detecting HPV-16 specific hybridization) we used 0.1X and 0.5 % SDS at 60

°C. The filter was dried at room temperature and exposed at -70 °C to kodak XAR-5 X ray film with Cronex intensifing screens. Films were developed with the aid of a kodak X-0 mat film automatic processor. We considered positive in this preliminary assay, ten samples of cervical carcinoma, one blood sample (S6) and the sample of condyloma acuminata.

Southern blot analysis (from gel electrophoresis to autoradiography in fig. 1) was carried out in order to gain more insight into the nature of the hybridizing samples. Genomic DNAs (10 ug) were digested overnight with an excess of the endonuclease EcoRI or PstI, concentrated by ethanol precipitation and electrophoresed through a 1.0 % agarose gel prepared in 40 mM Tris (pH 8.0) 4 mM sodium acetate and 1mM EDTA. The DNA in the gel was denatured, neutralized and transferred to nitrocellulose filters as described by Southern (36). The filters were then baked and pretrated with the prehybridization solution and then were hybridized as above. The filters were washed twice in 2X SSC and 0.5 % SDS for 30 min at room temperature, and depending of the stringency conditions desired, they were then either washed twice in the same solution at 50 °C for 30 min each time (low stringency), or twice in 0.1 X SSC and 0.5 % SDS at 60°C, also for 30 min each time (high stringency). After these washes the filters were subjected to autoradiography as above.

Under low stringency conditions (40°C below the melting temperature), we detected positive signals in eleven (54%) of the twenty samples of cervical carcinomas, and in one of the six samples of healthy women. When we rewashed the filters in high stringency conditions (0.1 X SSC at 60°C), seven samples retained the signal, including the sample of the normal woman.

The number of DNA copies of the hybridizing HPV sequences was obtained by comparing the hybridization intensity of the positive samples with the signal generated by pHPV-16 DNA standards also included and run in parallel in the agarose gel. We estimated the hybridization signals comming from the positive samples to have values between 1 and 10 copies per cell.

Six out of the eleven positive samples, equivalent to 30% of the twenty samples of cervical carcinomas, gave still signal under high stringency washing canditions, and thus were classified as positives to HPV-16. These results agree with those observed in U.S.A. (7, 9, 14, 17), Mexico City (21) and in other countries (18, 10, 19, 11). However this HPV-16 frequency in our cervical samples is lower than those reported for Italy (23), England (13, 15), Panama (22) and Brazil (20).

It is very likely that the signal lost when the stringency conditions were rised may be revealing the presence, in the rest of the samples, of papillomavirus types 18, 31 and 35.

When we analyzed the data in relation to the stage of the disease, we found a higher percentage of HPV DNA in the first two stages. On the other hand, our results obtained with respect to the number of DNA copies did not show a correlation with the stage of the ailment.

The presence of one positive sample in the six specimen from healthy women was observed. Although at present we have not investigated further this finding, similar results have been obtained by other investigators (11, 27, 28, 29, 30).

Of the positive samples, we selected nine and digested them with the endonuclease PstI. The samples which showed positive signal in the dot blot experiments were included in this batch of DNAs. The DNAs in the filters were hybridized with the radio-labeled pHPV-16 DNA and washed in low stringency conditions. As results, we observed similar multiple bands resembling the expected pattern for Pstl digestion of HPV-16 DNA; however we also observed differences in band sizes in 6 samples.

One thing that we would like to point out from the results, is the heterogeneity of bands observed in the positive samples from those theoretically expected. We believe that this variation can not be explained by partial digestion of our samples, because we used an excess of restriction enzyme units (calibrated by testing several concentrations of enzyme). In addition we saw no signal that could resemble the three conformational forms of undigested extrachromosomal HPV DNA. The only possible explanations are unusually high polymorphism or integration of the DNA of HPV in the genome of the infected cells.

Finally, we detected a signal in the dot blot experiment from one patient's blood sample. Due to this, we decided to continue studying this sample by cutting its purified DNA with the endonuclease Pstl. Performing Southern blot analysis, we detected a band of approximately 3.9 kb. When the stage of sickness in which the patient might be in was analyzed, it was found to belong to the third stage, which meant that the patient presented expansion of the sickness. Nevertheless, since this was observed with only one of our patients under study, we believe that more patients will be needed before making conclusions out of this observation.

With the increasing number of newly reported papillomavirus types, the reverse-blot technique (26) and the polymerase chain reaction (43) may be considered as better alternatives for the early detection of papilloma virus in carriers using cervical scrapping.

CONCLUSIONS

By Dot blot and Southern blot hibridization techniques, we detected HPV signal in fifty four percent of the cervical samples. The signal in six of them, the equivalent to thirty percent of the twenty samples of cervical carcinomas, was specific of HPV-16.

While we found an inverse correlation between the stage of the disease and the incidence of HPV DNA, no correlation was found between the number of DNA copies of HPV and the stage of the disorder.

The finding of DNA sequences of HPV-16 in one sample of peripheral blood of a patient with cervical carcinoma, may have important implications in the study of the development of this disease.

On the basis of the high percentage of HPV positivity of the cervical carcinomas samples analyzed, we have now included as one of our inmediate goals, to correlate our findings with the type of responses to therapeutic strategies as well as with the evolution that this disease may have in our patients.

ACKNOWLEDGMENTS

We would like to express our gratitude to Dr. Patricio Gariglio and Dr. Grady Saunders for gifts of materials. Our gratitude also extends to all those members of both U.L.I.E.G. and those pertaining to Dr. Patricio Gariglios' laboratory, that helped us during this work.

M.L.G.G. thanks to National Council of Science and Technology (CONACYT) of the Mexican Goverment for support through thesis scholarship.

This project was supported by grants from CONACYT and Fondo de Estudios e Investigación Ricardo J. Zevada.

H.A.B.S. would like to thank the Faculty of Medicine of the Universidad Autónoma de Nuevo León for their continuous support.

REFERENCES

1.- R. Peto, Introduction: Geographic Patterns and Trends. "Viral Ethiology of Cervical Cancer". R. Peto and Zur Hausen Harald, eds., Cold Spring Harbor Laboratory, New York. (1986).

2.- A. Singer and P. French. Natural History and Epidemiology of Cervical Carcinoma. "Cancer of the Uterine Cervix". D. C. H. Mc Brien and T. F. Slater. eds., Academic Press. London (1984).

3.- J. Rodríguez-Peral. Carcinoma cervicouterino en el Noroeste de México. Aspectos clínicos. Rev. Med. I. M. S. S. 21: 183 (1983).

4.- M. P. Vessey. Epidemiology of Cervical Cancer: Role of Hormonal Factors, Cigarette Smoking and Occupation. "Viral Ethiology of Cervical Cancer". R. Peto and Zur Hausen Harald, eds., Cold Spring Harbor Laboratory, New York. (1986).

5.- M. Dürst, L. Gissmann, H. Ikenberg and H. zur Hausen. A papilloma virus DNA from a cervical carcinoma and its prevalence in cancer biopsy samples from different geographic regions. Proc. Natl. Acad. Sci. U. S. A. 80: 3812. (1983).

6.- M. Boshart, L. Gissmann, H. Ikenberg, A. Klenheinz, W. Scheurlen and H. zur Hausen. A new type of papilloma virus DNA, its presence in genital cancer biopsies and in cell lines derived from cervical caner. EMBO. J. 3: 1151 (1984).

7.- A. T. Lorincz, W. D. Lancaster, R. J. Kurman, A. B. Jenson and G. F. Temple. Characterization of Human Papillomaviruses in Cervical Neoplasias and Their Detection in Routine Clinical Screening. "Viral Ethiology of Cervical Cancer". R. Peto and Zur Hausen Harald, eds., Cold Spring Harbor Laboratory, New York. (1986).

8.- A. T. Lorincz, A. P. Quinn, W. D. Lancaster and G. F. Temple. A New Type of Papillomavirus Associated with Cancer of the Uterine Cervix. Virology 159:187 (1987).

9.- R. Dekmesian, X. Chen, T. Kuo, N. Ordoñez, R. L. Katz. DNA Hybridization for Human Papillomavirus (HPV) in Cervical Lesions. Arch Pathol. Lab. Med. 111:22. (1987).

10.- M. Dürst, L. Gissmann, H. Ikenberg and H. zur Hausen. A papilloma virus DNA from a cervical carcinoma and its prevalence in cancer biopsy samples from different geographic regions. Proc. Natl. Acad. Sci. U. S. A. 80: 3812. (1983).

11.- D. W. M. Millan, J.A. Davis, T.E. Torbet and M. S. Campo. DNA sequences of human papillomavirus types 11, 16, and 18 in lesions of uterine cervix in the west Scotland. Br. Med. J. 293:93. (1986).

12.- H. Shirasawa, Y. Tomita, K. Kubota, T. Kasai, S. Sekiya, H. Takamisawa and B. Simuzu. Detection of human papilloma virus type 16 DNA and evidence for integration into the cell DNA in cervical dysplasia. J. Gen. Virol. 67: 2011. (1986).

13.- J. M. Scholl, E. M. Kingsley-Pillers, R. E. Robinson and P. J. Farrell. Prevalence of human papilloma virus type 16 DNA in cervical carcinoma samples in East Anglia. Int. J. Cancer 35: 215. (1985).

14.- M. Fukushima, T. Okagaki, L. B. Twiggs, B. A. Chark, K. R. Zachow, R. S. Ostrow and A. J. Faras. Histological types of carcinoma of the uterine cervix and the detection of human papilloma virus DNA. Cancer Res. 45: 3252. (1985).

15.- J. C. M. Macnab, S. A. Walkinshaw, J. W. Cordiner and J. B. Clements. Human papillomavirus in clinically and histologically normal tissue of patients with genital cancer. N. Engl. J. Med. 315:1052. (1986).

16.- H. Lehn, P. Kriegy, G. Saver. Papilloma virus genomes in human cervical tumors: analysis of their transcriptional activity. Proc. Natl Acad. Sci. U. S. A. 82: 5540. (1985).

17.- R. S. Ostrow, D. A. Manias, B. A. Clark, T. Okagake, L. B. Twiggs and A. J. Faras. Detection of Human Papillomavirus DNA in Invasive Carcinomas of the Cervix by In Situ Hybridization. Cancer Res. 47:649. (1987).

18.- H. Yoshikawa, T. Matsukura, E. Yamamoto, T. Kawana, M. Mizuno and K. Yoshike. Ocurrence of human papilloma virus types 16 and 18 DNA in cervical carcinomas from Japan: age of patients and histological type of carcinomas. Jpn. J. Cancer Res. 76: 667. (1985).

19.- Y. Tsunokawa, N. Takebe, S. Nozawa, T. Kasamatsu, L. Gissmann, H. zur Hausen, M. Tereda and T. Sugimura. Presence of human papilloma virus type 16 and type 18 DNA sequences and their expression in cervical cancer and cell lines from japanese patients. Int. J. Cancer 37: 499. (1986).

20.- D. J. Mc Cance, A. Kalache, K. Ashdown, L. Andrade, F. Menezes, P. Smith and R. Doll. Human Papilloma virus types 16 and 18 in carcinomas of the penis from Brazil. Int. J. Cancer 37: 55 (1986).

21.- P. Gariglio, R. Ocadiz and R. Sauceda. Human papiloma virus DNA sequences and c-myc oncogene alterations in uterine cervix carcinoma. Cancer Cell. 5:343 (1987).

22.- S. S. Prakash, W. C. deBritton and W. E. Rawls. Herpes simplex virus types 2 and human papilloma virus type 16 in cervicitis, dysplasia and invasive cervical carcinoma. Int. J. Cancer 35: 51. (1985).

23.- D. DiLuca, S. Pilotti, B. Stetanon. A. Rotola, P. Momimi, M. Tognon, G. de Palo, F. Rilke and E. Cassai. Human papilloma virus type 16 DNA in genital tumors; A pathological and molecular analysis. J. Gen. Virol. 67: 583. (1986).

24.- Y. K. Tomita, T. Kasai, S. Sekiya, H. Takamizawa and B. Simuzu. Detection of Human Papillomavirus DNA in Genital Warts, Cervical Dysplasias and Neoplasias. Intervirology 25:151 (1986).

25.- C. P. Crum, M. Mitao, R. U. Levine and S. Silvestein. Cervical papilloma viruses segregate whithin morphologically distinct precancerous lesions. J. Virology. 54: 675. (1985).

26.- H. K. Fife, R. E. Rogers and B. W. Zwickl. Symptomatic and Asymptomatic Cervical Infections with Human Papillomavirus During Pregnancy. J. Infect. Dis. 156:904. (1987).

27.- A. Schneider, H. Kraus, R. Schuhmann and L. Gissmann. Papilloma virus infection of the lower genital tract: detection of viral DNA in gynecological swabs. Int. J. Cancer 35: 443. (1985).

28.- D. H. Webb, R. E. Rogers and K. H. Fife. A One-Step Method for Detecting and Typing Human Papillomavirus DNA in Cervical Scrapr Specimens from Women with Cervical Dysplasia. J. Infect. Dis. 156:912. (1987).

29.- P. G. Toon, J. R. Arrand, L. P. Wilson and D. S. Sharp. Human papillomavirus infection of the uterine cervix of woman without cytological signs of neoplasia. Br. Med. J. 293:1261. (1986).

30.- M. F. Cox, C. A. Meanwell, N. J. Maitland, C. Blackledge, C. Scully and J. A. Jordan. Human papillomavirus Type 16 homologous DNA in Normal Human Ectocervix. Lancet. 8499:157. (1986).

31.- K. Seedorf, G. Krammer, M. Durst, S. Suhal and W. G. Rowekamp. Human Papillomavirus Type 16 DNA Sequence. Virology 145:181. (1985).

32.- D. B. Clewel Nature of Col E₁ plasmid replication in Escherichia coli in the presence of chloramphenicol J. Bacteriol 110:667 (1972).

33.- A. P. Feinberg and B. Vogelstein. A Technique for radiolabeling DNA
     restriction Endonuclease fragment to high specific activity. Anal.
     Biochem. 132:6. (1983).

34.- F. C. Kafatos, C. W. Jones and A. Efstatiadis. Determination of
     nucleic acid sequence homologies and relative concentrations by a dot
     hybridization procedure. Nucl. Acids Res. 7:1541. (1979).

35.- D. T. Denhardt. A membrane-filter technique for the detection of
     complementary DNA. Biochem. Biophys. Res. Commun. 23:641-646. (1966).

36.- E. M. Southern. Detection of specific sequences among DNA fragments
     separated by gel electrophoresis. J. Mol. Biol. 98: 503. (1975).

37.- H. W. Jones and G. S. Jones. "Tratado de Ginecología de Novak,"
     Interamericana. Mexico (1985).

38.- K. E. Davies,"Human genetic diseases". IRL PRESS. Oxford (1986).

39.- P. L. Iversen, J. E. Mata and R. N. Hines. Rapid isolation of both
     RNA and DNA from cultured cells or whole tissues with a benchop
     ultracentrifuge. Biotechniques 5:521. (1987).

40.- E. Seto, and T. S. Benedict-Yen. Detection of Cytomegalovirus
     infections by means of DNA isolated from paraffin-embedded tissues
     and dot hybridization. Am. J. Pathol. 127:409. (1987).

41.- C. Gautreau, C. Rahuel, J. Carton and G. Lucotte. Comparison of Two
     Methods of High-Molecular-Weight DNA Isolation from Human Leucocytes.
     Anal. Biochem. 134:320. (1983).

42.- R. A. Gatti, P. Concannon, W. Salser. Multiple Use of Southern Blots.
     Biotechniques. 1:148 (1984).

43.- D. K. Shibata, N. Arnheim and J. Martin. Detection of Human Papilloma
     virus in Paraffin-Embedded tissue using the polymerase chain reaction
     J. Exp. Med. 167:225. (1988).

POPULATION DYNAMICS OF AIDS: BASIC CONCEPTS

Marco V. José

Director of Epidemiology, Centro de Investigaciones
sobre Enfermedades Infecciosas, Instituto Nacional
de Salud Pública. Fco. de P. Miranda 177, 5° piso,
Unidad Plateros 01480, México, D.F.

Jesús Kumate

Vice-Minister of Health, Subsecretaría de Servicios
de Salud, Lieja 7, 1er Piso, Cuauhtémoc, 06696,
México, D.F., México

INTRODUCTION

Considerable progress related to the molecular, gene-
tic,[1,2,3,] clinical[4] and epidemiological[5,6] aspects of acquired
immunodeficiency syndrome (AIDS) has been achieved in a very
short period of time. However, the current knowledge about the
epidemiological characteristics of the virus transmission and
persistence within human communities seems to be not well understood. In the absence of immediate help of effective drugs or
vaccines, prevention is the only available strategy. Yet many
politicians behave as if AIDS were just a temporary problem
without considering the cyclic nature of most infectious diseases and disregarding that the epidemic has not even clearly reached its peak incidence. It is true that the population-level
data which is available is indeed from a recent epidemic and
the time scales of many of the epidemiological parameters are
of the order of years.

The elaboration of mathematical models, even simple ones,
may permit us to make some predictions about the course of
the infection and to understand more about the dynamic nature
of the disease. This is important since AIDS disease is not
only infectious but it is highly lethal a combination of characteristics almost unknown in the recent history of pandemics.

In this chapter we present the basic notions of relatively
simple mathematical models about the population dynamics of
AIDS with which reliable long term predictions can be made.
We emphasize that although an accurate modelling of the spread
of AIDS is difficult because of the large number of parameters
required, even these simplified models provide useful concepts
about the dynamical nature of the diseases. The chapter is

organized as follows: the first part begins with a survey of
the basic epidemiological parameters which are necessary to
describe the population dynamics of AIDS. We next formulate a
basic mathematical model with which we get an approximate estimate of the latent period. A predicted pattern about the course
of the epidemic considering the heterogeneity in sexual activity
is illustrated and a relationship between the total fraction
eventually infected and peak values of incidence is obtained.

Basic Epidemiological Concepts

One of the most important questions that epidemic models
can address is: What are the necessary conditions for an
epidemic to become established? The central concept of
epidemiological models is the intrinsic reproductive rate of
the infection usually denoted by R_0[6,7]. The term R_0 is defined
as the average number of secondary infections produced by one
infected individual in a wholly susceptible population.[6,8]

Considering AIDS as a sexually transmitted disease (STD),
we have[8]

$$R_0 = \beta c(t)D, \qquad (1)$$

Where β is the average probability that a given susceptible will acquire the infection from an infected individual;
$c(t)$ denotes the average rate at which NEW sexual partners are
acquired and is not simply the expected number of partners
$E(i)$, but rather is[8,9]

$$c(t) = \frac{E(i^2)}{E(i)} = E(i) + \frac{Var(i)}{E(i)} \qquad (2)$$

Here $E(i^n)$ is the nth moment of the distribution in the
number of sexual partners, i, and $Var(i)$ is the variance of
that distribution; the term D denotes the average duration of
infectiousness; this assumes that on average, a single value
can characterize all infectives regardless of whether they will
go on the get full-blown AIDS; then the rate at which infectives move out from the infectious class is $v=D^{-1}$.

It is still not known why only a fraction of those infected do develop full-blown AIDS. Let f be the fraction of infecteds who go on eventually to develop full-blown AIDS; whether all those infected with the human immunodeficiency virus
(HIV) are moving towards AIDS at different rates while other
never will, is not still known. Among the unknowns are such
factors as the duration of the latent period denoted $L= \mu^{-1}$,
i.e. the time that elapses from initial infection to the point
where an infected person is infectious to other susceptibles.
Whether the presence of detectable levels of virus infection
indicates infectiousness is not clear at presente.

In contrast, the incubation period s is defined as the
interval between the point of acquisition of the infection
and the point of appearance of symptoms of full-blown AIDS.

Mathematical Model for a Homogeneous Epidemic

We first consider a very simple mathematical framework which mirrors the dynamics of AIDS as a STD of homosexual men mixing homogeneously. The same concepts are involved for the groups at risk (e.g. drug-related cases, heterosexual transmitted AIDS or perinatal AIDS).

Even with this oversimplified framework we will show how various observed factors influence the dynamics of disease persistence and spread.

Let us consider a closed population of fixed size N in which the density of susceptibles, infectious and AIDS-patients at time t are denoted by X, Y and Z; then the total population size is $N=X+Y+Z/f$.

We ignore AIDS-related deaths since the effects of mortality are likely to show up only after most people have been infected. Recruitment is also ignored. We assume that the population is closed on the time scale of the epidemic.

The model is[10]

$$\frac{dx}{dt} = -\lambda X \tag{3}$$

$$\frac{dy}{dt} = \lambda X - vY \tag{4}$$

$$\frac{dz}{dt} = fvY \tag{5}$$

Here, λ is the so-called force of infection, and as in other STD, is the probability that infection is acquired from any new partner; then λ is defined as

$$\lambda = \beta \ c(t)\left(\frac{Y}{N}\right) \tag{6}$$

The most common data we have is about the initial rise (approximately exponential) of the incidence of diagnosed cases of full-blown AIDS. Thus in the early stages of the epidemic the number of susceptibles is approximately equal to N, this is $X \simeq N$, and equations (4) and (6) give

$$\frac{dY}{dt} = [\ \beta \ c(t) - v\ ]\ Y \tag{7}$$

That is, the incidence of infection, and thence the incidence of diagnosed AIDS (i.e. vY), is expressed as

$$Y(t) = Y(o) \ \ EXP \ \ [(\beta c(t) - v)t] \tag{8}$$

Similarly, in the early stages of the epidemic, if there are c(t) cases at time t, there will be (Ro-1) c(t) cases and interval of time D later and therefore

$$\frac{dc(t)}{dt} = \frac{(Ro-1)}{D} c(t)dt \tag{9}$$

Integrating equation (9) gives

$$c(t) = c(o) \; EXP\left(\frac{(Ro-1)}{D}\right)t \tag{10}$$

Now let Λ be the compound interest rate of increase in the number of diagnosed cases in the initial phases in which there is roughly exponential growth, then

$$\Lambda = \frac{a}{t_d} \tag{11}$$

Where t_d is the doubling time and the constant $a=ln2\simeq0.7$. Thus we can identify either from equations (8) or (10) the exponential growth rate, Λ, as being related to Ro by

$$\Lambda = \frac{(Ro-1)}{D} \tag{12}$$

Therefore the initial doubling time, t_d, either for infection (as revealed by seropositivity) or for diagnosed cases of full-blown AIDS is approximately

$$t_d = \frac{a.D}{Ro-1} \tag{13}$$

Alternatively we can write

$$Ro = 1 + D \tag{14}$$

The latent period

Let us now include a latent class of individuals, denoted by H. Let μ be the rate of transition through this latent class. Then the average duration of the latent period is $L = \mu^{-1}$. The basic set of equation now become

$$\frac{dX}{dt} = -\lambda X \tag{15}$$

$$\frac{dH}{dt} = \lambda X - \mu H \tag{16}$$

$$\frac{dY}{dt} = \mu H - vY \tag{17}$$

346

$$\frac{dZ}{dt} = fvY \qquad (18)$$

Again this set of equations reduces to a relatively tri-vial set of linear equations when we assume that $X \approx N$ and we obtain

$$\frac{dH}{dt} = \alpha Y - \mu H \qquad (19)$$

Where $\alpha = \beta c(t)$. Simple manipulation of equations (17) and (19) leads to a second-order linear equation in Y (Ricatti equation) of the form

$$Y'' + b_1 Y' + b_2 Y = 0, \qquad (20)$$

where $b_1 = (v + \mu)$ and $b_2 = \mu(v - \alpha)$. Equation (20) has the standard solution

$$Y(t) = A \, EXP \, (\Lambda_1 t) + B \, EXP \, (\Lambda_2 t), \qquad (21)$$

Where A and B are determined by the initial conditions $Y(0) = b_1$ and $Y'(0) = b_2$ and $A = (b_1 \Lambda_2 - b_2) / (\Lambda_2 - \Lambda_1)$ and $B = (b_2 - b_1 \Lambda_1) / (\Lambda_2 - \Lambda_1)$.

The rate parameters Λ_1 and Λ_2 are determined by the quadratic equation

$$\Lambda^2 + (\mu + v)\Lambda + (v - \alpha) = 0 \qquad (22)$$

The two solutions of equation (22) are

$$\Lambda_{1,2} = \frac{1}{2} \left\{ -(\mu + v) \pm \left\{ (\mu + v)^2 + 4\mu(\alpha - v) \right\}^{\frac{1}{2}} \right\} \qquad (23)$$

There will be only one positive value of Λ provided $\alpha > v$, otherwise, both Λ_1 and Λ_2 will be negative, and the infection will not be able to sustain itself. Conversely if $\alpha > v$, the positive route will dominate and eventually lead to exponential growth rate given by Λ_1. This means that the doubling rate initially observed is approximately related to the initial value of α ($\alpha = \beta c(0)$) by the relationship

$$\Lambda = -\frac{1}{2}(\mu + v) + \frac{1}{2} \left\{ (\mu + v)^2 + 4\mu(\alpha - \mu) \right\}^{\frac{1}{2}} \qquad (24)$$

The eradication criterion corresponds to changes in sexual habits such that

$$c(t) \rightarrow x_c \, c(o) \qquad (25)$$

Where x_c is a critical fraction of susceptible individuals. Thus the infection will be unable to maintain itself if $\alpha < v$; that is, decline follows if $x < x_c$ with

$$x_c = \frac{v}{\alpha} \qquad (26)$$

347

Then the critical fraction of susceptible individuals is

$$x_c^{-1} = 1 + \frac{\Lambda(\Lambda+\mu+v)}{\mu v} \qquad (27)$$

Notice that once one includes a latent class we have a quasi-exponential growth.

Considering that $D^{-1} = v$, $L^{-1} = \mu$ and $Ro = x_c^{-1}$ we get from equation (27)

$$Ro = (1+\Lambda D)(1+\Lambda L) \qquad (28)$$

Estimates of Ro are between 5 and 10, $D \cong 5$ years and $\Lambda \cong 1$ year^{-1} [8], therefore the latent period should range from few days up to 8 months.

We want to highlight the fact that Ro is between 5 and 10 and this means that according to observed patterns each infected individual is infecting, on average, 5 or more susceptible individuals.

Mathematical Model for a Heterogeneous Epidemic

We will now consider the fact that there is a distribution of numbers of sexual partners which may be characterized at time t by the probability for a randomly chosen individual to have i sexual partners per unit time, $F(i)$. We can then divide the total at risk population, N, into subgroups, N_i whose members have on average i sexual partners per unit of time, then[8,9]

$$N_i = N.F(i) \qquad (29)$$

The number of susceptibles, infectious and AIDS-patients in the ith class are X_i, Y_i and Z_i, respectively; the probability of acquiring infection is taken to be i number of partners times λ. Thus equations (2) and (4) generalize to

$$\frac{dX_i}{dt} = -i\,\lambda\,X_i \qquad (30)$$

$$\frac{dY_i}{dt} = i\lambda X_i - vY_i \qquad (31)$$

The force of infection, λ, depends on β and on the probability that a partner is infectious; weighting partners by their degree of sexual activity, we have

$$\lambda = \beta \sum_i iY_i \Big/ \sum_i iN_i \qquad (32)$$

A computer simulation of the course of the epidemic of AIDS in a closed population of homosexually active males (N=831) utilizing the model described by equations (30) and (31) is shown in Figure 1. Assuming a threshold value of Ro=6 and a rate of acquiring new partners of c(t)=1, the duration of all the epidemic would be of about 25 years (curve of infecteds) and the population of susceptibles (curve of susceptibles) would be scanty after 12 years of initiated the epidemic. Note that there is a slower rate of appearance of new cases once the epidemic reaches its highest incidence. In this computer simulation a small value of c(t) was considered (estimates of c(t) ranges from 1 to more than 50 partners per year[11]) and the incubation period was not taken into account. A larger value of c(t) would reduce the rate of appearance of the cases. According to equation (2) both the mean and variance of sexual activity contribute to determine the magnitude of the epidemic within a community. Highly promiscous individuals are quickly infected and therefore the less active individuals may remain uninfected for longer periods of time. Thus, programs targetted to modify the behaviour of the most active individuals would reduce substantially Ro, eventually to less than unity.

The incubation period would augment the duration of the epidemic, i.e., there would be a slower decline of cases after the peak incidence.

Total Fraction Infected and Peak Incidence

The epidemic of AIDS may currently be at or near its peak incidence. Thus, a relationship between the total fraction likely to get full-blown AIDS, I, and the peak value of incidence, P, may be useful.

Fo a homogeneous population in which all the individuals have the same epidemiological characteristics I depends only on Ro according to[6]

$$I = 1 - EXP(-IRo) \qquad (33)$$

Unless Ro is close to unity, which is a priori unlikely, the peak incidence in a relatively homogeneous population obeys the relationship

$$P \sim \frac{f}{D} \qquad (34)$$

The peak incidence is expressed as a fraction of the total population at risk.

Considering the basic equations (3) and (4) note that Y is stationary when $\lambda X = vY$, this is when Rox=1, where x=X/N. Dividing equation (3) into equation (4) we get

$$\frac{dY}{dX} = -1 + \left(\frac{N}{Rox} \right) \qquad (35)$$

Integrating from t=0 to t=t gives

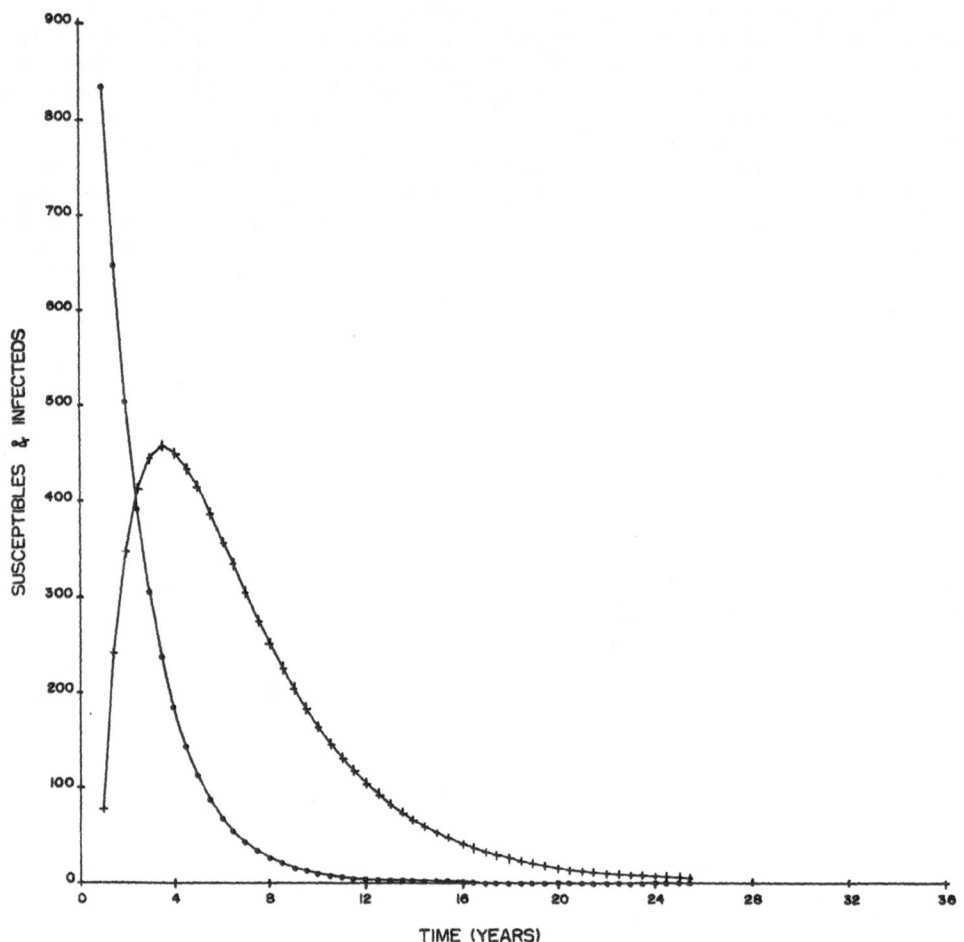

Figure 1. Prediction of the dynamics of susceptibles
(•) and infecteds (+) utilizing the model described by
equations (30) and (31) which assume heterogeneity in the
sexual activity of a closed population of homosexual. The
values of the parameters are: $v=0.17$; $R_o=6$; $c(t)=1$; $\lambda=0.5$;
the initial number of susceptibles is $X(.) = 831$.

$$Y(t) = N - X(t) + \left(\frac{N}{Ro}\right) \ln\left(\frac{X}{N}\right) \qquad (36)$$

The total fraction of infecteds reaches its maximum when Ro = N/X, thus

$$I_{MAX} = \left(\frac{Y}{N}\right)_{MAX} = 1 - [1+lnRo] / Ro \qquad (37)$$

We have now a relationship between I_{MAX} and P expressed as

$$I_{MAX} = P \left(\frac{f}{D}\right)^{-1} \qquad (38)$$

If we take into account the incubation period of duration S, then the relation in equation (34) is modified to

$$P \sim \frac{f}{D+S} \qquad (39)$$

The heterogeneity in sexual behaviour can cause further modifications, resulting in the approximate generalization

$$P \sim \left(\frac{f}{D+S}\right) \left(\frac{(E(i))^2}{E(i^2)}\right) \qquad (40)$$

Alternatively equation (40) can be written as

$$P \sim \frac{f}{(D+S)\ (1+CV^2)} \qquad (41)$$

That is, the homogeneous result expressed by equation (39) is reduced by a factor $[1+(\text{coefficient of variation})^2]^{-1}$

CONCLUSION

The kind of simple models presented here should not serve as a substitute for a detailed analysis of particular life histories of individual communities. Nevertheless it is the thinking underlying these models that may assist health decision makers for the adoption of population policies in efforts to ameliorate the negative effects of AIDS epidemic. Given the many uncertainties sorrounding the epidemiology of the disease it would be improper to draw more than the general implications of the results presented here. Predicted patterns of AIDS epidemic, (as the one shown in Figure 1), based on relatively simple assumptions may serve as a rough guide for grasping the nature of observed facts. It is noteworthy to mention that

AIDS disease has come to stay.

The latency period is an important factor which determine the time required to observe heterosexual transmission. The fact that the interval between initial infection and the beginning of infectiousness is up to 8 months implies that individuals DO NOT generally become infected many years after their partners' infection. Thus observable transmission from female to male to female may occur even in the absence of large numbers of infectious women into the population.

Estimates of the initial doubling times of the epidemic is related to the basic reproductive rate of the infection, Ro (see eq. (13)). One of the major unknowns is the proportion of infecteds who go on to develop full-blown AIDS and the heterogeneity in sexual behaviour. These two parameters may be inferred from information about the peak incidence of the epidemic (see eq. (41)).

Epidemic models must now focus in determining which changes in sexual behaviour (or drug use) would have the greatest impact in halting the spread of the infection.

Acknowledgment

Marco V. José thanks helpful advises from Prof. Robert M. May.

REFERENCES

1. Barre-Sinoussi, F., Chermann, J.C., Rey, F., Nugeyre, M.T., Chamaret, S., Gruest, J., Dauguet, C., Axler-Blin, C., Vezinet-Brun, F., Rouzioux, C., Rozenbaum, W., Montagnier, L., Isolation of a T-lymphotropic retrovirus from a patient at risk for acquired immune deficiency (AIDS), Science, 220:868-871 (1983).
2. Gallo, R.C., Salahuddin, S.Z., Popovic, M., Sheaver, G.M., Kaplan, M., Haynes, B.F., Palker, T.J., Redfield, R., Oleske, J., Safai, B., White, G., Foster, P., Markham, P.D. Frequent detection and isolation of cytopathic retroviruses (HTLV-III) from patients with AIDS and at risk for AIDS.
3. Hahn, B.H., Shaw, G.H., Taylor, M.E., Redfield, R., Markham, P.D., Salahuddin, S.Z., Wong-Staal, F., Gallo, R.C., Parks, E.S., Parks, W.P., Genetic variation in HTLV-III/LAV over time in patients with AIDS for at risk for AIDS, Science, 232:1548-1553 (1986).
4. Centers for Disease Control: Classification system for human T-lymphotropic virus type III/Lymphadenopathy-Associated virus infections. Ann. Intern. Med. 105:234-237 (1986).
5. Lui, K.J., Lawrence, D.N., Morgan, W.H., Peterman, T.A., Haverkos, H.W., Bregmann, D.J., A model-based approach for estimating the mean incubation period of transfussion-associated acquired immunodeficiency syndrome, Proc. Nat. Acad. Sci., 83:3051-3055 (1986).
6. Kermack, W.O., Mckendrick, A.G., A Contribution to the

      mathematical theory of epidemics, Proc. Roy. Soc. London, 115A:700-721 (1927).

7.   Bailey, N.T.J., The Mathematical Theory of Infectious Diseases, Macmillan, New York, ed. 2, 1975.

8.   May, R.M., Anderson, R.M., Transmission dynamics of HIV infection, Nature, 326:137-142 (1987).

9.   May, R.M., The Search for Patterns in the Balance of Nature: Advances and Retreats, Ecology 67:1115-1126 (1986).

10.   Anderson, R.M., Medley, G.F., May, R.M., Johnson, A.M.; A Preliminary Study of the Transmission Dynamics of the Human Immunodeficiency Virus (HIV), the Causative Agent of AIDS, IMA J. Math. Applied Med. Biol. 3:229-263 (1986).

11.   Carne, C.A., Weller, I.V.D., Sutherland, S., Cheinsong-Popov, R., Ferns, R. B., Williams, P., Mindel, A., Tedder, R., Adler, M.S., Rising prevalence of human T-lymphtropic virus type III (HTLV-III) infection in homosexual men in London. The lancet i:1261-1262 (1985).

STRUCTURE-BASED TARGETING OF DRUGS TO HIV PROTEINS

Frederick A. Eiserling, David S. Eisenberg, Susan Horvath+
and Gary Fujii

Molecular Biology Institute, University of California
Los Angeles, CA 90024
+California Institute of Technology, Pasadena, CA 90061

INTRODUCTION

During the past two years a major effort has begun in the USA, France, Germany, U.K. and other countries to develop programs of rational drug design in the fight against Acquired Immunodeficiency Syndrome (AIDS). Rational drug design is based on the idea that the detailed biological and structural analysis of the life cycle of the pathogen, the Human Immunodeficiency Virus (HIV) can locate those sensitive events where drugs can block that cycle, and that specific drugs can be developed through a detailed knowledge of the structure of these sensitive components at atomic resolution. This is also known as structure-based drug targeting. The life cycle of the virus has been described in detail elsewhere (for example, Fauci, 1988). The obvious sensitive steps are adsorption, entry, uncoating, reverse transcriptase activity, integration, the five or six regulatory gene functions, protease activity, and the steps in virus assembly. Figure 1 shows the viral structure, and labels several identified components.

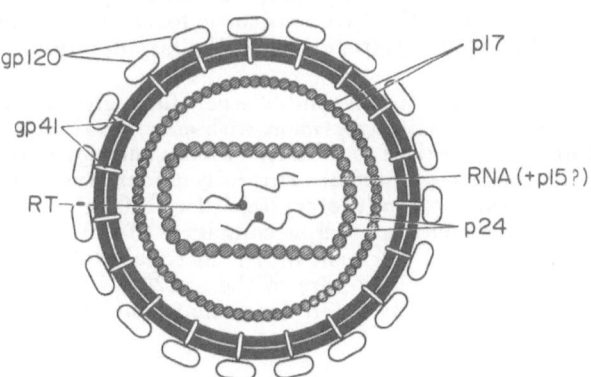

Fig. 1. Model of the structure of HIV based on work by Gelderblom (1987) and others.

The U.S. National Institutes of Health has a number of programs designed to combat the AIDS epidemic, primarily in the National Institute of Allergy and Infectious Diseases (NIAID) and the National Cancer Institute. About a year ago, a new program was announced in the Institute of General Medical Sciences for a

"Program For Structural Studies Related to AIDS", under the direction of Dr. Marvin Cassman. That program now supports the activities of eight to ten research groups dedicated to providing a structural basis for the design of drugs. This program is in addition to the one already underway in the NIAID called the National Cooperative Drug Discovery Groups for the treatment of AIDS. Together, there are at least 30 major research groups, including x-ray crystallographers, nuclear magnetic resonance spectroscopists, synthetic organic chemists, computer-aided molecular design chemists, molecular biologists and virologists, working in this area, involving both universities and biotechnology industries. The purpose of this communication is to describe the nature of our HIV-related projects, and some preliminary progress, from our Structure Group at the University of California, Los Angeles.

OVERVIEW

One of our main projects uses a combination of x-ray diffraction and electron microscopy to study the structure of the HIV envelope surface glycoprotein, gp120, and to examine the envelope (env) gene products gp120 and gp41 by sequence analysis methods. Our collaborator, Dr. L. Lasky of Genentech, has engineered a soluble, active analog of HIV glycoprotein gp120, and isolated it in highly pure form in large amounts. The HIV glycoprotein coat gp120 contains roughly 450 amino acids, with 24 asparagine-linked N-glycosylation sites, and an N-terminal signal-sequence-like domain. The HIV gp120 is derived in vivo from a gp160 precursor by cleavage toward the C-terminus to produce gp120 and the transmembrane gp41. Lasky has created a C-terminus several amino acids from the gp41 cleavage site, and has substituted 25 amino acid residues from the herpes simplex virus glycoprotein D sequence at the N-terminus of gp120, to give a secretion signal.

This reengineered protein is biosynthesized in chinese hamster ovary cells and secreted constitutively. The system produces copious gp120 (15-20 mg/liter of culture), which is purified to 99% purity on a monoclonal affinity column. The molecular mass of the glycoprotein consists of 53,000 Da of amino acids, a nearly equal mass of neutral mannose, and about one fifth the mass of sialic acid residues.

The gp120 protein is soluble and Lasky has shown that even when iodinated, it binds to the cellular T4 (CD4) lymphocyte receptor with a high affinity. This is a a major biological function and shows that the reengineered protein retains the correct fold of the viral surface glycoprotein. The protein also retains the 17 invariant Cys residues that form 8 invariant disulfides, and this probably accounts for the finding that the protein is stable to heating. On isoelectric focusing gels, this gp120 displays several bands, suggesting some heterogeneity in the glycosylation.

The gp120 molecule is extreme soluble when fully glycosylated and does not crystallize in any useful way. After treatment with enzymes that partially deglycosylate the protein it has been possible to produce ordered aggregates that have potential for further structural analysis. These are currently under study in our laboratories. Projects are also underway to produce crystals of gp120 in complexes with monoclonal antibody (Fab) fragments, as well as complexes of gp120 with the cellular T4 (CD4) receptor or a functional fragment of T4, in collaboration with Dr. L. Hood at Caltech. A knowledge of the structure of the gp120-T4 receptor complex would be especially useful in designing targeted drugs that could block attachment of the virus.

Another sensitive step in the virus life cycle is the reverse transcription of viral RNA into DNA, and integration of the newly synthesized double strand DNA into the host genome as a provirus. Successful blockage of this step would inhibit virus reproduction. This is a promising approach since host cells do not contain reverse transcriptase and inhibition could be highly specific for the viral protein. The most successful nucleoside analog AZT, has serious toxicity and more effective treatment is needed. Dr. Richard Dickerson's group has begun a collaborative effort with Drs. Edward Arnold and Stephen Hughes to produce and crystallize HIV reverse transcriptase. The first protein samples have arrived and crystallization efforts are underway.

The goals of this project are to study crystallographically the specific interactions between drugs and macromolecules, in particular short oligomers of DNA, in order to learn how a specific drug molecule "reads" a DNA sequence or a binding site on a protein, with the goal of being able to design new drugs that can be targeted narrowly to a particular macromolecule. The plan is to solve the crystal structure of one or more reverse transcriptases from retroviruses, including but not limited to that from HIV, in order to understand the structure and function of what appears to be an evolutionarily related family of DNA polymerases. The enzymes are being studied alone and in complexes with potential inhibitors and abortive substrate analogues. The main goal is to use the knowledge of drug specificity, and of reverse transcriptase structure, to design synthetic drug molecules capable of blocking the reverse transcriptase activity of HIV, and to prevent the replication of the AIDS virus in vivo.

The theoretical basis of how drugs bind to proteins requires further study and development. Dr. Douglas Rees is working on these problems. Recognition and specificity in macromolecular binding are achieved through complementary interactions between the molecular surfaces of the associating molecules. Rational drug design for the treatment of AIDS must be based on identification of structural features of the components of HIV which will allow effective targeting of drugs against the virus. Dr. Rees is looking at the identification of target zones on proteins. These zones are regions of proteins which are promising sites for directing specific ligands. His analysis is based on characterization of the surface roughness of proteins. Previous work suggests that tight binding sites on proteins have atomically rough surfaces. New approaches to establishing the interaction region between two molecular surfaces ("docking") are being developed. The project also involves an evaluation of the energetics of interactions between surfaces. Effects on protein–ligand binding energetics of surface hydrophobicity, roughness, and membranes are being characterized. The theoretical results will be applied to problems in drug design.

Dr. David Sigman is examining the direct and specific chemical attack of phenanthroline copper complexes on controlling sites of the HIV genome. The 1,10-phenan-throline-cuprous complex with hydrogen peroxide as a co-reactant makes single-stranded nicks in RNA and DNA by oxidatively degrading the ribose moiety. The scission chemistry is independent of the base attached to the ribose moiety but depends on local sequence. The reaction efficiently inactivates many membrane enveloped DNA and RNA viruses. The antiviral activity can be attributed to degradation of the genome as well as inhibition of RNA and DNA polymerases due to the formation of oligonucleotide products with 3'phosphomonoester termini. Dr. Sigman's laboratory has developed the capability of targeting this nuclease activity using oligonucleotide and DNA binding proteins as carriers. He is using three approaches to develop antiviral agents based on this nuclease activity. First, the intrinsic reactivity of an important control region of HIV will be examined with the unsubstituted complex to characterize its structure and to determine if pharmacologically active DNA ligands can enhance the nuclease activity. Secondly the procaryotic Ara C protein has been transformed into a nucleolytic agent because there is a sequence in the viral genome which closely matches a known binding site of this protein. Thirdly, Dr. Sigman is directing these nucleolytic agents to viral transcripts by linking 1,10-phenanthroline to complementary nonhydrolyzable methyl phosphonate oligonucleotides. These nucleolytic activities, which have been targeted to an HIV control region, are being directed to T-lymphocytes by attachment to HIV envelope protein and peptides, and anti-receptor antibodies.

Together, these projects are not moving forward at as rapid a pace as we would like, in part because of the extreme scarcity of HIV proteins. Although many ingenious ways have been found to make HIV products, in bacteria, yeast and cultured animal cells, at some point the HIV proteins will have to be purified from the virus itself, which at present is a daunting task.

SEQUENCE ANALYSIS OF THE HIV env PROTEIN

The HIV surface glycoprotein is made as a M_r 160,000 precursor protein that is cleaved, possibly during assembly, into a M_r 120,000 surface protein and a M_r 41,000 transmembrane envelope protein. Following the determination of the HIV genome nucleotide sequence in a number of laboratories in 1985, details of the env gene sequence and its protein products became available. We have concentrated our analysis of the env gene sequence on that of Muesing, et al. (1985), since the gp120 protein we study has been provided by Drs. L.A. Lasky and T.J. Gregory from Genentech. Studies by several groups have shown that there are conserved and highly variable regions of the envelope gene product (Starcich, et al. 1986; Kowalski, et al. 1987). This variability could account for the difficulty in producing a successful antiviral immune response in the host.

A method developed several years ago (Eisenberg, et al. 1982) permits the analysis of amino acid sequences to measure the helical hydrophobic moment, the tendency of an alpha helix to seek the surface between a polar and an apolar phase. Alpha helices having one polar cylindrical face and one apolar face are termed "amphipathic" or "amphiphilic". They tend to seek the boundary between an apolar phase (air, the protein interior or a membrane surface) and an aqueous phase (water or cytoplasm). The reason is that these helices can lie parallel to an interfacial boundary, with their apolar side chains forming hydrophobic interactions with the apolar phase, and their polar and charged side chains interacting with the strong dipoles of water. Some surface-seeking helices, such as melittin from bee venom, function in the processes of membrane lysis and membrane fusion. Recently, D. Eisenberg and M. Wesson (unpublished) have estimated the helical hydrophobic moments for 4,400 protein sequences (about 1 million residues). Among those proteins with the largest moments were two segments from the HIV gp41 region, suggesting that these two segments are alpha-helical, amphiphilic, and may be membrane–associated. Segment 1 of three strains of HIV is located near the C-terminus at residues 839–846 of the sequence of Muesing, et al. (1985), while Segment 2 is located at residues 776–785. Another sequence of interest, although not at all as amphiphilic as Segments 1 or 2, is a sequence identified by Gallaher (1987) as being conserved among a large number of viruses that induce membrane fusion. For reference, we identify this potential fusion peptide sequence as Segment 3, located at residues 519–524. Because of the potential interest in these peptides as having potential membrane–related functions in HIV replication, further studies were pursued as described below.

VESICLE FUSION RESULTS

To test the properties of these peptides, sequences including segments 1, 2, and 3 were identified and prepared synthetically at Caltech using solid phase synthesis chemistry (Bruist, et al. 1987) The sequences are Segment 1: 829–848; Segment 2: 774–792; Segment 3: 510–534. All of the peptides were somewhat insoluble, and Segment 3 peptide has proved to be completely insoluble. Lipid vesicles (liposomes) were prepared using either Distearoylphosphatidyl choline or Dipalmitoylphosphatidyl choline and sonication, followed by purification on a Sephadex G50–80 column. The purified vesicles were diluted into phosphate buffered saline (pH 7.4) or citrate buffered saline (pH 4.0), and the peptides added to final concentrations of 5 mg/ml. The preparations were incubated at room temperature and examined by electron microscopy.

At neutral pH there was no effect of either peptide segments on liposome structure. The 100 nm vesicles remained in their original state. However, when incubated at pH 4.0, both Segment 1 and 2 induced the transformation of the vesicle preparation into large flat sheets several micrometers in diameter (Figs 2,3). The conversion of small, spherical liposomes into large sheets implies a substantial surface orienting property for each of these peptides. The small protein from bee venom, mellitin, also lyses and transforms these liposomes, but does not induce sheet formation. Rather, the vesicles both fuse into larger spheres and aggregate into grapelike clusters.

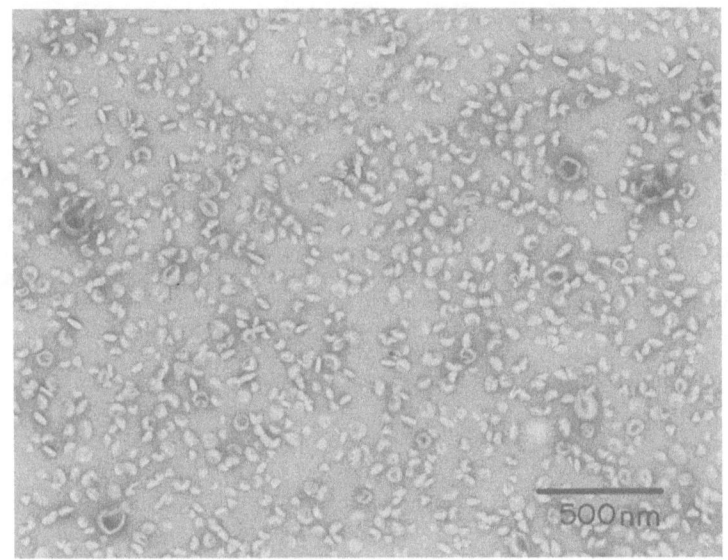

Fig. 2. Electron micrograph of lipid vesicles plus "peptide I" (see text) prepared at pH 7.0. The liposomes are intact and separate.

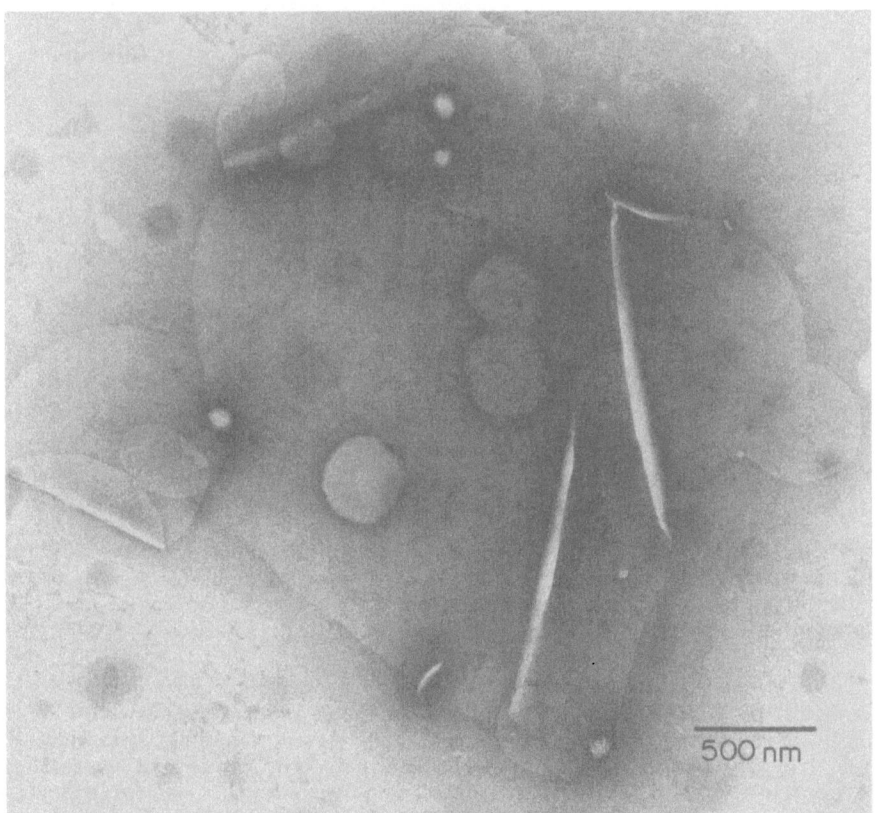

Fig. 3. Electron micrograph of the large, flat sheets that remain after incubation of the same lipid vesicles as shown in Fig. 2, but prepared at pH 4.0. Few if any original sized liposomes remain.

GP120 OLIGOMER STRUCTURE - A PUZZLE

The structure of the HIV virion is only slowly beginning to emerge. A careful study by Gelderblom, et al. (1987) has primarily used embedding and sectioning methods combined with immuno–electron microscopy. Although the resolution is rather low (5.0–7.0 nm), several important features have emerged. These are summarized in Fig. 1. Gelderblom, et al. (1987) have proposed that the glycoprotein knobs are arranged on the virion surface in a T=7 icosahedral surface lattice, and that pentameric and hexameric clustering could be seen, based on counting of the surface knobs seen in various views by sectioning. Dimensions of the knobs are consistent with such an arrangement as related to the virus surface. Based on this arrangement there would be 72 knobs/virion, and a thin section through the center would be expected to show 18–20 knobs around the periphery of the particle. The dimensions reported for the knobs (8-9 x 15 nm) are quite large, and are nearly the same as those known for the oligomeric protein glutamine synthetase (Almassy, et al. 1986).

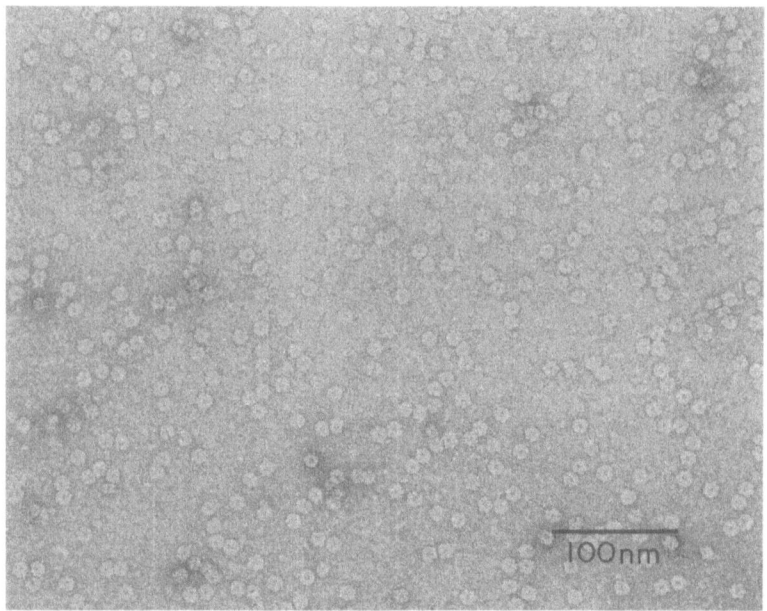

Fig. 4. Electron micrograph of a mixture of glutamine synthetase and monomeric gp120 expressed from CHO cells as described in the text. The large, clearly visible molecules of glutamine synthetase are mixed with a larger number of small, fuzzy dots (monomeric gp120).

The oligomeric state of gp 120 is unknown, and suggestions have been made that it might be trimeric based on analogy to the influenza HA protein (Wiley, et al. 1984). If the above dimensions are correct, however, it is clear that there could be more than three copies of gp120 per oligomer since glutamine synthetase contains 12 polypeptide chains of M_r 55,000 within the same volume. It will be important to determine the volume of the surface knobs by other means such as frozen–hydrated and unstained scanning transmission electron microscopy to confirm these dimensions (and to measure the mass within the molecular envelope). Since there is also an extraordinarily high degree of glycosylation in gp120 it is not possible to predict either the shape or the oligomeric state from the amino acid sequence or from the appearance of stained structures in sectioned material. It is interesting that none of the expression systems for gp120 (bacterial, yeast or CHO cells) has produced oligomeric gp120 – only monomers are made, even when glycosylation appears normal. This suggests the possibility that gp41 in the viral envelope may be involved in directing formation of the correct oligomer, since none of the env gene expression systems include the highly hydrophobic gp41 sequence. The solution to this puzzle will be important in determining which protein structural features are important in maintaining subunit interactions, and which could be available for attack by specific blocking drugs.

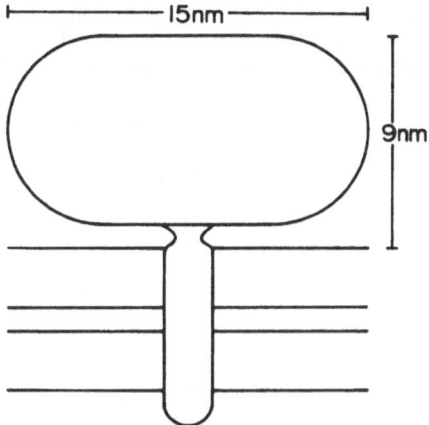

Fig. 5. Dimensions of an HIV surface knob from thin section electron microscopy. The volume of the envelope surface glycoprotein knob from this method could accommodate a number of copies of the monomeric gp120. There is no information at present as to the number of copies within the knob, and the solution to this problem is an important goal of structural research on HIV.

REFERENCES

Almassy, R.J., Janson, C.A., Hamlin, R., Xuong, N-H., and Eisenberg, D., 1986, Novel subunit-subunit interactions in the structure of glutamine synthetase, Nature, 323:304.

Bruist, M.F., Horvath, S., Hood, L.E., Steitz, A.A., Simon, M.I., 1987, Synthesis of a site–specific DNA–binding peptide, Science, 235:777.

Eisenberg, D., Weiss, R.M., and Terwilliger, T.C., 1982, The helical hydrophobic moment: a measure of the amphiphilicity of a helix, Nature, 299, 371.

Fauci, A., 1988, The human immunodeficiency virus: infectivity and mechanisms of pathogenesis, Science, 239:617.

Gallaher, W.R., 1987, Detection of a fusion peptide sequence in the transmembrane protein of human immunodeficiency virus, Cell, 50:327.

Gelderblom, H.R., Hausmann, E.H.S., Özel, M., Pauli, G. and Koch, M.A., 1987, Fine structure of human immunodeficiency virus (HIV) and immunolocalization of structural proteins, Virology, 156:171.

Kowalski, M., Potz, J., Basiripour, L., Dorfman, T., Goh, W.C., Terwilliger, E., Dayton, A., Rosen, C., Haseltine, W., and Sodroski, J., 1987, Functional regions of the envelope glycoprotein of human immunodeficiency virus type 1, Science, 237:1351.

Starcich, B.R., Hahn, B.H., Shaw, G.M., McNeely, P.D., Modrow, S., Wolf, H., Parks, E.S., Parks, W.P., Josephs, S.F., Gallo, R.C., and Wong-Staal, F., 1986, Identification and characterization of conserved and variable regions in the envelope gene of HTLV-III/LAV, the retrovirus of AIDS, Cell, 45:637.

Wiley, D.C., Wilson, I.A. and Skehel, J.J., 1984, The hemagglutinin membrane glycoprotein of influenza virus, in "Biological Macromolecules and Assemblies," F. A. Jurnak and A. McPherson, eds., Wiley–Interscience, New York.

ON THE DEVELOPMENT OF NEW CHEMOTHERAPEUTIC AGENTS IN THE COMING DECADE

Paul O. P. Ts'o

Division of Biophysics
The Johns Hopkins University
Baltimore, Maryland

The major advances in scientific research which currently have a significant impact on medicine can be generally described in two areas: 1) application of instrumentation such as laser beam surgery and NMR imaging, etc. and 2) the understanding of biology and medicine at the molecular and atomic levels. For the development of new therapeutic agents, the second area certainly is the focal point of attention. Thus, the major impact of the molecular sciences on the medical and pharmaceutical industries is the topic of this short essay.

Among the major advances in biomedical research and molecular biology is the understanding of genes and gene expression as the fundamental parameters of the normal and pathological states of living systems, including those of the pathogens such as viruses. Most viruses, though not all, have been or are being analyzed for the nucleic acid sequences of their genes. Indeed, there is a big push to sequence the entire human genome as a contribution to the fundamental understanding of human biology and medicine.

The increase in our knowledge of genes and their functions coincides with rapid and powerful developments in nucleic acid chemistry. Advances in nucleic acid organic chemistry have led to the facile synthesis of nucleic acids several hundred nucleotides in length, and advances in nucleic acid physical chemistry provide the means to study the secondary structure and the interaction of nucleic acids.

Therefore, a major challenge in biomedical research is to design new agents which specifically modify and modulate the biological expression of certain important genes in living cells and in the host. The first requirement is specificity, i.e. one must be able to build an agent which specifically recognizes one gene or one gene product from the other genes or other gene products. This is accomplished through the Watson-Crick complementary base pairing scheme. After all, this is the basic mechanism which preserves the identity and specificity of genes in the replication of cells.

Nucleic acid physical chemistry indicates that such specific interactions from about 12-20 units of bases in length will be sufficient for stability at human body temperature (37°C) and will have a sufficient degree of uniqueness for base-base recognition.

Once we have synthesized an agent which can recognize and interact with a specific gene or gene product, this agent must have the ability to enter the living cells unmodified before the agent can exert its therapeutic function. However, living cells are guarded by membranes protecting against the intrusion of exogenous genetic materials. In fact, the strategy of a virus is to interact with the cell membrane through its specific component in order to gain the entry of viral genetic materials. Therefore, the first requirement of this new therapeutic gene-specific agent is its ability to pass through the membrane of a living cell and remain intact.

The second requirement for this new therapeutic agent is to bypass the defense system (the nucleases) of the host cell. Nucleases exist in serum and intracellularly, serving both as scavengers or as guards to destroy foreign nucleic acids. The new therapeutic agent must be immune to nucleases.

In addition to the requirements for specific recognition, an increased affinity in the interaction between the sequence-specific agent and the target nucleic acid without sacrificing specificity is highly desirable. Thus, if a generalized, attractive force is added to the agent, then such an agent may bind to the target nucleic acid more strongly but not necessarily more discriminatingly. This increase in affinity, when coupled with a decrease in selectively, may defeat the original objective, since a non-discriminating interaction would most likely induce strong toxic effects. Parenthetically, a single-stranded nucleic acid not only interacts with the complementary nucleic acids in the base pairing scheme described above, but will also interact with enzymes, proteins, and perhaps other cellular components. Thus, the addition of an oligonucleotide into the cell can lead to other non-specific and unpredictable side effects through protein-nucleic acid interaction.

Matagen, An Acronym for "Masking Tape for Gene Expression"

Matagen, the structure of which is shown in Figure 1, is an oligonucleoside methylphosphonate in which an oxygen atom of the phosphorus

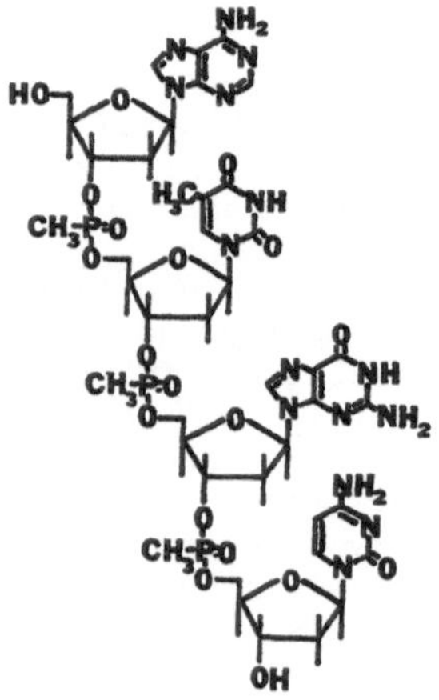

d–$A_pT_pG_pC$

Fig. 1. Structure of Matagen, an oligonucleoside methylphosphonate

backbone is replaced by a methyl group. This nonionic oligonucleotide analog affords the least structural variation from the naturally occurring nucleic acid. Without changing the size of the molecule, this modification removes the negative charge from the molecule. Now this nonionic Matagen can penetrate a mammalian cell membrane because of its lipophilic property. Matagen is also totally resistant to all known nucleases owing to the modification of the backbone and the removal of the negative charge. In addition, Matagen forms complexes with the target complementary strand of nucleic acid of a higher stability, due to the removal of the negative charge repulsion at each unit of the building block. The increase in the affinity/stability of these nonionic analogs in duplex formation still allows the analog to retain a high degree of specificity, since the removal of the charge repulsion is done per unit of the building block. The increase in enthalpy for binding is about 1.5 Kcal per base pair, or approximately one-quarter of the enthalpy in the duplex formation (ΔH~6 Kcal/base pair) as compared to the duplex formed between two strands of negatively charged nucleic acids. Since this compound is not charged and has a modified backbone, there is no indication Matagen will interact with a protein or any other cellular component.

With the above properties, Matagen can bind to a single-stranded nucleic acid target, thereby interfering with, or masking the function of this target nucleic acid in living cells. So far, this compound does not attack the genes (DNA in double-stranded form), but can suppress the function or the formation of the gene products in nucleic acids. This type of molecule can be used to control gene expression by masking the gene products at the level of transcription (message) of the gene. Thus, this family of molecules has been termed "Matagen, an acronym for "Masking Tape for Gene Expression".

It should be noted that Matagen is not a compound but a _family_ of compounds. Each Matagen is synthesized with a specific sequence of 10 to 20 units in length, and will block the transcription of a targeted gene and therefore mask the expression of this targeted gene.

At present, sequence-specific Matagen can function in two ways. First, Matagen can function as a masking tape of the messenger RNA, thereby blocking the translation of such a message, resulting in the blockage of protein synthesis. Little or no protein of this gene can now be produced in the cells. Examples of this study can be found in the successful experiments in suppressing specifically the synthesis of globin in reticulocytes[1], as well as in the inhibition of Vesicular stomatitis virus (VSV) in mouse L cells[2].

The second mode of action of Matagen concerns the inhibition of splicing of the immature messenger RNA with the understanding that the unspliced messages are rapidly destroyed in the nucleus. An example of this action of Matagen is its ability to inhibit herpes simplex virus function and replication in simian and human cells[3]. In this study, the splice junction of the immediate early pre-mRNA 4 and 5 of HSV-1 was masked by the addition of a sequence-specific Matagen, leading to the loss of viral replication in living cells. The highly specific action of this particular Matagen is well-documented in this paper. It should be noted that application of Matagen inside living cells can block the function of a gene, but will not destroy that gene.

The synthesis of Matagen was first done by a method analogous to the phosphotriester method[4] and more recently by the phosphoamidite approach using a DNA synthesizer, with yields above 95% per addition of nucleotidyl units. Thus, for the synthesis of a 12- to 14-mer, the isolated yield can be 15-60% depending on the sequence (Miller, unpublished data; and Ref. 5). In addition, this molecule can be characterized for its purity and further defined through analysis of its sequence[4].

The inhibitory action of Matagen, i.e. forming a complex with complementary target nucleic acids, is exerted through physical interaction whereby the association process dominates at low levels of binding but the dissociation process becomes dominant when the extent of binding increases. From a theoretical standpoint, such a binding process approaches a 100% level asymptomatically. In other words, the physical interaction process becomes less and less efficient at higher levels of interaction. In addition, when one tries to increase further the high level of interaction by increasing the concentration of Matagen, the selectivity in binding begins to suffer. Therefore, it is imperative to develop a second generation of Matagen in which the physical binding will lead to a chemical interaction.

Derivatized Matagen

In addition to the need to improve the efficiency of binding at a high level of interaction, it is also desirable to eliminate permanently the target nucleic acid through a chemical reaction.

Currently, this objective has been achieved successfully in vitro by attaching psoralen, a photocrosslinking group, to Matagen[5] (Figure 2). The Matagen will first interact physically with the target nucleic acid by forming a physical duplex according to base-pairing specificity, and

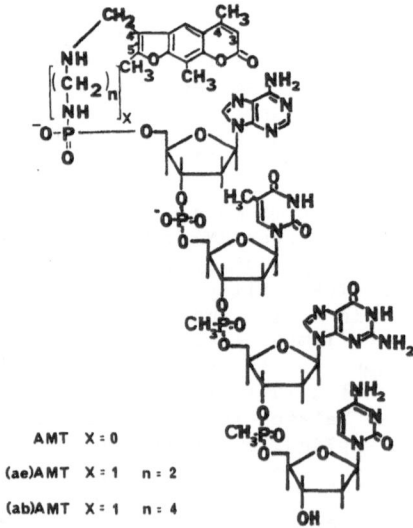

AMT   X = 0

(ae)AMT   X = 1   n = 2

(ab)AMT   X = 1   n = 4

Fig. 2. General structure of deoxyribonucleoside methylphosphonates derivatized with 4'-(aminoalkyl)-
4,5',8-trimethylpsoralen.

then form a covalent crosslink with the target nucleic acid upon activation of the chemically-linked psoralen using near-UV irradiation[5]. Preliminary results indicate that psoralen derivatized Matagen has at least three advantages over the underivatized Matagen: 1) The effective concentration has been reduced by 20-50-fold; 2) The timing of intervention in the gene function (such as a viral gene) in the host cell can now be controlled as desired. This advantage will give additional insight for the biological system leading to a greater knowledge for therapeutic application; and 3) Derivatized Matagen will provide the scientific evidence of the specificity of Matagen interacting only with the target nucleic acid inside living cells or animals.

Research is currently in progress to develop chemically reactive groups which provide further specificity as well as efficiency without the requirement of photoactivation. This will undoubtedly form the third generation of Matagen.

Requirements for Utilizing Matagen as a Therapeutic Agent

From the above description we can summarize the requirements for utilizing Matagen as a therapeutic agent.

1. One must know the molecular biology of the diseases or the pathogens.

2. One must know the sequences of the genes involved in the disease or in the virus or pathogen.

3. One must know the conformation of the targeted nucleic acid sequences if one is going to attack the single-stranded form of a messenger RNA. One must know which region of the messenger RNA is the most vulnerable.

4. One must know how to synthesize a sequence-specific Matagen efficiently and at low cost, and derivatize the Matagen to be an effective and sequence-specific chemical reagent.

5. One must know how to deliver Matagen to the host with emphasis on the possible additional specificity of the organ.

6. Finally, one must continue to develop new Matagens and new approaches to control genes in double-stranded DNA form.

As for the limitations of Matagen, the following points should be noted.

1) The approach of Matagen is fundamentally based on suppression, removal, surgery, etc., and is ideal for use against diseases caused by external pathogens such as viruses. This approach is also applicable in the pathological states where undesirable or deleterious genes can be suppressed or excised in order to bring about a cure. However, this approach is powerless when a new gene is needed to correct the pathogenic state.

2) Matagen is incapable of penetrating certain pathogens whose thick cell walls form a physical barrier. For example, Matagen cannot penetrate gram negative bacteria such as E. coli. However, Matagen is expected to be effective against

367

microplasma, as well as certain parasites such as trypanosomes.

3) It is both difficult and inefficient to expect by a direct, chemical attack to eliminate the infectious virus in the host by more than 99%, or a 2 log unit in reduction. Therefore, Matagen can be very effective in stopping virina but may not be capable of eliminating all of the targeted virus. It is necessary to mobilize the immune defense system of the host in order to eliminate the remaining viruses or pathogens or cancer cells in the host.

Comparison of Matagen With Other Forms of Conventional Therapeutic Agents

Currently, most therapeutic agents work on the enzyme level or the membrane level. Nucleoside analogs have been used extensively as antiviral drugs. One of the most well known of these analogs is acyclovir, which is an analog of guanosine and is used as an anti-herpes drug. Acyclovir has a molecular weight of approximately 400, and such a simple molecule cannot store too much information in order to have a high degree of specificity in its therapeutic actions. In contrast, Matagen is 12 to 20 nucleotides in length and has a molecular weight of about 6000-8000. Therefore, Matagen can have 12 to 16 times more structural information than a nucleoside analog. In essence, this is analogous to the comparison of two computers, one with a 4K core capacity and the other with a 64K core capacity. Moreover, a nucleoside analog works on the level of enzymes, usually taking advantage of the differences between the host enzyme and the viral enzyme, thereby differentially inhibiting the viral enzyme. However, there are many more copies of viral enzymes in cells than there are copies of the viral genome or even the gene message. For example, there may be one million viral enzyme molecules, but only 1000 to 10,000 viral genetic messages and perhaps only 1-10 copies of the viral genes.

Clearly, in attacking the viruses in living cells, the targets for Matagen are the viral genetic messages which are far less in number. Therefore, there is a greater chance of success. However, the targets for nucleoside analogs are the enzyme molecules, which are much more in number and therefore, the inhibition would require much higher concentrations of the drug. More importantly, the interference or attack on the virus by Matagen, or on any host gene, is based on solid knowledge of nucleic acid sequences, while the development of nucleoside analogs as therapeutic agents is based on trial and error or guesswork. For this reason, viral strains resistant to acyclovir can be easily developed and then the whole program of drug development for another acyclovir substitute for this resistant mutant has to be repeated over again. Using Matagen, one can add several Matagens in a therapeutic mixture to attack the viral genome from several positions of the viral genes simultaneously. Therefore, it is statistically impossible that several point mutations of a viral genome will take place simultaneously along the viral genome in order to escape the masking action exerted by all these Matagens in combination. Thus, it will be very difficult to develop resistant viral strains to Matagen, in contrast to acyclovir. Furthermore, if a resistant virus to Matagen is uncovered, one can simply isolate such a resistant strain and analyze the new base sequence of this resistant strain. Another new Matagen can then be synthesized with a new sequence that will inhibit this resistant viral variant.

Currently there is a major advantage of the nucleoside analog, i.e. it is much cheaper to be synthesized on a large scale using current technology. However, the technology to streamline the synthetic procedure of Matagen and make it cost-effective will soon be developed. Furthermore, it is theoretically possible to develop a chemically reactive Matagen so that a much smaller amount is needed in comparison with the amount of nucleoside analog needed. For instance, if a current daily dose of 200 mg acyclovir as an anti-herpes drug can be replaced by a 10 mg daily dose of Matagen, then the economic factor will be much more favorable for Matagen.

There is an additional interesting point: when employing Matagen as a therapeutic agent, one can utilize all the findings from studies in molecular biology and molecular medicine about viruses and other diseases, particularly about the genes involved in these situations. In other words, all the progress in the knowledge of analyzing nucleic acid sequences in molecular biology and molecular medicine can lead to the construction of Matagens as therapeutic agents targeted toward diseases.

One more important point should be mentioned. Matagen is not a biotechnological product which depends on the synthesis of this compound by microorganisms or living systems. Matagen is developed and synthesized through chemistry, and therefore in many ways is a chemical approach based on the knowledge of biology and molecular biology. Matagen is not dependent on biological systems for its production.

Modern Advances in Immunology in the Study of Biological Response Modifiers and in the Understanding of the Use of Nucleic Acids as Therapeutic Agents

Currently advances in molecular immunology are slowly unravelling the immune system or the the self-defense system of humans developed during the course of evolution. Clearly, from the time of the emergence of the human race to the dawn of medicine, humans have survived solely on their own ability to defend against pathogens. Our current knowledge of the immune defense system reveals that this system is indeed very complex, perhaps involving scores of cell types and hundreds of factors interacting with each other. The system has a memory and many cybernetic regulatory cycles. After all, it is the heart of the survival of every species.

Factors termed "biological response modifiers" have been identified as part of the immune defense system. One such factor produced by the host is interferon, which was perceived to be the main factor excreted by the host in the defense of viral infection. Therefore, it was thought that once interferon could be synthesized, the problem of viral infection would be solved. Subsequently, this hope was extended to the use of interferon as a growth regulator, particularly in the inhibition of cancer growth and malignancy. Since these biological response modifiers were produced by the host, it was perceived erroneously that these factors would produce no toxic side effects in the host when applied exogenously. Therefore, the main challenge was to synthesize these factors, which are produced normally and only in the human body, on a commercial scale.

As a triumph of modern biochemistry and genetic engineering, interferon was indeed purified and crystallized, and its genes were isolated and put into vectors. Interferon is now being produced on a large scale for commercial use. However, it was soon found to have substantial toxic effects when administered to patients. This toxic effect was originally thought to be due to "impurities" in the preparation of the interferon. In retrospect, this is not surprising at all, since these factors produced by the host are powerful biological response modifiers. Administration of these factors to a human being certainly would upset the delicate balance of the regulatory system, leading to undesirable toxic side effects.

Furthermore, it was soon discovered that there was not just one kind of interferon, but three major categories of interferon, with several and even scores of sub-species in each category. These many species of interferon are not identical in their actions, and at present we have no idea how they interact. Nor do we know which species of interferon the physician should introduce to treat a particular patient, or what the dose should be for a particular patient. It has now become clear to a growing population of physicians and scientists, that in view of the complexities of the human immune defense system, the application of one or two of these factors in a generalized dose range to a patient is very difficult to be successful. If too small an amount is administered, the system would probably self-regulate itself and cancel the effect in a cybernetic manner. If an excessive amount is introduced, the balance of the regulatory system could be thrown off course, leading to an adverse effect.

Slowly, the promise of successfully utilizing these factors, including interferon, began to be elusive. Simply put, it is not a question of knowing how to make them, but rather a question of how to use them.

In light of this background, the function of a double-stranded RNA (dsRNA) provided a very exciting prospect. Research in the early 1960's uncovered that challenging human cells with many viral intermediates, most of which are dsRNA's, would produce one or several species of interferon. It was soon discovered that the shape and the size of the dsRNA is a factor for recognition, and not the sequence in the dsRNA. For example, polyriboadenylic acid plus polyribouridylic acid duplex (poly A·poly U) as well as polyinosinic acid plus polycytidylic acid (poly I·poly C), are powerful interferon inducers, even though obviously these synthetic polynucleotide complexes do not contain any genetic information. The requirement of the shape and conformation of molecules is quite strict, however[6]. The polynucleotide with the deoxynucleoside backbone or 2' O-methylriboside backbone, does not have this function. Simply put, the molecule has to be a dsRNA molecule, and not a dsDNA molecule, or DNA-RNA hybrid molecule, or a poly 2' O-methylribonucleotide duplex molecule, etc. It was further demonstrated that a small structural variation can be tolerated and that dsRNA can still retain its main biological function, i.e. the substitution of one 2' O methylriboside residue in about one turn of the ribose nucleic acid helix (about 10-12 nucleotides in length)[6].

During the early days, when the production of large quantities of interferon seemed so remote, a dsRNA such as poly I·poly C was considered as a possible substitute for interferon. Subsequently, it was discovered that while this dsRNA is a very powerful interferon inducer, it is highly toxic, one side effect being the induction of host antibodies against the dsRNA upon a continuous administration of dsRNA to the host[7,8]. Research on the biological action of dsRNA soon uncovered additional fascinating information[9]. Double-stranded RNA not only is a powerful interferon inducer, but also is an obligatory activator of two of the important enzyme systems in the host which are induced by interferon. Double-stranded RNA thus serves as an obligatory activator of 2'-5' oligo A synthetase which synthesizes 2'-5' oligo A from ATP in the presence of dsRNA. The 2'-5' oligo A in turn is an activator of endonuclease. Double-stranded RNA also serves to activate a kinase system, which plays an important role in the antiviral action in cells. Further findings indicate dsRNA has a very broad spectrum of action on cells, although this action is yet to be well understood. One example is that the inhibitory action of a mismatched dsRNA on HIV infection in cloned human T cells cannot be removed by the antibody against α-interferon[10]. Current data suggest dsRNA and interferon interact with each other, but are not equivalent to each other. Double-stranded RNA has a very large and profound effect on cells, which is not directly related to interferon action, although dsRNA is a powerful inducer of all types of interferon(s), as well as many types of cytokines.

Due to its toxicity, dsRNA was not being used extensively as a therapeutic drug. As a result, in the late 1960's researchers at Johns Hopkins began to look in another direction[11]. The new approach involved preparing a special kind of dsRNA which could stimulate the immune defense system or the cytokine system, but then can be destroyed readily by the host after the stimulation was initiated. This research led to the discovery of a class of mismatched dsRNAs, a typical example of which was poly I·poly $C_{12}U$. On the average, there is one uridylic nucleotide residue in the polycytidylic acid. Since uracil does not form a base pair with hypoxanthine, poly I.poly $C_{12}U$ is a mismatched dsRNA which, on the average, contains one mismatched base pair (I·U) out of 12 matched base pairs (I·C). This molecule functions as a nearly perfectly matched poly I·poly C, but is much more sensitive to degradation by nucleases. Poly I·poly $C_{12}U$ was termed as Atvogen at Johns Hopkins, an acronym for "Anti-Viral Onco-Gen".

It has been shown clearly that the mismatched poly I·poly $C_{12}U$ is far less toxic than poly I·poly C, and that it induces no antibodies from the host[7,8,12]. Recently, in a very crucial experiment, it was demonstrated that upon repeated injections up to 100 times into a rat during a time period of one year, the effectiveness of activation of NK cells upon injection by this molecule remains the same[13]. In other words, repeated injections over a year up to 100 times does not induce a hyporesponse state in the animal; the animal responds to the first injection in a manner and magnitude similar to its response to the 100th injection.

Conceptually, application of mismatched poly I·poly $C_{12}U$ to the host is very different than the administration of one interferon or one biological response modifier to the host. In the former case, a stimulus is introduced. The host is now asked to respond by producing all the factors which would be produced naturally by the host. If the host is capable of doing so, the immune system of the host will be enhanced. Since the stimulus is not a live virus, the enhanced or activated immune system of the host can be used to defend or to attack other targets such as other viruses or cancer cells, within the host. In other words, the host has been stimulated by a generalized challenge which would soon disappear, and the host would now be left with an active immune defense state. Thus, all the immune regulatory factors are being produced, hopefully in the proper proportion, by the host upon such a generalized stimulus. In contrast, in the latter approach, usually one of the immune regulatory factors such as interferon-α or interferon-γ or one of the interleukins or tumor necrosis factors, is administered to the patient. In this case, the regulatory system and the immune system are being perturbed with the hope that the proper response can be obtained.

In view of the complexity of the immune defense system, perhaps it is a long time off that a cocktail of all the factors in proper proportion can be prepared and administered to a patient in order to produce a desirable effect. Currently, there is little scientific basis that addition of one factor at a set dose can lead to desirable effects across a large population of patients. Therefore, the approach of administering a stimulus (the mismatched dsRNA) appears to have a much better chance of being successful than administering one factor at a time to the patient. This is perhaps one reason why therapeutic treatment using biological response modifiers has not been as successful as was originally anticipated.

Problems in Clinical Trials with Biological Response Modifiers

In addition to the difficulties outlined above, there is another major conceptual problem with the application of biological response modifiers, such as interferon, as therapeutic agents. So far, application of biological response modifiers in the treatment of disease usually leads to a partial response in the patient population: for instance, one-fifth to one-third of the patients may give a satisfactory response, while others may not. This result is not conducive to the further development of the biological response modifier as a therapeutic agent, particularly since the reason for the lack of response is unknown. This difficulty can be traced to two major problems, as described below.

In the conventional approach of using drugs to treat disease, the host is being considered basically as a "container" in which the virus or the cancer cells are harbored. The drug is then introduced in the "container" and the drug then pursues the pathogens or pathogenic cells. It is logically assumed that the higher the amount of drug introduced to the "container", the more effective the treatment, as long as the "container" can retain these therapeutic agents. The host, as the "container", presumably takes a passive role, and therefore a population of "containers" is assumed to have similar properties after normalization for size (per body weight in kilograms) or the surface (per surface area in cm^2). Thus, the concept of "maximum tolerable dose" is the guiding principle. One would therefore administer the highest concentration of drug which the host, or "container", is able to contain (i.e. tolerate).

In administering a biological response modifier for treatment, such as interferon, or administering an immune stimulus, such as mismatched dsRNA, it is clear that these agents do not attack the virus or the cancer cells in the body per se. They merely activate the defense system of the host to attack the virus. In this mode, the host now assumes an active role. It is the host, not the drug, that attacks the virus. In a complex cybernetic cycle of regulation, clearly an overdose of this type of drug may even reduce the effectiveness of the drug by suppressing or derailing the host's defense system through a feedback reaction. Therefore, it is important to obtain a maximally effective dose (or optimum dose) and not the maximum tolerated dose, as described previously. Clearly, the questions of dose-response becomes far more complicated.

The second concept in a conventional drug treatment is that all patients will respond more or less the same to this conventional drug, as would be the case for all the passive "containers". Therefore, one can do massive, randomized, double blinded trials in order to find the maximum tolerated dose (which presumably is the most effective dose) for a large population as a homogeneous group. In contrast, if each individual is supposed to wage his/her own individual battle against the virus or the cancer cell, then a generalized treatment dose may have little meaning. It is clear that a test or a screening system has to be developed for the responsiveness of each individual patient who is to receive this immune stimulus or immune enhancing factor. Some will respond, while others may not; some will respond to a certain dose, while others will respond to a different dose.

Thus, the question of treatment has to be coupled tightly to the testing/screening of the response of each individual, a procedure which is much more sophisticated.

It appears that in order to utilize the patient's own immune system in a natural way, the use of appropriate tests coupled with a general but brief stimulus intermittently, appears to be most logical. If a patient is able to respond to this harmless stimulus, then the success of such a therapy can also be assumed for many diseases curable by the immune self-defense system. After all, this immune system has been solely responsible for preserving the human race in the past.

Conclusion

Advances in our knowledge about nutrition, the discovery of antibiotics, and the development of vaccines have greatly reduced human suffering from disease and have increased the longevity of human life. New knowledge in molecular biology and molecular medicine (particularly in genetics, virology, and immunology) has provided a basis of new information for the design and preparation of therapeutic agents, particularly for viral infections and neoplasia. The knowledge about genes and their functions as well as the development of nucleic acid chemistry provide a unique opportunity to design innovative therapeutic agents to suppress or to modify gene function. Matagen, a family of nonionic oligonucleotide analogs of 12 to 16 residues in length, belongs to this new generation of therapeutic agents. Matagens are synthesized in a sequence-specific manner and each one is complementary to a target nucleic acid, thereby masking the function of the target nucleic acid inside the living cells. Practical problems including the technology of large scale synthesis, the question of pharmacokinetics, and the issue of drug delivery, etc. remain to be solved. However, there are no foreseeable obstacles to the successful development and the use of this family of anti-messenger RNA or antigene drugs.

As our knowledge in immunology and virology increases, we begin to understand the complexity and the power of the immune self-defense system which has preserved all the species thus far against pathogens. We have attempted in utilizing this knowledge to develop immuno enhancement agents as a therapeutic modality. One of the best approaches is to provide the host with a brief, harmless but powerful stimulus in an intermittent manner to enhance the secretion of all the immunoregulatory factors by the host in proper proportion, so that the host's immunodefense state remains alert and strong. Cells in the host have a special ability to recognize double-stranded RNA, which often represents the size and shape of a viral replicating intermediate. Upon sensing the presence of a dsRNA, the immune system inside the cells in the host begins to be activated for the defense against the perceived attack. Thus, dsRNA is a powerful stimulus for the immune system.

The agent proposed is a mismatched dsRNA, poly I·poly $C_{12}U$ (Atvogen), which has been tested extensively in clinical trials. In this treatment approach, the individual patient is called upon to fight the virus or cancer cell directly and actively. Therefore, the information on the host response is needed to design an effective dose for a particular disease or the particular stage of a disease and this information can only be obtained by an intelligent screening of individual patients. Without coupling testing and treatment as the basis for the dose strategy, the success of utilizing the immune response through exogenous addition of factors or stimuli will be much too limited to become useful.

It should be noted the approach of Matagen and the approach of Atvogen (poly I·poly $C_{12}U$) are complementary and are both based on the science and technology of nucleic acids. The complete elimination of the viruses or cancer cells left behind after treatment by Matagen most likely will depend on the action of the host immune defense system. Therefore, a combination of Atvogen treatment after Matagen treatment appears to be very logical. In addition, Atvogen can be used as a prophylactic agent to prevent the onset of disease or to reduce the severity of the disease. In such a manner, Matagen can again be more effective against a relatively minor infection. Therefore, it appears that these two treatments can be used sequentially, depending on the patient and the disease.

Finally, success in this area may even have implications in the problem of aging, since neoplasia and the loss of immunodefense against infection are the final challenges confronting humans in the latter part of their lifespan.

Additional references on Matagen and Atvogen are also available[14-18].

Acknowledgement

The research conducted at The Johns Hopkins University is a collaborative effort, with special acknowledgement to Professor Paul Miller in the area of Matagen research, and Professor Paul Lietman in the area of Atvogen research. Special thanks is also given to Ms. Dorothy Lindstrom for the preparation of this manuscript.

References

1. A. Murakami, K. Blake and P. Miller, Characterization of sequence-specific oligodeoxyribonucleoside methylphosphonates and their interaction with rabbit globin mRNA, Biochemistry 24:4141 (1985).

2. C. Agris, K. Blake, P. Miller, M. Reddy, and P. Ts'o, Inhibition of Vesicular stomatitis virus protein synthesis and infection by sequence-specific oligodeoxyribonucleoside methylphosphonates, Biochemistry 25:6268 (1986).

3. C. Smith, L. Aurelian, M. Reddy, P. Miller, and P. Ts'o, Antiviral effect of an oligo (nucleoside methylphosphonate) complementary to the splice junction of herpes simplex virus type I immediate early pre-mRNAs 4 and 5, Proc. Natl. Acad. Sci. USA 83:2787 (1986).

4. P. Miller, M. Reddy, A. Murakami, K. Blake, S.-B. Lin, and C. Agris, Solid-phase synthesis of oligodeoxyribonucleoside methylphosphonates, Biochemistry 25:5092 (1986).

5. B. Lee, A. Murakami, K. Blake, S.-B. Lin and P. Miller, Interaction of psoralen-derivatized oligodeoxyribonucleoside methylphosphonates with single-stranded DNA, Biochemistry 27: 3197 (1988).

6. J. Greene, J. Alderfer, I. Tazawa, P. Ts'o, S. Tazawa, J. O'Malley, and W. Carter, Biochemistry 17:4214 (1978).

7. W. Carter, J. O'Malley, M. Beeson, P. Cunnington, A. Kelvin, A. Vere-Hodge, J. Alderfer, and P. Ts'o, An integrated and comparative study of the antiviral effects and other biological properties of the polyinosinic-polycytidylic acid duplex and its mismatched analogues. III. Chronic effects and immunological features, Molec. Pharm. 12:440 (1976).

8. P. Ts'o, J. Alderfer and J. Levy, An integrated and comparative study of the antiviral effects and other biological properties of the polyinosinic acid-polycytidylic acid and its mismatched analogues, Molec. Pharm. 12:299 (1976).

9. J. Greene and P. Ts'o, Double-stranded RNA and its analogs: The prospects and the promise of the first nucleic acid therapeutic agent, in: Clinical Application of Interferon and Their Inducers, D. Stringfellow, ed., Marcel Dekker, Inc., New York (1986).

10. J. Laurence, J. Kulkosky, S. Friedman, D. Posnett, and P. Ts'o, Poly I·poly $C_{12}U$-mediated inhibition of loss of alloantigen responsiveness and viral replication in human CD4 + T cell clones exposed to human immunodeficiency virus in vitro, J. Clin. Invest. 80:1631 (1987).

11. W. Carter, P. Pitha, L. Marshall, I. Tazawa, S. Tazawa and P. Ts'o, Structural requirements of the $rI_n·rC_n$ complex for induction of human interferon, J. Mol. Biol. 70:567-587 (1972).

12. W. A. Carter, I. Brodsky, M.G. Pelligrino, H.F. Henriques, D.M. Parenti, R.S. Schulof, et al., Clinical, immunological, and virological effects of ampligen, a mismatched double-stranded RNA, in patients with AIDS or AIDS-related complex. Lancet, i:1286 (1987).

13. D. Nolibe and M.N. Thang, Stimulation of natural killer cytotoxicity by long-term treatment with double-stranded polynucleotides without induction of hyporesponsiveness, Cancer Immunol. Immunother.:66 (1988).

14. P. Miller, C. Agris, L. Aurelian, K. Blake, S. Glave, S.-B. Lin, A. Murakami, M. Reddy, C. Smith, S. Spitz and P. Ts'o, Matagen (Masking Tape for Gene Expression): A family of sequence-specific oligonucleoside methylphosphonates, in:

"Molecular Mechanisms of Carcinogenic and Antitumor Activity," C. Chagas and B. Pullman, eds., Pontifical Academy of Science, Vatican City (1986).

15. P. Miller and P. Ts'o, A new approach to chemotherapy based on molecular biology and nucleic acid chemistry: Matagen (Masking Tape for Gene Expression), in: "Anti-Cancer Drug Design", Vol. 2, The Macmillan Press Ltd., London (1987).

16. L. Aurelian, C. Rinehart, M. Wachsman, M. Kulka, and P. Ts'o, Augmentation of natural immune defense mechanisms and therapeutic potential of a mismatched double-stranded polynucleotide in cutaneous herpes simplex virus type 2 infection, J. Gen. Virol. 68:2831 (1987).

17. P. Ts'o, P. Miller, L. Aurelian, A. Murakami, C. Agris, K. Blake, S.-B. Lin, B. Lee and C. Smith, An approach to chemotherapy based on base sequence information and nucleic acid chemistry. Matagen (Masking Tape for Gene Expression), in: "Annals of the New York Academy of Sciences, Vol. 507, Biological Approaches to the Controlled Delivery of Drugs" (1987).

18. P. Miller and P. Ts'o, Oligonucleotide inhibitors of gene expression in living cells: New opportunities in drug design, in: Annual Reports in Medicinal Chemistry, Vol. 23, in press (1988).

INVESTIGATION OF METABOLIC DISORDERS
IN PATIENTS AND ANIMAL MODELS USING
MAGNETIC RESONANCE SPECTROSCOPY

Gregory S. Karczmar and Michael W. Weiner

Magnetic Resonance Unit, Veterans Administration
Medical Center and Depts. of Medicine and Radiology
University of California, San Francisco, C.A.

INTRODUCTION

This chapter will review the current status of in Vivo
Magnetic Resonance Spectroscopy (MRS), with emphasis on the
clinical applications of MRS. The results of MRS studies of
muscle function, tumor metabolism, and other applications
will be discussed. Techniques for obtaining MR signals from
selected volumes of tissue and/or selected metabolites will
be described.

BASIC PRINCIPLES

Although current techniques for clinical diagnosis
provide important information concerning anatomy and
function, they provide little or no direct information
concerning tissue metabolite levels and energy metabolism.
Magnetic Resonance (MR) makes it possible to measure levels
of important metabolites directly and non-invasively in Vivo.

Magnetic Resonance Spectroscopy (MRS) detects signals
from a number of nuclei which are naturally present in
biological tissues. These include protons (^{1}H), phosphorus
(^{31}P), sodium (^{23}Na), and nitrogen (^{14}N). Nuclei which are
not naturally present such as ^{13}C and ^{19}F can be introduced
in labeled molecules (i.e. ^{13}C labeled glucose) The signals
from these nuclei are sensitive to their chemical
environment, and in general signals from different
metabolites occur at different frequencies; thus MRS can be
used to identify and quantitate levels of a variety of
metabolites.

The most commonly detected nucleus in in Vivo studies is
^{31}P (1). In general, MRS can detect phosphorus containing
metabolites which are present in concentrations of greater
than 0.5 mM. Figure 1 illustrates a typical ^{31}P spectrum
obtained from the brain of a human volunteer. The spectrum
was obtained at 2T using a Philips Gyroscan Spectrometer, as
are all other spectra of human subjects presented here.

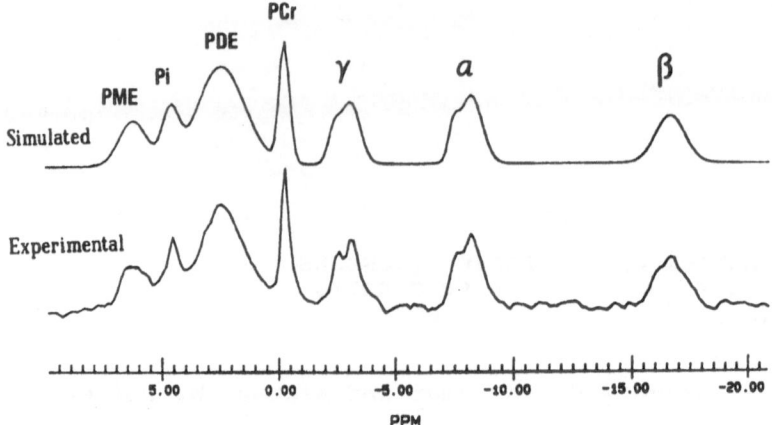

Fig. 1. ^{31}P MRS spectrum of the brain of a human volunteer, obtained using the ISIS technique. A) Experimental data. B) Computer generated fit of the data. The spectrum was obtained from a 65 ml volume in 20 minutes. (From Roth et al.)

(The method used to obtain this spectrum is discussed below). Signals are detected from the α, β, and gamma ATP phosphates, phosphocreatine (PCr), inorganic phosphate (P_i), phosphodiesters (PDE), and phosphomonoesters. The resonance frequency, or chemical shift (shown on the X axis) of each signal identifies the chemical species. Furthermore, the chemical shift of the P_i resonance is a measure of pH, and the chemical shift of the β ATP is a measure of intracellular magnesium concentrations. The Y axis is in arbitrary intensity units.

The extremely high concentration of water protons in tissue, close to 100 $\underline{M}$, makes Magnetic Resonance Imaging possible, but interferes with observation of ^{1}H resonances of other metabolites which are present in millimolar concentrations. Nevertheless, MRS techniques have been developed which allow observation of metabolites such as n-acetylaspartate, creatine and phosphocreatine, lactate, glutamine, and glutamate by ^{1}H MRS (2).

The intensities of signals detected by MRS are influenced by a large number of factors. These include the magnetic field strength, longitudinal (T_1) and transverse (T_2) relaxation times of the signals, the sensitivity of the detector (which is often spatially inhomogeneous), and the gains of various amplifiers. Roth et al. have developed a method based on the use of standards, T_1 measurements, and analysis of data using computer programs, which allows calculation of absolute concentrations of metabolites from the observed intensities (3). As discussed below, determination of absolute concentrations can be a powerful tool for biochemical and clinical studies.

USE OF MRS TO MONITOR ENERGY METABOLISM IN VIVO

The chemicals which are detected by MRS are vital to cellular energetics. ATP is synthesized in the mitochondria by oxidative phosphorylation, and in the cytoplasm by glycolysis. P_i and intracellular pH are important modulators of ATP synthesis. Fluorine containing probes are currently being developed which will allow MR measurements of tissue O_2 tension (4). Lactate can be detected in tissues with a high rate of glycolysis. Thus MRS can be used to detect disorders which affect energy metabolism. Anoxia or ischemia inhibit oxidative phosphorylation, and diminish ATP levels. Furthermore, if oxidative production of ATP is slowed, glycolysis may be stimulated, and MRS may detect increased production of lactic acid, and a decrease in tissue pH.

A related potential of MRS is its application to the study of thermodynamics in vivo. Energy derived from the hydrolysis of ATP drives most biochemical reactions, either directly or indirectly. The energy derived from ATP hydrolysis, i.e. $\Delta G(ATP)$ is given by:

$$\Delta G(ATP) = \Delta G^O + RT*ln([ATP]/[ADP]*[P_i]). \qquad (1)$$

ATP and P_i intensities measured by MRS, can be translated into absolute molar concentrations (3). In brain and muscle, the free ADP concentration can be calculated from MRS data if the equilibrium constant for the creatine kinase reaction, and the concentration of creatine are known. The concentration of free Mg^{+2} and pH also influence $\Delta G(ATP)$, and these can also be estimated from MRS spectra, as described above. Thus $\Delta G(ATP)$ can be calculated from MRS data for a variety of tissues, and under a variety of conditions, and these data can be used to study thermodynamic control of biochemical processes.

In general, it appears that the $\Delta G(ATP)$ in normal tissues is considerably above the energy requirements of sodium transport and muscle contraction. In other words, the energy available for cell work functions is far in excess of energy requirements. Under these circumstances, the enzymatic reactions of cellular work are probably limited kinetically rather than thermodynamically. However, a variety of physiological and pathological conditions increase the rate of ATP utilization, or inhibit the rate of ATP generation, leading to a fall of $\Delta G(ATP)$. Common examples of these conditions are muscular exercise, ischemia, brain convulsions, or hypoxia. As the $\Delta G(ATP)$ approaches ΔG for a particular work function, the reaction may no longer proceed. A number of investigators are working to determine whether MRS can be used to monitor these changes in $\Delta G(ATP)$, and thus predict changes in the ability of a tissue to do work. It should be emphasized however that use of MRS to calculate $\Delta G(ATP)$ is based on a number of assumptions. One of these is that the metabolite concentrations and pH in the tissue studied are homogeneous, so that $\Delta G(ATP)$ is also homogeneous.

Successful application of MRS for in Vivo studies depends on the ability to obtain signals from selected volumes of tissue, while suppressing signals from surrounding tissue. In Vivo MRS spectra are usually obtained using an inductor tuned to the proper frequency, which is known as a surface coil. The surface coil is a flat loop of wire which provides some spatial resolution because it is sensitive to signals which are no more than one coil diameter away. Further resolution is obtained by applying a magnetic field gradient across the sample, which causes nuclei to resonate at a frequency which depends on their spatial location. Excitation applied at the proper frequency can then be used to selectively excite nuclei within a single volume of interest.

A class of localization techniques known as chemical shift imaging techniques make it possible to acquire signal from of a number of regions of tissue simultaneously. Gradients can be used to define the region in one, two, or three dimensions. Figure 2 shows the result of applying one such method, known as the surface coil rotating frame experiment, to obtain ^{31}P spectra of an exteriorized mouse kidney (5). Two regions of the kidney can be distinguished; the cortex, which contains relatively high ATP, and low P_i, and the medulla, in which ATP is lower. Clearly, application of these techniques in Vivo can allow good discrimination of different regions of tissue. The size of the volumes which can be resolved is determined primarily by sensitivity. In high field magnets (i.e. 5 - 9 Tesla) which can accommodate small animals, spectra can be obtained from as little as 50 mgs of tissue.

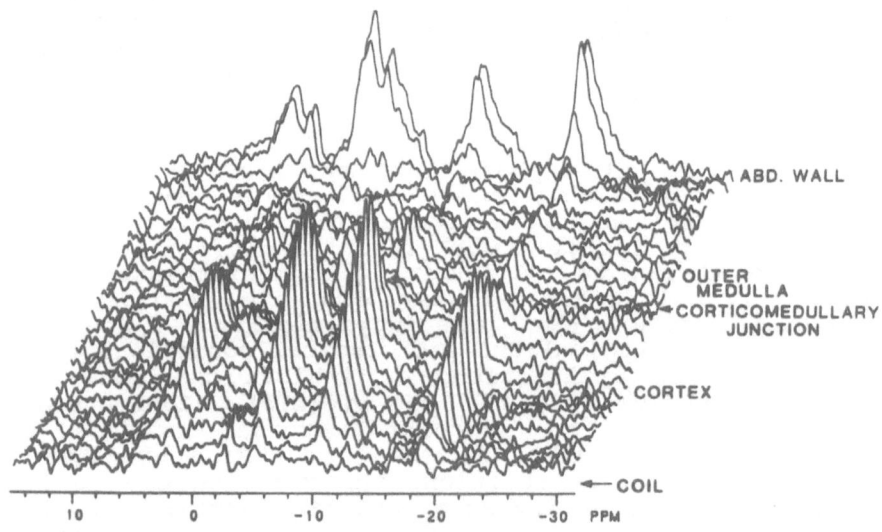

Fig. 2. One dimensional ^{31}P chemical shift image of the exteriorized mouse kidney, obtained in a 7 T. Chemical shifts are given along the X axis, the distance from the detector is given along Y, and the intensities of the signals are given along Z. (From Bogusky et al.)

However, in the lower field magnets (2 T or lower) which are large enough to accommodate human subjects, the volumes which can be discriminated are no smaller than 0.5 gms at the surface of the body near the detector. In deeper regions of tissue, for example in the kidney, volumes of at least 50 mls are needed for adequate sensitivity. Although techniques for spatially resolved spectroscopy are constantly improving, the fundamental insensitivity of MR will continue to sharply limit resolution.

In the remainder of this chapter application of the principles described above to study a number of different organs and tissues will be reviewed.

CANCER

One of the most promising clinical applications of MRS is in the design of cancer therapy. Glickson (6,7) and Griffiths (8) and their coworkers first characterized ^{31}P MRS spectra of rodent tumors in Vivo. Tumors are generally found to have low ATP/P_i and PCr, and high pH; this is believed to reflect poor perfusion and a high rate of glycolysis. Tumors often have high concentrations of phosphomonoesters and diesters relative to the tissue of origin; this may reflect abnormally high rates of membrane synthesis and degradation. Glickson et al. (6,7) and Griffiths et al. (8) and others also demonstrated that ^{31}P MRS spectra of experimental tumors showed rapid changes after initiation of a variety of therapies including chemotherapy, radiation, and hyperthermia (9). These results have been confirmed and extended by other investigators.

Figure 3 shows results of a study, performed in our laboratory (10) by Shine and coworkers, of the effect of Tumor Necrosis Factor (TNF) on a murine methylcholanthrene-induced (Meth) sarcoma. Within 120 minutes after administration of TNF, dramatic decreases in ATP and PCr, and increases in P_i were observed. These metabolic changes were correlated with the earliest histological appearance of thrombosis and hemorrhage.

MRS has also been used to assess the effect of metabolic inhibitors on rat rhabdomyosarcomas. Studies performed in our laboratory by Arbeit and coworkers demonstrated that 2-deoxyglucose, a selective inhibitor of glycolysis, reduced tumor ATP concentrations by 50% or more (Figure 4). The spectra also showed increased intensity in the PME region due to accumulation of 2-deoxyglucose-6-phosphate. Insulin produced a selective and significant decrease in ATP, and an increase in P_i in these tumors (11). Metabolism in other tissues appeared unaffected. These results suggest that these tumors depend heavily on glycolysis for production of ATP, and are unable to use alternative substrates when access to glucose is blocked.

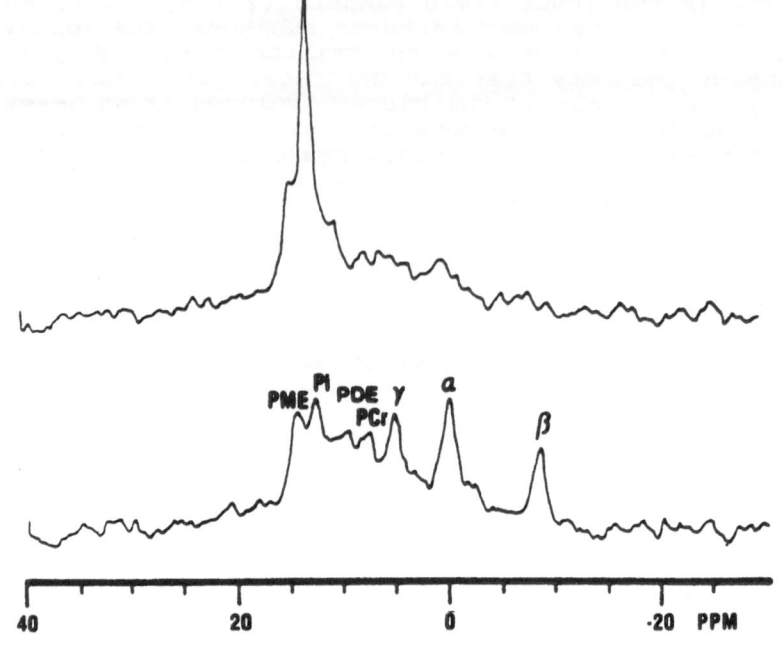

Fig. 3. ^{31}P spectra showing effect of TNF on a murine
Meth-a sarcoma. (From Shine et al.)

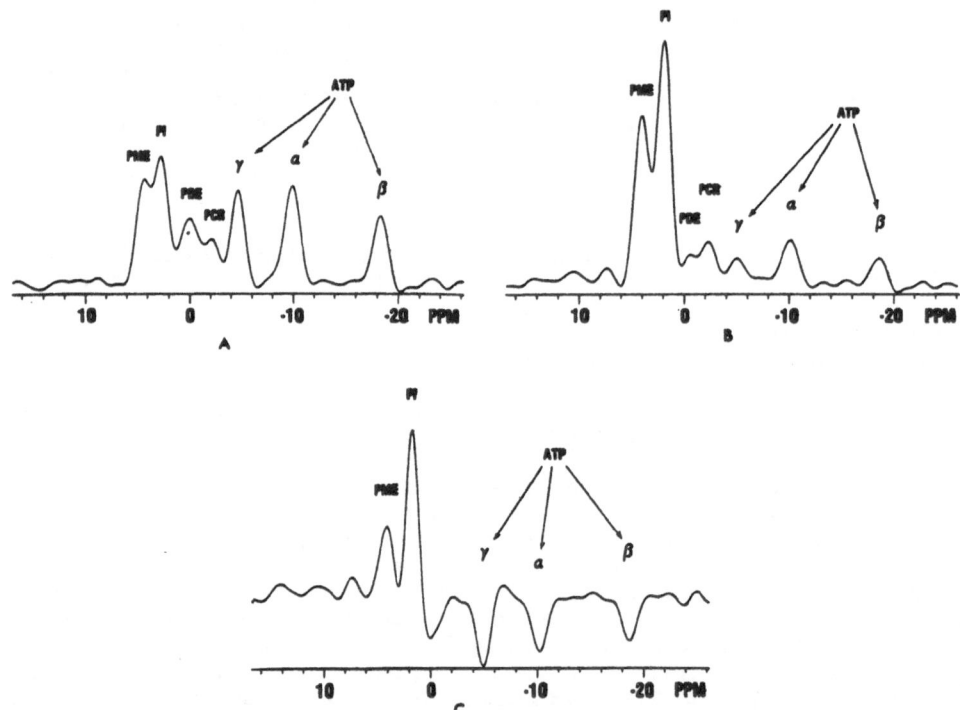

Fig. 4. ^{31}P MRS spectra showing the effect of 2DG on
rat rhabdomyosarcoma.   A) Before 2DG, B)
after 2DG, and C) difference.

One dimensional chemical shift images can be used to detect metabolic heterogeneity within tumors (12,13). Spectra of a rat rhabdomyosarcoma, obtained at 2 T using the surface coil rotating frame experiment, (Figure 5) show two distinguishable tumor zones. Tissue at the surface of the tumor has high levels of ATP, some PCr, and relatively low P_i. In a deeper region within the tumor, ATP/P_i and PCr are lower than at the surface.

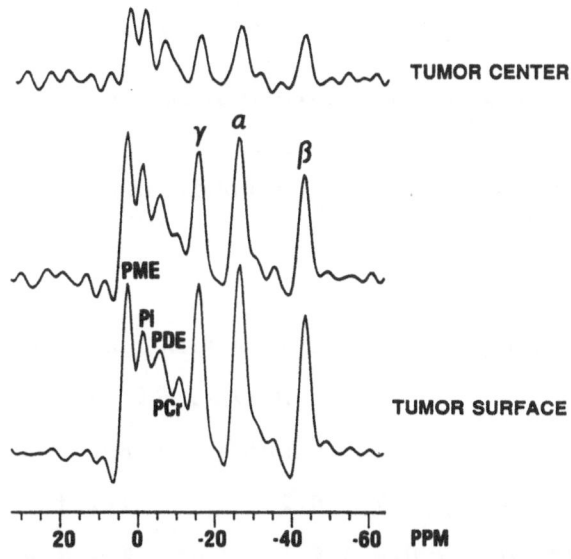

Fig. 5. One dimensional ^{31}P chemical shift image of a rat rhabdomyosarcoma. The upper spectra come from the surface of the tumor nearest the coil, and the lower spectra come from skeletal muscle beneath the tumor. (From Karczmar et al.)

These observations support the hypothesis that superficial regions of the tumor are able to maintain a high rate of oxidative phosphorylation because they are well perfused, while deeper regions are relatively hypoxic, and may rely on glycolysis as a source of energy. These techniques may also reveal spatial heterogeneity of the response of tumors to therapy.

Few studies of human tumors have been reported to date. In general these studies suggest that in superficial tumors outside of the brain, the PME/ATP ratio is higher than in normal tissue, and that this ratio decreases as tumors respond to treatment.

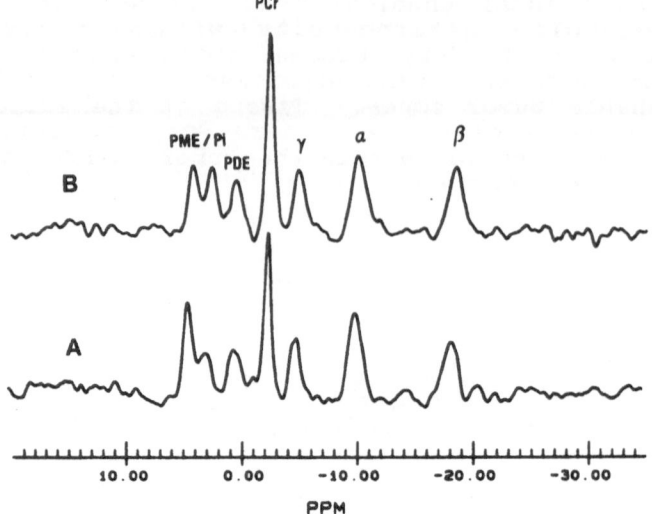

Fig. 6. ^{31}P Spectra of a lymphoma in a human subject before (A) and after (B) treatment with methotrexate.

Table 1. Comparison of responders and non-responders

PME/β-ATP:

	Responders	Non-responders
Control	1.7 ± 0.2	1.1 ± 0.1
Treated	0.8 ± 0.2	1.7 ± 0.3
%change	−55	+50

Studies performed on superficial human tumors in this laboratory support these findings (14). Figure 6 shows spectra of a lymphoma which responded to therapy, before (A) and after (B) treatment with methotrexate. Table 1 shows that in tumors which responded to therapy, as determined by tumor size, and other clinical tests, PME/β-ATP decreased, while in tumors which did not respond to therapy, PME/β-ATP increased.

Studies performed on brain tumors in this laboratory by Hubesch et al. (15) suggested that relative concentrations of tumor metabolites detected by ^{31}P NMR are not significantly different from those of normal brain tissue. However, calculations of absolute concentrations from MRS data showed that the absolute concentrations of metabolites in brain tumors are much lower than in normal brain.

MRS may ultimately provide an early indication of response of tumors to treatment, which can be used as a guide by physicians. In addition MRS may become an important experimental tool for the design of new therapy.

The first use of MRS to study intact tissue was a study of excised muscle (17). Subsequently, experiments have been performed on animal muscles, and more recently on normal (18) and diseased (19) human muscles. Muscular exercise results in depletion of PCr, increased P_i, and tissue acidosis. These changes follow characteristic time courses in normal subjects which can be used as a basis for comparison with data from patients with muscle diseases. Radda et al. were able to diagnose Mcardle's syndrome using MRS, by demonstrating that the patient's muscles did not acidify during exercise, reflecting an inability to break down glycogen to fuel glycolysis (20).

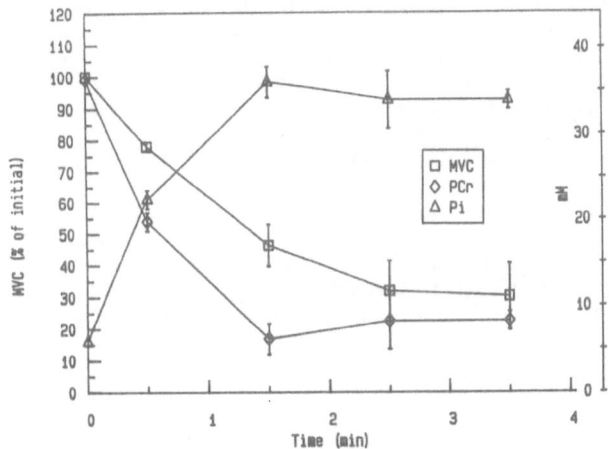

Fig. 7. Effects of 4 minute sustained exercise of the human adductor pollicis on PCr, P_i, and MVC. (From Miller et al.)

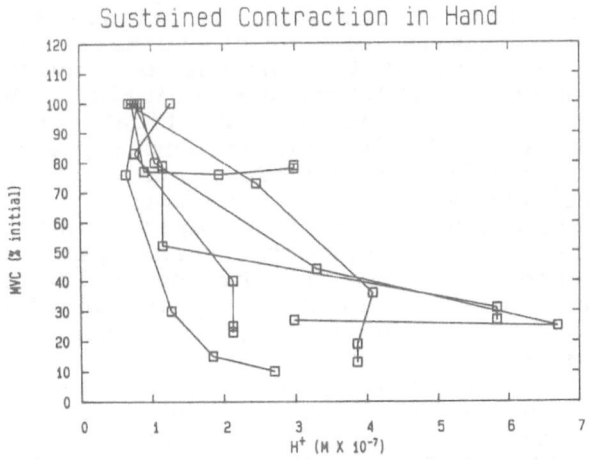

Fig. 8. Plot of [H^+] versus MVC during exercise of the human adductor policis. (From Miller et al.)

An important area of MRS research is the metabolic basis of muscular fatigue (21). Previous work has suggested that a fall in tissue pH is responsible for fatigue although the evidence is circumstantial and controversial. Miller et al., in this laboratory (22), performed experiments to test this hypothesis.

Figure 7 shows the effects of 4 minute sustained exercise of the adductor pollicis on PCr, P_i, and maximum voluntary contraction (MVC), which is an indicator of muscle fatigue. Exercise produced a rapid fall of PCr and increase in P_i, as fatigue increased. ATP concentrations were not significantly altered, because ATP is buffered by PCr. Figure 8 shows a good correlation between increasing H^+ and decreasing MVC at high pH, emphasizing the role of H^+. Dawson et al. (23) recently proposed that $H_2PO_4^-$ may be a direct inhibitor of actinomycin cross-bridge formation, and suggested a role of this compound in muscular fatigue. $H_2PO_4^-$ can be calculated from pH and P_i, measured by MRS. Results obtained by Miller et al. demonstrate a good correlation between decreasing MVC and increasing $H_2PO_4^-$. To further investigate the role of $H_2PO_4^-$ in human muscular fatigue, Miller and associates studied the effects of aerobic and anaerobic exercise on the human adductor pollicis and tibialis anterior. The relationship between H^+ and MVC, and $H_2PO_4^-$ and MVC was not significantly different in the two muscle groups, during the two types of exercise. This constant relationship between $H_2PO_4^-$ and fatigue supports the concept that $H_2PO_4^-$ is an important determinant of muscle fatigue.

HEART

Early studies demonstrating the feasibility of measuring various phosphates in perfused hearts were performed by Jacobus (24), Gadian (25) and their coworkers. It was quickly demonstrated that anoxia or ischemia results in a rapid depletion of PCr, accompanied by a rise in P_i and an acid shift of the P_i peak due to tissue acidosis. MRS studies on the hearts of intact animals have been more difficult to perform, but a number of investigators have reported studies with laboratory animals. Work in this laboratory has used ^{31}P MRS to investigate changes of high-energy phosphate metabolism during coronary occlusion in the intact pig. Camacho, Lanzer, Toy, and colleagues used ^{31}P MRS to investigate effects of acute coronary occlusion (26). Coronary occlusion caused a rapid depletion of PCr accompanied by a rise of P_i, without any change of ATP. The fall of pH, calculated from the chemical shift of the P_i peak was slower than the fall of PCr. Reduction in contractility produced by coronary occlusion, monitored by ultrasonic crystals, was much more rapid than changes detected by MR. However, the insensitivity of MR requires a long acquisition time for each MR spectrum. Thus, rapid changes in metabolites are not easily detected. Schaefer and colleagues used ^{31}P MRS to investigate metabolic alterations produced by partial steady state occlusion of the pig coronary artery. Progressive reduction of coronary blood flow lead to a progressive decrease in PCr and an increase in P_i. Endocardial blood flow, monitored by radioactive microspheres, directly correlated with alterations of PCr/P_i. These findings are consistent with the hypothesis that reduction of myocardial contractility is directly related to alterations of energy metabolism. Diminished contractility may be due to a decrease in $\Delta G(ATP)$ or changes of phosphate and/or H^+.

MRS studies of human heart were first reported by Bottomley (27) and subsequently by Blackeledge (28). Bottomley et al. reported that myocardial infarction in man was associated with partial depletion of PCr, and prolonged elevation of P_i. These early studies of human heart used MRS techniques which provided high spatial resolution in only one dimension. MRS spectra of human heart with 3-dimensional resolution were obtained in our laboratory using the ISIS technique (29,30). The spectrum shown in Figure 9 was obtained from a 60 ml volume in the apex of the heart, in 35 minutes. To minimize motion artifacts, data acquisition was gated using EKG; data was acquired only at peak systole. Preliminary results of Schaefer et al. obtained using these techniques suggest that spectra obtained from patients with cardiomyopathies show unusually high intensity resonances in the PDE region, with a PDE/PCr ratio of more than twice control.

Because myocardial ischemia and infarction are common causes of morbidity and mortality, it can be expected that MRS will be widely used in experimental animals and eventually human patients.

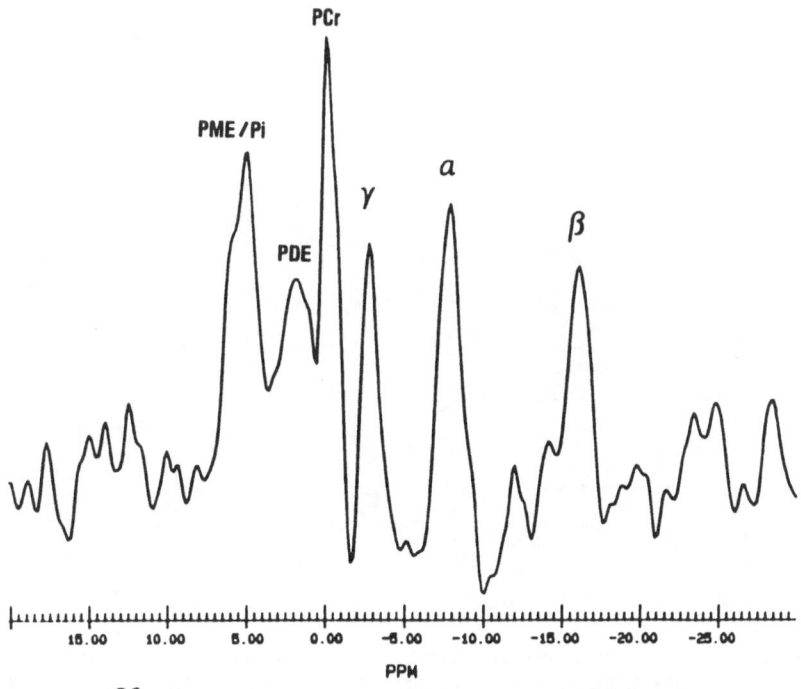

Fig. 9. ^{31}P spectrum of a 60 ml volume in the apex of the heart was obtained in 35 minutes.

Chance et al (31) first obtained ^{31}P MR spectra from the head of an anesthetized mouse. Subsequently, Ackerman et al. (32) used a surface coil to study the brain of a living rat. Thulborn et al. first studied the effects of ischemia produced by carotid occlusion. Ischemia produced the expected loss of PCr and ATP, a rise of P_i, and acidification (33). Since this pioneering work, many investigators have used MRS to study the effects of ischemia, seizures, hypoglycemia, anoxia, anesthetics, metabolic inhibitors and other agents which affect brain metabolism (34).

Roth and coworkers (3) in this laboratory used the ISIS technique to obtain ^{31}P MRS spectra from the human brain (Figure 1). These workers calculated absolute concentrations of metabolites in brain, and decreases in these concentrations in brain tumors (see above).

Figure 10A shows a ^{31}P MR spectrum of human liver obtained by Meyerhoff et al. (35) using the ISIS technique. The MR image (Figure 10B) shows the volume from which a ^{31}P spectrum was obtained.

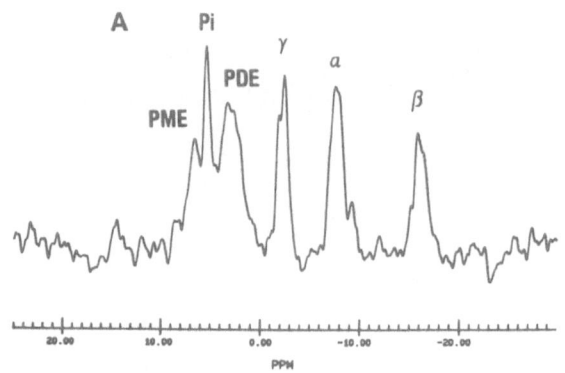

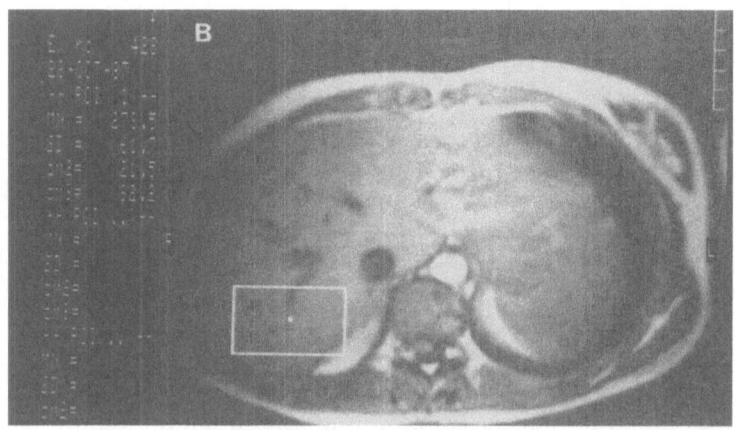

Fig. 10. A ^{31}P spectrum of human liver acquired in 15 minutes (A) and an image (B) showing the box from which the ISIS spectrum was obtained. (From Meyerhoff et al.)

Meyerhoff et al. recently observed that the
concentrations of various phosphate compounds were markedly
reduced in patients with the diffuse liver diseases of
alchoholic hepatitis and cirrhosis. Several investigators
have studied the effects of the sugar fructose on hepatic
energy metabolism in animals and man. Radda and coworkers
(36) used this approach to detected abnormalities of fructose
metabolism in several volunteers in whom fructose intolerance
had not previously been diagnosed.

In Vivo MRS studies of the human orthotopic kidney
present a considerable challenge because of its deep location
and respiratory motion. Transplanted kidneys are closer to
the surface, but signals from thick overlying muscle can
contaminate spectra. Boska et al. (37) have used the ISIS
method to study human orthotopic and transplanted kidneys in
this laboratory. Spectra of orthotopic kidney (Figure 11A),
and transplanted kidney (Figure 11B) can be identified
because of the unusually high intensity in the PDE region,
which may be due in part to urine phosphate, and the lack of
PCr. Preliminary results of Boska et al. suggested that PME
is elevated in kidney transplants relative to normal kidneys.
This may be related to transplantation, or may reflect
hypertrophy in transplanted kidneys.

An ultimate goal of MRS studies of transplanted kidneys
is to predict rejection of the transplant, and to distinguish
between rejection and cyclosporin toxicity.

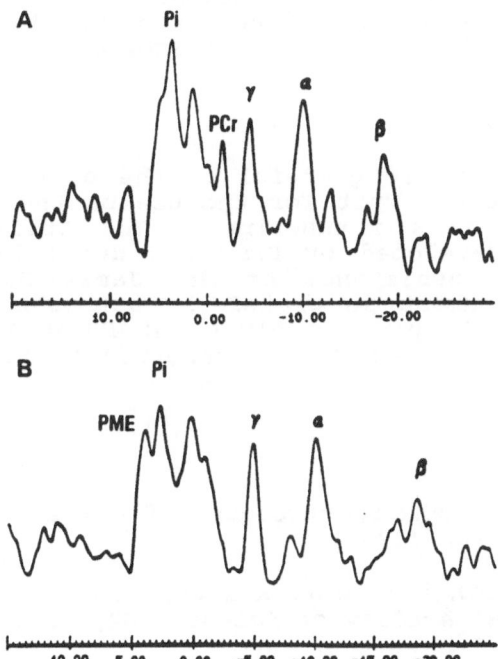

Fig. 11. Spectra of orthotopic kidney (A), and
transplanted kidney (B) can be identified
because of the unusually high intensity in
the PDE region, which may be due in part to
urine phosphate, and the lack of PCr. (From
Matson et al.)

PROBLEMS AND CONCLUSIONS

This chapter has emphasized the capabilities of MRS for biochemical and clinical research. However, a number of technical and interpretive problems limit the value of this technique. The major problem is sensitivity: MRS signals from tissue metabolites are extremely weak and living tissue generates large amounts of noise. Therefore MRS signals must be obtained from relatively large volumes of tissue, given currently available magnetic field strengths. MRS is at present relatively insensitive to tissue heterogeneity which exists at several levels: intracellular heterogeneity (mitochondria, nucleus, cytosol, endoplasmic reticulum), intracellular versus extracellular heterogeneity, and zonal heterogeneity (eg. epicardium v.s. endocardium). In addition to heterogeneity of metabolite levels, there may be heterogeneity of relaxation times, magnetic field susceptibility, and other factors which affect MR signals. MRS detects only mobile metabolites, metabolites which are tightly bound to macromolecular structures are not detected.

Despite these difficulties, the ability of MRS to noninvasively monitor a number of compounds which play an important role in tissue metabolism, with some degree of spatial discrimination, make it an important tool for the investigation of the pathophysiology of common conditions in human subjects and animal models of human disease. In the near future, our ability to obtain, understand, and apply information from MRS should improve dramatically. Ultimately, this new technique for clinical investigation may become an important diagnostic resource.

ACKNOWLEDGEMENTS

The authors are grateful to the staff of the UCSF/VAMC Magnetic Resonance Unit for the use of their data, and help in preparing this manuscript. The Philips spectroscopy package was developed by Drs. Jan den Hollander and Peter Luyten. The assistance of Mr. James Buchanan with the Gyroscan instrument is gratefully acknowledged. This work was supported in part by NIH Grant R01AM33293, and Philips Medical Systems. G.S.K. is supported by NIH fellowship F32 AM07770-01.

REFERENCES

1) R.K. Gupta, NMR Spectroscopy of Cells and Organisms, CRC Press, Boca Raton (1987)

2) H.P. Hetherington, M.J. Avison, R.G. Shulman, Proceedings of the National Academy of Sciences 82, 3115 (1985)

3) K. Roth, B. Hubesch, S. Naruse, J. Gober, T. Lawry, M. Boska, G.B. Matson, and M.W. Weiner, (Abstract) Society of Magnetic Resonance in Medicine 2, 608 (1987)

4) D. Eidelberg, G. Johnson, D. Barnes, P.S. Tofts, D. Delpy, D. Plummer, and W.I. McDonald, Magnetic Resonance in Medicine 6, 244 (1988)

5) R.T. Bogusky, M. Garwood, G.B. Matson, G. Acosta, L.D. Cowgill, and T. Schleich, Magnetic Resonance in Medicine 3, 251 (1986)

6) L.M. Schiffer, P.G. Braunschweiger, J.D. Glickson, W.T. Evanochko, T.C. Ng, Annals of the New York Academy of Science, 6481 (1981)

7) W.T. Evanochko, T.C. Ng, M.B. Lilly, A.J. Lawson, T.H. Corbelt, J.R. Durant, J.D. Glickson, Proc Natl Acad Sci USA 80, 334 (1983)

8) J.R. Griffiths, A.N. Stevens, R.A. Iles, R.E. Gordon, D. Shaw, Bioscience Reports 1, 319 (1981)

9) J.M. Maris, B. Chance, in Magnetic Resonance Annual, ed. Herbert Y. Kressel, Raven Press, New York, pp. 213 (1986)

10) N. Shine, M.A. Palladino, J.S. Patton, A. Deisseroth, G. Karczmar, G. Matson, and M.W. Weiner, Journal of Clinical Investigation, submitted

11) J.M. Arbeit, J.B. Toy, M.W. Weiner, (Abstract) Society of Magnetic Resonance in Medicine 1, 64 (1987)

12) G.S. Karczmar, M.W. Weiner, and G.B. Matson, Journal of Magnetic Resonance 71, 360 (1987)

13) G.S. Karczmar, T. Lawry, G.B. Matson, and M.W. Weiner, NMR in Biomedicine, submitted

14) G.S. Karczmar, J. Poole, M.D. Boska, D.J. Meyerhoff, B. Hubesch, K.Roth, G.B. Matson, J. Arbeit, F. Valone, and M.W. Weiner, (Abstract) Society of Magnetic Resonance in Medicine in Medicine, submitted (1988)

15) B. Hubesch, D. Sappey-Marinier, K. Roth, E. Sanuki, J.E. Hodes, G.B. Matson, and M.W. Weiner, (Abstract) Society of Magnetic Resonance in Medicine, submitted (1988)

16) D.T. Twieg, D. Meyerhoff, J. Gober, M. Boska, B. Hubesch, S. Schaefer, K. Roth, D. Sappey-Marinier, and M.W. Weiner, (Abstract) Society of Magnetic Resonance in Medicine, submitted (1988)

17) D.I. Hoult, S.J.W. Busby, D.G. Gadian, G.K. Radda, R.E. Richards, and P.J. Seeley, Nature 252, 285 (1974)

18) D.J. Taylor, P.J.Bore, P. Styles, D.G. Gadian, and G.K. Radda, Molecular Biology in Medicine 1, 77 (1983)

19) E.D. Ross, G.K. Radda, D.G. Gadian, G. Rocker, M. Esiri, and J. Falconer-Smith 304, 1338 (1981)

20) G.K. Radda, Science 233, 640 (1986)

21) R.G. Miller, D. Giannini, H.S. Milner-Brown, R.B. Layser, A.P. Koretsky, D. Hooper, and M.W. Weiner, Muscle and Nerve, in press

22) M.R. Boska, M. Moussavi, R. Miller, R. Layser, and M.W. Weiner, (Abstract) Society of Magnetic Resonance in Medicine 2, 446 (1986)

23) M.J. Dawson, K.J. Brooks, D. Mcfarlane, and S.J. Smith, (Abstract) Society of Magnetic Resonance in Medicine 2, 447 (1986)

24) W.E. Jacobus, G.J. Taylor, D.P. Hollis, and R.L. Nunnally, Nature 265, 756 (1977)

25) D.G. Gadian, D.I. Hoult, G.K. Radda, P.J. Seeley, B. Chance, and C. Barlow, Proceedings of the National Academy of Science USA 73, 4446 (1976)

26) B.J. Toy, P. Lanzer, A. Camacho, M. Valenza, J. Gober, G.S. Karczmar, E. Botnivick, B. Massie, G.B. Matson, and M.W. Weiner, (Abstract) Society of Magnetic Resonance in Medicine 2, 1018 (1987)

27) P.A. Bottomley, R.J. Herfkens, L.S. Smith, S. Brazzamano, R. Blinder, L.W. Hedlund, J.L. Swain, and R.W. Redington, Proceedings of the National Academy of Science USA 82, 8747 (1985)

28) M.J. Blackledge, B. Rajagopalan, R.D. Oberhansli, N.M. Bolas, P. Styles, and G.K. Radda, Proceedings of the National Academy of Science 84, 4283 (1987)

29) R.J. Ordidge, A. Connelly, J.A.B. Lohman, J Magn Reson 66, 283 (1986)

30) G.B. Matson, D.T. Twieg, G.S. Karczmar, T.J. Lawry, J.R. Gober, M. Valenza, M.D. Boska, and M.W. Weiner, Radiology, in press

31) B. Chance, Y. Nakase, M. Bond, J.S. Leigh Jr., and G. McDonald, Proceedings of the National Academy of Science USA 75, 4925 (1978)

32) J.J.H. Ackerman, T.H. Grove, G.G. Wong, D.G. Gadian, and G.K. Radda, Nature 283, 167 (1980)

33) K.R. Thulborn, G. du Boulaym and G.K. Radda, Journal of Cerebral Blood Flow and Metabolism 1, 580 (1981)

34) J.W. Prichard and R.G. Shulman, Annual Review of Neurosciences 9, 61 (1986)

35) D.J. Meyerhoff, G.B. Matson, G.S. Karczmar, M.D. Boska, D.R. Rockey, M.W. Weiner, Radiology, submitted

36) R.D. Oberhanslii, G.J. Galloway, D.J. Taylor, P.J. Bore, G.K. Radda, British Journal of Radiology 59, 695 (1986)

37) M.D. Boska, D.B. Twieg, G.S. Karczmar, D.J. Meyerhoff, D. Sapey-Marinier, G.B. Matson, and M.W. Weiner (Abstract) Society of Magnetic Resonance in Medicine, submitted (1988)

ELECTROROTATION OF BLOOD CELLS:
MEDICAL APPLICATIONS

Roland Glaser, Marcel Egger,
Cornelia Pritzen and Heiko Ziervogel

Sektion Biologie, Bereich Biophysik
Humboldt-Universität
Berlin, GDR

INTRODUCTION

The use of dielectric measurements as a tool for studying cellular parameters even for diagnostic reasons is a well established technique of continuing importance. Usually the passive electrical properties of tissues and cell suspensions are measured by the electrical impedance method (10). Recently electrorotation was introduced as a method which allows the measurement of these properties in single cells (1,2,7,8). In contrast to cell electrophoresis, which allows the determination of surface charges of cells by the measurement of cell movement in a constant DC-field, in case of electrorotation the spin of cells in a rotating high frequency field is measured in dependence of the rotation frequency on the applied field.

Differences between the electrical properties of the cells in relation to their environment are the basic conditions for electrorotation. Biological cells in the simplest case can be considered as spheres containing an electrolyte medium with high electric conductivity (about 0.2 S/m) and a dielectric constant near water (about 50) surrounded by a membrane with low conductivity (1 uS/m) whose dielectric constant is significantly different from that of water (about 8). Electrorotation was found to be attributed to polarization phenomena resulting from charge separations on these dielectric boundary layers. This polarisation effect can be described by a resulting electrical dipole. At low frequencies, this dipole follows the vector of the rotating field without delay, but with increasing frequency a definite angle of delay occurs, leading to a permanent torque of the cell. The direction of this torque in relation to the spin of the field depends on the nature of the induced dipole, and therefore on the frequency of the applied field and the dielectric properties of the cell. In the low frequency range an "anti-field rotation", and at higher frequencies a "co-field rotation" takes place. In any case the rotation speed is much slower than the rotation frequency of the field. The torque of the cells therefore can easily be measured by microscopic observations.

EXPERIMENTAL TECHNIQUES

Usually electrorotation is measured in a small chamber, mounted on a glass slide under a microscope. This chamber contains four electrodes made of platinum, silver or stainless steel. They have a mutual distance of several millimeters, depending on the size of the observed objects and the necessary field strength. The electrodes are driven by sinusoidal waves or pulses with a definite phase shift. This generates an electric field in the chamber with a field strength of several kV/m, rotating in a frequency range between 1 kHz and 35 MHz.

The spin of individual cells is measured in the simplest way by microscopic observation using a stop watch. One has to take particular care to minimise the mechanical friction of the rotating objects. The best way is to observe particles in free suspension. In some cases, however, cells can be measured even after sedimentation of glass or other surfaces.

For electrorotation measurements the cells have to be suspended in solutions with low conductivity. Usually isotonic sucrose solutions are applied with a low content of cations.

EXAMPLES OF ELECTROROTATION MEASUREMENTS

FIG 1A indicates an example for a typical electrorotation spectrum of a human erythrocyte as a very simple cell. The measured rotation speed of the cell in arbitary units (rad/s, rotations/s, rotation speed in relation to the maximal rotation etc.) is plotted versus the logarithm of the applied field frequency. In some cases a parameter is used, called "rotation". This is the rotation speed divided by the square of the applied field strength. This parameter is usually in the order of 10^{-8} to 10^{-7} $rad.m^2s^{-1}v^{-2}$ (see Fig. 3). It has the same significance for electrorotation as "mobility" for cell electrophoreses.

In the simplest case, this electrorotation spectrum can be fitted by a theoretical curve, representing a single-shell model (A spherical cell with homogeneous cytoplasma surrounded by a single membrane). The two characteristic frequencies (f_1, f_2) and the corresponding rotations (N_1, N_2) contain the basic information about the cell. Knowing the cell radius, the outside conductivity and the dielectric constant of the solution, one can calculate the internal conductivity of the cell, their membrane conductivity and the membrane capacity with the help of this model (7, 8).

All changes in these parameters are reflected by the shape of the whole electrorotation spectrum. For example the first characteristic frequency indicates the membrane conductivity. If the cell becomes leaky, the value of N_1 decreases dramatically. This can be used to indicate, whether cells reseal fully after electric break down (3, 6).

Fig. 1B indicates an electrorotation spectrum of a human peripheral lymphocyte with a large nucleus. This spectrum

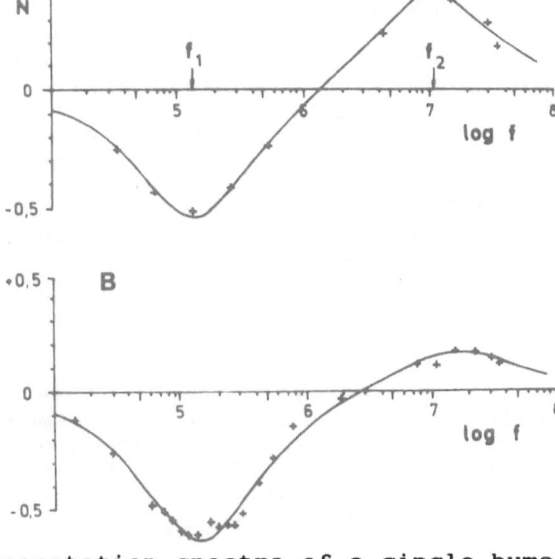

Fig. 1. Electrorotation spectra of a single human erythrocyte
(A) and a peripheral blood lymphocyte (B). The plotted
curves indicate functions corresponding to the single-
-shell model fitted to the points. In case A the model
fitts the points well. In case B, systematic deviations
occure in the region of the anti-field rotation,due to
the influence of the dielectric properties of the
nucleus.

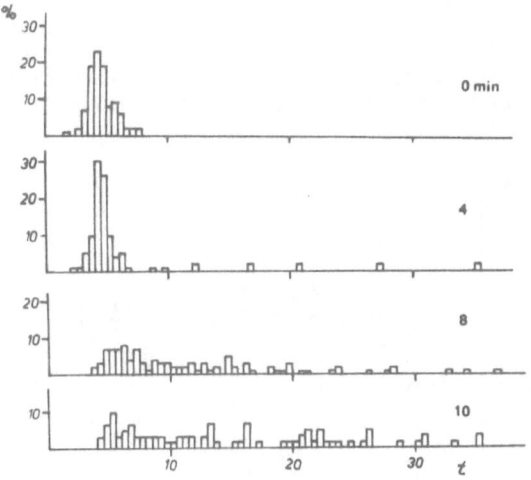

Fig. 2. Interaction of influenza virus (A/PR 8/34) with human
erythrocytes as reflected in the anti-field rotation of
erythrocytes. Abscissa: time in sec. for one full revolution of the cell, ordinate: relative number of cells
in the corresponding classes (in %). The graphs show the
gradual diminuition of the number of cells with high
rotation speed (resp. low rotation time) as a function
of the incubation time (t). (Experimental conditions:
1.7 µg virus protein per ml, pH=5.2, T=310 K).

shows significant deviations in the anti-field rotation from the fitted curve according to the single cell model. We could indicate, that in this way the electrical behaviour of the nucleus is directly measurable (11). It is not possible to evaluate cellular parameters from multi-shell objects in the same direct way as in case of the single-shell objects. In any case, however, important qualitative conclusions are possible. So we were able to characterize membrane events following the mitogenic stimulation. An enhanced cell membrane conductivity was detected for the whole cell population immediately after addition of the mitogen. But only successfully stimulated cells remain in this status, whereas other cells which were unaffected by the stimulation were measured to return to a situation of low membrane conductivity, comparable to that before stimulation (11).

Fig. 2 demonstrates experiments on virus-cell interactions. In this case influenza virus was added to suspensions of erythrocytes at 273 K for 5 minutes (process of virus adsorption). Than the temperature was rised up to 310 K and the rotation speed of individual cells was measured near the first characteristic frequency (f_1 in Fig 1A).

At the beginning of the experiment the rotation speed (or as a reciproc - the time for one full rotation) was distributed statistically. After several minutes at 310 K an increasing number of cells shows a slower rotation speed. As we could demonstrate in a recent paper (8), not only the rotaion speed, but also the first characteristic frequency shifts as a result of virus - cell interaction. The advantage of this method is the possibility to follow the individual behaviour of single cells. The change of the electrorotational parameters (f_1, N_1) is the result of the increase of the membrane permeability of red cells after fusion with the virus and the subsequent decrease of the internal conductivity. This corresponds to the observations of Pasternak et al (9).

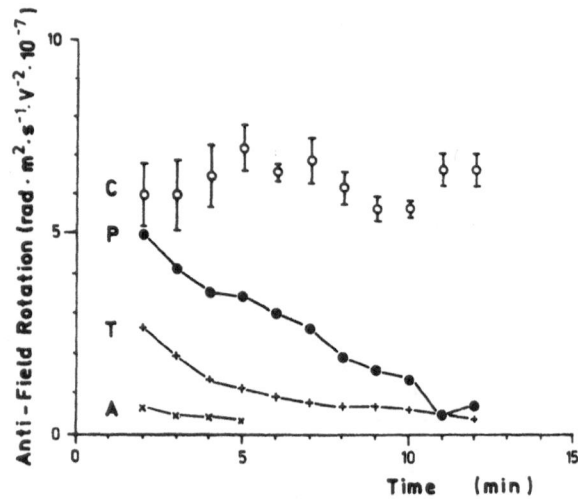

Fig. 3. Activation kinetics of platelets as reflected in the anti-field rotation (C - control, P- PAP-activated, T - thrombin activated, A - activated by A 23187-ionophore)

As a third example, the possibility of measuring the activation of platelets will be demonstrated (Fig. 3). The electrorotation spectrum of platelets can be fitted by the single-shell model (4, 5). There is a clear correlation between the serotonin release (measured with ^{14}C-serotonin) and the decrease of rotation near the first characteristic frequency of the platelets. The receptor-coupled exocytotic release reactions induced by different activators were not only characterized by their significant rotation decrease but also by their fast kinetics

There are no qualitative differences between the electrorotation kinetics of ionophore- and receptor coupled exocytotic release reactions of activated platelets. A dramatic decrease of the anti-field rotation within the first two minutes was observed applying inducers such as thrombin or the calcium-ionophore A 23187. On the contrary, adrenalin shows only a slow decrease of rotation. This could be a further hint towards the existence of different exocytotic pathways at least in the case of platelets.

This example shows that it is possible to follow the activation kinetics of individual thrombocytes with the use of electrorotation measurements. In this way it is also possible to investigate pathological platelets, especially platelet membrane receptor defects. This is helpful especially in cases, like the May-Hegglin-syndrome, where the number of platelets is limited.

SUMMARY

In this short rewiev we tried to demonstrate that the electrorotation is a new and useful method to indicate changes of physical properties of individual cells. A full survey on the application of this method in biological, medical, pharmacological as well as biotechnological research was given in a previous paper (8). The method is relatively young and its application therefore not fully discovered. Electrorotation allows to follow changes in single cells without damaging them. This means a great advantage in some biotechnological questions, where it is necessary to select cells with special properties for further cultivation. The disadvantage of the method is the fact, that all properties are measured under conditions of low ionic strength solutions. Many cells resist this treatment for a long time. The aim of this paper was to direct the interest to this method to check possibilities of its useful application.

REFERENCES

1. Arnold, W.M. and Zimmermann,U., Rotating-field induced rotation and measurement of the membrane capacitance of single mesophyll cells of Avena sativa, Z. Naturforsch. 37c:908 (1982).
2. Arnold, W.M. and Zimmermann,U., Rotation of isolated cell in a rotating electric field, Naturwiss. 69:297 (1982).
3. Engel, J., Donath, E., Gimsa, J., Electrorotation of red cells after dielectric breakdown, studia biophysica (in press)
4. Egger, M., Donath, E., Spangenberg, P., Ladhoff, A.-M.,

Gimsa, J., Glaser, R., Till, U., Human platelets electrorotation behavior. Influence of activators and cytoskeleton modifications, Biochim. Biophys. Acta (in press)

5. Egger, M., Donath, E., Ziemer, S., and Glaser, R., Electrorotation - A new method for investigating membrane events during thrombocyte activation. Influence of drugs and osmotic pressure, Biochim. Biophys. Acta 861:122 (1986).

6. Fuhr, G., Mller, Th., Wagner, A. and Donath E., Electrorotation of oat protoplasts before and after fusion. Plant and Cell Physiol. 28:549 (1987).

7. Glaser, R. and Fuhr, G., Electrorotation of single cells - a new method for assessment of membrane properties, in: "Electrical Double Layers in Biology", M. Blank, ed., Plenum Press, New York (1986).

8. Glaser, R., Fuhr, G., Electrorotation - the spin of cells in rotating high frequency electric fields. In: M. Blank and E. Findl: Mechanistic Approaches to Interaction of Electric and Electromechanic Fields with Living Systems. Plenum Press, New York 271-290 (1987)

9. Pasternak, C. A., Bashford, C. L. Gray, M. A., Miklem, K. J., Ion permeability and membrane potential in viral disease. in: A. Pullman, V. Vasilescu, L. Packer (eds.) Water and Ions in Biological Systems, Plenum Press, New York 1985, 483-490

10. Pethig, R., Kell, D.B., The passive electrical properties of biological systems: their significance in physiology, biophysics and biotechnology. Phys. Med. Biol. 32: 933 (1987)

11. Ziervogel, H., Glaser, R., and Schadow, D., Heymann, S.: Electrorotation of lymphocytes - the influence of membrane events and nucleus, Bioscience Reports 6: 973 (1986)

MOLECULAR DYSFUNCTIONS IN LEAD POISONING: AN IN VITRO MODEL

Diana Reséndez-Pérez[1], F. J. Sánchez-Anzaldo[2], Hugo A. Barrera-Saldaña[1,3] and Salvador Said-Fernández[4]

[1]ULIEG, Depto.de Bioquímica, Facultad de Medicina, Universidad Autónoma de Nuevo León, Monterrey, N. L.,[2] Laboratorios Clínicos de Puebla, Puebla, Pue., [3] International Center for Molecular Medicine, Monterrey,N. L. [4] División de Biología Celular, UNIBIN, IMSS, Monterrey, N. L., México

INTRODUCTION

Lead intoxication is known as saturnism. This agent, in the majority part of people, slowly produces intoxication symptoms, according to their internal concentration[1]. Bones having a great lead affinity and constitutes a saturable defense mechanism[2]. The main clinical features of lead poisoning are the following: arterial hypertension[3], vascular smooth muscle spasm[4], gastrointestinal disorders, occasional drowsiness, headache, irritability, and paralysis of certain striated muscle groups. Kidneys, hemopoietic system, and bones are also affected by lead: in severe chronic intoxication cases interstitial nephritis normocytic normochromic anemia, other erythrocytic disorders[3,4], arthritis and gout can be observed[7].

The most recent studies in the lead intoxication field have focused on determining lead toxic effects on particular enzymes from metabolic and biosynthetic routes, and knowing their relation with the above mentioned lesions and dysfunctions. The first well defined biochemical disorders produced by lead were those concerning the hematopoietic system[8], in which the main effect of lead is the inhibition of two enzymes from the biosynthetic route of hemo group: the τ-aminolevulino acid dehydratase (ALA-D) and the heme synthetase[8]. The inhibition of the former results in a δ-aminolevulinic acid (ALA) accumulation, first in the extracellular space and later in the urine, with which it is excreted (ALA-U). The consequence of the heme synthetase inhibition is a protoporphyrin-zinc (PPZ) accumulation in the erythrocyte, as an effect of the incorporation of Zn^{++} instead of Fe^{++} into the protoporphyrin molecule. The possible inhibiting mechanism of ALA-D is by binding Pb^{++} directly to one of several of its -SH groups, which modify this enzyme's active site. The heme synthetase inhibition mechanism has not yet been completely defined[4].

GENETIC IMPLICATIONS

Lead and its derivatives are considered as a potential producer of damage in experimental animal and human genetic material: In 1963, Dingwall-Fordyce and Lane[9] found significant differences in malignant neoplasm frequency between a group of workers in a car storage battery factory and its respective control group. It has also been reported that lead increases miscarriage frequencies[10], produces developmental malformations[9-13] and has infertility, mutagenic and carcinogenic effects[14]. There is evidence that indicate that lead produces damage to genetic material: it was observed that the chromosomes from a lead intoxicated patient were bound to each other by a long and thin filament[15]; and in mice leukocytes chromosomic breach like alterations and chromatid rupture increased frequencies were found[16]. In peripheral blood cultures of lead exposed laborers and in lymphocytes which were cultured in a medium containing different lead acetate concentrations, a significant increase in chromosomic and chromatid damage was reported (for a review see ref. 17). It has been proposed, notwithstanding, that the above and several other genetic material damages are not exclusively due to lead, but are the result of a concerted action of lead and several other factors, as the patients age, individual susceptibility, lead exposition lengths, dietary calcium and simultaneous exposition to cadmium[17].

A MODEL FOR LEAD-EFFECT HUMAN DNA-TRANSCRIPTION INVESTIGATION

In spite of the relationship of lead to congenital abnormalities and other relevant implications, relatively little attention has been paid to this metal's effects on the molecular mechanism of human genetic expression. With this idea in mind, we recently developed an in vitro model which permits the investigation of lead's effect on human DNA transcription, which is an important and complex component of gene expression. The placenta was selected as the study subject, considering that this organ meets several favorable characteristics for a such purpose: (i) it is an abundant material; (ii) under appropriate conditions, its genetic expression mechanism can be kept intact, even several weeks after its obtention; (iii) its use, as opposed to other human organs, does not implicate either damage for donors or serious ethical limitations; (iv) placental misfunction is a risk element for fetus development[18], (iv) it is very active in its synthesis of diverse peptide hormones[18], which implies a very active biosynthetic process, where DNA transcription plays a relevant role[19], and (v) this organ is directly related to the previously mentioned congenital defects imputed to lead intoxication. The model consists of (i) an enriched placental nuclear fraction, as human DNA dependant RNA polymerase and DNA source, and (ii) an appropriate mixture of reagents for RNA synthesis. (i) The former are obtained as follow: human placentae are collected immediately after cesarean sections or normal deliveries, which immediately are washed with a cold sodium citrate buffer, pH 7.0, and the cotyledons are separated and used as the source of nuclei. These latter are obtained by homogenizing the cotyledons and filtering and centrifuging the homogenate. In order to standardize the DNA concentration in the RNA synthesis reaction mixtures, the number of nuclei and its DNA content is determined in each preparation. In a typical experiment, 50 μl volumes were used, which contain approximately 1.15×10^6 nuclei and 15 μg of DNA.(ii) the reaction mixture is composed of HEPES, as pH buffer; B-mercaptoethanol, as reducer; KCl and $MgCl_2$, as salts required by the RNA polymerase; triphosphate nucleosides as substrates; glycerol; and [^{3}H]-UTP as a radioactive label. When the above are incubated at 29°C for 150 min with variable nuclei dose, an asymptotic function is obtained, with a maximum [^{3}H]-UMP incorporation of 3.2 pmoles at the end of the incubation. The supplementation of the previously described RNA synthesis model with variable concentrations of $PbCl_2$ produces a clear RNA synthesis inhibition with a relatively low doses: 50% with 5 μM and 90% with 30 μM; a gradual recuperation of about 30% with respect to the [^{3}H]-UMP incorporated into the controls occurs with 78 μM to 300μM. Higher dosages reinitiate the inhibition, until 100%, with 2.25 mM $PbCl_2$. With the same reaction mixture similar results can be obtained by using _Escherichia coli_ total extracts instead of the placental nuclear fraction. Forty µM of lead chloride inhibits 50% of the bacterial RNA synthesis, a recuperation of 80% is observed between 0.2 and 0.6 mM, and a final 80% inhibition occurs with 2.25 mM PbCl$_2$. From the above, it can be concluded that the human placental and _E. coli_ RNA polymerase activities are totally inhibited by lead chloride and that the former is more sensitive than the latter to the lead toxic effect. Our results with the _E. coli_ total extracts, including the recuperation with intermediate doses agreeing with the observations of Hoffman and Niyogi[20] who used a purified _E. coli_ RNA polymerase. In addition, in the previously mentioned study it was found that Pb^{++}, Cd^{++}, Co^{++}, Cu^{++} and Mn^{++} stimulated RNA synthesis initiation, which could explain the mitogenic activity imputed to these metal intoxicants. Popenoe and Schmaeler[21] found an 80% to 90% inhibition of the human DNA polymerase with 10 µM lead nitrate.

The possible mechanism for human placental RNA synthesis inhibition could be the direct interaction of lead with the RNA polymerase, or the interaction of this metal with the DNA template or the triphosphate nucleoside substrates. The first possibility is that which has been described for the greater part of heavy metals. These reversibly bind to -SH groups and change the active tridimensional enzymic structure[22]. This mechanism was described by Novelo and Stirpe in 1969[23] in rat liver isolated nuclei, who found that the inhibitory effect of Hg^{++}, Cd^{++} and Ag^{++} upon the enzymic activity was reversed with the supplementation of the transcription system with cystein or mercaptoethanol. The cadmium inhibitory effect was reversed only after the nucleotide concentration in the reaction mixture was augmented 3-fold. This latter result suggests that the inhibition could be also due to this metal's interaction with the RNA polymerase substrates. The DNA polymerase also is inhibited by the above mentioned metals, but it did not occur when the assayed mixtures were preincubated with EDTA, or other compounds with free -SH groups. In these experiments a noncompetitive lineal inhibition was observed. That inhibition is not dependent on DNA as a template or on triphosphate nucleosides; which suggests the direct interaction of metals with the enzyme[21]. On the other hand, Eichhorn and Shin detected that the metallic ions interact with nucleic acids, synthetic polynucleotides and triphosphate nucleosides, and change their melting temperatures, maximum absorption point and steric conformation. These metals, therefore, could inhibit the nucleic acid polymerases, by diminishing the template or the substrate availability, by making triphosphate nucleoside metal complexes[24].

PERSPECTIVES

In the peripheral blood of chronically lead exposed persons this metal reaches levels of 2.5 µM. That concentration is capable of inhibiting 25% of the RNA synthesis in the above described DNA transcription model. If both nucleic acid classes could be affected _in vivo_ with similar dosage as those affecting the RNA synthesis _in vitro_, the direct consequence could be the apparition of defective genetic material. The tissues of an organism in gestation, which are in an very active reproduction and differentiation processes, would be specially sensitive to the toxic effect of an agent that affects the genetic expression mechanism, which could explain, at least in part, the congenital malformations in animals[11] and humans imputed to lead intoxication[10,13]. There exist several defense systems in the animal cells that counteract the toxic effect of lead and other heavy metals. The main mechanisms of heavy metal neutralization are the following: (i) intranuclear inclusion bodies (INIB), described in lead intoxicated rat kidney; it has been suggested that the formation of these proteic corpuscles are induced by lead, and that its specific function is to chelate this metal[23]; (ii) a cadmium binding protein (CdBP), found in cadmium intoxicated rat liver[24]; (iii)the metallothionein (MT)[27], from mice liver and kidney and a hamster ovary cell line, which has properties similar to those of INIB and CdBP, and make complexes with Cd^{++}, Zn^{++}, Cu^{++} or Hg^{++}; these metals induce an increase in the MT-I-gene transcription velocity in different tissues, which suggests a heavy metal detoxification autogenous mechanism[28]. Thus, some congenital defects could be the result of a combination of the lead toxic effect upon the genetic expression mechanism and failures in the metal neutralization autogenous systems.

We believe that an in vitro model as the one developed by us could be useful to define the possible relationship between the effect on transcription of toxic compounds and specific cellular dysfunctions induced by lead. This could be achieved, for example, by analyzing the in vitro transcription and translation of genetic coding for enzymes related to congenital abnormalities.

ACKNOWLEDGMENTS

We thank The English for Graduate Studies in Biomedicine, section of The Modern Language Department, Faculty of Medicine, Autonomous University of Nuevo León for stylistic suggestions.

REFERENCES

1. P. S. I.,Barry, A comparison of concentrations of lead in human tissues, Brit. J. Industr. Med. 32:119 (1975).
2. R. W. Baloh, Laboratory diagnosis of increased lead absorption, Arch. Environ. Health 28:198 (1974).
3. E. K. Silbergeld and J.J. Chisolm, Lead poisoning: altered urinary-catecholamine metabolites as indicators of intoxication in mice and children, Science, 192:153 (1976).
4. F. Piccinini, L. Favalli and M. C. Chiari, Experimental investigations on the contractions induced by lead in arterial smooth muscle, Toxicology, 8:43(1977).
5. P. B. Hammond, Exposure to humans of lead, Ann. Rev. Pharmacol. Toxicol. 17:197 (1977).
6. F. J. Sánchez-Anzaldo Chap.I. Bioquímica de la intoxicación por plomo, in "Intoxicación por plomo", G. Molina-Ballesteros, Subdirección General Médica, Jefatura de Servicios de Enseñanza e Investigación, Instituto Mexicano del Seguro Social, México, D. F. (1986).
7. G. D. Braunstein, Hypogonadism in chronically lead-poisoned men, Infertility 1:33 (1978).
8. M. R. Moore, The biochemistry of the porphyrins, Clinics in Hematology, 9:227 (1980)
9. I. Dingwall-Fordyce and R. E., A followup of lead workers, Brit. J. Ind. Med. 20:213 (1963).
10. D. G. Wibberley, A. K. Kheve, J. H. Edwards, D. I. Rusahton. Lead levels in human placentae from normal and malformed births, J. Med. Genet., 14:339 (1977).
11. M. R. Mcclain and A. B. Becker, teratogenicity, toxicity and placental transfer of lead nitrate in rats. Toxicol. Appl. Pharmacol. 31:72 (1975).
12. V. H. Furm and S. J. Carpenter, Developmental malformations resulting from the administration of lead salts, Exp. Molec. Pathol.7:208 (1967).
13. L. H. Needleman, M. Rabinowits, A. Levinton, S. Linn and S. Schoenba-

un, the relationship between prenatal exposure to lead and congenital anomalies. JAMA, 251:2956 (1984).

14. M. M. Varma, S. R. Soshi, A. O. Ademi, Mutagenicity and infertility following administration of lead subacetate to Swiss male mice, Experientia, 30:486 (1974).

15. M. Carfagna, U. Cocco and O. Elmino, Sulle associazioni cromosomiche mediante filamento nelle cellule somatiche umane, Atti. Ass. Genet. Ital. 12:276 (1967).

16. L. A. Muro and R. A. Goyer, Chromosome damage in experimental lead poisoning, Arch. Path. 87:660 (1969).

17. C. H. Leal-Garza and R. Garza-Chapa, Chap. III. Implicaciones genéticas, in "Intoxicación por plomo", G. Molina-Ballesteros, Subdirección General Médica, Servicios de Enseñanza e Investigación, Instituto Mexicano del Seguro Social, México, D. F. (1986).

18. P. Beaconsfield, G. Birdwood and R. Beaconsfield, The placenta, Scientific American 243:80 (1980).

19. H. A. Barrera-Saldaña, Transcriptional products of the human placental lactogen gene, J. Biol. Chem., 257:12399 (1982).

20. D. J. Hoffman and S. K. Niyogi, Metals mutagens and carcinogens affect RNA synthesis rates in a distinct manner, Science 198:513 (1977).

21. E. A. Popenoe and M. A. Schmaeler, Interaction of Human DNA polymerase beta with ions of cooper, lead and cadmium, Arch. Biochem. Biophys. 196:109 (1979).

22. L. A. Lenhinger, "Enzymes, Kinetics and inhibition," Worth Publishers Inc. New York (1975).

23. F. Novelo and F. Stirpe, The effects of copper and other ions on the ribonucleic acid polymerase activity of isolated rat liver nuclei. Biochem. J. 111:115 (1981).

24. G. L. Eichhorn and Y. A. Shin, Interaction of metal ions with polynucleotides and related compounds. XII. The relative effect of various metal ions on DNA helicity. J. Am. Chem. Soc., 90:7323 (1965).

25. R. K. Shelton and M. P. Egle, the proteins of lead induced intranuclear inclusion bodies. J. Biol. Chem. 257:11802 (1982).

26. A. H. Hidalgo, V. Koppa and E. S. Bryan, Effect of cadmium on RNA polymerase and protein synthesis in rat liver, FEBS Letters, 64:159 (1976).

27. G. Gick, S. K. McCarty Jr. and S. K. McCarty Sr., The role of metallothionein synthesis in cadmium- and zinc- resistant CHO K1 cells, Exptl. Cell. Res., 132:23 (1981).

28. M. D. Durnan and D. R. Palmiter, Transcriptional regulation of the mouse metallothionein I gene by heavy metals, J. Chem. 256:5712 (1981).

MOLECULAR BASIS OF CONTRAST IN MRI

T. F. Egan , H. E. Rorschach [1], and C. F. Hazlewood

Department of Physiology and Molecular Biophysics
Baylor College of Medicine, Houston, TX 77030
[1]Physics Department, Rice University, Houston, TX 77251

INTRODUCTION

Water is an essential component as well as the most abundant molecule in all living systems. In magnetic resonance imaging (MRI) water is fundamental to image formation. Furthermore, the physical-chemical properties of cellular and extracellular water provide major contributions to the inherent contrast found in each image. The physical properties of water as a solvent are subject to change when macromolecular structure and/or motion is altered. Thus, precise knowledge of water-macromolecular interactions is required for our understanding of information content in an image obtained by magnetic resonance techniques.

Many experimental techniques have been applied to the study of water in cellular and other interfacial systems (1-3). A subset of this group is those measurements yielding information on the molecular motions and dynamics of water; including dielectric relaxation (4), nuclear magnetic resonance relaxation (5,6), and quasi-elastic neutron scattering (7,8). Within the time and distance scales of molecular water motion probed by each of these techniques, distinct differences are observed between bulk water and water in biological systems. Analysis and interpretation of these experimental results has proved difficult. A detailed microscopic model of water interactions is not complete for bulk water and even less complete for the case of water interacting with macromolecules. The purpose of this paper is to report progress in the development of a physical picture of water-protein interactions, and the relationship of this interaction to the nuclear magnetic relaxation times of water in biological tissues.

BACKGROUND

Nuclear magnetic resonance has been and continues to be a primary source of information about both water and proteins in the biological environment. Measurements of NMR relaxation times of water in biological tissues in varying states of health and disease have provided the basis for important medical applications of NMR (9). As referred to above, the NMR relaxation times T_1 (longitudinal relaxation time) and T_2 (transverse relaxation time) are fundamental to image contrast. In particular, the proton Larmor frequency dependence of T_1 is especially sensitive to the physical properties of water in association with other molecular systems, including biological systems.

The NMR relaxation of pure bulk water is known well (10): a single relaxation mechanism correctly predicts the magnitude and Larmor frequency independence of T_1 . The relaxation mechanism (or mechanisms) for water in complex heterogeneous environments on the other hand is not well understood: the relaxation time T_1 is substantially reduced from the values for bulk water and also depends upon the Larmor frequency in the range where bulk water does not. One interpretation of these observations begins with the assumption that the total water population is partitioned into a number of environments or phases (two in the simplest case). Within each environment, all water molecules experience roughly the same interaction conditions. NMR and other studies have provided data suggesting that one environment is represented by a fraction of solvent water molecules which is closely associated with surfaces of solute components in solution (3,5,6,11-13), forming a hydration layer that surrounds each surface. In the biological system, the available surface area for hydration is large. The simplest two-phase model of the heterogeneous water system is the "bound-free" model; the "free" fraction has properties essentially unchanged from pure bulk water and the remaining fraction is hydration water. (The proportion of water in the "bound" phase (or environment) is the subject of debate although it is usually assumed small. More sophisticated multi-phase (12,13) and continuum phase (3) models also exist.) The macroscopic relaxation time T_1 results from relaxation mechanisms occurring in all of the water phases and is also influenced by mixing of water between the different states.

Analysis of NMR and quasi-elastic neutron scattering experiments in this two-state model indicates that the aqueous phase (solvent) water is different from pure bulk water. This difference in the physical properties of the solvent water fraction, although significant,is not sufficient alone to account for the large differences in T_1 relaxation times . Instead, the major contribution to water proton relaxation will be shown to result from the "bound" water protons and solute protons which produce relaxation in the "free" phase by exchange interactions.

MATERIALS AND METHODS

Proton NMR relaxation measurements have been conducted on a biological organism - cysts of <u>Artemia</u> (a primitive crustacean commonly called brine shrimp). The <u>Artemia</u> cyst is capable of withstanding long periods of extreme dehydration in an ametabolic state. Water introduction to the cyst initiates the development cycle and eventual emergence of larva from the acellular shell which surrounds the cysts (14). This unique organism is well described and is widely used as a model for studies of a large number of biological problems and, it is particularly well-suited for the study of the physical properties of water in living cells (15).

<u>Artemia</u> cysts were purchased from San Francisco Bay Brand (Newark, CA.). The cysts were cleansed and sorted by size (200µ average diameter). The cysts were hydrated in 0.1 M NaCl for a minimum of 12 hours at 2-4 C. This treatment produces a water hydration level of 1.3 grams of water per gram of dry solid (or ≈65% wet weight). Prior to NMR

measurements, the cysts are removed from the salt solution and briefly rinsed with water which is then quickly removed following the procedure of Clegg (16). NMR tubes were loaded with cysts and placed in the magnet. Sample temperature was regulated at 22 C.

T$_1$ NMR relaxation measurements were obtained over a Larmor frequency range greater than four decades. Conventional fixed field permanent, electro-, and superconducting magnets were used to cover the proton range between 10 and 500 MHz using the inversion recovery method to measure T$_1$. At lower proton frequencies between 0.01 and 50 MHz, a field-cycling magnet, or NMRD relaxometer (17), was used. The relaxation rate data, R$_1$ = T$_1^{-1}$, for the <u>Artemia</u> cysts is shown in Figure 1. At these high water concentrations, the NMR signal from the cysts is dominated by a single broad water peak.

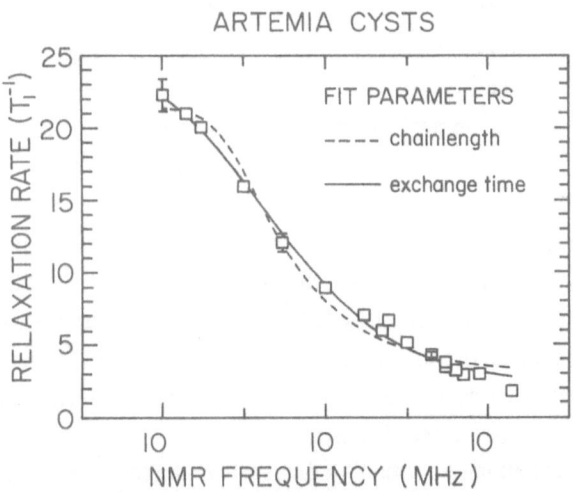

Figure 1. Water longitudinal relaxation rate data obtained for <u>Artemia</u> cysts. The dotted and solid lines are general least squares fits to these data using eq. (8). For the dotted line, the exchange rate (R$_T$ = 10^6 s^{-1}) and solvent / solute proton ratio (M = 2) where held constant. An amplitude term, the correlation time τ_{cmax}, and a constant relaxation rate term were allowed to vary. The values of these fit parameters are 6.8, 2.6 x 10^{-6} s, and 4.5 s^{-1}, respectively. In the case of the solid line, the correlation time τ_{cmax} is assumed to be large such that $\omega\tau_{cmax} \approx \infty$, and held constant (with M = 2 also). The exchange rate, an amplitude term, and a constant relaxation rate were allowed to vary, producing fit values of 22.7 s^{-1}, 11.2, and 3.3 s^{-1}, respectively (see text for details).

THEORY

According to BPP theory (10), the NMR relaxation times T$_1$ and T$_2$ are determined by statistical processes dependent on the small variations in the magnitude and direction of the local magnetic field (relative to the external NMR field) at each nuclear spin site due to fields originating from neighboring spins. These local magnetic field fluctuations are characterized by a correlation time, τ_c, determined by the motion of the neighboring spins and the molecules of which they are part, such as protons in water. For single phase, small molecule, homogeneous liquids, like water, a single short correlation time ($\tau_c = 10^{-12}$ s) due to rotational Brownian motion describes the fluctuating fields (from dipole fields in the case of protons). In terms of the amplitude of the mean square fluctuating field $\overline{h^2}$ and the gyromagnetic ratio γ, the relaxation rates R_1 and R_2 are expressed as:

$$R_1(\omega) = \frac{1}{T_1(\omega)} = \frac{2}{3} \gamma^2 \overline{h^2} \frac{\tau_c}{1 + \omega^2 \tau_c^2} \tag{1}$$

$$R_2(\omega) = \frac{1}{T_2(\omega)} = \frac{1}{3} \gamma^2 \overline{h^2} \tau_c + \frac{1}{2} R_1 \tag{2}$$

Many studies of NMR relaxation times, particularly those over a wide range of Larmor frequencies, have demonstrated that water in biological and heterogeneous systems is not described by eq. (1) with a single long correlation time (18). Instead, the frequency dependence of R_1 of water in biological systems suggests a distribution of long correlation times within a single aqueous environment. This correlation time distribution is usually different for each microenvironment or compartment within a cell and also different for macroenvironments (tissues and organs) within an organism. Because of this frequency dependence of R_1, in many instances, relaxation rate contrast is potentially greater at very low NMR frequencies, lower than the usual NMR imaging Larmor frequencies.

If $g(\tau_c)$ is a distribution of correlation times where the fraction of correlation times between τ_c and $\tau_c + d\tau_c$ is $g(\tau_c) d\tau_c$, then eqs. (1) and (2) can be generalized to

$$R_1 = \frac{2}{3} \gamma^2 \overline{h^2} \int_0^\infty \frac{\tau_c \, g(\tau_c) \, d\tau_c}{1 + \omega^2 \tau_c^2} \tag{3}$$

$$R_2 = \frac{1}{3} \gamma^2 \overline{h^2} \int_0^\infty \tau_c \, g(\tau_c) \, d\tau_c + \frac{1}{2} R_1 \tag{4}$$

The polymer dynamics model (19) postulates a relaxation mechanism in which a distribution of correlation times results from the thermally excited transverse modes of motion of chain-like macromolecules (proteins) or segments of macromolecules in solution. In this model a wave equation describes the rotational motion between adjacent pairs of spins along the chainlength. A different correlation time is associated with the decay constant for each of the heavily damped transverse normal mode excitations along the chain. For a wavelength $\lambda = 2\pi/q$, the correlation time is $\tau_c = Cq^{-2}$ (where C is a composition and structure dependent constant of the chain). The density of modes as a function of τ_c is the correlation time distribution;

$$g(\tau_c) = g(q) \frac{dq}{d\tau_c} = \frac{L}{2\pi} C \tau_c^{-3/2} \tag{5}$$

where L is the length of the chain (for 1-dimensional transverse waves). With this $g(\tau_c)$, eqs. (3) and (4) become

$$R_1 = \frac{2}{3} \gamma^2 \overline{h^2} \frac{CL}{2\pi} \omega^{-1/2} \int_{\omega\tau_{cmin}}^{\omega\tau_{cmax}} \frac{x^{-1/2}dx}{1+x^2} \tag{6}$$

$$R_2 = \frac{1}{3} \gamma^2 \overline{h^2} \frac{CL}{\pi} \left[\sqrt{\tau_{cmax}} - \sqrt{\tau_{cmin}} \right] + \frac{1}{2} R_1 \tag{7}$$

where the largest correlation time $\tau_{cmax} = C q_{cmax}^{-2}$ (and $\tau_{cmin} = C q_{cmin}^{-2}$).

In the polymer dynamics model, the source of long correlation times and therefore high relaxation rates in the proton spin system are the motions of large biological molecules, such as proteins. This relaxation mechanism is effective for those protons covalently bound to the protein as well as tightly bound water molecules. The relaxation rate of protein-associated protons, R_{1p}, can be transferred to the entire water environment through magnetic exchange interactions. In a two-phase exchange model, the water longitudinal relaxation rate, R_1, is

$$R_1 = \frac{1}{2}(\lambda_p + \lambda_s) - \frac{1}{2}\left[(\lambda_p - \lambda_s)^2 + 4 M R_T^2 \right]^{1/2} \tag{8}$$

where $\lambda_p = R_{1p} + MR_T$, $\lambda_s = R_s + MR_T$, and M is the ratio of solvent water protons to protein protons (17). In this notation R_{1p} and R_s are the intrinsic relaxation rates of "bound phase" protein protons and "free solvent phase" water protons respectively, and R_T is the cross relaxation rate for transfer of magnetization between phases. In the limit $MR_T \gg |R_{1p} - R_s|$, eq. (8) reduces to the usual fast exchange expression in which R_1 is the population weighted sum of the relaxation rate for each phase;

$$R_1 = \left(\frac{M}{1+M} \right) R_s + \left(\frac{1}{1+M} \right) R_{1p} \tag{9}$$

RESULTS AND DISCUSSION

In order to examine the behavior of the two-phase exchange model, eq. (8), we assume;

a) the intrinsic relaxation rate of the "free solvent phase" water, R_S, is constant and not very different from the pure water rate, $0.3 < R_S < 1.0$ s^{-1}, and

b) the intrinsic relaxation rate of the "bound" proton phase, R_{1p}, is given by eq. (6). First, for the case of fast exchange, we fix R_T at a large rate of 10^6 s^{-1}. The Larmor frequency behavior of eq. (6) depends upon the value of the integral (the lower limit is set to zero, for $\omega \ll \tau_{cmin}^{-1}$ which is valid for the NMR frequency range)

$$\int_0^a \frac{x^{-1/2} \, dx}{1 + x^2} = \frac{\sqrt{2}}{4} \ln\left[\frac{a + \sqrt{2a} + 1}{a - \sqrt{2a} + 1}\right]$$

$$+ \frac{\sqrt{2}}{2}\left[\tan^{-1}(\sqrt{2a} + 1) + \tan^{-1}(\sqrt{2a} - 1)\right] \qquad (10).$$

In the high frequency limit, $a \to \infty$ ($\omega\tau_{cmax} \gg 1$), the integral evaluates to $\pi/\sqrt{2}$ and $R_1(\omega) \propto \omega^{-1/2}$. At low frequencies, $a \to 0$ ($\omega\tau_{cmax} \ll 1$), the integral approaches $2\sqrt{\omega\tau_{cmax}}$ and R_1 is constant.

Treating τ_{cmax} as an adjustable parameter in R_{1p} (eq. (6)), a family of curves representing R_1 given by eq. (8) is plotted in Figure 2. The frequency dispersion occurs at $\omega = \tau_{cmax}^{-1}$ as in the BPP model with $\tau_c = \tau_{cmax}$, however the dispersion is much weaker: $\omega^{-1/2}$ instead of ω^{-2} characteristic of eq. (1). We emphasize that τ_{cmax} is a physical parameter in the polymer dynamics model directly related to the maximum chainlength in the protein macromolecules.

For fast exchange conditions, the low-frequency behavior of the relaxation rate R_1 frequency dispersion is solely determined by τ_{cmax} in the polymer dynamics model. However, if fast exchange conditions are not met, $MR_T \approx |R_{1p} - R_S|$, then the relaxation rate R_1 at low frequencies is primarily determined by the cross relaxation rate R_T. A second family of curves is plotted in Figure 3 using R_T as an adjustable parameter and fixing $\tau_{cmax} \approx \infty$ in each case. For $R_T = 10^6$ s^{-1} (fast exchange) the relaxation rate $R_1 \approx R_{1p} \propto \omega^{-1/2}$. At slower exchange rates the relaxation rate approaches R_T at low frequencies while at the high frequency range, the characteristic $\omega^{-1/2}$ dependence is evident.

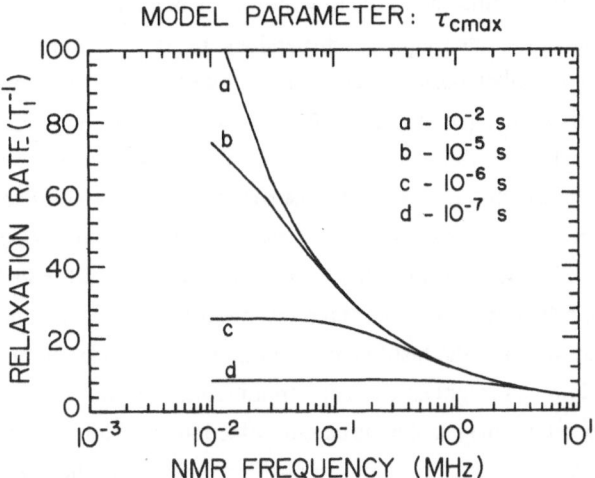

Figure 2. Family of curves representing variation in τ_{cmax}. The amplitude scaling of the relaxation rate is arbitrary but constant for each of the curves a - d. The exchange rate R_T was fixed at 10^6 s^{-1}. The variable M in eq. (8) was set to 1 and the non-frequency dependent constant rate contributions were fixed at 1 s^{-1}.

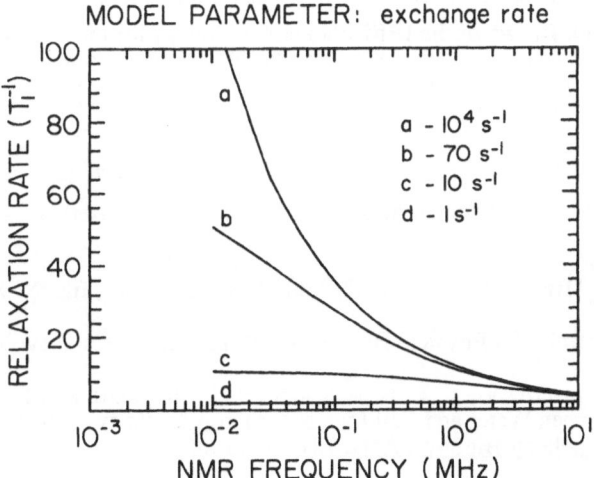

Figure 3. Family of curves representing variation in R_T. The amplitude scaling of the relaxation rate is the same as in Figure 2. The product $\omega\tau_{cmax} \approx \infty$. The variable M in eq. (8) was set to 1 and the non-frequency dependent constant rate contributions were fixed at 1 s^{-1}.

411

CONCLUSION

Comparison of Figures 2 and 3 indicates that two independent physical parameters of this model, τ_{cmax} and R_T, contribute to the low frequency behavior of R_1; each produces similar spin-lattice relaxation rate frequency dispersions. In addition, numerical fits to biological relaxation rate data using either τ_{cmax} or R_T as the fitting parameter are shown in Figure 1. Adequate fits are obtained by either approach. In the case of the Artemia cyst data, it is difficult to argue from a physical basis which is the appropriate parameter to use as a least squares fitting variable. The organism is a complex composite of many different proteins so that τ_{cmax} can only be regarded as an average correlation time determined over all possible chainlengths present. On the other hand, water exchange dynamics between "bound" and "free" water populations is also poorly understood, but if the low-frequency dispersion is due to cross-relaxation, then we are far from the fast-exchange limit. (Hydration studies are possible for cysts of the brine shrimp and might assist in determining the role of R_T.)

Another approach to this problem is careful selection of a biological system suitable for NMR study. Concentrated aqueous solutions of purified proteins might be especially appropriate systems. The class of amino acid polymers termed homo-polyamino acids fall into this category (20). The polyamino acids are polymerized forms of the amino acids available in several narrow ranges of molecular weights, corresponding to homogeneous, well-defined chainlengths. Polyamino acid conformation can also be controlled by the pH of the solution.

ACKNOWLEDGEMENTS

We acknowledge support from the R. A. Welch Foundation. We also acknowledge S. H. Koenig and R. D. Brown of the IBM Watson Laboratory for their assistance in the collection of data.

REFERENCES

1. W. Drost-Hanson and J. S. Clegg (editors), "Cell-Associated Water", Academic Press, Inc. New York (1979).

2. F. Franks (editor), "Biophysics of Water", John Wiley and Sons, Inc. New York (1982).

3. G. N. Ling, "In Search of a Physical Basis of Life", Plenum Press, New York (1984).

4. J. S. Clegg, S. Szwarnowski, V. E. R. McClean, R. J. Sheppard, and E. H. Grant, Interrelationships between Water and Cell Metabolism in Artemia cysts. X. Microwave Dielectric Studies, Biochem. Biophys. Acta. 721:458 (1982).

5. P. A. Bottomly, T. H. Foster, R. E. Argersinger, and L. M. Pfeifer, A Review of Normal Tissue Hydrogen NMR Relaxation Times and Relaxation Mechanisms from 1-100 MHz: Dependence on Tissue Type, NMR Frequency, Temperature, Species, Excision, and Age, Med. Phys. 11:425 (1984).

6. P. A. Bottomly, T. H. Foster, R. E. Argersinger, and L. M. Pfeifer, A Review of ^{1}H Nuclear Magnetic Resonance Relaxation in Pathology: Are T_1 and T_2 Diagnostic?, Med. Phys. 14:1 (1987).

7. E. C.Trantham, H. E. Rorschach, J. S. Clegg, C. F. Hazlewood, R. M. Nicklow, and N. Wakabayashi , Diffusive Properties of Water in Artemia as Determined from Quasi-Elastic Neutron Scattering Spectra, Biophys. J. 45:927 (1984).

8. H. E. Rorschach, D. W. Bearden, C. F. Hazlewood, D. B. Heidorn, and R. M. Nicklow, Quasi-Elastic Scattering Studies of Water Diffusion, Scanning Microscopy 1:2043 (1987).

9. D. D. Stark and W. G. Bradley, "Magnetic Resonance Imaging", The C.W. Mosley Co. St. Louis, (1988).

10. N. Bloembergen, E. M. Purcell, and R. V. Pound, Relaxation Effects in Nuclear Magnetic Resonance Absorption, Phys. Rev. 73:679 (1948).

11. G. D. Fullerton, V. A. Ord, and I. L. Cameron, An Evaluation of the Hydration of Lysozyme by an NMR Titration Method, Biochem. Biophys. Acta. 869:230 (1986).

12. G. D. Fullerton, in: "Magnetic Resonance Imaging", D. D. Stark and W. G. Bradley, editors. The C.W. Mosley Co. St. Louis., 36 (1988).

13. S. E. Koenig, in: "Water in Polymers", S.P. Rowland, editor. ACS Symp. Series No. 127:157 (1980).

14. J. S. Clegg, Interrelationships between Water and Cell Metabolism in Artemia cysts. XI. Density Measurements, Cell Biophysics 6:154 (1984).

15. J. S. Clegg, Properties and Metabolism of the Aqueous Cytoplasm and its Boundaries, Am. J. Physiol. 246:R-136 (1984).

16. J. S. Clegg, Interrelationships between Water and Cell Metabolism in Artemia cysts. Hydration-Dehydration from Liquid and Vapour Phases, J. Exptl. Biol. 61:291 (1982).

17. K. Hallenga and S. E. Koenig, Protein Rotational Relaxation as Studied by Solvent ^{1}H and ^{2}H Magnetic Relaxation, Biochemistry 15:4255 (1976).

18. J. M. Escanye, D. Canet, and J. Robert, Frequency Dependence of Water Proton Longitudinal Nuclear Magnetic Relaxation Times in Mouse Muscle at 20°C, Biochem. Biophys. Acta. 721:305 (1982).

19. H. E. Rorschach and C. F. Hazlewood, Protein Dynamics and the NMR Relaxation Time T_1 of Water in Biological Systems, J. Mag. Res. 70:79 (1986).

20. A. L. Lehninger, "Biochemistry", second edition, Worth Publishers, Inc., 130 (1975).

413

DISEASE DETECTION BY MR IMAGING: T1 and T2 RELAXATION TIMES

Luis E. Todd, Guillermo Elizondo, and Ralph Weissleder

Magnetic Resonance Unit., University Hospital
Monterrey, Nuevo Leon, Mexico

Tumor tissues and serum of tumor patients studied by proton nuclear magnetic resonance (NMR) show increased T1 and T2 relaxation times. These increases in T1 and T2 form the basis of tumor detection by magnetic resonance (MR) imaging. In the present work the dependence of the MR imaging signal on T1 and T2 relaxation times will be elucidated. The characteristic MR appearance of a variety of tissues will be discussed on a physiological basis.

Physical principles of relaxation times

The net magnetization vector of hydrogen protons aligns itself longitudinally to an externally applied magnetic field. If a radiofrequency (RF) pulse is subsequently applied, this magnetization vector is temporally transferred into another plane at an angle X. After cessation of the external RF pulse both longitudinal and transverse components of this magnetization vector return to equilibrium. The longitudinal magnetization is recovered by spin-lattice relaxation, and is expressed as the T1 value measured in milliseconds. T1 relaxation involves transfer of energy to the environment (lattice) from spin systems excited by the RF pulse. Transverse magnetization decays because of spin-spin relaxation which involves the loss of spin coherence among exited nuclei and is expressed as the T2 value. The relaxation mechanisms are considerably modulated by molecular motion, diffusion, flow, and magnetic field inhomogeneities.

Physiological principles of relaxation times

Molecular motion within a sample is best described by the correlation time (Tc), which is the average time between collisions of typical molecules within a sample. This molecular motion of molecules causes the local magnetic field to fluctuate and interact with the spin systems. However, only motion with a frequency equal or double to the Larmor resonance frequency of hydrogen contributes to proton T1 and T2 relaxation.

Rapid molecular motion (short Tc) such as in liquids is characterized by relatively small interaction with protons at the Larmor frequency; the resulting relaxation is therefore ineffective. Such rapid motion averages the local field almost to zero and equals T1 and T2 relaxation times. T1 and T2 are inversely proportional to Tc of the sample. T1 is minimum when Tc is equal to the inverse of the resonance frequency. Fat for example has a Tc close to the inverse of the resonance frequency and therefore has a short T1. Tc is long in macromolecules such as proteins, resulting in ineffective spin-lattice relaxation, and thus long T1. On the contrary, the low frequency motion of macromolecules contributes to spin-spin relaxation, resulting in shortening of T2. Because of the extremely long T1 and short T2 of macromolecules, no signal in MR arises directly from them. However, macromolecules can influence the magnetic relaxation of adjacent water molecules and are thus an important source of tissue contrast

MR tissue contrast

MR tissue contrast depends on one or more basic tissue parameters such as relaxation times, hydrogen density, chemical shift, magnetic susceptibility, and vascular parameters (perfusion, diffusion, blood flow). In conventional spin-echo and inversion recovery pulse sequences, T1 and T2 relaxation time differences between normal and abnormal tissues are the most important parameters determining tissue contrast. Hydrogen density usually has the least influence, as only fat and fluids differ significantly from other tissues. Chemical shift, magnetic susceptibility, and diffusion effects contribute little to the conventional MR images unless the pulse sequence design is altered.

Pulse sequences can be specifically designed and altered by the timing parameters TR, TE, TI, and the flip angle X to display contrast as the difference in T1 and T2.

Proton imaging

Hydrogen protons in water molecules are the principal source of MR signal in soft tissues. "Biological water" exists in at least two states: "free" and "bound". Free water, such as in extracellular fluid, has relatively unrestricted motion. Bound water, such as intracellular water, is bonded to hydrophilic groups of intracellular proteins and membranes. This interaction restricts the motion of water molecules and shortens T1 and T2 relaxation times. Many investigators favor the fast-exchange two-phase model. In this model, the free and bound portions of intracellular water are in rapid exchange with each other, and the relaxation rate of tissue therefore reflects a weighted average of its free and bound components.

Tissues with increased amounts of free water have longer T1 and T2 relaxation times. Such an increase in free water is present in various pathological conditions and forms the basis for detection of disease by MR imaging. For example, inflammation causes the extracellular water to increase (edema) and tumor cells are known to have increased intracellular water content

T1 relaxation times also correlate with the rate of cell division and the phase of mitosis, even when total cell water is constant. T1 of synchronized cell lines is shortest during the S phase, when the nuclear microstructure is complete, whereas T1 is longest during the M phase when chromosomes are most condensed and chromosomal surface area is therefore at a minimum. Tumor tissue is characterized by a more active cellular M phase and thus shows increased T1.

T2 relaxation times correlate with disintegration of cellular components. For example, dissolution of normal cellular structure (depolymerization) causes T2 to increase.

MR imaging of physiological tissues

In the following paragraphs a brief discussion of the relation of in vitro tissue relaxation times and in vivo MR imaging appearance is presented. The knowledge of the molecular composition of tissues helps to select the pulse sequences necessary to improve image contrast and allow tissue characterization.

Soft tissues

Much of the water in soft tissues (liver, pancreas, adrenals, muscle) is intracellular and therefore affected by hydrogen bonding to cellular macromolecules. As a result, cellular tissues have short T1 and short T2 relaxation times. Their spin-echo intensities are intermediate between fat and bone on both T1- and T2-weighted images.

Cellular tissues that are active in protein synthesis and therefore contain abundant endoplasmic reticulum (liver, pancreas) have an especially short T1 relaxation times. Muscle has a rigid matrix of actin and myosin that is less effective in shortening T1 relaxation than it is in shortening T2 relaxation. This explains why muscle has lower signal intensity on both T1- and T2-weighted images than most other soft tissues.

Soft tissues composed primarily of extracellular water (kidney, prostate, testis, spleen, venous plexus) have long T1 and long T2 relaxation times. As a result, these tissues show lower signal intensity on T1 weighted images and higher signal intensity on T2 weighted images than cellular tissues.

Pathological tissues

Increased amounts of extracellular water can be found in edematous tissues and neoplasms causing long T1 and long T2 relaxation times, low signal on T1-weighted images, and high signal on T2-weighted images. The ability to separate tissues on the basis by their extracellular or "free" water content is the most important advantage of MR.

Fat

Adipose tissues have a 10-15% higher proton density than other soft tissues resulting from characteristic CH_2 chains. Protons within CH_2 groups precess approximately 3ppm slower than protons in water, a phenomenon exploited in chemical shift techniques used to separate fat from water. However, T1 and T2 relaxation times still remain the most important contribution to MR image contrast. CH_2 protons in fat tumble at a rate close to the Larmor frequency. As a result, interaction with the environment is efficient, T1 relaxation is short, and signal intensity is high on T1-weighted images. Protons in fat have a shorter T2 than free water because of the larger size and longer correlation time of lipid molecules. Because of this shorter T2, the signal from fat decreases with respect to water on sufficiently T2-weighted MR images. However, fat may appear isointense or hyperintense relative to free water (urine, CSF) because of opposing T1 effects.

Mineral rich and collagenous tissues

Bone, calculi, and calcifications contain few mobile protons and consequently result in a small MR signal. Collagenous tissue (tendons, ligaments, fibrocartilague, or fibrosis) contain little water among the tight protein macromolecules, resulting in a long T1 relaxation time and therefore low signal intensity on T1-weighted images. The orientation of collagen fibers causes motion of water molecules to be anisotropic to the magnetic field, resulting in extremely short T2 relaxation times. As a result, collagenous tissues show a low signal intensity on T2-weighted images.

Iron rich tissues

Free iron atoms and hemoglobin degradation products show important paramagnetic and/or magnetic susceptibility altering properties. Hemoglobin in oxygenated blood has only one unpaired electron and is therefore a weak paramagnetic agent. Oxygenated blood has relaxation characteristics similar to free water, which explains the appearance of slow flowing blood in hemangioma and venous plexuses on T1 and T2-weighted images. Similar relaxation times also contribute to the difficulties in differentiating fresh hemorrhage -few hours old- from other fluids.

Deoxyhemoglobin in older hemorrhage is formed from oxyhemoglobin by chemical reduction. Deoxyhemoglobin is paramagnetic because of four unpaired electrons in the ferrous (Fe^{2+}) ion. However, the three dimensional structure of deoxyhemoglobin, sterically hinders water molecules to access the unpaired molecules and little effect on spin lattice relaxation (T1) occurs.

Deoxyhemoglobin diffusion within red blood cells creates zones of different magnetic susceptibility and field gradients within the cell. Conventional spin-echo techniques partially compensate for this inhomogeneities as a 180° refocusing pulse is applied in these techniques. However, losses in phase coherence are not corrected and a decreased signal is seen in acute hemorrhage on T2-weighted images. Imaging techniques which do not recompensate for field inhomogeneities (such as gradient echo sequences) will display acute hemorrhage as low signal intensity on both T1 and T2-weighted images.

With time, red blood cells of hemorrhage lyse, and deoxyhemoglobin is oxidized to its ferric (Fe^{3+}) form, methemoglobin. Methemoglobin is also paramagnetic, but its five unpaired electrons are more accessible to water molecules shortening the T1 relaxation times of nearby water molecules. Subacute (more than 7 days) hemorrhage therefore typically has a high signal intensity on both T1 and T2-weighted images.

In the presence of hemosiderin such as in chronic hemorrhage or hemosiderosis, pronounced spin dephasing and T2 shortening occurs due to magnetic susceptibility effect. MR signal intensity is low, even on T1-weighted spin echo images, because of the inherent T2 contrast on the spin echo technique.

Proteinaceous fluids

Proteins in aqueous solutions bind water and therefore enhance spin-lattice relaxation and shorten T1. The more proteinaceous the fluid, (synovial fluid, abscesses, and proteinaceous cysts) the shorter T1 relaxation times. In analogy, protein rich tumors such as neurofibromas, contain large amounts of extracellular water in a mucopolysacharide matrix. As a result, these tumors have similar relaxation times and signal intensity characteristics as proteinaceous fluids.

Summary

In summary, MR signal intensity reflect the physical characteristics of protons in the tissue being imaged. Fat, cellular tissues, extracellular and neoplastic water, and hemorrhage all have distinct signal characteristics and can be separated by proper pulse sequence selection. A better understanding of the biophysical mechanisms of relaxation, their relationship with neoplastic tissue, and the chemical information available from in vivo MR spectroscopy, will improve cancer detection and tissue characterization.

MAGNETIC RESONANCE RELAXATION TIMES AND IMAGING IN THE PATHOPHYSIOLOGY OF MUSCLES

L. K. Misra [1], E. E. Kim [1], C. F. Hazlewood [2], and L. W. Dennis [3]

[1] Division of Diagnostic Imaging, University of Texas M.D.Anderson Cancer Center, Houston, Texas, 77030. [2] Department of Physiology and Molecular Biophysics, Baylor College of Medicine, Houston Texas, 77030
[3] Exxon Research and Engineering, Baytown, Texas, 77522

INTRODUCTION

Magnetic resonance (MR) is one of the sophisticated techniques that are becoming enormously important in modern medical practice and research. Due to its ability to provide valuable information on any selected region of the human body in a noninvasive and risk free manner, MR has recently emerged as a powerful diagnostic modality for the detection and evaluation of a number of disorders. [1,2,3,4,5,6] Moreover, it is now the modality of choice in the diagnosis of the neurological diseases. [1,3-5] In general, the medical application of MR has revolved around two complimentary aspects. The first, and by far the widest application is concerned with the anatomic depiction of the living organism at any region of interest in the form of cross sectional images. It utilizes the protons, particularly those in water, and is referred to as magnetic resonance imaging (MRI). The second application involves acquisition of spectra which may provide quantitative assessment of major metabolites, particularly those containing phosphorus and protons. Significant insights into the biochemical alterations that are caused by disease may be observed by this technique, appropriately called magnetic resonance spectroscopy (MRS). Although MRS holds great promise for the future, progress in the field was delayed until high field spectrometers and localization techniques were developed. A discussion of MRS is outside the scope of this review; however, a number of excellent reviews, [7,8,9,10,11,12,13] published recently, summarize the rapid advancement that may expedite the realization of its potential in diagnostic medicine.

ADVANTAGES AND LIMITATIONS

MRI offers a number of advantages over other currently used diagnostic modalities. It is noninvasive and safe. Both of these properties allow repeated MRI scanning which may help in following the changes in pathophysiological processes and evaluation of the response to therapeutic intervention. The noninvasiveness and safety also make MRI a good diagnostic modality for pregnant women, neonates, and children. The capability to scan in multiple planes is an additional advantage of MRI. Simultaneous examination of MR images of axial, sagittal, and coronal sections, for example, enhances the possibility of detecting local lesions which may be missed in a single slice or plane. Furthermore, MRI provides excellent soft tissue contrast. Although the spatial resolution of MR images is generally very high, recent progress in its enhancement has led to remarkable improvements in the multislice imaging capability without prolonging the scanning time. [1,4]

Inherent superiority in soft tissue contrast gives MRI a distinct advantage over the other imaging modalities. The MR images present detailed anatomic information with remarkable clarity. Furthermore, contrast agents are not necessary to visually identify and localize most pathologic lesions in the MR images. The signal intensity of MR images depends on many parameters, such as proton density, relaxation times, motion, chemical shift, diffusion coefficients, and magnetic susceptibilities. The information content of MR images may be, however, suitably tailored by selectively increasing the contributions of one or more of these parameters.[1] This multiparametric dependence gives MRI a remarkable flexibility. It allows the optimization of imaging parameters to specific diagnostic situations and thus enhances the clinical utility of MRI. Furthermore, these MR parameters provide biophysical information through which insights into the pathophysiological status of tissues may be gained.[14]

MRI has certain limitations , and efforts to overcome some of them are showing varying degree of success. The major limitation concerns the relatively long time that is required to acquire the data. Apart from the discomfort of claustrophobia, the long scanning time causes artifacts from natural movement of organs, arterial pulsation, etc. Introduction of new techniques such as fast low angle shot (FLASH), echo-planar imaging, and cine MRI proved helpful in reducing both the scan time and the motional artifacts.[1] A second drawback is that the benefits of MRI could not be extended to persons with cardiac pacemakers and metal implants. Because of the involvement of multiple parameters, the interpretation of the MR images becomes complex.[1] Not only the features depicted, but also the instrumental settings used for scanning are important for image interpretation. Although the alterations in MR parameters show a high degree of sensitivity for detecting pathophysiological events, the specificity of these changes have not been satisfactorily determined.[1,15] Furthermore, the mechanism responsible for the changes in MR parameters are not clearly understood.[14] Such an understanding is essential in order for the full realization of the clinical potential of MRI. The MR parameters fail to detect calcification which is important for staging and evaluation of a number of disorders.[1] Finally, the purchase, maintenance, and operation of MR imagers is expensive and, therefore, MRI may not be a cost-effective means for routine diagnosis.[16,17,18]

FOCUS OF THE REVIEW

Since the surge in clinical application of MRI began in 1981, a wide range of pathologies has been extensively investigated. A number of excellent reviews have summarized the progress in evaluation of the abnormalities of central nervous system[19,20,21], cardiovascular system[22,23], kidneys[24], and liver.[25,26] Despite numerous advantages that were partially listed in the preceding section, the application of MRI to muscle disorders has been limited.[27,28,29,30,31,32] The primary purpose of this review is to summarize the studies on the pathology of muscles that were conducted in our laboratory and those of other investigators. We will also identify the aspects that are poorly understood and speculate on the possible approaches that may expand the application of MR techniques in evaluation of muscle functions.

ANIMAL MODELS

An extensive review of the published data by Bottomley, et al[33] revealed wide variation in the values of MR parameters. This has made meaningful interpretation of data difficult.[33] Although a number of factors have been implicated, the heterogeneity of biological tissue has been considered a major source for the reported wide variation in the MR parameters.[15,33] In an attempt to minimize this variation, we used well characterized genetic strains of animals in our investigations. Most of our studies were conducted in an avian model for the following reasons. Lines of chickens with a high degree of genetic homogeneity are available. Following a single artificial insemination, hens lay many fertilized eggs which may be saved and set for development such that closely related chicks, in numbers that are sufficient for meaningful experiments, could be hatched at about the same time. Embryos may be easily obtained at selected stages for ontogenic studies. The chicks are active immediately after hatching and are large enough to contain muscles adequate in size for MR measurements. Unlike other species, muscles of chickens show remarkable fiber homogeneity.[34,35] While pectoralis major (PM), and posterior latissimus dorsi (PLD) are almost entirely fast-twitch muscles, the anterior latissimus dorsi (ALD) and metapatagialis latissimus dorsi (MLD) are slow-tonic muscles.[34,35] These advantages make the chicken ideal for studying the effects of normal development and functional dichotomy on the MR characteristics of muscles.

Chickens are also suitable for evaluating the effects of pathologic development on the muscle MR characteristics. An autosomal gene causes muscular dystrophy in chickens. [34,35] The biochemical and histopathological changes in the dystrophic muscles of chickens are similar to those reported in the affected muscles of patients with Duchenne muscular dystrophy. [36] A major advantage of this animal model is that, unlike Duchenne muscular dystrophy, variations in the onset, course, and severity of myopathy are minimal due to the genetic homogeneity of the dystrophic chicken lines. Each dystrophic line has a genetically related normal line which serves as a control.[34,37] All fast-twitch muscles are affected in the dystrophic chickens, but the degenerative changes are invariably more severe in the PM than in the other fast muscles. [34,37] This intrinsic variation allows the study of the relationship of observed MR changes to the severity of myopathy. Availability of simple and highly reproducible performance tests helps in correlating the impairement in muscle functions with alterations in MR parameters.

The avian model is amenable to the modification of the phenotypic expression of the dystrophic gene. Depending upon the selection pressure used in their development, some dystrophic chicken lines show muscle hypertrophy while others undergo muscle atrophy.[34,37] This difference in the type of degenerative processes provides a means to examine the lesion specificity of altered MR characteristics of muscles.

In addition to dystrophic chickens, we have utilized the diabetic strain of mice to study the effects of hormonal abnormalities on the MR parameters of muscles. Also, we examined the relationship between age related physiologic changes in the hormonal states and MR parameters of muscles in the rat model. These well characterized models are widely used because of their similarity with comparable hormonal deficiencies in humans.

MR IMAGES

By its ability to scan in various planes and to display simultaneously the images of several thin slices, MRI shows unrivaled superiority in the evaluation of muscle anatomy. The muscle boundaries are sharply defined, primarily because of the high signal intensity of surrounding fat. [38,30] The type of pulse sequences used, however, determines to a large extent the contrast seen between the muscle tissue and its adjacent fat. [38] The use of inappropriate pulse intervals may even lead to the total loss of this intrinsic contrast. [38] A recent study discussed the disruption in muscle interfaces of the head and neck region that were revealed in the MR images, in relation to planning surgical and other therapeutic interventions.[31,39] We were able to visualize muscles that were less than half a gram in weight and 2 mm in thickness with adequate resolution.[40,41] The axial images, shown in Figure 1, revealed maximum anatomic details. The coronal and sagittal images were, however, useful in studying the course of muscles from their origin to their insertion. They also enhanced the accuracy for estimating the severity of muscle hypertrophy and atrophy. A recent study demonstrated the value of coronal MR images in detecting and localizing the abnormalities of muscles and tendons in patients with shoulder impairment syndrome. [31,39]

Anatomic information, even though useful, may have limited diagnostic values. Recently, Fisher et al [28] demonstrated the inadequacy of morphologic changes such as the contour, shape, and size for the reliable detection of a number of muscle pathologies. Increasing the contributions of relaxation parameters enhances the diagnostic value of MR images. The signal intensity (SI) of muscles and background noise may be electronically measured and their ratio (S/N) may encode useful information. Based on such S/N determinations, we distinguished slow and fast muscles in our recent study. [40] The S/N values of the slow muscles were significantly higher than those of the fast muscles in the MR images. This was an important finding which demonstrated the value of MRI in the characterization of normal tissues. Increases in SI of muscles were also reported in conditions such as fatty infiltration, atrophy, post surgical trauma, infection, and hemorrhage. [28] Although the changes in SI strongly indicate abnormality of muscles, their diagnostic value is limited by the lack of specificity. [28] Moreover, the information obtained from SI values may be inadequate for the interpretation of the severity of muscle damage.

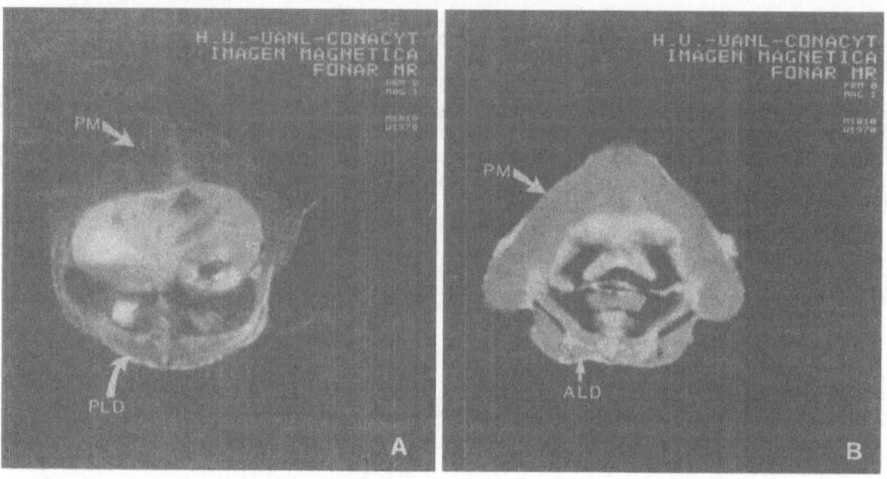

Figure 1. Axial section showing pectoralis major (PM), posterior (PLD), and anterior (ALD) latissimus dorsi muscles. T_1 weighted images of the chest with repetition time (TR) of 500 msec and echo time (TE) of 28 msec.

PROTON RELAXATION TIMES

Among the MR parameters, proton relaxation times show the highest degree of sensitivity for detecting pathophysiological events. [1,14,15,28] They basically represent the time required for protons to return to equilibrium following a radio-frequency excitation. [1,14] The two relaxation parameters generally used in MRI are: 1). the spin-lattice, longitudinal, or energy relaxation time, T_1 ; and 2). the spin-spin, transverse, or phase relaxation time, T_2. Both T_1 and T_2 relaxation times are considered as physical measures of the mobility of proton containing molecules, principally water and fat, in biological tissues. Although most of the reported T_1 and T_2 values represent in actuality the combined contributions of the protons of water and fat molecules, they are generally regarded as reflecting the physical properties of water, the predominant component of biological tissues. The resonances of water and lipid protons are separated in frequency by chemical shielding, and the resultant chemical shift differences makes it possible to measure the relaxation times of water and lipid protons separately.

The T_1 and T_2 values of tissues may be measured from MR images. The more frequently used imaging methods use the slope of the curve formed by only two data points to calculate the T_1 and T_2 values. The accuracy of these measurements may be difficult to assess. Since the diagnostic value of MRI ultimately depends upon reliable estimates of the T_1 and T_2 changes in pathophysiological events, it is important to build a reliable data base of the proton relaxation values of various normal and pathologic tissues. Accurate determination of the T_1 and T_2 values requires a large number of data points on the magnetization decay curve. Recently, the proton relaxation times, particularly T_1, of tissues have been measured in living organisms. Such noninvasive measurements have been referred to as the *in vivo* relaxation values. The proton relaxation times of excised tissues, the so-called *in vitro* relaxation times, were collected for the characterization of normal and abnormal tissues. In fact, the birth and development of the MRI modality itself may be justifiably attributed to the impact of the original *in vitro* T_1 measurements of Damadian. [42] In spite of the relative ease and increased accuracy of the acquisition of *in vivo* relaxation properties in recent times, the importance of *in vitro* T_1 and T_2 measurements has not diminished. They not only stimulate the extension of MRI applications to new pathologies, but also provide the basis for greater insights into the relaxation mechanisms. In this review we will discuss both the *in vitro* and *in vivo* proton relaxation times of muscles.

SENSITIVITY TO PHYSIOLOGICAL DIFFERENCES

Skeletal muscles were extensively used in the early *in vitro* MR investigations. The reported T_1 and T_2 values ranged from 92 msec to 1400 msec and 9 to 195 msec, respectively. [43] This wide variation was primarily due to the fact that the measurements were made at different frequencies. [33,43] Moreover, the selection of muscles for investigation was not based on their contractile properties. Since, coincidentally, mixed-fiber muscles were mostly used, the early investigations did not relate the relaxation characteristics of muscles to their composition and contractile properties.

The unique fiber homogeneity in several muscles of the chicken allowed us to examine the ability of the proton relaxation times to characterize different muscle types. We measured the relaxation times at four different frequencies (20, 30, 80, and 200 MHz) to determine the consistency of their relationship to the fiber composition of muscles. Data from a representative set of experiment are summarized in Table I. The T_1 and T_2 values of slow-tonic ALD and MLD muscles are significantly (P<0.01) larger than those of the homogeneous fast-twitch PM and PLD muscles. [44] The water content of fast and slow muscles are similar to each other (See Table II). This suggests that the T_1 and T_2 values depend not on the water content alone, but also on the physical state of water in muscles. Our results demonstrated that the T_1 and T_2 relaxation time sensitively reflect the differences in fiber type, biochemical composition, and contractile properties of the muscles. Similar results have been reported recently by Polak *et al* [45] They related the larger T_1 and T_2 values of predominantly slow soleus muscles of rats and rabbits compared to those of largely fast gastrocnemius muscles to their enlarged extracellular space. Earlier, Mardini *et al* found no difference in the T_1 and T_2 values of predominantly fast psoas muscles and slow soleus muscles of rabbits. [46] However, dietary manipulations such as limiting potassium or increasing cholesterol significantly reduced the T_1 and T_2 values of psoas muscles and helped in the differentiation of the fast and slow muscles.

Table I. Proton Relaxation Times of Muscles of Normal and Dystrophic Chicks

	T_1 (msec) Normal	Dystrophic	T_2 (msec) Normal	Dystrophic
Fast Muscles				
Pectoralis Major	500 ± 21	568 ± 18[#]	43 ± 3	60 ± 5[#]
Posterior Latissimus Dorsi	539 ± 15	578 ± 27[*]	45 ± 4	60 ± 5[#]
Biceps Bracii	530 ± 21	577 ± 14[*]	42 ± 3	53 ± 5[*]
Patagialis	553 ± 32	594 ± 16[*]	47 ± 4	59 ± 3[*]
Slow Muscles				
Anterior Latissimus Dorsi	563 ± 24	578 ± 28	61 ± 3	60 ± 4
Metapatagialis Latissimus Dorsi	551 ± 9	553 ± 28	63 ± 4	67 ± 10
Rudimentary Muscles				
Serratus Metapatagialis	523 ± 26	573 ± 49[*]	52 ± 4	58 ± 7[*]

Six Chicks each of normal (412) and dystrophic (413) lines were used at 10 weeks of age. The T_1 and T_2 values were measured on an IBM PC-20 NMR spectrometer at 20 MHz. The values are presented as mean P S.D. (Unpublished data).

* Significant at P < 0.05 and # Significant at P < 0.01.

Table II. Water and Fat Content of Normal and Dystrophic Muscles

	Percent Water		Percent Fat	
	Normal	Dystrophic	Normal	Dystrophic
Fast Muscles				
Pectoralis Major	76.39 ± 1.46	80.54 ± 1.52[#]	1.14 ± 0.55	3.03 ± 1.79[*]
Posterior Latissimus Dorsi	80.51 ± 2.80	81.33 ± 2.48	1.03 ± 0.32	5.12 ± 3.05[#]
Biceps Bracii	78.05 ± 2.15	80.35 ± 1.41[*]	1.83 ± 1.57	1.79 ± 1.72
Patagialis	79.33 ± 2.21	81.07 ± 1.1	2.15 ± 1.95	3.44 ± 1.58
Slow Muscles				
Anterior Latissimus Dorsi	82.39 ± 1.79	80.58 ± 2.07	4.41 ± 3.15	6.44 ± 4.72
Metapatagialis	82.93 ± 3.19	80.91 ± 4.34	6.34± 3.97	8.43 ± 2.50
Rudimentary Muscles				
Serratus Metapatagialis	80.04 ± 2.29	81.89 ± 2.45	3.10 ± 1.54	3.02 ± 3.71

Water content was determined by a gravimetric method. Fat was extracted by the Folch technique (*J. Biol. Chem.* 226:497 (1959). The values are presented as mean ± S.D. (Unpublished data).

* Significant at $P < 0.05$ and # Significant at $P < 0.01$.

In common with these and other investigations, we assumed the T_1 and T_2 values of muscles to be monoexponential in our early experiments. [44,47] Our subsequent studies revealed a nonexponential magnetization decay due to more than one relaxation component in muscles. [40,48,49] Our studies indicate that the routinely used monoexponential fits to data may not adequately reflect the biological complexity of muscles. Analysis of the multiexponential behavior increased significantly the ability of proton relaxation times to detect physiologic and pathologic differences between muscles. [40,48,49] When the short and long relaxation components were used for comparison, the magnitude of differences in the T_1 and T_2 values between the slow and fast muscles increased approximately two fold (Table III). The contribution of long relaxation components were generally higher in the slow muscles than in the fast muscles. If the hypothesis relating the long relaxation components to the extracellular water is correct [15,50] then the differences in the T_1 and T_2 values between the slow and fast muscles that we observed may be attributed to the contributions of the extracellular water.

We examined the effects of muscle activity on the T_1 and T_2 relaxation times. We stimulated the PM muscles of 3 week-old chicks using a 25 volt pulse with a duration of 2 msec and frequency which increased from 2 MHz to 16 MHz over a period of 2 hours. The electrical stimulation decreased the T_2 values by approximately 20%. The T_2 values for stimulated vs. unstimulated muscles were 78 ± 4 and 100 ± 12 msec for the long and 33 ± 2 and 40 ± 3 msec for the short components, respectively. The fraction contributing to the long and short components were approximately 70% and 24% respectively, and did not significantly change after electrical stimulation. The reduction in T_1 values was approximately 7%. These results demonstrate that the alterations in the proton relaxation times reflect the metabolic changes that occur in transient muscle activity. Rumeur *et al* studied the effects of intensive electrical stimulation on the T_2 relaxation times of fast muscles in rats. [51] Their results indicate a significant increase in the long T_2 components, which are at variance with our results. A recent study demonstrated the use of MRI in

Table III. Mono-exponential and Multi-exponential T_1 and T_2 Relaxation Times of Fast and Slow Muscles

Muscles	Monoexponential	Multiexponential	
	T_1	T_1 (Short)	T_1 (Long)
Pectoralis Major[*]	625 ± 15	183 ± 40	657 ± 23
Posterior Latissimus[*] Dorsi	660 ± 9	210 ± 32	700 ± 7[#]
Anterior Latissimus[**] Dorsi	744 ± 31[##]	322 ± 22[##]	1012 ± 113[##]
	T_2	T_2 (Short)	T_2 (Long)
Pectoralis Major[*]	56 ± 4	36 ± 3	124 ± 13
Posterior Latissimus[*] Dorsi	59 ± 3	40 ± 1	124 ± 16
Anterior Latissimus[**] Dorsi	99 ± 8[##]	41 ± 2	226 ± 13

The T_1 and T_2 values (in msec) were measured on a Bruker SXP spectrometer operating at 30 MHz and are presented as Mean ± S.D. (Unpublished data).

[*] Fast muscle   [**] Slow muscle

[#] Significantly different (P<0.01) from T_1 (Long) of pectoralis major.

[##] Significantly different (P<0.001) from the corresponding T_1 and T_2 values of pectoralis major and posterior latissimus dorsi muscles.

evaluating the effects of various forms of exercise on the muscle MR relaxation values. [52] These values appeared to reflect changes on the content and distribution of water. The study provided a new approach of using exercise to distinguish active and inactive muscles by MRI which is an important consideration in planning the extent of dissection in surgery.

The characterization of normal tissue by their relaxation times is a crucial step in the full exploitation of the clinical potential of MRI. Such data bases need further expansion. The modifying effects of diet, level and type of physiologic activity, various forms of conditioning and physical training, etc on the relaxation times of muscles are not fully understood. Extension of the application of MRI to sport and preventive medicine may depend on careful investigations of these and related aspects.

ASSESSMENT OF DEVELOPMENT AND GROWTH

Marked changes in the chemical composition and contractile functions of muscles occur during normal development. [53,54] Concomitant with these postnatal changes is a gradual reduction in the proton relaxation times of muscles which was shown to stabilize by 30 days of age in rats. [55] We examined the change in MR relaxation times of muscles during embryonic and neonatal development and subsequent postnatal growth. [56] We observed essentially no change in the T_1 values of PM muscles during the final stages of embryonic development. [56] A dramatic decrease (approximately 25%) in the muscle T_1 values appeared by 8 days of postnatal development. Subsequently, the rate of reduction declined and the T_1 values stabilized by 32 days of age. Similar age dependent decrease in the T_1 values of human muscles has been recently reported. [57] The correlation coefficient between the muscle T_1 values of 15 male volunteers and their age was as high as -0.91.

All of these studies measured the T_1 values from the composite proton signal. In a recent study, we determined the *in vivo* T_1 values of water and fat protons separately as a function of development. [32] The data is shown graphically in Figure 2. Consistent with our earlier *in vitro* studies, we observed a significant decrease in the water T_1 values of normal muscle during the first 2 weeks of postnatal development. The fat T_1 values increased during the development of muscles.

Basic information of the development-associated changes in the MR values, as summarized above, is valuable as a guideline for future clinical research. These results suggest the potential use of MRI in evaluating pre- and post-natal growth processes. MRI may be extended to possibly study the muscle weakness that is associated with disability, disuse, and senescence.

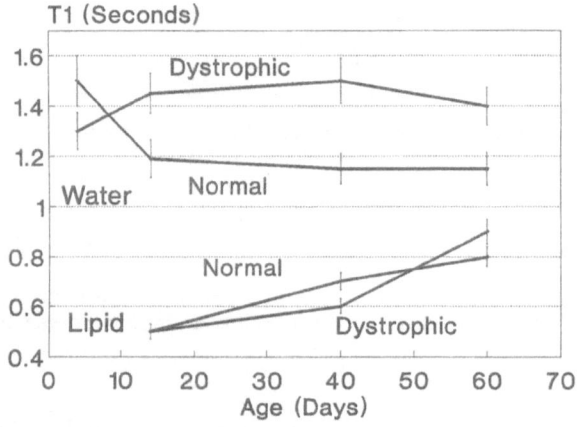

Figure 2. *In Vivo* T_1 values of normal and hypertrophic pectoralis major muscles in chicks as a function of age. Eight chicks in each group were measured. Results reported as mean ± S.D. (Narayana, *et al* [32]).

EVALUATION OF PATHOLOGICAL PROCESSES

Detection of Myopathy

Most pathological processes invariably alter the proton relaxation times of tissues. The T_1 and T_2 values are more reliable for detecting and evaluating lesions than the shape, size, contour, or proton density of muscles. [28] Fatty atrophy was shown to decrease the T_1 and increase the T_2 values of muscles. [28] Other pathological abnormalities such as postsurgical repair, intramuscular hemorrhage, and infections lead to an increase in both T_1 and T_2 values of muscles. [28] In a different study, localized changes in the proton relaxation times were found to reflect the severity of inflammatory reaction in various regions of muscles. [29] The MR relaxation times were observed to reliably detect intramuscular abscesses. [30] Administration of a gadolinium compound delineated the necrotic tissue center of these intramuscular abscesses from their cellular periphery. [30] Although the T_1 and T_2 values were elevated by tumorogensis in muscles, it was difficult to differentiate malignant and benign growth on the basis of relaxation time measurements. [30]

We studied the effects of degenerative changes on the muscle proton relaxation times in the avian model of muscular dystrophy. In our initial studies, we found a significant prolongation of the T_1 and T_2 values of dystrophic PM muscles. [58] The data obtained at 20 MHz is summarized in Table I and Table II. Similar data obtained at 30 MHz is shown graphically in Figure 3 and Figure 4.

The effect of dystrophy was more pronounced on the T_2 values than on the T_1 values of PM muscles. [58] The T_1 and T_2 values of heart, spleen, liver, brain, uterus, and aorta were similar between the normal and dystrophic muscles. These results indicated that the elevation in the T_1 and T_2 values were specific to the affected tissue and may be used to detect myopathy. A few years later Borghi *et al* [59] reported a wide variation in the relaxation values of muscles from patients with a variety of myopathies.

Evaluation of Severity

In the chicken model, the degenerative changes in the PM muscles are invariably more severe than those in the other affected fast muscles. [44,47,48] Results summarized in Table I and Table II show that the increase in the T_1 and T_2 values of dystrophic PM muscles were significantly higher than those of other affected fast muscles. These results suggest that the proton relaxation times are sensitive to pathological variations and the magnitude of increase in the T_1 and T_2 relaxation may be used to evaluate the severity of myopathy. Furthermore, the T_1 and T_2 values of dystrophic muscles resemble those of normal slow muscles. This is consistent with the similarities in the biochemical compositions between the dystrophic and slow muscles that has been demonstrated in a number of studies. [34,35,37] Results also suggest that high T_1 and T_2 values may not necessarily indicate a pathological event, but may reflect a preponderance of slow fibers in the muscle. [40] Therefore, caution should be exercised in the interpretation of relaxation data.

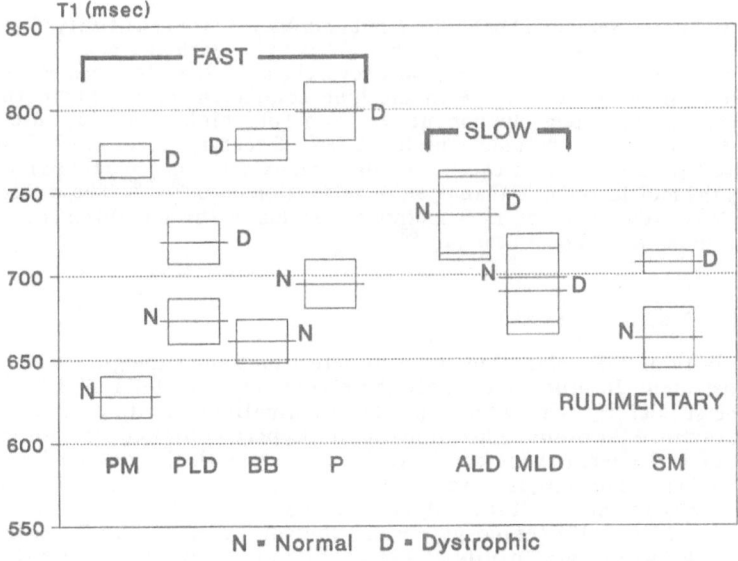

Figure 3. T_1 values of muscles from six chicks each of normal (412) and dystrophic (413) lines. Bruker SXP NMR spectrometer was used at 30 MHz. Muscles are in order as listed in Table I. (Unpublished data).

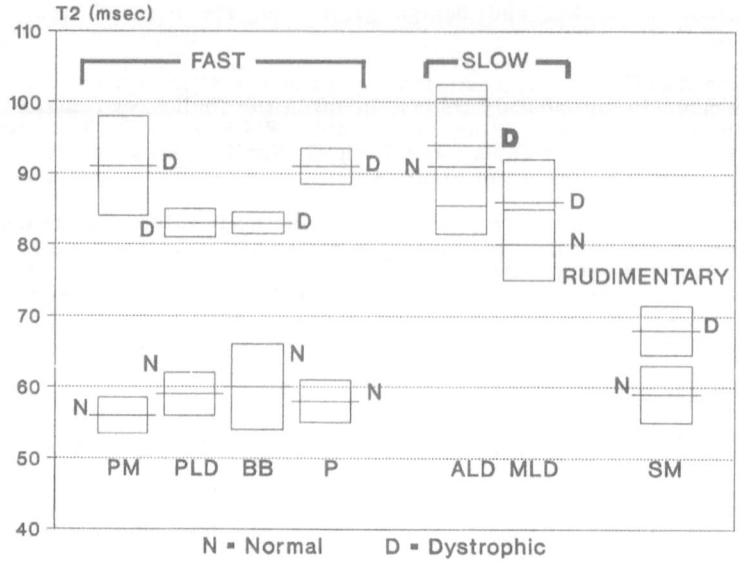

Figure 4. T_2 values of muscles of normal and dystrophic chicks. See caption to Figure 3 for experimental details. (Unpublished data).

Serratus metapatagialis (SMP) is an interesting muscle. Its original function was to maintain the tautness of skin during flight. Since chickens lost their ability to fly, SMP became vestigial. The mixed SMP muscle contains slow fiber which becomes predisposed to degenerative changes in the dystrophic chickens. Table I shows that the evolutionary disuse does not alter the T_1 and T_2 values of SMP muscles in normal chickens, while degenerative changes significantly prolongs its relaxation times.

Analysis for multiexponential behavior further enhanced the sensitivity of relaxation times in detecting and evaluating myopathy.[48,49,59, 60] Both the short and the long relaxation components increased in the most severely affected PM muscles. The other affected muscles showed an increase in the long component only. Furthermore, the relative increase in this long component was two fold higher in PM muscles (40%) compared to that of other affected muscles (20%). Recently, other investigators also reported improved detection and evaluation of various disorders, particularly those of liver, by using the multiexponential analysis of relaxation data.[60,61,62] These observations have led to the development of new methods that make the measurement of *in vivo* multiple relaxation components feasible.[63]

Differentiation of Muscle Lesions

The sensitivity of MR relaxation times to detect pathophysiologic events has been widely demonstrated. In order to enhance the effectiveness of MRI, it is important to investigate the specificity with which various abnormalities are diagnosed. With this objective we examined the proton relaxation times of hypertrophic and atrophic muscles.[64] The first set of experiments were conducted in three genetically related lines of New Hampshire Chickens. The data is summarized in Table IV. The dystrophic gene caused muscle hypertrophy in one line (Line 413) and atrophy in the other (Line 307). Chickens of the third line (Line 412) served as normal control. The T_1 and T_2 values of all dystrophic muscles were significantly elevated. More importantly, our results showed that the magnitude of increase in the relaxation values of atrophic muscles were significantly higher than those of hypertrophic muscles.[64] The water content of the

Table IV Proton Relaxation Times of Pectoralis Major Muscles of Normal and Dystrophic (Hypertrophic and Atrophic) Lines

Genotype	T_1 (msec) Short	Long	T_2 (msec) Short	Long
Normal (Line 412)	193 ± 12	784 ± 56	29 ± 11	102 ± 24
Hypertrophic (Line 413)	153 ± 57	1005 ± 21	41 ± 3	219 ± 22
Atrophic (Line 307)	179 ± 12	1135 ± 71	50 ± 3	258 ± 19

A Bruker SXP NMR Spectrometer was used to measure T_1 and T_2 values at 30 MHz. Five to six chickens from each genetic line were used. Results are presented as mean ± S.D. (Data from Misra et al [64]).

hypertrophic and atrophic muscles was similar. The fat content of atrophic muscles, however, was several fold higher than that of hypertrophic muscles. [64] Since fat is known to reduce the relaxation times of the composite proton signal, it cannot explain the greater increases in T_1 and T_2 values of atrophic muscles relative to the hypertrophic muscles. This suggests that yet undetermined factors, perhaps intrinsic to the lesion formation, may contribute to the differences in the relaxation times that were observed between atrophic and hypertrophic muscles. Despite the uncertainty of the underlying mechanism, our results indicate the potential of MR relaxation times to differentiate these two muscle lesions.

In the second set of experiments, we compared the water T_1 relaxation times of PM, PLD, and ALD muscles of unrelated lines 307 (New Hampshire) and 433 (White Leghorn). Their respective genetic controls were also included in the study. The data is presented in Table V. The dystrophic gene induces atrophy of PM and PLD muscles in these two different breeds of chicken, while ALD muscles remain unaffected. Atrophic changes in both genotypes increased the T_1 values of PM ($P<0.01$) and PLD ($P<0.05$) muscles in proportion to their known degree of severity. There were, however, no differences in the relaxation values of the corresponding muscles between the two atrophic lines. The T_1 values of normal muscles were also similar in chickens of the two control lines. When considered in the context of the data of the first experiment, these results show that the type of lesion, rather than the genotype, affected the T_1 values of muscles.

Finally, we compared the *in vivo* T_1 relaxation times of water and fat protons, determined separately in the hypertrophic and atrophic PM muscles. [65] Although both lesions elevated the T_1 values of muscles, the increase was significantly greater in the atrophic muscles (57%) than in the hypertrophic muscles (22%). The atrophic changes increased the fat T_1 values of muscles. In comparison, the fat T_1 values of the hypertrophic muscles remained essentially identical to those of the normal muscles. These results suggest that the fat T_1 values may be potentially important in enhancing the specificity in diagnosis of muscle pathologies.

Although these results are encouraging, additional studies are needed to investigate the specificity of MR relaxation parameters. Specificity remains the major shortcoming of MRI. Our studies indicate that the T_1 relaxation times, determined separately for the fat and water protons, may be helpful. The value of multiexponential analysis of relaxation data has not been adequately established. Recently developed techniques [63] for selectively suppressing the contribution of one or more relaxation components may be evaluated. The possibility of using contrast agents for enhancing the specificity of MR relaxation has not been fully explored. [66] An understanding of the biochemical composition of various lesions may form the basis for devising new strategies aimed at enhancing the specificity of MR relaxation times.

Table V The Spin Lattice Relaxation Time (T_1) of Severely, Moderately, and Unaffected Muscles in Two Genotypes

	Pectoralis Major	Posterior Latissimus Dorsi	Anterior Latissimus Dorsi
Normal White Leghorn (Line 003)	1.28 ± 0.06	1.32 ± 0.10	1.33 ± 0.07
Dystrophic White Leghorn (Line 433)	1.84 ± 0.18	1.55 ± 0.10	1.30 ± 0.02
Normal New Hampshire (Line 412)	1.27 ± 0.05	1.32 ± 0.08	1.32 ± 0.09
Dystrophic New Hampshire (Line 307)	1.81 ± .04	1.56 ± 0.04	1.32 ± 0.04

The T_1 results of water protons were measured on a Varian XL200 NMR spectrometer operating at 200 MHz. Results are presented as mean ± S.D. (Unpublished data).

Determination of Disease Progression

Among the available diagnostic modalities, MRI is uniquely suited for sequentially following the course of myopathy. The noninvasiveness, safety, and high sensitivity to pathophysiologic changes allow several repetitions of MR measurements. We measured the T_1 relaxation times of water and fat protons at approximately the same location of the most severely affected PM muscles of chickens at 4, 14, 40 and 60 days of age.[32] The results of this study are shown in Figure 2. Except for the 4 day-old chicks, the proton spectrum of the PM muscles contained water and fat peaks. An age related reduction in the water T_1 values of muscles was observed in the normal chickens. No such reduction was evident in the dystrophic birds. This suggests a failure of the maturation processes in the dystrophic chicks. Because of this failure, the water T_1 values of dystrophic muscles were higher than those of normal muscles at 4 days of age. The fat T_1 values of muscles were essentially identical in the 14 day-old normal and dystrophic chicks. These values of normal and dystrophic muscles increased comparably during the subsequent development period. In our earlier ontogenic studies, significant differences in the *in vitro* T_1 values between normal and dystrophic chicks were observed before the functional impairment was apparent.[56] A recent study on Duchenne muscular dystrophy patients showed an elevated T_1 relaxation time of affected muscles in the preclinical stages of the disease.[57] As the disease progressed, the T_1 values of the muscles declined in the patients. This decline was attributed to the infiltration of muscle tissue with fat. By 11-14 years of age, the fat almost totally replaced the muscles, and the measured T_1 values were identical with those of subcutaneous fat. In these studies, the composite proton signals containing the contributions from both water and fat were used for the T_1 measurements.[57] This inherent limitation of the methodology made study of the T_1 alterations of the actual muscle tissue difficult. It also precluded the examination of the advanced stages of disease. Our studies indicate that separate determination of water and fat T_1 values may largely overcome the difficulties of following the course of myopathy through stages that are complicated by high fat infiltration.[65] Decomposition of the relaxation data into the long and short components may also improve the staging and evaluation of disease progression.[48,59,60]

We examined the effects of natural recovery of function on the muscle relaxation values. The right sciatic nerve was crushed in rats. This resulted in an immediate functional impairment of their right leg, which induced atrophic changes in its gastrocnemius muscle. The muscle eventually recovered following the normal repair of the nerve. The gastrocnemius muscle from the left leg were used as internal controls. In addition, this study included gastrocnemius muscles from untreated rats. The data is

Table VI. Proton T_1 and T_2 Relaxation Times of Gastrocnemius Muscles of Rats Following Nerve Damage

	T_{1S} (msec)	T_{1L} (msec)	T_{2S} (msec)	T_{2L} (msec)
7 Days Post-Crushing				
Control	198 ± 16	684 ± 41	41 ± 8	147 ± 13
Atrophic	233 ± 21*	746 ± 28	52 ± 5*	189 ± 20*
Control lateral	184 ± 18	698 ± 49	46 ± 4	139 ± 18
28 Days Post-Crushing				
Control	189 ± 23	677 ± 36	48 ± 7	154 ± 19
Recovered	201 ± 19	694 ± 42	46 ± 8	171 ± 18
Control lateral	212 ± 27	687 ± 39	43 ± 5	166 ± 17

The right sciatic nerve was crushed for 1 minute in rats weighing 250 grams. The control lateral sciatic nerves were exposed but not crushed. Sample size was five in each treatment group. Results are presented as mean ± S.D. * Indicates a significant difference (P<0.05) from the other values. (Unpublished data).

summarized in Table VI. The T_1 and T_2 values of gastrocnemius muscles from a group of rats were measured at a time when the functional impairment and muscle atrophy were evident. These values were measured in a second group of rats after the normal function and morphology were fully restored. The T_1 and T_2 values of gastrocnemius muscles from the leg with the intact sciatic nerve were similar to those from the untreated rats. Transient atrophy increased the relaxation times of muscles. The T_1 and T_2 parameters returned to normal values once repair processes were completed. These results suggested that the T_1 and T_2 values accurately reflected the restoration of the structural and functional status of muscles.

The ability to detect early and follow the course of pathophysiological processes makes MRI a powerful diagnostic modality. Such an ability is valuable in the identification of developmental disorders. Early detection may help in instituting appropriate, early treatment so that damage may be minimized. Sequential follow up of pathological events has obvious advantages for modification of therapeutic approaches and prognosis. These advantages of MRI have been effectivly used in the diagnosis and management of neurological disorders.

Evaluation of Response to Therapy

The advantages discussed in the preceding section are equally relevant for exploring the potential of MRI in evaluating the response to therapeutic intervention. In order to examine this yet untapped potential, we conducted double-blind studies. After undergoing treatment and functional tests in Dr. Entrikin's laboratory at the University of California Medical School at Davis, the chicks were transported to us with their identification numbers coded. We measured the relaxation times of PM muscles of the chicks, randomly selected, following a three day adjustment period. In the first set of experiments, the response to corticosterone-21-acetate was examined. The data, summarized in Table VII, clearly shows that the treatment significantly improved the muscle functions of dystrophic chicks, as measured by the standard exhaustion score (ES) method. [67] This improvement was reflected in marked reduction in the T_1 values of PM muscles. While significant differences in the muscle T_1 values between untreated normal and dystrophic chicks were observed, corticosterone reduced the T_1 values essentially to the level found in the treated controls. An interesting observation in this study concerns

Table VII. Effects of Corticosterone-21-acetate on the Righting Ability and Muscle T_1 values of Normal and Dystrophic Chicks

Treatment[1]	Normal		Dystrophic	
	T_1 (msec)	ES[2]	T_1 (msec)	ES[2]
Control	680 ± 23[4]	24.3 ± 2.5	829 ± 32[4]	0.00
Corticosterone[3]	734 ± 15	20.3 ± 3.1	729 ± 47	12.0 ± 2.6

1. Number of chicks per treatment was 6. 2. ES = exhaustion score.
3. Corticosterone-21-acetate, 12 mg/Kg ,2-28 days of age, was injected IP.
4. Significantly different, P<0.01. (Data from Misra, *et al* [67]).

the hormone induced elevation of T_1 values in the normal muscles. This was consistent with our findings in rats that ovariectomy reduced the T_1 values of several organs, including skeletal muscles. [68] Extended treatment of a group of ovariectomized rats with estrogen analogs either restored or, more frequently, elevated the T_1 values of these organs. These T_1 changes occurred without the treatment affecting the water content of the organs.

In the second set of double-blind experiments, the response to dexamethasone (DEX) treatment was evaluated using MR relaxation times. [49]   DEX treatment markedly improved the ES's and reduced the T_1 and T_2 values of the PM muscles of dystrophic chickens. Decomposition of the relaxation data into two components further enhanced the treatment differences. For example, the T_{1S} and T_{1L} values of dystrophic muscles were 75% and 35% higher than the corresponding values of normal muscles. The monoexponential T_1 values of dystrophic muscles, in comparison, were elevated by only 13%. DEX treatment markedly reduced both T_{1S} and T_{1L} values. But T_{1L} values of muscles of treated dystrophic chicks were significantly higher than those of normal muscles. This accurately reflected the partial functional recovery that was evident in the DEX treated dystrophic chicks. Although DEX reduced the short and long T_2 components of dystrophic chicks, the effects were not as striking as those on the muscle T_1 values.

Our studies using the chicken and rat models demonstrate the potential value of MRI in pharmacological research. Sequential follow up of the drug response in the same set of animals reduces the effects of biological variations that may complicate the interpretation of data from several sets of animals. Simultaneously, this approach will also increase the cost-effectiveness of therapeutic trials. Extensive work in the future is, however, needed to establish the effectiveness of MRI in the screening and testing of pharmacological agents.

Insights at the Subcellular Level

Calcium metabolism is known to be impaired in the degenerative diseases of muscles. [69] In an earlier study we observed abnormalities in the synthesis and activity of calmodulin in dystrophic muscles. [70] Calmodulin is a cytosolic protein which plays a major role in the regulation of cellular calcium by the sarcoplasmic reticulum (SR) of the dystrophic chicks. Because of this functional importance, we decided to study the effects of myopathy-associated alterations on the proton relaxation times of SR. We separated the SR from the PM muscles of 6 week-old normal and dystrophic chicks by gradient centrifugation. The SR pellets were tightly packed for the T_1 determinations. The T_1 values of dystrophic SR (862 ± 27 msec) were significantly higher than those of normal SR (715 ± 29 ms). These results suggested that the functional and structural defect at the subcellular level may be investigated using MR relaxation times.

Table VIII. *In vivo* T_1 and BWF Values of Pectoralis Major Muscles in Chickens

	T_1	BWF (%)	% Water
Normal (Line 03)	1076	10.72	73.72
Atrophic (Line 433)	1660	4.44	80.80
Normal (Line 412)	1150	9.61	75.36
Hypertrophic (Line 413)	1400	6.63	80.21

T_1 values were measured at 80 MHz. BWF was calculated according to Matsumura, *et al.* [57] Water content was determined gravimetrically. (Data from Misra and Narayana [65]).

MECHANISMS OF MR RELAXATION TIMES

MR characterization of biological tissues depends upon relaxation processes whose mechanisms are not well understood. Several problems complicate the elucidation of the relaxation mechanisms. The published data, although extensive, show a large variation in the T_1 and T_2 values of tissues. [14,33] Many pathophysiological events may alter the content and distribution of water in tissues. Also, the heterogeneity of biological tissues may add to the complexity by giving rise to more than one relaxation component. In addition, relaxation times are known to vary with the frequency at which the measurement is made. A complete explanation of mechanisms should, therefore, account for hydration, multiexponential relaxation, and frequency dispersion. A number of models, mostly phenomenological in nature, have been proposed to explain the MR relaxation behavior in biological systems, yet a consensus on a model and relaxation mechanisms remain elusive. [14,33]

Zimmerman and Brittin proposed the "two fraction fast exchange" model which explains the observed T_1 and T_2 values as a weighted average of the relaxation times of bulk water and hydration water. [71] We tested their model in our early studies. [47,58] Our analysis suggested that the increase in the relaxation values of dystrophic muscles could not be explained on the basis of their water content alone. Fullerton *et al* proposed the fast proton diffusion (FPD) model in 1982, [72] essentially reformulating the Zimmerman-Brittin equations to include more than two distinct water phases , or compartments.

Table IX. Effect of Experimental Dehydration on T_1 and BWF Values of Muscles

	T_1	BWF (%)	% Water
Normal chickens (water *ad libitum*)	616	7.9	74.4
Hypertrophic chickens (water *ad libitum*)	687	6.8	76.4
Normal chickens (water withheld)	512	10.0	70.4
Hypertrophic chickens (water withheld)	547	9.2	69.2

T_1 values were measured at 20 MHz. Water content was determined gravimetrically. BWF was calculated according to Matsumura. [57] (Unpublished data).

According to the these models, binding of water to macromolecular surfaces diminishes the mobility of hydration water. This loss of motional freedom is reflected in a marked reduction in the relaxation times of biological tissues compared to bulk water. Using the equations given with the FPD model, the fraction of water bound to macromolecules , also called the bound water fraction (BWF), may be roughly estimated. Recently, Matsumura *et al* explained their observations of alterations in the T_1 values of muscles from patients with early stages of Duchenne muscular dystrophy in terms of changes in BWF. [57] They attributed these changes to both regeneration and degeneration of muscle fibers. Consistent with their observations, we also found the FPD model to be useful in explaining the difference between the atrophic and hypertrophic muscles (Table VIII).[65]

We further examined this relationship between the T_1 and BWF values by altering the hydration of muscles. In our initial experiments, summarized in Table IX, deprivation of drinking water to normal and dystrophic chicks for a period of 96 hours reduced the water content of PM muscles on average by 4 to 6 percent. [73] The reduction in T_1 values was on the order of 100 to 140 msec. The calculated BWF values were increased significantly by muscle dehydration. If the assumptions inherent in the FPD model are correct, then this data indicates that the changes in T_1 on muscle dehydration are due to changes in the amount and distribution of water in the tissues.

In subsequent experiments we used diabetic mice (C57/BL/Ks(db/db)) to study the effects of dehydration from naturally occurring diuresis on muscle tissue (Table X). Osmotic diuresis caused by hyperglycemia reduced the T_1 values of gastrocnemius muscles of diabetic mice compared to those of normal mice. We also examined the effects of experimentally induced diuresis. Low (10 mg/kg) and high (50 mg/kg) doses of lasix, which causes diuresis by inhibiting sodium resorption, was injected IP in 16 week-old normal mice (C57/BL/Ks). One hour later, the T_1 values of gastrocnemius muscles were measured. Low doses of lasix did not affect the muscle T_1 values. The T_1 values were, however, reduced by high doses of lasix compared to those of noninjected controls. Within the context of the FPD model, the calculated BWF of muscles following both the diabetes and lasix induced diuresis occurs at the expense of "free water." If direct measurements of "bound water" becomes available, then the FPD model can be tested directly. Such measurements would be valuable in testing the validity of interpretations based on the FPD model.

The origin and number of relaxation components in biological tissues are not precisely known. Some investigators associate these relaxation components with water, fat , and proteins in the tissue. [14,50] Others postulate that the relaxation components originate from various compartments in the tissue. [14,45,50] In general, the slow (long) and fast (short) relaxing components are assigned to the extra- and intracellular compartments. [14,50] A third group of investigators associates the slow component with

Table X. T_1 and BWF Values in Normal and Diabetic Mice

Mice Genotype	Lasix Dose (mg/kg)	T_1 (msec)	BWF (%)
Normal	None	691	6.8
Diabetic	None	625	7.8
Normal	10	686	6.7
Normal	50	624	7.8
Normal	None	697	6.7

16 Week-old C57/B2/Ks mice were used. Blood glucose levels were 151 ± 33 and 428 ± 81 mg/dl for normal and dystrophic mice, respectively. T_1 values were measured at 20 MHz. BWF was calculated according to Matsumura. [57] (Unpublished data).

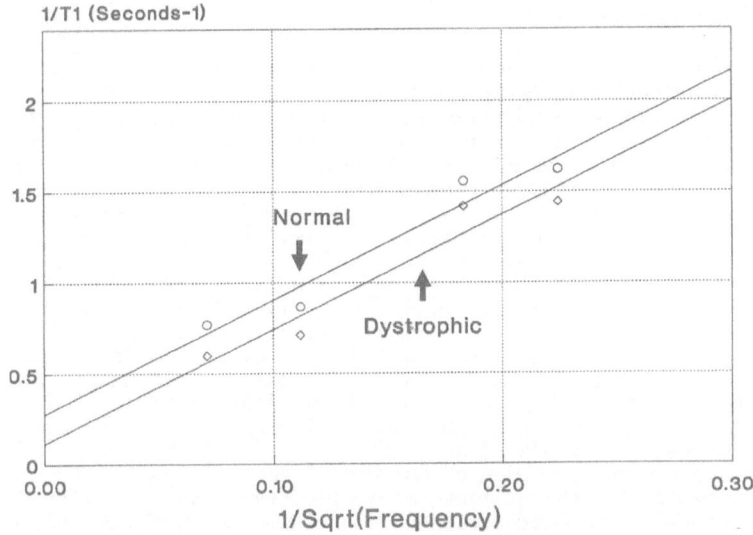

Figure 5. Frequency dependence of T_1 of normal and dystrophic pectoralis major muscles.

the vasculature, and the fast components with the interstitial and intracellular water. [51] On a physical basis, Mathur- de Vre attributes the multiexponential behavior to a) local heterogeneity in the state of tissue water, b) non-averaging of the local field fluctuations, and c) diffusion. [14] Cross relaxation between the bulk water protons and bulk macromolecular protons may cause multiexponential decay of magnetization in biological tissues. Additional work is needed to resolve the occurrence, origin, and mechanism of multiple relaxation components. This is important as consideration of various relaxation components appears to enhance the effectiveness of MRI in the detection, evaluation, and understanding of pathophysiological processes. Finally, theories have been advanced to explain the relaxation parameters and their frequency dependence based on first principles. These theories attempt to explain experimental results in terms of biophysical processes. Koenig and his associates proposed a single model to explain the proton relaxation behavior in the tissues. [74,75,76] Their extensive work, using the nuclear magnetic relaxation dispersion (NMRD) profile of homogeneous proteins solutions, diamagnetic proteins, metalloproteins, and normal and abnormal tissues, has recently been reviewed. [74,75,76] In their theory water is assumed to be relatively free to diffuse throughout the intracellular and extracellular structures of the tissue. Also, the NMRD profiles were postulated to reflect collisional and thermal processes. Recently a new model, the protein dynamics model, was proposed. [77] In this model, the water relaxation mechanisms depend on the excitations originating from Brownian motion of cellular protein segments. Such excitations are proposed to be regulated by the size, conformation, and mechanical properties of the cellular proteins and by the viscosity of water. This model allows for the evaluation of protein-water interactions as a mechanism for proton relaxation times. Very recently Schauer et al proposed another theory based on the premise that hydration water exhibits orientational order with respect to the protein surfaces. [78] Detailed discussion of these models is outside the scope of this review. Plans are underway to use these models to explain the mechanisms of relaxation phenomena in normal and pathological muscles. The data shown in Figure 5 shows an approximately linear relationship between $1/T_1$ and the inverse of the square root of the frequency of measurement. Such a relationship suggests that these theoretical models could be experimentally tested with an objective to gain insights into the relaxation mechanisms.

The early stages of development of MRI, spanning the decade from 1973 to 1983, recorded phenomenal success in the clinical applications of MRI. A number of pathologies, hither to either missed or diagnosed with uncertainty, began to be reliably detected by MRI. These encouraging developments led to an euphoria that MRI could resolve most of the problems in diagnostic medicine. As the number of MR imagers multiplied and the clinical experience increased worldwide, the shortcomings of MRI became apparent. This not only moderated the expectations from MRI to realistic levels, but also set the tone for future studies to overcome the identified shortcomings. The major shortcoming of MRI is the limited specificity. Therefore, the most challenging task for investigators in the field is to improve the specificity of MRI. A number of approaches are being pursued to accomplish this singularly important task.

The routine MR images provide predominantly qualitative information which is generally unreliable for differentiating various pathological states. Strategies for quantitating the biophysical and biochemical information that is encoded in images are, therefore, needed. In this context, computerized image analysis techniques that were originally developed by the United States National Aeronautical and Space Administration for interpretation of satellite images were evaluated in MRI. These techniques may use either the previously known biochemical and biophysical information on tissues in the "supervised" scheme, or multivariate statistical analysis in the "unsupervised" scheme to quantitate the information content of each pixel. The versatility of these techniques, based on fuzzy mathematical set theory, is remarkable. They are capable of analyzing several images, taken under different conditions, and thereby potentially improving the specificity. Being the most quantitative MR parameter, the relaxation times are fundamentally important for enhancing the specificity of MRI. The multiexponential behavior of relaxation times add to the complexities. Although a challenging task, the resolution of origin and mechanism of the various relaxation components may enhance the sensitivity and possibly provide some insights into the specificity of MR changes that are associated with various pathophysiological events. New techniques may evolve from these investigations and help further improve the diagnostic capabilities of MRI. Studies of Kroeker and Henkelman [79] reveal the potential of using continuous distributions of relaxation times in understanding biological processes.

Crucial to MR images is the contrast between different types of tissues on the one hand, and the contrast between the normal and pathological tissues on the other. This contrast depends predominantly upon the differences in the relaxation times of tissues. The values of relaxation times depend on the strength of the magnetic field of the MR imager. Our early studies suggested that the T_1 differentiation between the normal and dystrophic muscles was enhanced when measurements were made at a low frequency. [80] Crooks at al [81] discussed the effects of magnetic field strength in relation to signal to noise ratio, spatial resolution, section thickness, echo times, and cost effectiveness. The optimum field strength is not precisely known. Future measurements at various frequencies may provide definitive answers. In addition, such measurements may also be useful in evaluating the models that were proposed by Koenig [74], Rorschach and Hazlewood [77], and Schauer, et al [78] to explain the mechanism of relaxation processes. Another approach toward contrast enhancement involves administration of paramagnetic agents which decrease the relaxation times. [82] Our current information being limited, it is important to evaluate the effectiveness of contrast agents in diagnosis of muscle disorders by MRI. The distribution of contrast agents in the body, following administration, may be regulated by modifying its structure. By careful selection of contrast agents and proper design of experiments, it is possible, therefore, to gain insights into the origin and mechanism of the multiple relaxation components. The recent studies of Fleckenstein et al [52], suggest the possible value of exercise in enhancing the contrast differentiation between active and nonactive muscles. Mardini et al [46] demonstrated the feasibility of dietary manipulations for contrast enhancements. These physiological and nutritional means of contrast enhancement appear to be promising, and merit further evaluation.

Chemical shift imaging is a potentially important field of research. Fat infiltration is not an uncommon feature of most myopathies. Using the spectral separation, it is now possible to acquire the images of either water or fat protons. The strengths and limitations of the various techniques for chemical shift imaging has been recently discussed. [83,84] Our recent, preliminary study suggests that chemical shift imaging may enhance the specificity of MRI in the differentiation of muscle lesions and also the evaluation of disease progression. [85] Additional studies are, however, necessary to fully understand the clinical applications of chemical shift imaging in muscle disorders.

MRI should not be limited to protons only. Other nucleii such as phosphorus, carbon, and sodium may play a crucial role in the detection and evaluation of muscle abnormalities. The reviews of MRS cited earlier [7-13] contain the details of the phosphorus studies. Brooks et al [86] recently reported the feasibility of ^{31}P imaging of phosphocreatine which was valuable in detecting anoxia in muscles. Using ^{31}P MR, we found that dystrophy significantly increase the pH of PM muscles. [87] The feasibility of ^{13}C imaging of animal tissues has been reported. [88] Its application in the study of muscle disorders, possibly by administration of labelled ^{13}C metabollites, remains to be explored. Such studies may provide insights into the metabolic alterations that are associated with muscle abnormalities. Pathophysiological events are known to affect the distribution of sodium in muscles. [54] Our plreliminary studies showed that not only the concentration (31.0 ± 5.9 vs 9.2 ± 1.6 mM/Kg) but also the T_1 values (24 ± 1 ms vs. 20 ± 1 ms) of sodium are increased in the dystrophic PM muslces compared to the normal muscles. These results suggest the potential value of ^{23}Na imaging in the diagnosis of muscle disorders. Recent developments of multinuclear MR techniques make possible the simultaneous acquisition of proton and sodium, [89] and proton and phosphorus data. [90] In addition to those discussed, many more nucleii are NMR active. Further developments of multinuclear MRI is highly probable either through imaging of nucleii naturally present or of nucleii added as tracers or contrast enhancers. All of these developments point to enormous potential of MR techniques in the diagnosis, evaluation, and even understanding of the pathophysilogical events in muscles.

ACKNOWLEDGMENTS

We express our thanks to Drs. P. A. Narayana, G. Elizondo-Riojas, R. K. Entrikin, T. F. Egan, and P. T. Beall. Results obtained from collaborative studies with these colleagues have been discussed in this review. Our studies were partly supported by grants from the National Institute of Health (ROI AR38741) and Diabetes Research and Education Foundation.

REFERENCES

1. D. G. Taylor, R. Inamdar, and M. C. Bushell, NMR imaging in theory and practice, *Phys. Med. Biol.* 33:635 (1988).

2. J. D. R. Miller and L. Starr, Information explosion in radiology, *J. Can. Assoc. Radiologists* 39:33 (1988).

3. A. R. Margulis and L. E. Crooks, Present and future status of MR imaging, *AJR* 150:487 (1988)

4. R. E. Steiner, Magnetic Resonance Imaging: Its impact on diagnostic radiology, *AJR* 145:883 (1985).

5. A. R. Margulis and M. R. Fisher, Present clinical status of Magnetic Resonance Imaging, *Mag. Res. Med.* 2:309 (1985).

6. W. G. Bradley, Jr., I. R. Young, and R. L. Nunnally, Society of magnetic resonance in medicine meeting in Montreal: Impressions and comments, *AJR* 148:1030 (1987).

7. O. A. C. Petroff, Biological ^1H NMR spectroscopy, *Comp. Biochem. Physiol.* 90B:249 (1988).

8. A. M. Margulis, The promise of Magnetic Resonance Spectroscopy for medical diagnosis, *Invest. Radiol.* 23:253 (1988).

9. B. M. Hitzig, J. W. Prichard, H. L. Kantor, W. R. Ellington, J. S. Ingwall, C. T. Burt, S. I. Helman, and J. Koutcher, NMR spectroscopy as an investigative technique in physiology, *FASEB J.* 1:22 (1987).

10. S. R. Williams and W. G. Gadian, Tissue Metabolism studies *in vivo* by nuclear magnetic resonance, *Quarterly J. Exp. Physiol.* 71:335 (1986).

11. M. J. Avison, H. P. Hetherington, and R. G. Shulman, Application of NMR to studies of tissue metabolism, *Ann. Rev. Biophys. Chem.* 15:377 (1986).

12. G. K. Radda and D. J. Taylor, Applications of nuclear magnetic resonance spectroscopy in pathology, *Int. Rev. Exp. Path.* 27:1 (1985).

13. P. J. Bore, The role of magnetic resonance spectroscopy in clinical medicine, *Mag. Reson. Imaging* 3:407 (1985).

14. R. Mathur-De Vre, Biomedical implications of the relaxation behavior of water related to NMR imaging, *Br. J. of Radiol.* 57:955 (1984).

15. L. E. Crooks, N. M. Hylton, D. A. Ortendahl, J. P. Rosin, and L. Kaufman, The value of relaxation times and density measurements in clinical MRI, *Invest. Radiol.* 22:159 (1987).

16. R. G. Evens, R. G. Jost, and R. G. Evens, Jr., Economic and utilization analysis of magnetic resonance imaging units in the United States in 1985, *AJR* 145:393 (1985).

17. K. M. Peddecord, E. A. Janon, and J. M. Robins, Use of MR Imaging in an outpatient clinic, *AJR* 148:809 (1987).

18. D. A. Durick, M. L. W. Phillips, Diffusion of an innovation: Adoption of MRI, *Radiologic Technology* 59:239 (1988).

19. G. M. Bydder, J. M. Pennock, R. E. Steiner, J. S. Orr, D. R. Bates, I. R. Young, The NMR diagnosis of cerebral tumors, *Mag. Reson. Med.* 1:5 (1984).

20. G. M. Bydder, R. E. Steiner, I. R. Young, A. S. Mall, D. J. Thomas, J. Marshall, C. A. Pellis, N. J. Legg, The NMR diagnosis of cerebral tumors, *AJR* 139:215 (1982).

21. M. A. Johnson and G. M. Bydder, NMRI imaging of the brain in children, *Br. Med. Bull.* 40:175 (1984).

22. M. T. Mc Namara and C. B. Higgins, Cardiovascular applications of magnetic resonance imaging, *Mag. Reson. Imaging* 2:167 (1984).

23. R. E. Steiner, Nuclear magnetic resonance imaging of the heart and mediastinum, *Br. Med. Bull.* 40:191 (1984).

24. F. W. Smith, A. Reed, J. R. Mallard, J. M. S. Hutchison, D. A. Power, G. R. D. Catto, Nuclear magnetic resonance tomographic imaging in renal disease, *Diagnostic Imaging* 51:209 (1982).

25. F. M. Doyle, J. M. Pennock, L. M. Banks, M. J. McDonnell, G. M. Bydder, R. E. Steiner, I. R. Young, G. J. Clarke, T. Passmore, and D. J. Gilderdale, Nuclear magnetic resonance imaging of the liver: initial evidence, *AJR* 138:193 (1982).

26. M. Gronskill, R. M. Henkelman, P. Y. Poon, M. B Gunes N. Ege, *et al*, Magnetic Resonance imaging, computed tomography and radionuclide scintigraphy of liver metastases, *J. Can. Assoc. Radiologists* 39:3 (1988).

27. K. L. Moon, Jr., H. K. Genant, C. A. Helms, N. I Chafetz, L. E. Crooks, and L Kaufman, Musculoskeletal applications of nuclear magnetic resonance, *Radiology 147:161 (1983).*

28. M. R. Fisher. G. C. Dooms, H. Hricak, C. Reinhold, and C. B. Higgins, Magnetic resonance imaging of the normal and pathologic muscular system, *Mag. Reson. Imaging* 4:491 (1986).

29. C. R. Ling and M. A. Foster, Changes in NMR relaxation time associated with local inflammatory response, *Phys. Med. Biol.* 27:853 (1982).

30. H. Paajanen, W. Grodd, D. Revel, B. Engelstad, and R. C. Brasch, Gadolinium-DPTA enhanced MR Imaging of intramuscular abscesses, *Mag. Reson. Imaging* 5:109 (1987).

31. J. D. Reeder and S. Andelman, The rotator cuff tear: MR evaluation, *Mag. Reson. Imaging* 5:331 (1987).

32. P. A. Narayana, W. W. Brey, M. V. Kulkarni, and L. K. Misra, *In vivo* proton spin-lattice relaxation times of normal and dystrophic muscles, *Mag. Reson. Imaging* 4:153 (1987).

33. P. A. Bottomley, C. J. Hardy, R. E. Argersinger, and G. Allen-Moore, A review of ^{1}H nuclear magnetic resonance relaxation in pathology: Are T_1 and T_2 diagnostic?, *Med. Phys.* 14:1 (1987).

34. B. W. Wilson, W. R. Randall, G. T. Patterson, R. K. Entrikin, Major physiologic and histochemical characteristics of inherited dystrophy of the chicken, *Ann. NY Acad. Sci.* 317:224 (1979).

35. E. Cosmos, J. Butler, E. P. Allard, and J. Mazliah, Factors that influence the phenotypic expression of genetically normal and dystrophic muscles, *Ann. NY Acad. Sci.* 317:571 (1979).

36. J. R. Mendell, R. Higgins, Z. Sahenk, and E. Cosmos, Relevance of genetic animal models of muscular dystrophy to human muscular dystrophies, *Ann. NY Acad. Sci.* 317:409 (1979).

37. B. W. Wilson, R. K. Entrikin, J. Sketelj, G. T. Patterson, and W. R. Randall, Genetics, cell biology and pharmacology of inherited muscular dystrophy if the chicken, *in Muscular Dystrophy , Proceedings of the International Symposium on Muscular Dystrophy*, S. Ebashi, ed. University of Tokyo Press, Tokyo, pp. 3-17 (1983).

38. M. A. Foster, J. M. S. Hutchison, J. R. Mallard, M. Fuller, Nuclear magnetic resonance pulse sequence and discrimination of high- and low-fat tissues, *Mag. Reson. Imaging* 2:187 (1984).

39. W. D. Middleton, J. B. Kneeland, G. F. Carrera, J. D. Cates, G. M. Kellman, N. G. Campagna, A. Jesmanowicz, W. Froncisz, and J. S. Hyde, High-resolution MR imaging of the normal rotator cuff, *AJR* 148:559 (1987).

40. L. K. Misra, C. F. Hazlewood, L. E. Todd, G. Elizondo-Riojas, Magnetic resonance imaging of slow and fast muscles, *Am. J. Vet. Res.* (in press).

41. L. K. Misra and G. Elizondo-Riojas, MRI of muscle hypertrophy and atrophy, *Avian Dis.* (submitted).

42. R. Damadian, Tumor detection by nuclear magnetic resonance, *Science* 171:1151 (1971).

43. P. A. Bottomley, T. H. Foster, R. E. Argersinger, and L. M. Pfeifer, A review of normal tissue hydrogen NMR relaxation times and relaxation mechanisms from 1-100 MHz: Dependence on tissue type, NMR frequency, temperature, species, excision, and age, *Med. Phys.* 11:425 (1984).

44. L. K. Misra, W. Vijjeswarrapur, G. W. Parker, D. W. Bearden, C. F. Hazlewood, Effects of evolutionary disuse, genetic selection, and severity of disease on muscle T_1 and T_2, *Mag. Reson. Imaging* 4:128 (1986).

45. J. F. Polak, F. A. Jolesz, D. F. Adams, NMR of skeletal muscle: Differences in Relaxation parameters related to extracellular/intracellular fluid spaces, *Invest. Radiol.* 23:107 (1988).

46. I. A. Mardini, R. J. M. McCarter, and G. D. Fullerton, NMR relaxation times of skeletal muscle: dependence on fiber type and diet, *Mag. Reson. Imaging* 4:393 (1986).

47. L. K. Misra, S. R. Kasturi, S. K. Kundu, Y. Harati, C. F. Hazlewood, M. G. Luthra, W. S. Yamanachi, R. P. Munjaal, and S. R. Amtey, Evaluation of muscle degeneration in inherited muscular dystrophy by nuclear magnetic resonance techniques, *Mag. Reson. Imaging* 1:79 (1982).

48. T. F. Egan, D. W. Bearden, C. F. Hazlewood, L. K. Misra, Changes in the proton spin-lattice relaxation times in muscular dystrophy, *Soc. Mag. Reson. Med. Meeting, 1986*, Abstracts 2:1182.

49. L. K. Misra and R. K. Entrikin, Corticosteroid therapy in avian muscular dystrophy:Evaluation by magnetic resonance relaxation times, *Expt. Neurol.* (in press).

50. P. S. Belton and R. G. Ratcliffe, NMR and compartmentation in biological tissues, *Prog. in NMR Spectrosc.* 17:241 (1985).

51. E. Le Rumeur, J. De Certaines, P. Toulouse, and P. Rochcongar, Water phases in rat striated muscles as determined by T_2 proton NMR relaxation times, *Mag. Reson. Imaging* 2:267 (1987).

52. J. L. Fleckenstein, R. C. Canby, R. W. Parkey, and R. M. Peshock, Acute effects of exercise on MR imaging of skeletal muscle in normal volunteers, *AJR* 151:231 (1988).

53. A. Von Bezold, Untersuchungen uber die verteilung von vasser, organischer materie, und anorganischen verbindungen im thierreiche, *Zeit.Wissenchaften Zoologic* 8:487 (1857).

54. L. K. Misra, N. K. R. Smith, D. C. Chang, R. L. Sparks, I. L. Cameron, P. T. Beall, R. Harrist, B. L. Nichols, R. C. Fanguy, and C. F. Hazlewood, Intracellular concentration of elements in normal and dystrophic skeletal muscles of the chicken, *J. Cell. Physiol.* 103:193 (1980).

55. C. F. Hazlewood, B. L. Nichols, D. C. Chang, and B. Brown, On the state of water in developing muscle: A study of the major phases of ordered water in skeletal muscle and its relationship to sodium concentration, *Johns Hopkins Med. J.* 128:117 (1971).

56. L. K. Misra, P. A. Narayana, P. T. Beall, S. R. Amtey, and C. F. Hazlewood, Developmental changes in T_1 relaxation times of muscle water protons in muscular dystrophy, *Mag. Reson. Med.* 1:205 (1984).

57. K. Matsumura, I Nakano, N. Fukuda, H. Ikehira, Y. Tateno, and Y. Aoki, Proton spin-lattice relaxation time of duchenne dystrophy skeletal muscle by magnetic resonance imaging, *Muscle Nerve* 11:97 (1988).

58. D. C. Chang, L. K. Misra, P. T. Beall, R. C. Fanguy, and C. F. Hazlewood, Nuclear magnetic resonance study of muscle water protons in muscular dystrophy of chickens, *J. Cell. Physiol.* 107:139 (1981).

59. L. Borghi, F. Savoldi, R. Scelsi, and M. Villa, Nuclear magnetic resonance of protons in normal and pathological human muscles, *Expt. Neurol.* 81:89 (1983).

60. S. G. Wu, D. R. Courtney, C. F. Hazlewood, and L. K. Misra, Multiexponential behavior of proton T_1 relaxation times of slow, fast, and degenerating muscles, *Fed. Proc.* 2:A958 (1988).

61. R. Barthwal, M. Hohn-Berlage, and K. Gersonde, *In vitro* proton T_1 and T_2 studies in rat liver: Analysis of multiexponential relaxation processes, *Mag. Reson. Med.* 3:863 (1986).

62. D. Grucker, Y. Mauss, and J. Sterbel, Effect of interpulse delay on NMR transverse relaxation rate of tissues, *Mag. Reson. Imaging* 4:441 (1986).

63. K. R. Metz, P. J. Stankiewicz, J. W. Sassani, and R. W. Briggs, Pulse techniques for the suppression of individual components in multiexponential relaxation curves, *Mag. Reson. Med.* 3:575 (1986).

64. L. K. Misra, T. F. Egan, D. R. Courtney, L. Chen, D. W. Bearden, and C. F. Hazlewood, NMR relaxation times of muscles reflect modifications in expression of dystrophic gene, *Mag. Reson. Imaging* 5:15 (1987).

65. L. K. Misra, P. A. Narayana, *In vivo* T_1 characterization of genetically-induced muscle atrophy, *Mag. Reson. Imaging* (submitted).

66. H. F. Bennett, H. W. Swartz, R. D. Brown, and S. H. Koenig, Modification of relation of lipid protons by molecular oxygen and nitroxides, *Invest. Radiol.* 22:502 (1987).

67. L. K. Misra, R. K. Entrikin, and C. F. Hazlewood, Effects of therapeutic intervention in NMR relaxation times of dystrophic muscles, *Proc. 12th Int. Conf. Mag. Reson. Biol. System* p136 (1986).

68. P. T. Beall and L. K. Misra, Effects of estrogen status on NMR relaxation times in tissues of female animals, *Biophys. J.* 45:253 (1984).

69. R. M. Kawamoto and R. J. Baskin, Isolation and characterization of sarcoplasmic reticulum from normal and dystrophic chicken, *Muscle Nerve* 9:248 (1986).

70. R. P. Munjaal, J. R. Dedman, and L. K. Misra, Elevation of calmodulin in avian muscular dystrophy, *Cell Calcium* 6:481 (1985).

71. J. R. Zimmerman and W. E. Brittin, Nuclear magnetic resonance studies in multiple phase systems: Lifetime of a water molecule in an absorbing phase on silica gel, *J. Phys. Chem.* 61:1328 (1957).

72. G. D. Fullerton, J. L. Potter, and N. C. Dornbluth, NMR relaxation of protons in tissues and other macromolecular water solutions, *Mag. Reson. Imaging* 1:209 (1982).

73. L. K. Misra, P. A. Narayana, D. W. Bearden, T. F. Egan, R. P. Munjaal, and C. F. Hazlewood, Reduction in the proton NMR relaxation times of dystrophic muscles following functional improvement, *Mag. Reson. Imaging* 2:205 (1984).

74. S. H. Koenig and R. D. Brown, III, Determinants of proton relaxation rates in tissue, *Mag. Reson. Med.* 1:437 (1984).

75. S. H. Koenig and R. D. Brown III, The importance of the motion of water for magnetic resonance imaging, *Invest. Radiol.* 20:297 (1985).

76. S. H. Koenig, Theory of relaxation of mobile water protons induced by protein NH moieties, with application to rat heart muscle and calf lens homogenates, *Biophys. J.* 53:91 (1988).

77. H. E. Rorschach and C. F. Hazlewood, Protein dynamics and the NMR relaxation time T_1 of water in biological systems, *J. Mag. Reson.* 70:79 (1986).

78. G. Schauer, R. Kimmich, and W. Nusser, Deuteron field-cycling relaxation spectroscopy and translational water diffusion in protein hydration shells, *Biophys. J.* 53:397 (1988).

79. R. M. Kroeker and R. M. Henkelman, Analysis of biological NMR relaxation data with continuous distributions of relaxation times, *J. Mag. Reson.* 69:218.

80. L. K. Misra, M. G. Luthra, S. R. Amtey, G. Elizondo-Riojas, S. H. Swezey, and L. E. Todd, Enhanced T_1 differentiation between normal and dystrophic muscles, *Mag. Reson. Imaging* 2:33 (1984).

81. L. E. Crooks, M. Arakawa, J. C. Hoenningar, B. McCarten, J. Watts, and L. Kaufman, Magnetic resonance imaging: Effects of magnetic field strength, *Radiology* 151:127 (1984).

82. V. M. Runge, J. A. Clanton, C. M. Lukehart, C. L. Partin, and A. E. James, Jr., Paramagnetic agents for contrast-enhanced NMR imaging: a review, *AJR* 141:1209 (1983).

83. R. B. Buxton, G. L. Wismer, T. J. Brady, and B. R. Rosen, Quantitative proton chemical-shift imaging, *Mag. Reson. Med.* 3:881 (1986).

84. C. J. Hardy and C. L. Dumoulin, Lipid and water supression by selective ^{1}H homonuclear polarization transfer, *Mag. Reson. Med.* 5:58 (1987).

85. L. K. Misra, J. W. Frazer, C. F. Hazlewood, and L. W. Dennis, ^{1}H NMR spectra of normal and dystrophic muscles, *Soc. Mag. Reson. Med., Book of Abstracts* p553 (1987).

86. W. M. Brooks, J. Field, M. G. Irving, and D. M. Doddrel, *In vivo* Determination of ^{31}P spin relaxation times (T_1, T_2, T_{1rho}) in rat leg muscle. Use if an off-axis solenoid coil, *Magn. Reson. Imaging* 4:245 (1986).

87. P. A. Narayana, J. L. Delayre, and L. K. Misra, *In vivo* ^{31}P NMR studies of avian dystrophic muscles, *Mag. Reson. Med.* 3:549 (1986).

88. D. W. Kormos and H. N. Yeung, NMR imaging of ^{13}C in animal tissues, *Mag. Reson. Med.* 4:500 (1987).

89. S. W. Lee. S. K. Hilal, and Z. H. Cho, A multinuclear magnetic resonance imaging technique-simultaneous proton and sodium imaging, *Mag. Reson. Imaging* 4:343 (1986).

90. D. B. Sprague, D. G. Gadian, S. R. Williams, E. Proctor, and A. W. Goode, Intracellular metabolites in rat muscle following trauma: a ^{31}P and ^{1}H nuclear magnetic resonance study, *J. Roy. Soc. Med.* 80:495 (1987).

MAGNETIC RESONANCE IMAGING (MRI) OF THE

PITUITARY GLAND IN SHEEHAN'S SYNDROME

Jesús Z. Villarreal (1), Guillermo Elizondo (2), Rafael
Real (1), José G. González (1), and Homero Náñez (1)

1.- Department of Internal Medicine 2.- Department of Radiology
Facultad de Medicina U.A.N.L. Monterrey, Nuevo León México

INTRODUCTION

Hypopituitarism designates absence of Pituitary Hormones Secretion.
Patients with this Syndrome may lack one (isolated deficiency) or several
hormones (Panhypopituitarism). In the post-puberal age, it usually presents
as a long standing illness characterized by symptoms and signs of one or
several hormone deficiencies. At the onset, most patients develop symptoms
of prolactin deficiency (no lactation) followed by symptoms of Gonadotrophin
failure (amenorrhea, loss of libido, impotence). Depending on the nature
and extension of the pituitary lesion, patients may develop symptoms and
signs of Hypothyroidism and eventually Adrenal Insufficiency appears.[1]

Classically, the diagnosis has been based on clinical history and
direct measurement of the pituitary hormones on both, basal and stimulated
conditions. However, the syndrome can be produced by several different
pathologic lesions of the pituitary[2] and the etiologic diagnosis requires
visualization of the sella turcica or the pituitary gland itself.

Tumors constitute the most common cause of hypopituitarism.[3] These
lesions may arise from the pituitary (Prolactinoma, Growth Hormone-Producing
adenoma, ACTH - producing tumor, non-functioning tumor) or from parasellar
tissues (craniopharyngioma, meningioma, glioma). The gland may be infil-
trated by granulomatous (Tb, Mycosis, Sarcoidosis) or malignant (Leukemia,
Histiocytosis) disorders. Also, the pituitary my suffer from infarction.
The necrosis of the pituitary has been reported in pregnant diabetics,
pituitary tumors and post-partum (Sheehan's Syndrome)[3].

Since the original description of Sheehan's Syndrome[4] numerous cases
have been reported in the literature. In fact, it is considered one of the
most common causes of hypopituitarism[3]. It explained 38% of 59 cases of
hypopituitarism seen at our Institution over a two year period. Post-partum
pituitary necrosis is considered secondary to extensive thrombosis of the
pituitary vessels during or following delivery usually in association with
hemorrhage and hypotension[5]. However, at least one study, reported the
presence of antibodies against the pituitary in the sera of a group of
patients with this Syndrome[6].

Patients present with progressive symptoms and signs of hypopituitarism. After delivery most patients are unable to lactate and this is

followed by symptoms of hypogonadism and hypothyroidism. Depending on the time of diagnosis patients may develop hypoadrenalism.

For visualization of the pituitary and the sella, several radiologic methods have been used (skull x rays, hypocycloidal tomography, pneumoence-phalography and Angiography)[7]. All have limitations because they do not permit a direct imaging of the gland and some are invasive.

The development of non-invasive, high resolution technics, such as, computed Tomography (CT) and Magnetic Resonance Imaging (MRI) has simplified the study of patients with pituitary disorders.[8] Although, some studies have found less sensitivity for MRI as compare to CT in detecting microade-nomas,[9] MRI is considered today a very exact method for evaluation of intrasellar and juxtasellar lesions.[10,11,12] Furthermore, in contrast to CT, MRI is free from ionizing radiation, has multiplanar capacity and excellent soft tissue characterization.[13] Finally it has no known hazardous biological effects.[14,15]

Until recently, the only radiological finding in Sheehan's Syndrome was a normal to reduced sella turcica size on plain skull films described by Meador and Worrell in 1966.[16] In 1983, Fleckman et al, reported for the first time radiologic documentation of Sheehan's Syndrome using computed tomography (CT).[17] They described an empty or partially empty sella of normal size in 11 of 13 women with post-partum hypopituitarism. One of their patients showed a normal CT of the pituitary and only 7 of 13 patients had documented vascular collapse during labor and some denied post-partum complications.

We present here the results of MRI and endocrine studies in twelve patients with documented Sheehan's Syndrome. To the best of our knowledge this is the first study of MRI in this entity

PATIENTS AND METHODS

Twelve women seen at our institution with history of obstetric complications and documented hypopituitarism were studied. Their clinical data are Shown in Table 1.

Their age ranged 36-59 yr. and the time elapsed between the obstetric accident and diagnosis ranged 6-25 yr with an average of 16.1 yr. Four of the patients had more than 4 pregnancies, five patients had between 2-4 pregnancies and three women had their obstetric complication on her first delivery. In all but one, the obstetric accident was hemorrhage. In one patient there was a combination of sepsis and hemorrhage.

ENDOCRINE TESTS

For documentation of hypopituitarism the patients had performed the following tests: Basal serum samples were obtained for measurement of Thyroxine (T4), Resin T_3 Uptake (RT$_3$U), Thyroid Stimulating Hormone (TSH), Prolactin (PRL), Follicle Stimulating Hormone (FSH), Luteinizing Hormone (LH), Cortisol and Growth Hormone (hGH). After this, stimulation of the pituitary was done by injecting Thyrotropin-Releasing Hormone, (TRH 200 mcg I.V.) and Gonadotrophin Releasing Hormone (GhRH 100 mcg I.V.) in all patients. In 9 patients Insulin - induced hypoglycemia (0.1 u/kg I.V.) was used for stimulation of cortisol and hGH. Additional samples were taken at 15', 30', 60', 90' and 120' after stimulation. We used for analysis only maximal responses.- Hormonal measurements were performed by Radioimmunoassay using comercial kits from Diagnostic Products. (L.A. Cal).

446

Table 1. Clinical data of twelve patients with
Sheehan's Syndrome

Patient	Age (yr.)	Evolution (yr.)*	Pregnancies	Complication
1.-	39	17	1	Hemorrhage + Sepsis
2.-	59	25	3	Hemorrhage
3.-	42	13	4	Hemorrhage
4.-	40	15	1	Hemorrhage
5.-	37	16	3	Hemorrhage
6.-	45	10	6	Hemorrhage
7.-	50	16	9	Hemorrhage
8.-	42	6	4	Hemorrhage
9.-	55	21	9	Hemorrhage
10.-	36	16	1	Hemorrhage
11.-	48	23	4	Hemorrhage
12.-	49	16	13	Hemorrhage

* Time elapsed between obstetric accident and MRI Study

MRI STUDIES

MRI images were obtained with a 0.3 T permanent magnetic imaging system (FONAR B 3,000 Melville, N.Y.), with a 25.6 cm. diameter head coil. Sections were 3.5 mm thick with 1.5 mm intersection gap, acquired on 256 x 256 matrix (1 mm pixels) and interpolated to 512 x 512 matrix for measurement and photography. T 1 weighted spin echo (SE) 500/28/4 (repetition time (TR) mesc/echo-Time (TE) mesc/number of excitations) was used as pulse sequence to obtain saggital and coronal images in all patients.

For cuantification of the pituitary gland we used the following scale: 0 complete absence of pituitary tissue; + when less than 25% of the sella volume was occupied by pituitary tissue; ++ 25-50% occupation; +++ 50-75% occupation and ++++ when more than 75% of the sella was occupied by the pituitary gland.

As controls we used MRI studies of the pituitary in 20 healthy, non-pregnant women aged 20-30 yr. A study of age and parity matched normal women was considered unwarranted.

RESULTS

Endocrine studies documented hypopituitarism in all patients. All of them had secondary hypothyroidism (Table 2). Although patient No. 7 had a basal TSH of 10.5 mu/ml and patient No. 3 a basal level of 9.7 mu/ml, all patients showed a lack of response to TRH administration. Ten patients had basal prolactins below 5 ng/ml, again, all showed a lack of respose to TRH.

Table 3 summarizes results of GnRH testing and insulin induced Hypoglycemia. Basal FSH and LH levels ranged 1.5-15 mIu/ml and 0.0-22.0 mIU/ml respectively. Although the majority of these measurements are considered within normal limits for menstruating women in follicular phase, 6 of our patients were in post-menopausal age and all were amenorrheic. Besides, there was no significant response to LHRH administration.

Insulin-induced hypoglycemia was provoked in 9 patients. Cortisol levels were low without response to hypoglycemia in five patients (2, 3, 5, 8 and 9). Patients 10 and 12 had normal basal levels with adequate response to stimulation. Patients 6 and 11 only had basal determinations and both were normal. Growth Hormone levels were tested in 8 patients. In all, basal levels were low without response to hypoglycemia.

Table 2. Thyroid Function and TRH tests in Patients with Sheehan's Syndrome

Patients	T_4 mcg/dl	RT_3I (%)	TSH (uu/ml)	PRL (ng/ml)
			B – S	B – S
1.–	4.3	16.3	5.1–8.2	13.0–15.6
2.–	5.1	28.8	5.0–5.4	ND–ND
3.–	2.0	27.0	9.7–12.1	4.0–6.4
4.–	4.5	16.0	1.6–3.2	1.8–2.8
5.–	1.0	25.0	4.1–7.2	-------
6.–	4.8	29.0	6.0–7.2	5.2–11.4
7.–	3.0	26.0	10.5–14.6	ND–0.6
8.–	5.4	20.9	0.5–0.7	ND–0.4
9.–	0.8	17.8	4.2–7.0	ND–ND
10.–	0.3	17.8	1.2–1.9	1.9–2.2
11.–	0.6	12.5	1.7–2.6	5.0–8.3
12.–	0.3	17.8	1.2–1.9	1.9–5.9
NORMAL	(4.5–12.5)	(25–35)	($<$ 6.0–*)	($<$ 20–*)

B Basal level S Maximal response of the hormone after TRH (200 ucg I.V.) administration ND Not detectable * A double increase or more above basal level is considered a normal response to stimulation.

MRI RESULTS

The studies in the control group showed the normal appearance of the pituitary[13] in all subjects. The gland had an homogeneous signal intensity and was slightly hyperintense relative to brainstem. It had a flat or slightly concave superior border on coronal sections, with the pituitary stalk in the midline. Posterioinferior intrasellar fat pad was seen hyperintense in comparison with the pituitary (Fig. 1 A, B).

Table 3. GnRH test and response of cortisol and Growth Hormone to insulin induced Hypoglycemia.

Patient	FSH(mIU/ml)	LH(mIU/ml)	Cortisol(Mcg/dl)	hGH(ng/ml)
	B–S	B–S	B–S	B–S
1.–	11.0–14.0	22–6.0	----------	--------
2.–	3.8–6.1	0.0–2.2	1.7–4.2	0.8–0.9
3.–	1.5–3.2	2.1–2.6	3.6–3.6	0.5–0.5
4.–	14.0–14.0	13.0–6.0	14.0–19.0	1.3–1.9
5.–	1.8–2.4	2.5–3.4	5.1–6.7	--------
6.–	3.0–4.2	2.0–3.8	20.5–	0.7–0.5
7.–	14.0–14.0	10.0–14.0	----------	--------
8.–	4.9–5.4	2.7–2.2	4.3–3.9	0.6–0.7
9.–	6.6–8.4	6.6–15.5	0.4–2.3	0.0–0.9
10.–	0.9–12.0	11.6–15.1	11.2–27.0	0.0.0.0
11.–	10.0–16.0	4.6–12.0	10.4–	--------
12.–	9.6–12.0	11.6–15.5	11.2–32.4	0.0–0.3
NORMAL	(3–20/*)	(5–15/*)	(5–25/*)	($<$ 5.0/*)

B Basal levels S Maximal level post-estimulation * An increase of at least double above basal is considered normal.

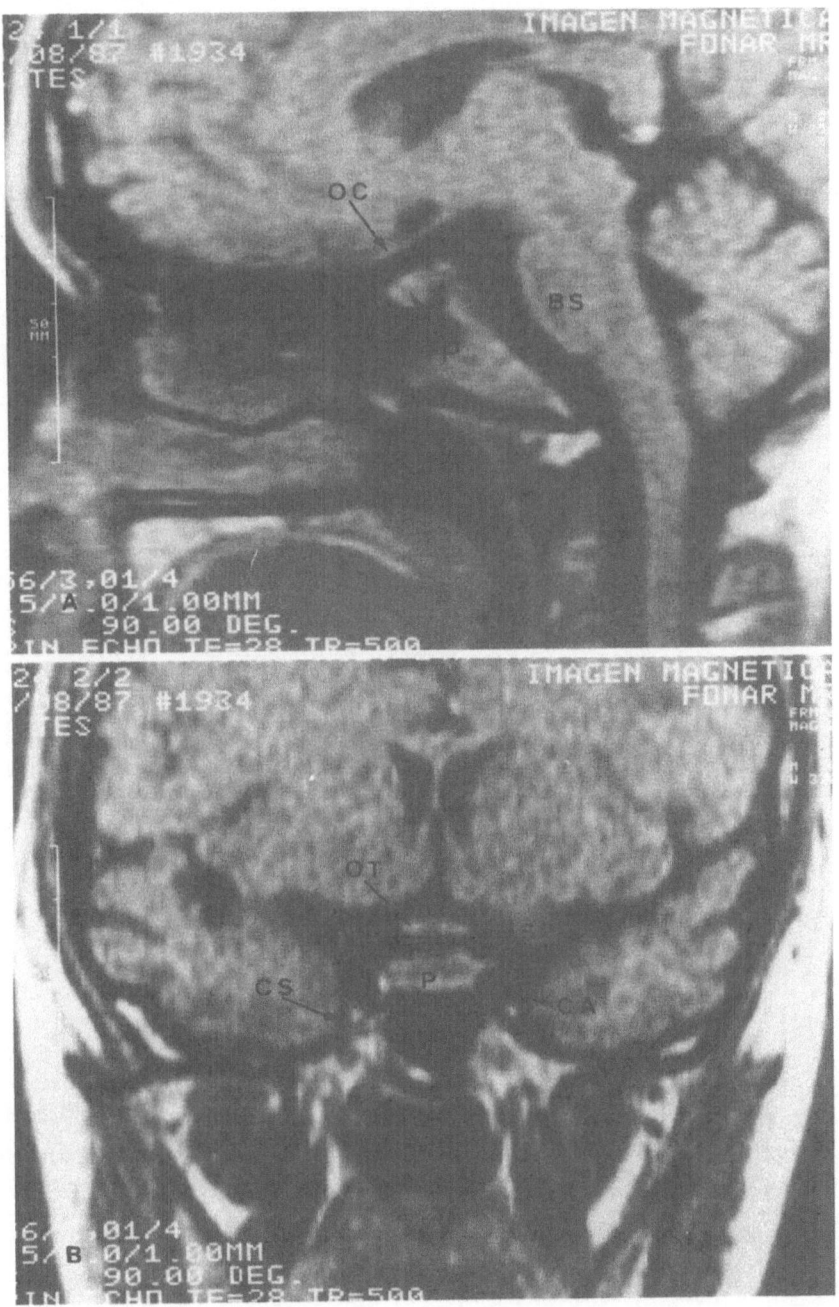

Figure 1. Normal pituitary gland. (A) Sagittal section. (B) Coronal view.
P Pituitary. BS Brain stem. OC Optic chiasm. PS Pituitary
stalk. CA Carotid artery. OT Optic tract. CS Cavernous
sinus.

The parasellar structures, including optic chiasm, suprasellar cistern, cavernous sinuses and carotid arteries were visualized and had normal signal intensities and anatomic characteristics. Normal pituitary diameters were: Vertical 6.0 + 0.6 mm, Transverse 12.4 + 1.1 mm and Anteroposterior 7.9 + 1.5 mm. Pituitary volume calculated as described by Dichiro[18] was 300 $\pm$ 60 mm[3].

Table 4 summarizes MRI findings in patients. None of them showed the characteristic, previously described, normal pituitary imaging. In fact, all showed a complete or almost complete absence of pituitary tissue. Eight patients had less than 25% of the sella volume occupied by glandular tissue. This was irregular, ill defined and very difficult to cuantificate with precision. Four patients (7,8,9,10) showed complete absence of the gland. Still it was possible to observe the pituitary stalk in all patients and the presence of the intrasellar fat pad, which looked hyperintense as described (Figures 2 and 3). Sella sizes were considered normal in all cases and were filled by cerebrospinal fluid.

No correlation was found between the amount of pituitary tissue visualized by MRI and the time elapsed between obstetric accident and imaging. No correlation could be made with endocrine function either.

DISCUSSION

In contrast to the study of Fleckman et al[17], none of our patients showed a normal appearance of the pituitary. However, the case reported in that study with normal CT, was a patient without history of vascular collapse during labor, but had severe headache and vomiting during the third trimester of pregnancy, making difficult to ruled out the presence of a pituitary tumor with infarction. Again, in contrast to Fleckman's study, we did not find any correlation between the amount of pituitary tissue and the time elapsed between the obstetric accident and MRI study. The earliest diagnosis in our study (patient 8) was made 6 years after the obstetric complication and this patient showed a complete absence of pituitary tissue.

Table 4.   MRI Findings in Sheehan's Syndrome

Patients	Time Interval (years)*	Amount of Pituitary Tissue
1.-	17	+
2.-	25	+
3.-	13	+
4.-	15	+
5.-	15	+
6.-	10	+
7.-	16	0
8.-	6	0
9.-	21	0
10.-	16	0
11.-	23	+
12.-	16	+

* Years elapsed between obstetric accident and MRI Study.

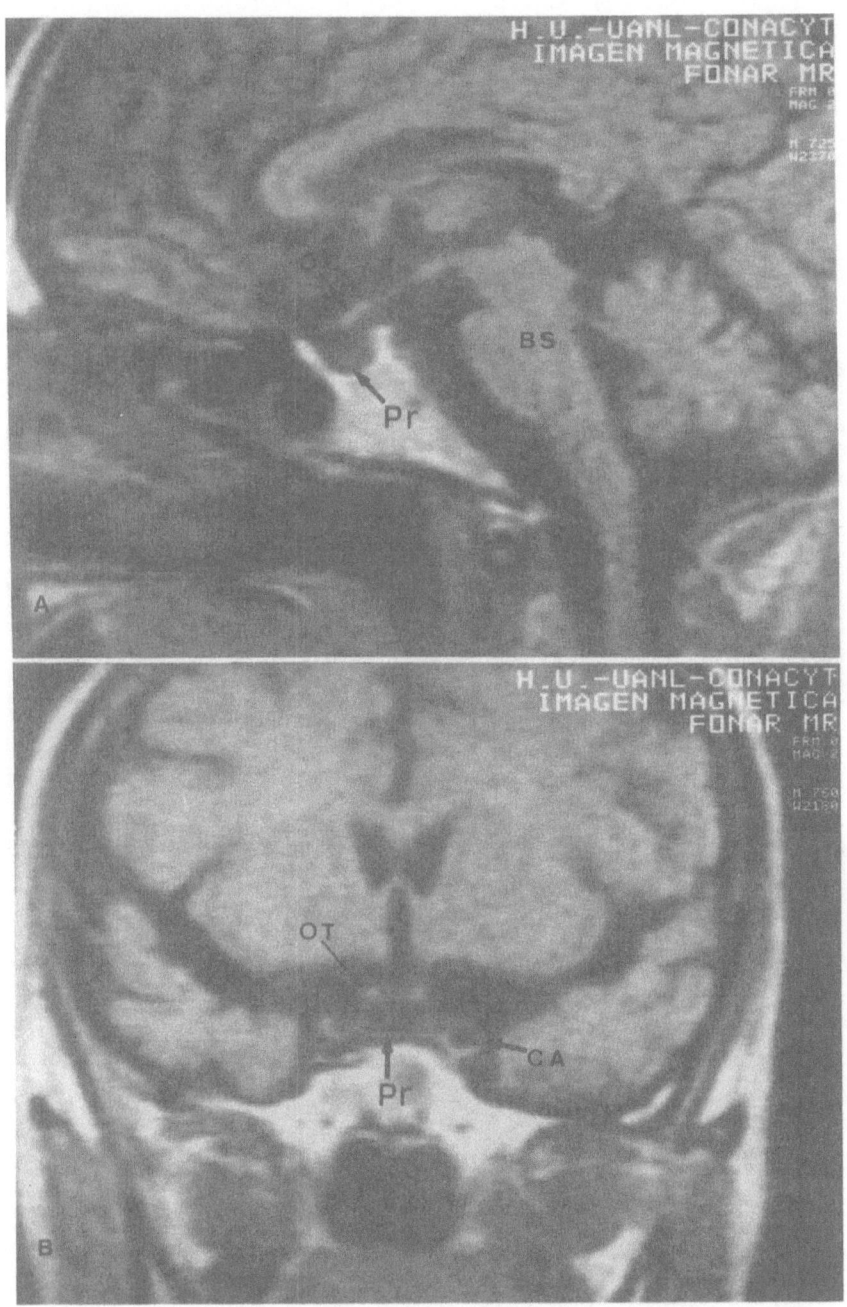

Figure 2. MRI study of patient No. 4. (A) Sagittal section. (B) Coronal
image. BS Brain stem. OC Optic chiasm. Pr Pituitary
remnants. OT Optic tract. CA Carotid artery.

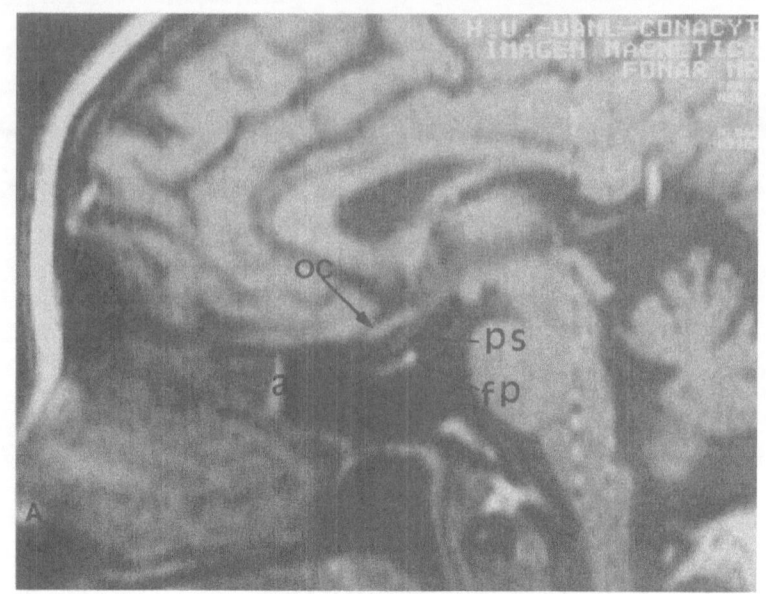

Figure 3 A.   Sagittal view of the patient No. 3.   OC Optic chiasm.
P.S. Pituitary stalk. fp Intrasellar fat pad. a Absence
of pituitary tissue.

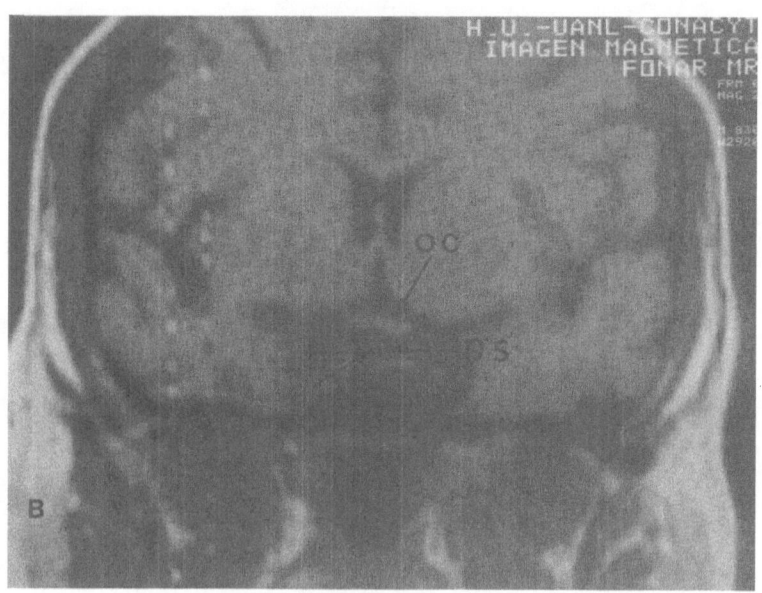

Figure 3 B.   Coronal section of patient No. 3.   OC Optic chiasm.
PS Pituitary stalk. a Absence of pituitary tissue

On the contrary, patient No. 2, whose diagnosis was made 25 years after her complicated labor showed some pituitary tissue within the sella. In fact, we have recently seen two patients with a 2 and 2 1/2 years history of Sheehan's Syndrome and on both, the MRI study showed a complete absence of pituitary tissue.

The reasons for these discrepancies may reside in the fact that in Fleckman's study, not all the patients had the classical history of Sheehan's Syndrome. Further, MRI permits a better characterization of soft tissues within the sella and the intrasellar fat pad and posterior pituitary may be better delineated.

The pituitary lesion in this syndrome is accepted to be necrosis due to thrombosis of pituitary vessels in the hyperplastic gland of pregnant women who suffer from hemorrhage and vascular collapse during labor.- No doubt exists about enlargement of the pituitary during pregnancy.[19,20] Recently, our group has shown, for the first time, in vivo documentation of this growth in a group of primigravid healthy women using MRI.[21] We found an increase of 136% in the size of the pituitary at the end of pregnancy. It is reasonable to think, that this enlarged gland, is very susceptible to decreases in blood perfusion. However, if this were the case, we would expect a disease of acute onset. On the contrary, Sheehan's Syndrome is a chronic an progressive illness. Some patients are able to lactate and even menstruate after the complicated pregnancy and symptoms may ensue many months or years after the obstetric accident. Engelberth et al,[6] reported the presence of antibodies against the pituitary in a group of women with post-partum hypopituitarism. Autoimmune hypophysistis has been confirmed in recent years.[22,23] An autoimmune process precipitated by partial infarction of the gland would be more likely to produce a progressive and chronic loss of pituitary secretion. Also it would explained the lack of correlation between time of evolution and pituitary imaging.- Immunological studies in these patients are warranted.

In contrast to what would have been expected, no correlation was found between endocrine function and the amount of pituitary tissue visualized by MRI. This is in accordance with Fleckman's report. The reason for this finding may be, that despite the high resolution of CT and MRI, are still insensitive methods to detect small clusters of pituitary cells in these patients. It is of interest the finding of slight increases in basal TSH levels in patients 3 and 7. Similar results have been reported previously and the reasons are unclear.[17,24]

In summary, Sheehan's Syndrome is a common cause of hypopituitarism. We have delineated the pituitary lesion in this entity using MRI. The method gives an excellent view of the pituitary gland and facilitates the differential diagnosis of hypopituitarism. Also, it will permit the study of patients at risk in order to obtain an earlier diagnosis and to gain a better understanding of its natural history.

ACKNOWLEDGMENT

We are indebted to Ludivina Segura for excellent secretarial assistance in typing the manuscript.

REFERENCES

1.- J.D. Wilson, D.W. Foster, Williams. Textbook of Endocrinology, W.B. Saunders Philadelphia (1985).

2.- P.O. Kohler, Clinical Endocrinology, Wiley Medical, New York (1986).

3.- G.T. Tindall, D.L. Barrow. Disorders of The Pituitary The C.V. Mosby Co. St. Louis (1986).

4.- H.L. Sheehan Post-partum necrosis of the anterior pituitary. J. Pathol. Bacteriol. 45: 189-214 (1937).

5.- H.L. Sheehan, Z.P. Stanfield, The Pathogenesis of post-partum necrosis of the anterior lobe of the pituitary gland. Acta Endocrinol. 37: 479-510 (1961)

6.- O. Engelberth, Z. Jezkova, Autoantibodies in Sheehan's Syndrome. Lancet 1: 1075 (1065).

7.- R.G. Ramsey, Neuroradiology, W.B. Saunders, Philadelphia (1987).

8.- S.H. Lee, K.C.V.G. Rao. Cranial Computed Tomography and MRI McGraw Hill, New York (1987).

9.- K.W. Pojunas, D.L. Daniels, A.L. Williams, V.M. Haughton, MR Imaging of Prolactin-secreting microadenomas. AJNR 7: 209-213 (1986).

10.- Kucharczyk, D.O. Davies, W.M. Kelly, G. Sze, D. Norman, T.H. Newton, Pituitary Adenomas: high-resolution MR imaging at 1.5 T Radiology 161: 761-765 (1986).

11.- B.C.P. Lee, M.D.F. Deck, Sellar and juxtasellar lesion detection with MR Radiology 157: 143-147 (1985).

12.- R. Oot, P.F.J. New, F.S. Buonanno, IL. L. Pykett, P. Kistler, R. Delapaz, K.R. Davis, J.M. Taveras, T.J. Brady MR imaging of pituitary adenomas using a prototype resistive magnet: preliminary assessment. AJNR 5: 131-137 (1984).

13.- B. Kaufman, Magnetic Resonance Imaging of the Pituitary gland. Radiol. Clin. North Am. 22: 795-803 (1984).

14.- R.D. Saunders, H. Smith, Safety aspects of NMR Clinical Imaging. Br. Med. Bull 40: 148-154 (1984).

15.- S. Wolff, L.E. Crooks, P. Brown, R. Howard, R.B. Painter Tests for DNA and chromosomal damage induced by nuclear resonance imaging. Radiology 136: 707-710 (1980).

16.- C.K. Meador, J.L. Worrell. The sella turcica in post-partum pituitary necrosis (Shehan's Syndrome) Ann Int. Med. 65: 250-264 (1966).

17.- A.M. Fleckman, U.K. Schubart, A. Danziger, N. Fleischer, Empty Sella of Normal Size in Sheehan's Syndrome Am. J. Medicine 75: 585-591 (1983).

18.- G. Dichiro, K.B. Nelson The Volume of the sella turcica. Am J. Roentgenol Radiol. Ther Nud Med. 87: 989-1098 (1962).

19.- R.M. Bergland, B.S. Ray, R.M. Torack. Anatomical Variations in the pituitary gland and adjacent structures in 225 human autopsy cases. J. Neurosurg. 28: 93-99 (1968).

20.- A.T. Rasmussen. Proportions of various subdivisions of normal adult human hypophysis cerebri and relative number of different types of cells in pars distalis, with biometric evaluations of age and sex differences and special consideration of basophilic invasion into infundibular process - Res. Nerv. Ment. Dis. Proc. 17; 118-150 (1938).

21.- J.G. González, G. Elizondo, D. Saldivar, H. Náñez, L. Todd, J.Z. Villarreal. Pituitary gland growth during normal pregnancy an in vivo-study using magnetic resonance imaging. Am. J. Med. (in press).

22.- S.L. Asa, J.M. Bilbao, K. kovacs, R.G. Josse, K. Kreines. Lymphocytic hypophysitis of pregnancy resulting in hypopituitarism: a distinct clinicopathologic entity. Ann Int. Med. 95: 166-171 (1981).

23.- D.S. Baskin, J.J. Townsend, C.B. Wilson. Lymphocytic adenohypophysitis of pregnancy simulating a pituitary adenoma: a distinct pathological entity. J. Neurosurg 56: 1480153 (1982).

24.- M. Shahmanesh, Z. Ali, M. Pourmand, I. Nourmand. Pituitary function tests in Sheehan's Syndrome. Clin. Endocrinol. 12: 303-311 (1980).

CANCER DETECTION BY MAGNETIC RESONANCE IMAGING

David D. Stark

Department Radiology, Massachusetts General Hospital, Fruit Street, Boston, MA 02114

INTRODUCTION

Limitations of CT

The absolute accuracy of CT, MR, or other imaging techniques for the detection of hepatic metastases is difficult to determine (1-2). However, for the clinical task of screening cancer patients at risk for hepatic metastases, CT clearly outperforms scintigraphy and sonography and is the accepted "Gold" standard. Unfortunately, recent data from several centers suggest that the sensitivity (true positive fraction, TPF) of contrast-enhanced CT for detecting individual hepatic lesions is only 34-37%, and the sensitivity for identifying patients with one or more lesions is only 73-74% (1-2). Indeed, the false negative rates for CT account for most of the unexpected late cancer deaths in patients who have previously undergone "curative" excision of primary cancers.

No consensus exists as to what constitutes a technically adequate CT examination (3). While most centers routinely use iodinated contrast material to improve lesion detection, in patients with hypervascular metastases iodinated contrast material obscures lesions. Therefore, under ideal circumstances both noncontrast and contrast enhanced scans would be performed routinely. While accepting the inconvenience, cost, and risks associated with iodinated contrast material, experts disagree regarding the best technique for contrast administration. Most infuse 40g iodine over 2-3 minutes while performing rapid sequential scans with table incrementation through the liver (dynamic sequential bolus CT) (3). Infusion of an additional 20g iodine with repeat scanning after a 4-6 hour delay (delayed CT) may detect additional lesions in some cases. Other data suggest that lesion detection is maximized when contrast is administered via an hepatic artery catheter (CT angiography). Such attempts to improve the performance of iodinated contrast agents have been offset by increased toxicity and invasiveness.

A technical limitation common to all CT techniques is the sequential single slice data acquisition, requiring patient cooperation in breath holding to duplicate the exact level of inspiration for each sequential scan. Scan-to-scan variations in breath holding result in CT examinations that skip some anatomic levels while duplicating others (3).

MR vs. CT

The accuracy of MR imaging relative to CT for the diagnosis of liver metastases was recently measured in a randomized, controlled comparative study including 135 subjects: 57 with cancer metastatic to the liver, 27 with benign cysts of hemangiomas, and 51 normal controls without focal liver disease (1). The sensitivity of a one-hour MR examination, consisting of 4 pulse sequences, for detecting individual metastatic deposits was 64%, significantly greater than 51% for contrast enhanced CT (p <.001) (Fig. 1). The difference in sensitivity for identifying patients with hepatic metastases was less (MR 82% vs CT 80%). In clinical practice it is often more important to identify abnormal patients than to count individual lesions. For example, once the diagnosis of hepatic metastases is established, needless excision of primary pancreatic, breast, lung, and perhaps some colon cancers can be avoided. Furthermore, appropriate systemic chemotherapy is begun as soon as possible. Therefore, the most important task in hepatic imaging is to distinguish normal patients from abnormal ones (those with one or more liver lesions).

Surgical treatment of hepatic neoplasms is increasingly common, particularly for metastatic colorectal cancer. When staging hepatic disease is clinically relevant, the superiority of MR for detecting individual metastatic deposits and the improved delineation of intrahepatic vascular and segmental anatomy is valuable in treatment planning (2). In planning resection of primary or metastatic hepatic neoplasms MRI can substitute for ultrasound, CT, and angiography.

Diagnostic techniques used to screen large numbers of patients must be specific as well as sensitive. For example, since most patients do not have liver metastases, it is important not to have a high false positive rate (low specificity). In patients without hepatic metastases, the specificity of MR was 99% vs 94% for CT, and this difference was statistically significant at p <.05 (1). Analysis of the performance of MR and contrast enhanced CT using receiver operating characteristic (ROC) curves shows MR to be a better test than CT for radiologists with varying degrees of overreading and underreading (1) (Fig. 2).

Pulse Sequence Performance

Pulse sequence performance, assessed by the ability of radiologists to detect lesions, has confirmed the validity of the contrast-to-noise ratio (CNR) as a quantitative index of lesion conspicuity (1,2,4). In clinical practice, significant differences are found between different MR pulse sequences for detection of individual lesions. The sensitivity of both T1 weighted SE (64%) and IR (65%) pulse sequences was significantly (p <.001) greater than either the TE 60 msec (43%) or TE 120 (43%) T2 weighted pulse sequences, Indeed, pulse sequence performance assessed by the ability of a radiologist to detect lesions was quite similar to the rank order of CNR performance and pulse sequence performance predicted by image contrast theory (1,2,4).

Field Strength

These comparative data are valid for pulse sequence performance on one particular machine (Technicare 0.6T). Similar results have been reported from other centers using similar techniques on a Picker 0.5T system (5). The relative efficacy of pulse sequences will vary considerably at higher field strengths. However, the CNR methodology is generally suitable for determining the best pulse sequence for any machine and for any clinical application.

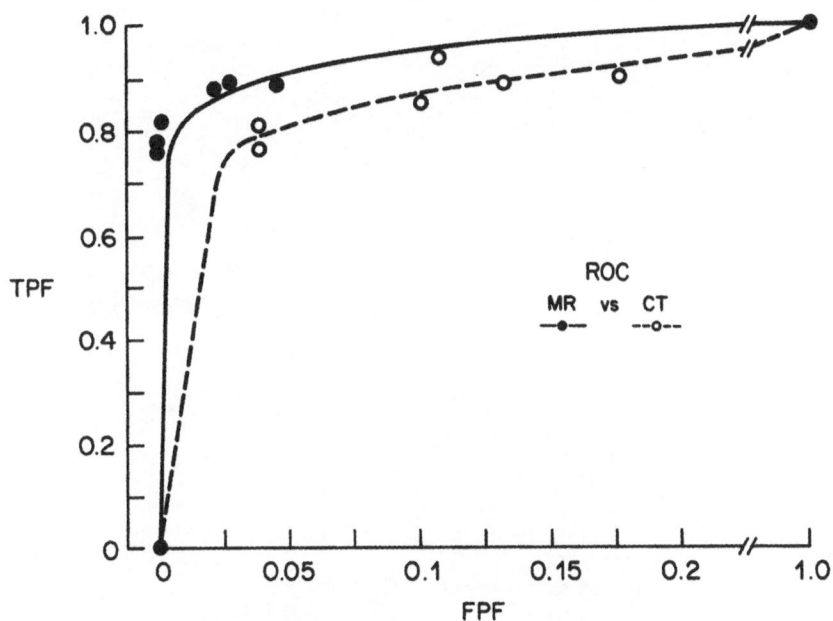

FIGURE   1.   Receiver   operating   characteristics   (ROC)   curves for the
diagnosis of metastatic liver   cancer. The   true positive
fraction is   shown on the ordinate and the false positive
fraction on the abscissa. need CT. The area under the MR
curve is greater than the area under the CT curve; therefore, MR was superior for this test group. Reproduced from
Ref. 1.

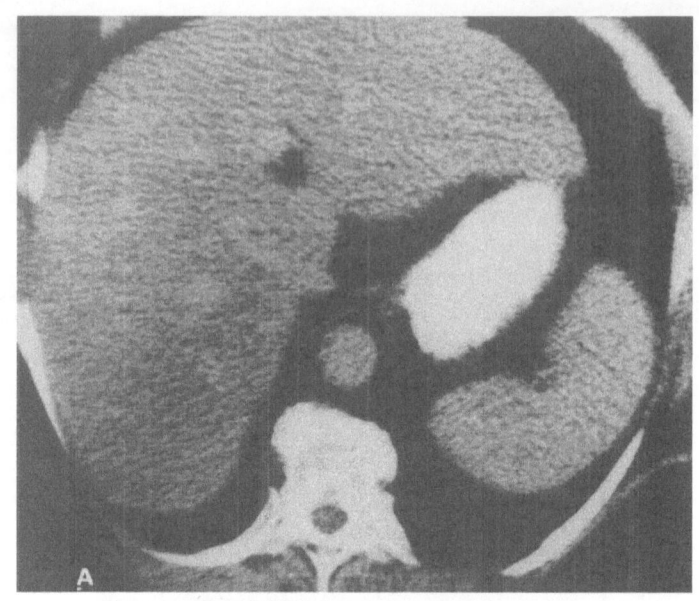

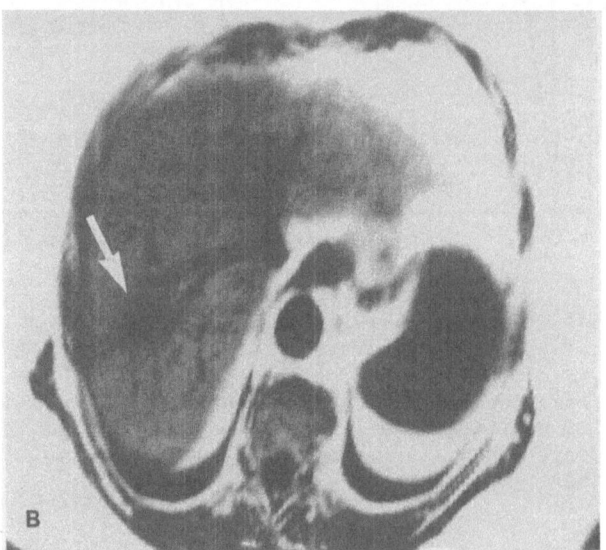

FIGURE   2.   Metastatic colon cancer.
(A) CT scan with dynamic bolus contrast technique. The scan
was reported as normal and no abnormality can be seen, even
in retrospect.
(B) Due to a rising CEA titer, MR was performed the same
week. T1 weighted SE 260/15 shows an hepatic metastasis
(arrow).

Preliminary work by Foley et al reported in more detail later in this symposium indicates that the T1 weighted pulse sequences most effective at low field strengths (0.02-1.0 Tesla) are less effective with high field strength (>1.0 Tesla) systems (6). This becomes evident on inspection of high field images which show a loss of T1-dependent image contrast (2). In theory, this may be due to convergence of tumor and liver T1 relaxation times at high field. T2 relaxation times show little or no variation with increasing field strength.

Calculations alone will not reliably predict pulse sequence performance since signal-to-noise ratios (SNR) as well as CNR are as dependent upon image noise as they are upon inherent tissue contrast. It is the noise, or graininess, of an image that obscures small lesions or lesions with reduced contrast. Due to the sensitivity of MR imaging to motion, it is motion-related ghost artifacts that are the major determinant of image noise (7). Therefore, independent of any theory of pulse sequence performance, it is the unique response of a given pulse sequence, field strength, and machine to motion that determines image quality (8). Analysis of SNR and CNR values for different machines indicates that the choice among optimized sequences is less important than matching optimal sequences to effective motion suppression techniques (9).

At low and midfields, signal averaging has been combined with short TR/short TE spin echo imaging. Alternatively, ordered phase encoding has been successfully combined with T1 weighted (long TR) inversion recovery imaging. At high field, ordered phase encoding (8) and gradient moment nulling (10) have been effectively combined with long TE T2 weighted pulse sequences.At the current time it is not known whether performance of T2 weighted pulse sequences at high field can equal the performance of contrast enhanced CT scanning or the performance of T1 weighted pulse sequences at mid field.

The Spleen: An Internal reference Phantom

To standardize comparison of different imaging systems and pulse sequence techniques, a readily available contrast phantom that models inherent tissue characteristics (proton density and relaxation times) as well as physiologic motion is required. The spleen of normal volunteers is an excellent model of liver cancer for several reasons: (1) the spleen has relaxation times (T1 and T2) as well as proton density similar to liver metastases; (2) the spleen is similar to cancer in containing little or no MR-observable lipids; (3) the spleen is located on the same transverse sections as the liver; and (4) the spleen is a large homogeneous organ from which reproducible image intensity data can be measured. The similarity of normal spleen to liver cancer has been confirmed on clinical images and it has been demonstrated that pulse sequence performance assessed using the spleen-liver CNR can substitute for cancer-liver CNR as a reliable predicter of pulse sequence performance in the detection of liver cancer (2,11).

For all clinical tasks, pulse sequence optimization pertains to the average of a group of volunteers or patients. Obviously, biological variation of either cancer or spleen will cause some differences in signal intensity in individual cases. CNR measurements are never used to establish a diagnosis in an individual patient. This parameter is, however, the appropriate way to quantitate the results obtained with technique for comparison with another. Note that no single imaging technique will ever be the best technique for all patients; our goal is to identify technique(s) that will be the best in the greatest number of patients. The use of the spleen in normal volunteers as a reference phantom is applicable to all pulse sequences, all machines, all field strengths and will greatly facilitate comparative evaluations of the CNR performance of different machines.

MR Image Quality

Just as CNR can be visually (qualitatively) assessed by inspection of liver-spleen contrast, the signal-to-noise ratio (SNR) can be qualitatively assessed by inspecting an image for "graininess". Adequate image quality is dependent upon freedom from motion related ghost artifacts and sufficient SNR to sharply delineate three hepatic veins and secondary branches of the portal venous system.

Unfortunately, manufacturers have not emphasized the development of body imaging techniques, and the performance of different machines varies widely. Without intensive efforts to optimize motion artifact suppression and pulse sequence performance for a given machine, routine MRI of the liver cannot be recommended. Another practical consideration is the superiority of CT scanning for examination of the peritoneum and retroperitoneum. In our experience, the failure of MR to image small pancreatic masses and the inability to distinguish retroperitoneal adenopathy from small bowel is a significant clinical limitation. When extrahepatic pathology is a consideration, the small incremental improvement in accuracy for detection of hepatic lesions may not compensate for lower accuracy of MR for detection of pathology outside the liver. However, MR can replace hepatic ultrasound and scintigraphy except when cost and availability are dominant factors.

Currently, most centers use MR as a problem-solving modality when conventional tests are inadequate or produce conflicting information. Certainly, for differential diagnosis of the two most common benign conditions that mimic cancer, focal fatty infiltration or cavernous hemangioma, MR is well established as the procedure of choice (2).

Differential Diagnosis: Tissue Characterization

Cavernous hemangioma is the most common benign hepatic neoplasm. Experience with imaging techniques such as sonography, MR, and angiography suggests that the world-wide incidence of hepatic cavernous hemangioma may be as high as 15% (2,12). Unfortunately, cavernous hemangiomas are often incidently discovered on abdominal sonograms, CT scans, or scintigrams obtained for other reasons. Since these lesions mimic the appearance of cancer, additional tests are then necessary.

Sonography has little or no tissue specificity since approximately 40% of cavernous hemangiomas have areas of decreased or mixed echogenicity mimicking the common sonographic appearance of cancer. Furthermore, cancer can also be uniformly be hypoechoic and match the appearance of "typical" cavernous hemangiomas. CT scanning has been advocated as a test for noninvasive differential diagnosis of cavernous hemangioma. However, as described by Freeny earlier in this symposium, the specialized single slice CT examination for cavernous hemangioma diagnostic in only half the cases. Scintigraphy with technetium labeled erythrocytes (blood pool study), is also a specialized study, the cost of which must be added to the cost of the original imaging test which disclosed the cavernous hemangioma in the first place. Although hypervascular metastases and hepatomas can retain labeled erythrocytes and mimic the appearance of cavernous hemangiomas, blood pool examinations are generally quite reliable for lesions larger than 3 cm in size. Single photon emmision computed tomography (SPECT) may allow evaluation of smaller lesions.

Cavernous hemangiomas are essentially lakes of slowly flowing blood

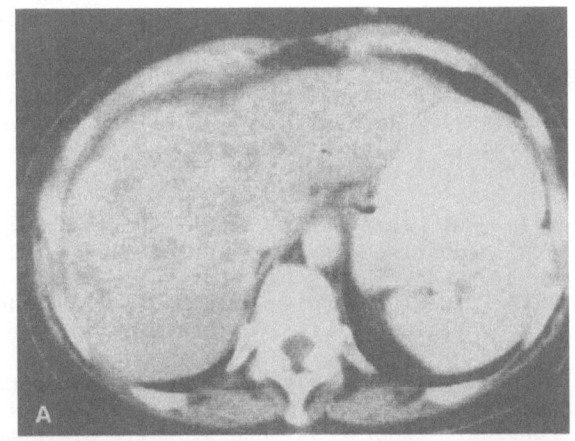

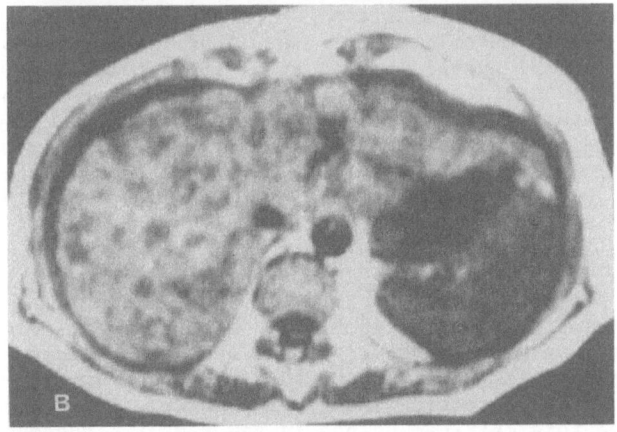

FIGURE   3.   Alcoholic liver disease in a patient with breast cancer.
(A) CT scan shows ascites which, in the presence of
abnormal plasma coagulation studies, precludes percutaneous
needle biopsy to distinguish fatty infiltration from
diffuse cancer.
(B) MRI T1-weighted (SE260/15) image shows extensive
hepatic infiltration with dark (long T1 tissue) nodules.
Note that adipose tissue is bright (short T1 tissue). The
MR findings are diagnostic of diffuse cancer, subsequently
proven at autopsy.

whose MRI features differ significantly from solid hepatic neoplasms (12). Scince fluids have extremely long T2 relaxation times, they differ from solid tumors on MR images performed with long TR/long TE T2 weighted techniques. Research studies have shown that T2 relaxation times of cavernous hemangiomas (150 ± 45 msec) are longer than primary or metastatic liver cancer (78 ± 32 msec) (2,12).

Although many cavernous hemangiomas show histologic evidence of septation or scarring, these regions do not interfer with the correct MRI diagnosis (13). Fibrous strands are small enough to undergo volume averaging and be dominated by the surrounding blood, and the "scar" of cavernous hemangioma in the liver has a high water content and long T2 (2,13). However, recently, it has been confirmed that echo delays of at least 100 msec, typically 120 or 180 msec, are required to consistently diagnose cavernous hemangioma by MRI (13).

The most useful MRI feature in the clinical diagnosis of cavernous hemangioma is visual comparison of the signal intensity of the hemangioma to that of adjacent liver, which has a very low signal intensity due to its short T2, and the signal intensity of other fluid collections such as cerebrospinal fluid, gallbladder bile, or gastric contents.

Pitfalls in the MR diagnosis of cavernous hemangioma exist when hypervascular metastases such as metastatic islet cell neoplasms are studied. These unusual tumors may occasionally mimic the appearance of cavernous hemangioma. More often, hypervascular tumors and necrotic sarcomas or cystic neoplasms of the liver show morphologic features diagnostic of cancer. Cancers frequently show heterogeneity, rings, or solid satellite tumors, features not seen in cavernous hemangioma (2). Fortunately, hepatoma rarely, if ever, mimics the MR appearance of cavernous hemangioma. When signal intensity and morphologic features are evaluated using an SE 2400/60,120,180 triple echo sequence, an overall accuracy of >95% can be achieved for the differential diagnosis of cavernous hemangioma and liver cancer in a representative hospital population (2,13).

At the present time, techniques for imaging the liver are not sufficiently transferable from machine to machine to allow easy introduction when fewer than 5 or 6 abdominal examinations are conducted each week. With higher volume, it becomes worthwhile to carefully optimize pulse sequences for lesion detection and develop experience with tissue characterization. Once this critical volume is reached, it becomes apparent that MRI efficiently serves to detect, stage, and characterize hepatic neoplasms better than any other single diagnostic method. Therefore, MR can be more cost effective than a battery of more expensive, more time consuming, or more invasive techniques. Future improvements in image quality for the upper abdomen as a whole, availability of gastrointestinal contrast agents, and development of tissue-specific magnetic contrast agents will ensure an expanding role for hepatic MR imaging in clinical practice.

REFERENCES

1. Stark DD, Wittenberg J, Butch RJ, Ferrucci JT: MR detection of hepatic metastases: a randomized, controlled comparison with CT. Radiology 1987;165:399-406.
2. Stark DD: Liver; Chapter 38 In: Magnetic Resonance Imaging, eds. Stark DD, Bradley WG, Mosby, St. Louis, 1988, pp934-1054.
3. Zeman RK, Clements LA, Silverman PA, et al: CT of the liver: a survey of preliminary methods for administration of contrast material. AJR 1988;150:107-109.

4. Stark DD, Wittenberg J, Edelman RR, et al: Detection of hepatic metastases by magnetic resonance: analysis of pulse sequence performance. Radiology 1986;159:365-370.
5. Reining JW, Dwyer AJ, Miller DL, et al: Liver metastasis detection: comparative sensitivities of MR imaging and CT scanning. Radiology 1987;162:43-47.
6. Foley WD, Kneeland JB, Cates JD, et al: Contrast optimization for the detection of focal hepatic lesions by MR imaging at 1.57. AJR 1987;149:1155-1160.
7. Stark DD, Hendrick RE, Hahn PF, Ferrucci JT: Motion artifact suppression by fast spin echo imaging. Radiology 1987;164:183-191.
8. Wood ML: Thoracic and abdominal motion artifacts; Chapter 33 In: <u>Magnetic Resonance Imaging</u>, eds. Stark DD, Bradley WC, Mosby, St. Louis, 1988, pp792-801.
9. Hendrick RE, Stark DD, Weissleder R, Foley WD, Kneeland JB: Maximizing liver lesion detection: a comparison of pulse sequence performance at different magnetic field strengths. Radiology 1987;165(P):182.
10. Haacke M, Bellon E: Artifacts; Chapter 8 In: <u>Magnetic Resonance Imaging</u>, eds. Stark DD, Bradley WG, Mosby, St. Louis, 1988, pp138-160.
11. Stark DD, Wittenberg J, Middleton MS, Ferrucci JT: Liver metastases: detection by phase-contrast MR imaging. Radiology 1986;158:327-332.
12. Stark DD, Felder RC, Wittenberg J, et al: Magnetic resonance imaging of cavernous hemangioma of the liver: tissue specific characterization. AJR 1985;145:213-220.
13. Rummeny E, Stark DD, Weissleder R, et al: Differential diagnosis of hepatic hemangioma and metastases by MR imaging: quantitative and qualitative criteria. Radiology 1987;165(P):147.

TREATMENT EVALUATION OF AMEBIC LIVER ABSCESS BY MR IMAGING

Guillermo Elizondo, Ralph Weissleder,
David D. Stark and Luis E. Todd

Massachusetts General Hospital, Harvard Medical School
Boston, MA 02114, and Hospital Universitario UANL
Monterrey, NL, México

A major goal in medicine is to know the biomolecular and biophysical processes of the human body in normal states; how these processes are altered in the pathological conditions; and how the body by itself or by medical treatment revert the abnormal situation to a normal state.

A wide number of techniques are employed actually to study the molecular changes in pathological states. Among them, Nuclear Magnetic Resonance (NMR) spectroscopic techniques have been used extensively in the last 40 years as a tool to differentiate between benign and malignant processes [1-5]. Furthermore, in the last decade, NMR has become an important diagnostic imaging technique, preserving most of the advantages of spectroscopy, that is, alterations in imaging represent alterations in the molecular composition and/or behavior of the organ in study.

In proton Magnetic Resonance Imaging (MRI) the signal intensity of a particular tissue depends of its proton density, its longitudinal and transverse relaxation times (T1 and T2, respectively), flow, and other parameters such as the presence of paramagnetic compounds (Fe, Mn, Mg, etc.), magnetic susceptibility, etc. In general, all pathological states, benign or malignant, have long T1 and T2 relaxation times, which is seen on imaging as areas of low signal intensity in images depending of T1 (T1-"weighted"), or as areas of high signal intensity on images depending of of T2 (T2-"weighted"). As a result, the specificity of MRI for differentiation between benign and malignant conditions is low, even when in most cases the sensitivity of MRI for differentiating between normal and pathological states is higher than with other imaging modalities.

When studying liver masses, we observed that untreated amebic liver abscesses had the same signal intensity characteristics described for other benign or malignant tumors in both T1 and T2-weighted images. However, after treatment amebic liver abscesses developed characteristic morphologic and signal intensity changes that allowed for differentiation among the all others hepatic tumors. We decided to conduct a prospective study to evaluate the utility of MRI in characterizing amebic liver abscess and its response to treatment. The experience and results of this study form the base of the present report.

Introduction

Amebic liver abscess (ALA) is an important medical problem in areas such as Mexico, the southwestern United States, and developing countries [6,7]. Approximately 3500 cases of invasive amebiasis are reported annually in the United States [8]. In addition, ALA is the most important complication of amebiasis and is present in 94% of fatal cases [6].

While negative serologic studies can exclude invasive amebiasis, imaging of the liver is necessary to distinguish patients with ALA from those with uncomplicated intestinal amebiasis [7,9]. Previous reports have described the detection of ALA with scintigraphy [10], sonography [11,12], and computed tomography (CT) [13,14]. Although the gradual resolution of ALA over 2 to 20 months has been observed with sonography and scintigraphy, imaging findings are not clinically helpful in monitoring early therapy [12,15]. We undertook this study to evaluate the MR characteristics of untreated ALA and compared these findings with ALA after medical and surgical treatment.

Materials and Methods

Patients. The diagnosis of ALA was confirmed by sonography and positive amebic serum antibody titers in 17 patients, three female and 14 male, ranged in age from 4 to 106 years. After the initial MR exam, all patients received metronidazole (500 mg q8h IV for 10 days) and dehydroemetidine (1 mg/kg q24h IM for 10 days). Twelve of 17 patients recovered without complications. Two patients with multiple abscesses (six and four, respectively) underwent laparotomy with surgical abscess drainage. Two other patients had a solitary abscess drained percutaneously. These invasive procedures were performed because of the large size of these ALA on the initial sonography examination. One patient died of septic shock following intrahepatic intravasular rupture of the ALA five days after medical treatment was begun. An autopsy was performed on this patient and liver specimens were obtained for gross pathologic and histologic correlation. Surgical specimens and aspiration material from the percutaneous drainage were also examined pathologically. All four surgically treated patients received the same antibiotic regimen as medically treated patients. No evidence of ascites, pleural effusion, or other complications of invasive amebiasis were detected clinically or by the diagnostic imaging studies.

Imaging. All patients had a pretreatment sonographic examination performed with a real time 3.5 MHz transducer. Patients were typically scanned in supine and left lateral decubitus positions. The CT studies were obtained using intravenously administered contrast media.

MR imaging was performed on a permanent magnet imaging system operating at 0.3 Tesla, corresponding to a proton resonance frequency of 13.8 MHz. Patients were imaged in a solenoidal coil with an aperture of 55 cm. Transverse and coronal multislice images with a slice thickness of 7 mm and a 3 mm interslice gap were acquired in all patients. Images with relative T1-weighting (T1W) were obtained using a spin echo (SE) pulse sequence with a TR of 300 msec and a TE of 16 msec (SE 300/16) and/or an SE 500/28. T2-weighted (T2W) images were obtained using an SE pulse sequence with a TR of 2000 msec and a TE of 84 msec. MR imaging was performed prior to therapy in all patients and repeated in 16 surviving patients 10 to 14 days after medical treatment was started. Additional studies were performed in 2 patients 4 days after initiation of treatment, in 3 patients after 30 days, and in another 3 patients after 40 days of treatment.

Results

Untreated abscess. A total of twenty-nine ALA were diagnosed sonographycally in 17 patients. Both T1W and T2W MR images detected all 29 ALA seen by sonography. One patient had six, one patient had four, two patients had three, and 13 patients had one ALA each. The average abscess cavity size was 6 cm (range 3-12 cm). Twenty-three ALA were found in the right and six in the left hepatic lobe. In patients with multiple abscesses, ALA tended to be located centrally, near the porta hepatis. In 2 patients, seven centrally located ALA surrounded the porta hepatis. Twenty-two of the 29 ALA were located peripherally and 4 of these protruded beyond the normal contour of the liver surface.

T1W images showed untreated ALA as hypointense regions with sharply defined margins corresponding to the sonographically defined cavity wall. The abscess cavity, devoid of normal hepatic vasculature, showed heterogeneous regions of hyper- and isointensity in 10 of 17 patients. In the other 7 patients the abscess cavity was homogeneously hypointense relative to the hepatic parenchyma. Heterogeneous and homogeneous ALA did not exist concurrently in any patient. All ALA showed a mass effect by compression of vessels or by protrusion beyond the normal contour of the liver surface if located peripherally.

Prior to treatment, incomplete concentric rings could be seen at the abscess cavity margin in only 3 of 17 patients. A inner ring bordering the low intensity cavity was isointense to normal liver and was distinguishable only in regions where an outer hypointense ring separated the abscess cavity from adjacent morphologically normal liver parenchyma .

On T2W pulse sequences the untreated ALA cavity were hyperintense relative to adjacent normal liver. The hyperintense region had ill defined margins and was larger than the cavity defined by sonography and the T1W images. This zone of hyperintensity extended peripherally from the lesion towards the liver capsule in a geographic, wedge-shaped fashion in 15 of 17 cases. This large, geographic hyperintensity corresponded to morphologically intact liver tissue which was indistinguishable from uninvolved liver by sonography or T1W MR images. In 11 of the 17 patients incomplete hypointense rings could be identified bordering the actual abscess cavity.

The surgical findings in four patients and autopsy study in another confirmed that the heterogeneous, hypointense central region seen on T1W images corresponded to necrotic debris, hemorrhage and liquefied material in the abscess cavity. These findings are consistent with previous pathologic studies of untreated ALA which demonstrated a central zone of cavitary necrosis surrounded by a very thin fibrous capsule [6]. Actually, the wall of untreated or partially treated ALA consists of a 3-7 mm zone of compressed and partially necrotic liver tissue [6,16] corresponding to the inner ring, isointense to liver tissue. This injured tissue is separated from viable hepatic parenchyma by a 1-3 mm zone of fibroblasts and macrophages [6,16] corresponding to the peripheral hypointense rings. The wedge-shaped hyperintense zone on T2W images corresponds to histologically normal liver tissue, without inflammatory cells. However, it is well known that histology is not a sensitive method for identifying hepatic edema, and furthermore, it has previously been shown that MR can detect hepatic edema earlier than histology [17].

<u>Treated Abscess</u>. When compared to the pretreatment images, the T1W images of medically treated patients showed a uniform decrease in signal intensity of the abscess cavity, and in all patients the cavity became homogeneous. This appearance of fluid signal intensity is similar to that of simple hepatic cysts [18]. In addition, as the cavity matured, the wall became sharply defined and the concentric rings became very prominent on both T1W and T2W images in all patients. The rings seen on T1W and T2W images were proven to be identical by electronic superimposition of the individual image pixels. As a result, the hypointense ring seen on T1W images corresponded exactly to the hypointense ring of T2W images. Peripheral to the hypointense ring, the broad hyperintense zone on pre-treatment T2W images corresponding to edema was now reduced to a narrow ring, hyperintense to liver. The previously larger hyperintense zone of edema was now isointense and morphologically indistinguishable from normal liver.

The follow-up MR images showed that none of the ALA cavities had decreased in size within the first two weeks of medical treatment, despite abscess wall maturation and resolution of hepatic edema. During this period marked symptomatic improvement including defervescensce and decreased hepatic tenderness was evident in 12 of the 16 patients. In 4 patients with large ALA on the initial sonographic exam, MR scans were performed prior and posterior to surgical intervention. After surgical drainage of two ALA both cavities collapsed and the mass effect on adjacent liver was diminished. Interestingly, surgical treatment did not alter maturation of the abscess wall and during subsequent antibiotic therapy the concentric rings became more prominent as usually seen in ALA responding to medical treatment. Intracavitary gas-fluid and/or fluid-fluid levels were seen in all 4 cases of surgically treated patients.

Even when we could not obtain pathologic material for correlation with MR images in the 12 patients responding to medical treatment, we were able to correlate our MR findings with published pathologic data [6,15,16], surgical and drained specimens, and an autopsy. The histologic resolution of amebic liver abscesses has been studied in detail and is consistent with the pathologic findings in our one patient who died after 5 days of treatment.

ALA responding to antibiotic therapy develop several distinct histologic zones. The center of the lesion undergoes liquefactive necrosis, been the finely granular eosinophilic liquid material usually lost in pathologic processing. Intensely inflamed granulation tissue rims the necrotic center of the abscess. Trophozoites of

Entamoeba histolytica, neutrophils, and macrophages adhere to the inner aspect of this zone at the interface with the necrotic core. This zone of granulation tissue corresponds to the inner ring on the T1W images, appearing isointense relative to normal liver. Peripheral to this granulation tissue there is a band of type I collagen with variable thickness, which in our partially treated patient measured 5-7 mm. The histologically proved rim of collagen corresponds to the hypointense ring on both T1W and T2W images. Outside this collagenous wall, hepatic tissue showed a mild inflammatory response with sinusoidal congestion, interstitial edema, and hepatocellular compression atrophy. On T2W MR images this corresponded to the outer hyperintense ring. With maturation of the abscess wall increased amounts of bile and some focal deposits of hemosiderin may be demonstrated histologically at the base of the granulation tissue and inner rim of the collagenous wall [16]; in our case very little bile or iron was seen.

Discussion

Detection of ALA is readily accomplished by imaging modalities such as sonography, scintigraphy, and CT, mainly because of the large size of symptomatic lesions. In previous imaging studies the decrease in size of ALA over 2 to 20 months time has been reported as a favorable response to treatment, corresponding pathologically to resorption of the abscess cavity [12,15]. However, clinical information alone has been relied upon to determine the early response to therapy, when the decision of surgical intervention is most critical. Our data suggest that diverse histologic changes occurring in healing ALA can be identified by MRI: 1) liquefaction of the abscess cavity corresponding to breakdown and homogenization of necrotic hepatic tissue and hemorrhagic material is seen as early as 4 days after initiation of effective medical treatment, 2) maturation of the abscess wall is shown by the appearance of prominent concentric rings, and 3) hepatic edema and its resolution can be effectively monitored by T2W images.

We found that the formation of the concentric rings after treatment seen on MRI correlates histologically with the maturation of the abscess wall [16]. The low signal intensity ring is of particular interest as it appears hypointense on both T1W and T2W post-treatment images. This MR feature is characteristic of tissues with low spin density (air,bone), short T2 relaxation times (hemosiderin, collagen) [19], rapid motion such as flowing blood, and in the case of normal uterine myometrium, the nature of the dark band is unknown [20]. Our histologic correlation suggests that the hypointense ring most probably corresponds to collagen. Although small amounts of stainable tissue iron have been described around the collagenous wall [16], this finding was not present in our autopsy specimen. Alternatively, rapid capillary blood flow within the collagenous tissue might account for presence of the hypointense ring.

A rim of enhancement in ALA has previously been observed by scintigraphy using 99mTc-DISIDA and has been attributed to an increased amount of radionuclide in the compressed hepatic parenchyma surrounding the abscess cavity [10]. Cholescintigraphic rim enhancement may be a relatively specific but insensitive finding as it was only present in 53% of the ALA in one study [10]. On CT, ALA may also show a rim which enhances after administration of contrast agents due to peripheral hyperperfusion [21]. This rim enhancement is quite non-specific as it may occur in metastases and bacterial liver abscesses. Detection of hepatic edema has not been possible using sonography, scintigraphy, or CT. In fact, even histology appears to be less sensitive than MRI for early detection of hepatic edema [17].

The evaluation of hepatic abscesses by MRI has been recently described by our group [22]. This experience suggests that standard MR imaging techniques are sufficiently sensitive to detect ALA. Our most significant finding is that MR imaging seems to be capable of following the response of ALA to treatment. During treatment the pattern of liquefaction of the abscess center, formation of concentric rings, and the resolution of hepatic edema may be useful in distinguishing responsive disease from treatment failures, potentially allowing changes in antibiotic therapy or surgical intervention.

References

1. Damadian R. Tumor detection by nuclear magnetic resonance. Science 1971; 171: 1151-1153.

2. Goldsmith M, Koutcher JA, Damadian R. Nuclear magnetic resonance in cancer, XII; applications of NMR malignancy index to human lung tumors. Br J Cancer 1977; 36: 235-242.

3. Damadian R, Zaner K, Hor D, DiMaio T. Human tumors detected by nuclear magnetic resonance. Proc Natl Acad Sci USA 1974; 71: 1471-1473.

4. Saryan LA, Hollis DP, Economou JS, Eggleston JC. Nuclear magnetic resonance studies of cancer. IV. Correlation of water content with tissue relaxation times. J Natl Cancer Inst 1974; 52: 599-602.

5. Beall PT, Hazlewwod CF, Rao PN. Nuclear magnetic resonance patterns of intracellular water as a function of HeLa cell cycle. Science 1976; 192: 904-907.

6. Brandt H, Tamayo RP. Pathology of human amebiasis. Human Pathology 1970; 1: 351-385.

7. Balasegaram M. Management of hepatic abscess. Curr Probl Surg 1981;18: 284-340.

8. Morbidity and Mortality Annual Supplement, Atlanta Center for Disease Control, US Public Health Service , Department of Health Education and Welfare, 1960-1974.

9. Palmer RB. Changes in the liver in amebic dysentery. Arch Path 1938; 25: 327-335.

10. Remedios PA, Colletti PM, Ralls PW. Hepatic amebic abscess cholescintigraphic rim enhancement. Radiology 1986;160: 395-398.

11. Ralls PW, Colletti PM, Quinn MF, et al. Sonographic findings in hepatic amebic abscesses. Radiology 1982;149:123-126.

12. Ralls PW, Quinn MF, Boswell WD, et al. Patterns of resolution in successfully treated hepatic amebic abscess : Sonographic evaluation. Radiology 1983; 149: 541-543.

13. Terrier F, Becker CD, Triller JK. Morphologic aspects of hepatic abscesses at computed tomography and ultrasound. Acta Radiologica Diagnosis 1983; 24:129-137.

14. Halvorsen RA, Korobkin M, Foster WL, et al. The variable CT appearence of hepatic abscesses. AJR 1984;141: 941-946.

15. Sheehy TW, Parmley LF, Johnston GS. Resolution time of an amebic liver abscess. Gastroenterology 1968; 55: 26-34.

16. Lushbaugh WB, Kairalla AB, Hofbauer AF, Sequential histopathology of cavitary liver abscess. Arch Pathol Lab Med 1980;104: 575-579.

17. Stark DD, Bass NM, Moss AA, et al. Nuclear Magnetic Resonance imaging of experimentally induced liver disease. Radiology 1983;148: 743-751.

18. Stark DD, Felder R, Wittenberg J, et al. Magnetic resonance imaging of cavernous hemangioma of the liver : tissue specific characterization. AJR 1985;145: 213-220.

19. Fullerton GD, Cameron IL, Ord VA. Orientation of tendons in the magnetic field and its effect on T2 relaxation times. Radiology 1985;155: 433-435.

20. Lee JKT, Gersell DJ, Balfe DM, et al. The uterus : In vitro MR-anatomic correlation of normal and abnormal specimens. Radiology 1985; 157:175-179.

21. Baert AL, Wackenheim A, Jeanmart L. Abdominal computed tomography. New York : Springer. 1980. pp.114-116.

22. Elizondo G, Weissleder R, Stark DD, et.al. Amebic liver abscess: Diagnosis and treatment evaluation with MR imaging. Radiology 1987; 165: 795-800.

PLANNING THE CLINICAL RESEARCH PROTOCOL TO UNDERSTAND THE RELATIONSHIP

BETWEEN CELL FUNCTION AND DISEASE

Luis E. Cañedo and Guillermo Elizondo

Department of Biochemistry, School of Medicine, Universidad Autónoma de Nuevo León and the International Center of Molecular Medicine, Monterrey, Nuevo León, México

INTRODUCTION

The clinical research protocol is the document in which the investigator expresses in advance the research problem that he or she is interested in. This document must be written with great care, since it reflects the comprehension that the investigator has of the state of the art and the place of his particular problem in the proposed area of research. When the investigator writes the protocol, he organizes his thoughts about the research problem: how clear is the question?, are the methods consistent with the objectives?, how is the answer to be handled?. Through the protocol, the investigator can realize if the project can be done and if he or she has the time and the money to carry it out. A wealth of information about how to write scientific articles have been published (1-7). However, a short and comprehensive review about how to plan the various kinds of clinical investigations is non existent. The present article is an attempt to describe the necessary guidelines to organize the clinical research protocol; provides information to understand the methodological structure of the original research articles found in modern medical publications, and offers references to the pertinent literature.

THE STATEMENT OF THE RESEARCH PROBLEM

The title. The planning of the project should start with a tentative title. The first word should indicate part of the content of the proposed research. The title should be short and informative, clearly indicating the object of the study to be investigated (8). The antecedents refers to the relevant published literature for the project under consideration and represents the state of art of the research area. The necessary search to be carried out should be intelligent, critical and up to date. The references employed must be identified in the text with a number corresponding to the bibliography provided at the end of the protocol (9-14). The objectives of the study should be something worth striving for. They must be derived as a logical conclusion of the antecedents and address important medical questions; therefore, all efforts and resources of the investigator must be directed toward these goals. The objectives of clinical research should be part of any of the following four major areas: a) The study of health. These studies use diagnostic procedures to describe normal values in human beings that define health. Deviations from these values are used as parameters to identify the diseased status. Some of the most important normal values are: physical and laboratory measures of the anatomical, histological, biochemical and physiological constants; studies of growth and development; psychological standards; and socio-economical issues. b) The study of disease. This type of research includes: the study of deviations from the normal values at macroscopic, microscopic and molecular levels; the description of the natural history of disease and its variants; the process and mechanisms of disease; studies of the etiology of disease; morbidity and mortality rates; and socio-economical issues. c) The assessment of the diagnostic procedures used to define health and disease. These type of studies may include: the definition of the measurable property of a particular disease and the procedure to be carried out; the establishment of the particular value of a variable that distinguishes normal from the diseased status; the sensitivity, specificity and predictive value of a test; and the study of costs. d) The evaluation of preventive and therapeutic measures. These type of studies use procedures designed to compare the outcome of treatments such as: diets, medical advice, psychotherapy, drugs, vaccines, surgery, anesthesia, interventional radiology, radiotherapy, risk-benefit ratio, and socio-economical issues. After classifying the main objective of the investigation under one or more of the four major areas of clinical research, the secondary objectives should be indicated (15). This perspective helps to point out where the contribution to the clinical science is to be made. The hypothesis are propositions or suppositions tentatively accepted to explain certain facts. These ideas not only provide the basis for further investigation, but, they also guide observation and experiment because they suggest what to observe and how to observe it. If the project has an a priori hypothesis to be tested, it should be clearly described in the protocol (16,17). Once the problem is stated free from ambiguity, the previous information can be used by the investigator to determine the proper methodological design or designs for his or her project. See Table 1.

CRITERIA FOR THE SELECTION OF THE STUDY DESIGN

Observational or Experimental. Man can gather scientific information from nature using two major procedures: observation and experimentation. Hence, the first criterion to select a proper methodological design is to classify each of the study objectives as observational or experimental (see table 2). If the investigator describes or measures without modifying the event as a it occurs in nature, the procedure is classified as observational (18). For example, study of normal growth by measuring the weight and height of normal growing children. If the researcher deliberately modifies one or more of the variables of the phenomenon under investigation, and describes or measures the event before, during and/or after the premeditated intervention, the procedure that will be utilized to reach the objective of the study is stated as experimental (18-21). For example, growth of children before and after the administration of somatostatin. Cross-Sectional or Longitudinal. The second criterion mentioned in table 2, takes into consideration the number of times that the same information is obtained from the same individual or population. When this information is obtained only once from the same subject, the study is called cross-sectional (18,21-23). An objective of the longitudinal studies is the study of change in relation with time. To reach this goal, the same information is obtained several times from the same subject (18,24-26). Both types of studies mentioned above, are independent with respect to the length of time. There are cross-sectional studies that conclude in very short time; an example of this would be, a report of an interesting finding in a X-Ray film. Others can last several weeks, for

Table 1. Methodological Designs used in Clinical Research Protocols.

Observational Designs	Experimental Designs
A. Surveys	A. Internally controlled
1. Descriptive	1. Self-controlled
a. Retrospective (33)*	a. Before-after
b. Prospective (34)	i. Single group (113)
2. Comparative	ii. Two or more groups(114)
a. Retrospective	b. Cross-over
i. Simultaneous (35)	i. Single group (83)
ii. Sequential (36)	ii. Paired organs (93)
b. Prospective (37)	iii. Two or more groups (81)
B. Cohorts	2. Parallel (115)
1. Forward Descriptive	B. Externally controlled
a. Retrospective (43)	1. Parallel (80)
b. Prospective (44)	C. Basic Experimental Designs
c. Retrospective-prospective (45)	
2. Forward Comparative	
a. Retrospective (46)	
b. Prospective (47)	
c. Retrospective-prospective (48)	
3. Backward Descriptive	
a. Case-history (49)	
4. Backward Comparative	
a. Case-control (50)	

In the present medical literature it is possible to identify two necessities. The first is a requirement for guidelines to organize original research articles (6,7). The second is a recognized need for an improved classification system for the medical literature (See Appendix 1 for details) (5,15,18,21,25). The present classification scheeme was used to identify the methodological designs from a sample of 90 original research articles selected by the authors from the current medical literature in a non-randomized way, and 100 original research articles selected by use of a table of random numbers from chapters of well known textbooks of medicine (233-240). Editorials, letters, reviews, and other non original research articles were excluded. In this preliminary attempt, it was possible to identify those articles with a single methodological design as the ones referred in this table; as well as, examples of articles with variants of one design: like a descriptive survey with sequential comparisons of variables (241); with simultaneous comparison of variables (242); and simultaneous-sequential comparative surveys (243). Studies with more than one design, like a cohort forward prospective design and a case control design (41); a cohort forward prospective design followed by a comparative survey (244); a retrospective survey and two independent clinical trials (245); a cohort forward study followed by an experimental animal model (246). Case reports with (130) and without (247) methodological design; as well as papers with omission in reporting the methods used (248). (* Numbers in parenthesis correspond to cited bibliography).

example a cross-sectional survey in a city, oriented to know the proportion of women under a program of birth control. Similarly, a longitudinal study can last minutes. For example, a study of the response to synthetic thyrotropin-releasing hormone in man (27); or years, as in the Framingham Study Group (28). Retrospective or Prospective. The third criterion of table 2 defines the temporal period from were the information needed to reach the objective will be obtained. By this yardstick, the study is considered retrospective when the information of interest for the present project was gathered in the past for a different purpose, and the data can be retrieved from charts, films or notebooks (18,29,30).

Table 2. Structure of the Methodological Designs used in Clinical Research Protocols.

Column groups: **A** = OBSERVATIONAL / EXPERIMENTAL; **B** = CROSS-SECTIONAL / LONGITUDINAL; **C** = RETROSPECTIVE / PROSPECTIVE; **D** = DESCRIPTIVE / COMPARATIVE; **E** = CAUSE-EFFECT / EFFECT-CAUSE. The upper-right diagonal is labelled METHODOLOGICAL CRITERIA / METHODOLOGICAL DESIGN.

OBSERVATIONAL	EXPERIMENTAL	CROSS-SECTIONAL	LONGITUDINAL	RETROSPECTIVE	PROSPECTIVE	DESCRIPTIVE	COMPARATIVE	CAUSE-EFFECT	EFFECT-CAUSE	METHODOLOGICAL DESIGN
X		X		X		X				
X		X			X	X				
X		X		X			X			
X		X			X		X			SURVEYS *
X			X	X		X	X			
X			X		X	X	X			
X			X	X	X	X	X			
X			X	X			X	X		
X			X		X		X	X		
X			X	X	X		X	X		COHORTS *
X			X	X		X			X	
X			X	X			X		X	
	X	X		X			X	X		CLINICAL TRIALS *, SURGICAL TRIALS *, BASIC EXPERIMENTAL DESIGNS *

(*) Study of individual cases, small groups or representative samples.

The time from which such information is obtained could be a fixed period in the past, for example diabetics registered in the hospital between 1980 and 1983 or up till the present (1980-1988). In contrast to the retrospective study, in the prospective design, the information relevant to the study will be captured in the future to fulfill the research objectives (18,26). For example: a study of the fetal growth during the following nine months of pregnancy. Descriptive or Comparative. The fourth criterion of table 2 defines whether the research objective is oriented to contrast an a priori hypothesis or not. If the goal of the study is only to give an organized account of the phenomenon under investigation, the project is descriptive (21-23). Frequently, only one population or event is observed and the association between the variables can be described to support a hypothesis; however, the study is not designed to contrast it. In the comparative studies, a comparison is made between two or more populations or events to contrast one or more a priori hypothesis (15,18-21). Cause to

Table 3. Survey Designs commonly found in Clinical Research Protocols.

METHODOLOGICAL CRITERIA		PAST	PRESENT	FUTURE
DESCRIPTIVE	RETROSPECTIVE	[2,3] or	$\frac{1}{3}$|	
	PROSPECTIVE		$\frac{1}{3}$| or	[2,3]
COMPARATIVE	RETROSPECTIVE (Simultaneous)	[2,3 A] [2,3 B] or ⋮ [2,3 N]	$\frac{1}{3}$| $\frac{1}{3}$| $\frac{1}{3}$|	
	RETROSPECTIVE (Sequential)	[2,3 A] [2,3 B]···[2,3 N] or $\frac{1}{3}$| $\frac{1}{3}$| $\frac{1}{3}$|		
	PROSPECTIVE		$\frac{1}{3}$| $\frac{1}{3}$| $\frac{1}{3}$|	[2,3 A] [2,3 B] ⋮ [2,3 N]

(1) moment in time, (2) period of time, (3) one sample, (AB...N) study groups.

Effect or Effect to Cause. The fifth and final criterion of Table 2 defines the epidemiological strategy of the study, and is specific for the longitudinal studies (26,30). In the studies of cause to effect, the subjects are selected on the basis of input variables and the direction of the acquisition of the information in relation with time is forwards (26,30). For example: people exposed and others not exposed to a risk factor, are followed up to determine the proportion of the effect in each group. In the studies of effect to cause, the subjects under study are selected on the basis of output variables (results or outcome), and the information is captured from subsequent to antecedent events (29,30). For example: patients with or without a stroke, are traced backwards for a few years to know the proportion in which each group was exposed to hypertension, considered as the putative risk factor. Every objective of the research project has to be classified under the five methodological criteria mentioned above. The type of the design needed to reach the objective comes as a result of its combination (See Table 2). One single protocol can be structured with several methodological designs.

SURVEYS

Any single observation or investigation of the facts about a medical related situation may be called a survey. Survey methods are well-grounded in statistical theory and detailed references are available (31,32). A survey can either be performed at essentially one moment in time in the course of an event (instant survey) or, for a fixed period of time (survey for a specific period). If the survey is carried out in one population, the study is descriptive (33,34). If two or more populations are surveyed and compared during the same period of time (35), or if two different samples from the same population or two or more populations are surveyed and compared, at different periods of time (36), the study is defined as comparative. Considering the time factor, if the survey is made with information collected in the past for another purpose, the survey is retrospective (33,35,36). Those performed with information that will be captured in the future are called prospective (34,37). Since any survey has to be planned in advance, the ones carried out at the present time can be considered as prospective surveys. Applications of this type of study design include: normal values of physical and laboratory measurements; differences of normal values among healthy individuals with different characteristics; study of specific processes and mechanisms of disease; population screening; morbidity, mortality, risk and probability measures; evaluation of diagnostic, preventive or curative procedures; and socio-economical issues. Survey designs commonly found in clinical research protocols are depicted in Tables 1 and 3.

COHORTS

The cohort design is oriented to search for cause-effect relationships (26,30). In the medical literature, the cohort studies (26) are considered as the antithesis of the case history (38) and case-control studies (39,40). These prospective and retrospective methods for studying associa-tion in medicine traditionally are classified from the point of view of the epidemiological strategy, i.e.: sampling by input variables (cause-effect), or sampling by output variables (effect-cause) (30), or classi-fied from the direction of the acquisition of the information in relation with time (Forwards or Backwards) (26,29). However, the cohort studies as well as the case history and case control studies, can also be considered as longitudinal studies of a group of people with similar characteristics (41); hence, both can be classified under this criterion. When this occurs, the inconsistency in the definition of the term cohort (42) or the need to consider each of these designs as separate categories (30,38-40), are eliminated without loosing reality in the design description. This perspective simplifies the structure of the clinical research designs (see Table 2); therefore, in the present classification scheme, the case history and case control designs are included in the cohort design. This term is defined as the longitudinal observation of a group of people with some characteristics in common. The methodology for carrying out a cohort study comprehends similar considerations as those used in survey planning (31,32). This type of study consists of several surveys that are carried out at different times with the same patient or population sample, in order to study change in relation with time.

The cohort-forward designs

In the cohort-forward study the individual, the group, or the groups of people are selected on the basis of input variables (causes) and then followed forward in time to study its future history in search of an outcome (26,30). If the cohort-forward study has as an objective a description of the changes that occur in the individual or in the popula-tion, the design is descriptive (43-45). If the history of two or more individuals or groups are followed with the aim to compare some of its characteristics, the study is comparative (46-48). In relation with time, the cohort-forward studies are retrospective if the study is carried out among a group of people whose past can be traced, using records with information gathered in previous time for other purposes. The group or groups are sampled on the basis of events that happened in the past (input variables), and their evolution is followed for a limited period of time or all the way up to the present (43,46). When planning a cohort forward retrospective studies the investigator must be assured that the records of the individuals forming the group are complete for the period of the

Table 4. Cohort Designs commonly found in Clinical Research Protocols.

DESIGN DESCRIPTION		PAST	PRESENT	FUTURE
COHORT FORWARD DESCRIPTIVE	RETROSPECTIVE	I →		
	PROSPECTIVE		I →	
	RETROSPECTIVE – PROSPECTIVE	I ────────→		
COHORT FORWARD COMPARATIVE *	RETROSPECTIVE	I A → ; I,4 B → ; ⋮ ; I,4N →		
	PROSPECTIVE			I A → ; I,4B → ; ⋮ ; I,4N →
	RETROSPECTIVE – PROSPECTIVE	I A ──────→ ; I,4B ──────→ ; ⋮ ; I,4N ──────→		
COHORT BACKWARD DESCRIPTIVE	CASE HISTORY	← 2,3		
COHORT BACKWARD COMPARATIVE *	CASE CONTROL	← 2,3 ; ← 4		

(1) a sample selected by input variables, (2) a sample selected by out put variables, (3) cases, (4) controls: can be more than one group, (*) internal o external controls, (AB...N) study groups, (────▶) in search of an outcome (forwards), (◀────) in search of a cause (backwards).

study. The cohort-forward studies are prospective if the individual, the group, or the groups of people, are organized in the present, and their future history is studied (44,47). In a cohort forward prospective study, the initial state of the units (patients, cells, values) must be characterized, and in addition to this, the future history of the units should be followed by measuring in sequence the variables of interest until the effects or outcome is reached. For this reason the investigator must be sure, by careful selection of the cohort, that the majority of the individuals forming the group will be maintained during the length of time in which the project will take place. In some projects, the information obtained in the past is only part of the information needed because the same patients whose records have been studied, will be followed in the future. These designs are classified as cohort-forward retrospective-prospective. In this alternative, the cohort groups can be formed with the existing patients followed by the localization and examination of their records before the future observation of the groups begin. In other instances, the records are studied first, and on the basis of this analysis, the cohort groups are formed and followed up (45,48). Applications of the cohort forward designs include: studies of normal values of growth and development; evolution of individual differences; the natural history of disease; study of the process and mechanisms of disease; estimation of prognostic values; risk measurements; evolution of diagnostic procedures; and socio-economical issues.

The cohort-backward designs

In these type of studies an individual, a group, or more groups of people with some characteristics in common, are sampled from the population on the basis of output variables, (results of an outcome or effect) and the search for the putative causes is pursued backward in time (29,30). If the study is carried out with one individual or population and a description of their history is carried out from subsequent to antecedent events, the study structure is called case history (38). i.e.: patients studied to search for the cause of the disease by backward pursuit of antecedent characteristics (49). If the study is based on a sample of two people or two or more groups, one selected by the presence of the effect (cases) and the other by its absence (controls); and, both are followed towards earlier times into the past to determine the degree of exposure of each sample to the putative cause, the study is called case-control (39,40). In these studies each case should be matched for those factors known to be important with one or more controls of similar characteristics with the case. An example of this would be, to study the association between the use of oral-contraceptives and the risk of ovarian cancer. The cases are defined as women with histologic diagnosis of ovarian cancer and the controls are selected from the general population to provide a stratified random sample of women without ovarian cancer (50). Applications of the cohort-backward design includes a description of the natural history of disease, a study of the process and mechanisms of disease, studies on the etiology of disease, and socio-economical issues. Cohort designs commonly found in clinical research protocols are shown in Tables 1 and 4.

THE CLINICAL TRIAL

This design is oriented to evaluate the effectiveness of one or more medical treatments. The therapy is administered to a preselected group of patients that are observed under controlled experimental conditions, and the outcome of such experiments, are used to elucidate specific medical problems (51-61).

The phases of the clinical trial

The preclinical phase consist of an extensive laboratory and clinical experimentation performed in animals to determine the pharmacokinetics of the drug; the structural and physiological toxicity in target organs; the nature and reversibility of toxic effects and the maximum dosage without toxicity. Before any drug is tested in man, its rationale and characteristics should be described in detail (62,63). Drug studies in man are divided into four phases. The first time that a new drug therapy is administered to a human being is considered as the first phase of the clinical research. The purpose of this first phase is to define the following initial aspects of the drug: safety and tolerance; the pharmacokinetics and bioavailability; efficacy; interactions and metabolism (64-66). When the drug appears promising for certain patients, the early therapeutic trials initiated will form part of the second phase of the clinical research. This phase is oriented to test whether the new drug is effective for one or more clinical indications; to obtain an initial estimate of the risk/benefit ratio; to determine the efficacy relative to available drugs and other methods of management; to study the pharmacokinetics of the drug; and to resolve the interactions between body fluids and/or tissues at different concentration of the drug (65-67). Comparisons of the new drug with the standard treatment or placebo among large group of patients, define the third phase of clinical research. This phase is oriented to obtain valid data on the clinical benefits and statistical evidence of efficacy and safety, as well as the incidence of adverse drug reactions. This information is focused to fulfill the requirements of the federal agencies to obtain authorization for marketing the drug (65,66,68). Phase four is oriented to overcome the limitation of premarketing evaluation of drugs in the following major areas: adverse reactions, efficacy issues, new drug applications and risk/benefit ratio (65,66,69). After the study phase has been defined, the following items must be specified in the protocol: Presentation form of the drug, precautions pertaining to the drug handling, route of administration, dosage, duration of therapy, procedures to be carried out if toxicity develops, other concomitant treatments. If a placebo is going to be used, its specifications should be included in the protocol.

Procedures to secure an unbiased comparison among the study groups

When the patients are homogeneous with respect to the characteristics to be compared, for example, in chronic stable diseases. They can be directly assigned to independent study groups by some random procedure. This procedure can be used when all the individuals that will be participating in the study are present before the start of the study (tangible population) or when they are going to be incorporated in the future, for example, patients that will be arriving to the hospital by a procedure not controlled by the investigator (conceptual population). If the patients are heterogeneous with respect to the variables to be compared, for example, in chronic variable diseases. It is necessary to look for essential systematic differences in the individuals being compared in order to integrate homogeneous sub-groups or blocks previous to random assignment. One study can have as many blocks as necessary. When it is needed to study the same problem among individuals with different characteristics, i.e. different ages, disease status or socio-economical issues, the investigator could arrange the participants in several strata. The outcome among these sub-groups will be compared at the end of the study.

The particular symptom or disease of interest to the study should be identified and/or understood before any treatment is attempted. Symptoms can be induced in healthy volunteers or studied in well defined patients. Depending on the disease characteristics, these can be classified as acute, sub-acute, chronic stable, or chronic variable (58): This classification scheme helps to point out the procedure to secure an unbiased comparison of the groups. For example: in some acute diseases open designs are appropriate: in chronic stable diseases, a parallel non stratified study could be used; while in a chronic variables disease, blocking and stratification is mandatory.

The randomized controlled trial is when the assignment of patients to treatments or treatments to patients is done by some random procedure i.e.: table of random numbers. In this type of study, the subjects, within the constrains of the experimental design, have the same opportunity to be allocated to the treatment or to the control group. This procedure eliminates the bias that could result from the assignment of treatments, balances covariates among the groups, and guarantees the validity of the statistical test of significance used to compare the treatments (70-73). The non randomized controlled trial, is when the subjects that will receive the treatment under study, as well as the controls, are allocated by means of some non-random process. i.e.: patient condition. The pros and

485

cons of randomized vs. non randomized controlled designs have been reviewed (74-79).

The control of the psychological bias: This criterion is oriented to control the psychological bias of the patient produced by his expectations or by his apprehension regarding the new treatment, the standard treatment, or the placebo. At the same time a control must exist over the evaluator's bias based on his preconceived ideas about the new drug, the existing treatments, or the placebo since this attitude could influence his appreciation of the outcome measurements to be gathered. These pitfalls can be avoided by the use of the following techniques. The open design. This technique can be used when the treatment outcome cannot be modified by the doctor or the patients attitude. An example of this, is when the treatment is administered to a patient with an acute life-threatening condition who has not responded to the existing therapy (80). The blind techniques. In the single blind technique the doctor knows which patient is getting which drug but the patient ignores which product he or she is receiving (81). In some cases, the evaluator is the one who has to be blinded in order to be able to measure objectively the outcome of the treatments (82). In the double blind technique, neither the doctor nor the patient knows which treatment is being administered to whom at which stage of the study (83-85). The double blind double placebo technique, is used when it is not possible to match the drugs physically or in dosage (86,87). And finally, the triple blind technique, in which the doctor, the patient and the external evaluator are blinded with respect to which treatment was administered to whom at which stage of the study.

Techniques to reduce the time of the trial

In some studies due to ethical, economical or other reasons, it is desirable that the experimental study be carried out in the minimum possible time. In these cases, the following designs should be taken into consideration: The cross-over design. In this strategy each patient receives two or more treatments in sequence and the outcome of each treatment is compared with the status of the patient before and after each treatment (15,81,83,88-93). The 2 x 2 factorial design. This design uses the same experimental set up to investigate four independent variables with the same number of individuals required to study only two (20). The validity of such studies depends on several a priori assumptions, that should be specified for each case (51,94). The sequential medical trial designs (95,96) are methods used to analyze the therapeutic trials using a continuous (95) or discrete (96-99) monitoring plan. The statistical assessment of the data under this program, offers the possibility of an early evaluation of the outcome of the trial. In other words, it allows for decisions to be made with respect to if a treatment is better, worse, or similar to the other; and, on this basis, stop the trial. The multicentric trial design (100-103) allows the possibility of obtaining a large number of patients within a short period of time. This design, is useful in studies of diseases with low prevalence, when long treatments are evaluated or, when it is advisable to compare the feasibility of a proposed treatment among several institutions.

THE SURGICAL TRIAL

The surgical trials are oriented to define new surgical or interventional treatments for human beings. These experiments diverge from drugs, in the sense that medicaments are standard treatments independent of any special skill or experience that the physician may have. In contrast, surgical procedures are not standard treatments because the skill and experience or each surgeon is unique and he or she are always part of the treatment (104).

The phases of the surgical trials

It is possible to differentiate four study phases in the evolution of a surgical procedure, before it can reach the maximum level of "standardization" i.e. that the procedure is well defined and reproducible in the hands of any well trained surgeon. The pre-surgical phase starts with the theoretical invention of a new surgical procedure to solve a medical problem. The procedure is first attempted in adequate animals models. If the evaluation of the experiments performed on the animals, shows that the procedure is better than the previous treatments, then the reproducibility of the procedure must be guaranteed. The conclusion of this phase should include the definition of a new surgical approach to solve a specific medical problem. The surgical phase I has as a main objective to demonstrate the usefulness of the surgical procedure in human beings. For some procedure the pre-surgical phase may not be possible and the procedure has to be initiated directly on human beings in any case, to reach the objective of surgical phase I, the first patient or group of patients cannot be selected by random procedures; because, in order to demonstrate the usefulness of the treatment, the surgeon has to give himself the best opportunity to succeed within the standard ethical limits. If under these conditions the procedure shows to be better in his or her hands than the previous medical or surgical treatments, a full description of the preliminary results and benefits of the new procedure has to be detailed. Once this is done, the surgeon then has to define the surgical procedure describing the techniques as easily and reproducible as possible in order that it can be followed through by other surgeons. The procedure defined in surgical phase I has to be taught to new surgeons who will be working in different environments. Performing the above, surgical phase II begins. This expanded phase has as an objective to balance out the covariants (individual skill, experience, team competence, quality of the institution, patients with different characteristics). Through this process, techniques, materials and apparatus employed, will evolve making the surgical procedure more independent from the skill and experience of a particular surgeon. During this phase, different type of patients are exposed to the procedure in order to define the ideal candidates. This process continues until it can be successfully performed in a reproducible way by other well trained surgeons. At this point the surgical approach that the new operation introduced, is considered at the maximum stage of "standardization". If under these conditions the surgical procedure shows to be superior than the existing previous treatments, the operation is ready for the surgical phase III. The objective of this last phase, is to compare the "standard" procedure with new techniques, materials, apparatus, as well as, with other surgical procedures of similar degree of evolution, placebo and other medical treatments. Its application to other medical conditions can also be investigated. The efficacy, sporadic adverse reactions and risk/benefit ratio are also defined in this phase. These comparisons can be made using random procedures if the ethical limitations allow for it. So as one can observe, once the process has been "standardized" in phase II, its evolution is not halted, rather, it continues through phase III.

The randomized controlled surgical trial

Although the same procedures to secure as unbiased comparisons among the study groups used in clinical trials could be considered in the case of the surgical trials. There exists a controversy in the medical literature in relation to when a randomized controlled trial can be used for surgical trials (104-110). In our view, these designs are appropriate for "standard" treatments only. i.e.: surgical phase III.

This classification scheme stands for the procedure that will be used to compare the outcome of the treatment. See Table 1. These studies can be divided into two main groups. The internally controlled trials. In this type of design, the control group or control values are contained within the same experimental design (15,88,111,112). The two major categories that conform this group are: 1. the self-controlled and 2. the parallel designs. In the self-controlled design, each patient acts as its own control (88,112). This category can be further subdivided in a) the before-after trial design in which each patient receives only one treatment and the outcome is evaluated by comparing the status of the patient before and after the treatment (15,88,113,114), and b) the cross-over design previously described (15,81,83,88-93). In the parallel designs, two or more patients or two or more groups are compared. Each patient of the same group receives only one treatment, and the outcome among the different patients or groups are compared (112,115). The second major category of the clinical and surgical trials consist of the Externally controlled designs, these schemes use a parallel design to evaluate the effectiveness of a procedure using as comparisons, historical controls, population based results or, prior published reports on the same subject (80,116-118). Examples of the most common designs of clinical and surgical trials is depicted in Tables 1 and 5.

THE BASIC EXPERIMENTAL DESIGNS

These designs are focused to contrast a hipothesis, very often, however, it is necessary to invent the experimental procedure or apply previously made designs to answer a new question in both cases the emphasis is made on the experimental procedure, as well as, in the observation of the natural fact. The studies included in this category are those focused to describe reality in man using basic laboratory procedures i.e.: the isolation of a new active peptide or the description of dinamic changes in heart physiology, as well as studies carried out in experimental models different than man i.e.: healthy animals, alternate biological species with similar diseases, cells, bacteria, virus, apparatus. These models are ideal representations of reality appropriate to test a hypothesis or to answer an specific research question, and are used to extract relevant information to support or explain a clinical fact (119). The basic medical research includes a broad variety of designs. However, most of them can be included on the following model categories:

Adequacy of the model to represent reality (120)

Iconic Models are those in which one or several properties of reality are preserved in the model generally with a change in scale, lower or higher for better representation. For example: the anatomic relationship of a part of the human body could be represented with the same properties in a 3-D plastic model or a plate in a book. Likewise an atomic model of a DNA molecule is an iconic model of the atomic distribution in the real molecule. Analogue Models. In these models one or several properties of reality are represented with a different property in the model. For example: the distribution of a disease in a country can be represented by a map, and the areas of high prevalence can be identified with a color. Likewise an animal can be used as an analogue model to perform a surgical procedure that is aimed to be performed in man. Symbolic Models. In these models, the variables, constants and their relationships for a particular natural phenomenon are represented by formulas that can be handled using the mathematical logic. For example: blood flow, partial pressure laws.

Table 5. Experimental Designs commonly found in Clinical Research Protocols.

DESIGN DESCRIPTION	PRESENT	FUTURE
BEFORE – AFTER TRIAL (one group).	S 1 [2 A] 3 C	
BEFORE – AFTER TRIAL (two or more groups).	S 1 [2 A] 3 C / S 1 [2 A] 3 C	
CROSS – OVER (paired organs).	S 1 [2 D, A] 3 C / S 1 [2 D', A] 3 C	
CROSS – OVER (single group).	S 1 [2 A] 1 [2 B] 3 C	
CROSS – OVER ✳ (two or more groups).	S 1 [2 A] 1 [2 B] 3 C / S 1 [2 B] 1 [2 A] 3 C	
PARALLEL ✳ ✳ (internal control). (two or more groups).	S 1 [2 A] 3 C / S 1 [2 B] 3 C	
PARALLEL (external controls).	S 1 [2 A] 3 C / [4]	

(S) start of experiment, (A) new treatment, (B) standard treatment or placebo, (C) end point, (D,D') can be two different areas of a single organ i.e.: skin or two matched paired organs i.e.: teeth, limbs, kidneys, lungs, (1) washout period, (2) treatment period, (3) follow-up period, (4) comparisons are made to data external to the current study, (5) treatments can be assigned to patients by randomm or non random procedures (67), (✳) latin square (75) is a variant of this designs. Patients can be assigned to treatments by random or non random procedures (69), (✳✳) stratified (46), factorial (16,47,76) and sequential clinical trials (77) are variants of this design. Patients can be assigned to treatments by random or non random procedures (70).

Capacity of the model to summarize empirical observations. (121).

An empirical model, summarizes the results of a particular group of observations. These models are valid as hypothesis to explain the behavior of a similar set of observations but it is risky to generalize its properties beyond the point representing the empirical observations. For example the E coli lactose operon or the alcohol metabolism in liver cells. A theoretical model, is an abstraction of the knowledge obtained from different empirical models. The scientific premises of these models abstract the general laws that regulate the behavior of real systems. For example the theory of natural selection or the structure of DNA.

COMMON METHODOLOGICAL GUIDELINES FOR OBSERVATIONAL AND EXPERIMENTAL STUDIES: MATERIAL AND METHODS SECTION

Variables and measuring scales

The variables are characteristics of the clinical phenomenae that may have different values among different individuals or observations (88,122,123). There is a tendency to transform the qualitative values into numerical concepts. This capacity depends on the technological advance. Within this limitation, the investigator should list those variables that he considers appropriate to be included in the protocol. Next, he has to select or design the best procedure or procedure to measure the outcome of the variables and imagine all the possible results and its interpretation. After this is carried out, every variable must be critically scrutinized in order to assure that its inclusion in the protocol is truly essential. This selection depends on the investigator creativity and experience. A good investigator chooses critical variables. A broadly used classification (124) is employed to determine the degree of precision of measuring a variable. The categorical variables refers to those values of qualitative nature, like skin color. This variable has different values: white, yellow, brown, black, which represent categories into which observations can be classified. These variables can be measured using the following two scales: a) the nominal scale uses names to determine the pertenence of a measured characteristics to a category, excluding all others (example: white skin). The values of this scale do not relate qualitatively or quantitatively. b) In the ordinal scale, the values are quantitatively related and the observations are compared to each other and placed in order. Example: Pain: light, moderate, intense. The numerical variables include those characteristics of quantitative nature, for instance, blood glucose, whose values are numerical scores. These values are measured within the following two scales: a) the interval scale, where the unit of measure is of the same size throughout the entire range of possible values of the variable. Using this scale it is possible to take any two scores and find out by how much they differ from each other. For example: temperature in degrees centigrades. b) The second scale is the ratio scale, which has the same characteristics as the interval scale, plus a true zero point. Weight, height, time, volume and pressure can be measured in this manner. In order to achieve the first step towards how to analyze the results obtained, the investigator should be able to examine his or her set of data and determine to which scale they belong to. (125).

Description of the target population and sample units

The target population should be described. Indicating whether the source of the statistical unit (patients, animals, cells) is ubicated in time and space (tangible); or not, (conceptual). The procedure that will be used to select the sample units must be pointed out, identifying the inclusion and exclusion criteria (32,126).

The design of the sample

Unlike a census, frequently the information of clinical research studies is gathered only from a small sample of people in order to learn something about the population from which the sample has been drawn. In statistical language, the term sample is most often used to describe a method for gathering information from a small number of individuals, randomly selected from a population. This is done so that each individual in the population has a known chance of selection. That is, a <u>representative sample</u> (32). These types of studies are useful when the objective of the project is to obtain a statistical profile of the population. However, in medicine, several problems limit the possibility to obtain such a sample. Frequently, the statistical units are not selected by the physician rather the physician is selected by them (127). In some study designs, the number of patients has to be small as in phase I studies. Other investigations in which the number of patients is small, is for instance, cases of rare illness (128); or when the importance of the outcome makes it illogical to delay the decision to communicate the news until a representative sample is obtained (129). In any case, this is not a limitation factor since the objective of clinical research is to obtain valid scientific knowledge and this can be gathered from a single patient. For instance, if a patient survives a disease with one hundred percent mortality by the use of a specific treatment, this fact would be considered enough evidence to support the treatment (130). Therefore, clinical research can be done with one patient, a group of patients or a representative sample. The validity of the information obtained has to be determined in each case.

<u>Errors to avoid during the selection of the sample</u>. When a representative sample should be conceived in the planning stages of the clinical research protocol, it is important to avoid errors during the sampling procedure. The sampling frame, should be clearly defined, as well as the process to identify and locate the sample units, and the sampling methods that will be used (simple, systematic, stratified, cluster) (32). In order to avoid the bias during the selection of the sample units, random procedures should be employed, i.e. table of random numbers (32).

<u>The sample size</u>. How many patients are needed in order to exclude random variation as an explanation of the differences that will be observed in the study? To answer this question the following criteria should be considered. <u>The magnitude of the difference expected</u>. This value is determined by the difference that will be clinically meaningful for those variables included in the protocol, large differences require less patients than those needed to detect small ones. <u>The variation of the characteristics to be detected in the population sample</u>. If the difference from patient to patient with respect to the characteristic to be measured is large more patients will be needed. If the population is more uniform in this regard, less patients are necessary (53). <u>The prevalence of the events searched for</u>. If the characteristic to be measured is frequently present, fewer patients will be required to measure it. If it is rare the sample should be large. <u>The exactnes of the measuring criteria</u>. If the investigator wants to measure a variable with great accuracy the sample needed could be larger than the one required for studies with less demand for precision. In addition, the following criteria have to be considered in comparative studies. <u>The null hypothesis</u>. This is a procedure used to compare the behavior of two populations by analyzing the results observed in the randomly selected samples. This can be summarized as follows: The investigator wants to know if treatment A is better than treatment B in the population of patients (Y). As a first step, he specifies the difference (D) that he regards as important between the two treatments. Then, the statement is reformulated in the form of the null hypothesis i.e.: treatment A = treatment B + D. Afterwards, he sets the error risk of rejecting the null hypothesis when it is true (risk α) or accepting it when it is false (risk β), associated to the statistical test previously identified. Once a decision has been done, the sample size is determined and by doing this, the power of the test $(1-\beta)$ is indirectly indicated. The acceptable limits as to how large or how small a risk should be, is a

value judgment. However, it is desirable to reduce the consequences of making false conclusions by setting small risk values. This can be done by increasing the number of patients. After the sample size has been settled, the subjects are selected and assigned randomly to the two treatment groups afterwards the responses are observed. If the value of the test statistic used to asses the data is higher than the value of the tables at the corresponding degree of freedom, the null hypothesis is rejected and the investigator can conclude that the treatment A is better than the treatment B, with a true difference in efficacy of a least D (131). If the calculated value of the significance test (132-134) is less than the value of the tables, the null hypothesis is accepted and the investigator concludes that there is no significant difference between the two treatments. The disease and treatment studied. When a better treatment is urgently needed for a grave disease. It is important to reduce the possibility of missing a valuable treatment by setting an small value for β risk (135,136). Alternatively it is desirable to minimize the consequences of accepting a new treatment when it is not really effective. In this case the acceptable limit for an α risk should be small.

Procedures to calculate the sample size.   To separate real from random effects, formulas, tables, nomograms and computer programs are available, to estimate the sample size needed for descriptive or comparative studies where the expected value or the differences among the groups have been well stablished (125,137-139), the same procedures are available for studies of chronic diseases or therapeutic trials where long periods of observation are required to detect small differences among the groups (140-149).

Description of the procedures and the data forms to collect the required information

Once the information needed for the study has been identified, the instruments to collect the data must be decided upon. If the information is going to be obtained through diagnostic procedures, i.e.: magnetic resonance imaging or laboratory tests, the required procedure must be described in detail. If the information is going to be captured through written questions, they have to be properly designed and pretested. Alternatives must be developed for controling response errors. Specific forms should be designed to point out the inclusion and exclusion criteria, to obtain the informed consent, clinical history, laboratory data and special studies, patient evolution, adverse effects, to define the criteria for evaluating the effect of treatment, and to summarize the results previous to analysis of the data (126, 150-152).

Ethical issues

In every step of the research proposal the investigator is faced with fundamental ethical problems. An appropriate justification of the procedures to be carried out should be included in the clinical research protocol. However after considering the research problem, its designs and the methodological guidelines the clinical researcher has to reconsider the ethical implications of the research proposal as a whole. The questions that should be thought are more conspicuous in the experimental than in the observational studies, some of the most important are: Is the investigator competent? (153); Are the eligibility requirements ethical? (126,154,155); How is the patient recruitment going to be carried out? (156-161); How many patients will be used for the study? (126,162); When it is necessary to obtain the informed consent? (163,164); How are the patients and participants in the investigation going to be protected from dangerous procedures (pathogens, mutagens, radioactive material)? (165-171); Are the physiological, diagnostic or therapeutic procedures ethical?

(172,173); When it is ethical to do research with children, pregnant women and impaired people? (174-179); When to use a random assignment? (180,181); When to use placebos? (182,183); When and how to stop a clinical trial? (98,184,185). The ethical values of the investigator (186-189) as well as the ethics of human experimentation and the safeguards to protect human beings, have been described in the Declaration of Helsinki in 1964 and the Declaration of Tokyo in 1975, as well as in other more recent documents (190-197).

Instructions to all personnel who will be part of the project

Professional and non professional personnel who will participate in the study should be informed of their responsibilities in each stage of the research project. Addresses and telephone numbers of key persons who can be contacted at any time must be annotated on the protocol. The instructions to the subjects that will participate in the study, the procedure to be followed with them, their restrictions, as well as their rights have to be described (125).

COMMON METHODOLOGICAL GUIDELINES FOR OBSERVATIONAL AND EXPERIMENTAL STUDIES: RESULTS SECTION

It should be stressed that the medical judgement of the data is an essential part of any clinical or surgical research study. Signs, symptoms and syndromes, are more useful if they are understood having an integrative view of the patient, within the framework of the study design. When it is necessary, the uncertainty of clinical decisions can be analyzed through the systematic approach of decision analysis (198-202). Some studies require statistical analysis to asses some of the clinical results, and the investigator should determine how and to what extent the statistical methods will be useful for his particular study (123,203-209). It should be pointed out that statistical significance does not necessarily mean that it is clinically important. If the investigator has any doubts, he or she should search for first rate statistical advice.

Selection of the statistical procedure to assess the data

When some of the conclusions of the study will be based upon an assessment of the data collected, the statistical methods that will be employed to make such estimation, must be described (150,151,203-208). A study carried out in a representative sample of original research articles from a general medical journal, showed that most part of the original research articles use only a few categories of statistical procedures: descriptive statistics, t-tests, analysis of contingency tables, non parametric tests, epidemiologic methods, correlation and simple regression (209). The decision to define the statistical method that will be used for the data analysis, is facilitated by considering the following methodological factors: Statistical methods to assess the data in descriptive studies. Graphical procedures such as histograms, percentile curves or scatter plots, are frequently used. Summary statistics as average ($\bar{X}$) and standard deviation (s) are employed to demonstrate variability among individuals with continuous variables; media and percentile ranges are used to describe proportions with asymmetrical distributions; the standard error of the mean ($s/\sqrt{n}$) is utilized to verify the variability of the sample mean (210-212). Statistical methods to assess the data in Comparative Studies. Formal statistical analysis should be preceded by adequate description and presentation of the data. The process of how this can be accomplished should be included in the protocol i.e.: graphical methods, summary statistics, mathematical transformations or tables (204). The interpretation of the results requires frequently the use of a null

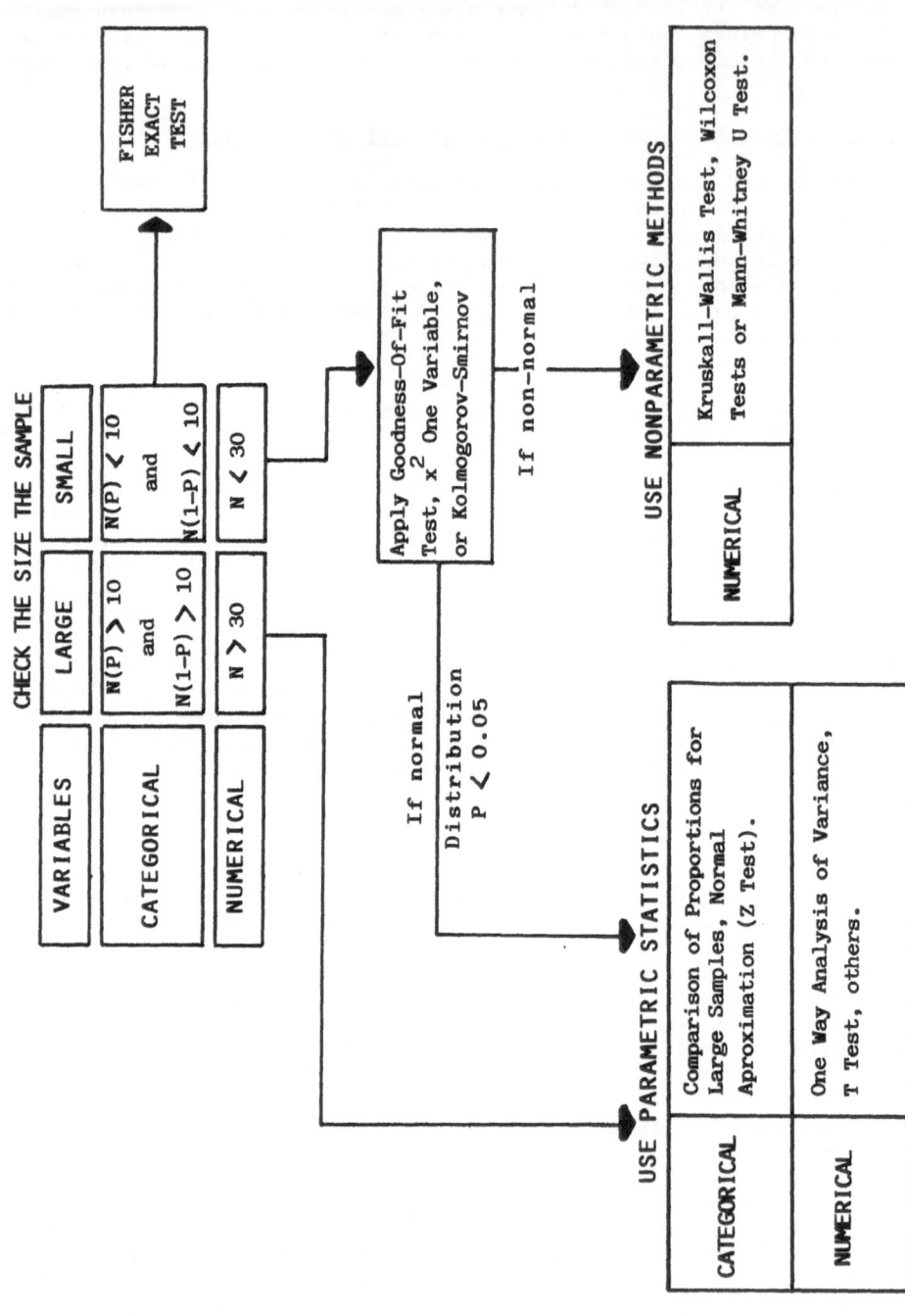

Fig. 1. Rule of Thumb to select the Statistical Methods to analyze the Data.

494

hypothesis. The practical problems that emerge when this theory is used in designs of complex structure such as the clinical trials have been pointed out (213). If an association among variables will be searched for in a randomized trial, the statistical procedure, correlation, chi-square or regression analysis, should be mentioned (204,214). In comparative studies, a rule of thumb to determine the statistical method to be used considers: The procedure used to measure the variables (79,122-124), the size of the sample (125,137-149) and their probability density functions (215). A summary of how to combine them is described in Fig. 1. Statistical methods associated to the study design. Each design has specific statistical procedures whether observational or experimental. For example, surveys rely more on descriptive methods, while cohorts use more non-parametric test (209). In retrospective studies chi-square test are used for the analysis of the observed association between the disease and the putative risk factor (216-218); in prospective studies incidence, and relative risk formulae are employed (211-213). When using a parallel design (unpaired data), the two sample t test should be used for continuous variables, chi-square test for categorical variables, and the Mann-Whitney test for free distribution data. An alternative to the above would be, a self controlled trial; before-after or cross-over (paired data) where the paired t test, the McNemar's test and the paired Wilcoxon test, will be used instead (204). Statistical methods associated to the clinical research area. The study of health, uses the statistical methods previously described for descriptive and comparative studies. The study of disease uses in addition: prevalence, incidence (219,220); risk measurements: relative risk and attributable risk (204,220); prognostic factors (216,222-227); odds ratio (228), log odds, and measures of association (204). To evaluate a diagnostic procedure, sensitivity, specificity, and predictive value (229,230) are considered. In treatment evaluation, t and chi-square are used as a significance test for the null hypothesis. Survival curves or life tables are used when estimates of the probability of survival as a function of time is needed (51-55,151,203,204). Common procedures may not be described but more complex methods deserve some explanation in order that they will be comprehensible to the protocol readers.

References

Complete data for each reference should be given. The reference style according to the National Library of Medicine, including the abbreviation of journal titles, are appropriate (2-4).

GUIDELINES TO IMPLEMENT THE PROJECT

Logistics

Once the structure of the clinical research protocol is complete, the estimated duration of the project should be in harmony with the intricacy of the investigation; and, the relationship between time and activities should be estimated. This is achieved by displaying in a flow chart, the proper temporal and logical sequence of the events that are expected for each stage of the study. This process is helpful to constrain the time of the study to the limits stipulated in the project (102).

Budget

If the investigator prepares the protocol for a research grant application, he must have a copy of the latest instructions of the granting agency, since policies and requirements change with time and with granting institution (231). All of them require that the investigator list all

participating human resources, biographical sketches for each principal
investigator, salaries and wages, travel expenses, facilities available
for the research project, equipment needed, materials and supplies, direct
and indirect costs.

COMMON FAILURES FOUND IN CLINICAL RESEARCH PROTOCOLS

When an evaluator reads a research proposal he will consider the
competence of the investigator reflected in his antecedents, as well as in
the document he has written. He will look if every one of the different
sections of the proposal has been carefully thought out. Some of the most
frequent shortcomings are (232): The investigator. The author of the
project is insufficiently trained to perform this particular research
problem; He does not evaluate objectively the role and experience of his
research associates; He does not know enough the research field nor he
describe adequately how he is going to obtain advice or collaboration from
more experienced investigators. The research problem. The antecedents of
the proposed research gives insufficient evidence to support the objectives or the hypothesis described; The research question is too general and
lacks specific objectives; The proposed research objectives are of little
importance and does not warrant the expenditure of human, physical or
economical resources on it; A retrospective or a pilot study is required
to propose a scientifically sound problem; The research problem or
procedures are ethically incorrect. The study design. The criteria used to
select the study design are incomplete; The type and number of methodological designs are inadequately conceived. The methodological guidelines.
The methodological guidelines to properly conduct an observational or
experimental design are incomplete or unclear to permit adequate evaluation; The clinical or surgical phases are weakly described or incomplete;
The controls to avoid the psychological bias and the partiality in the
treatment assignment; are inadequately defined; The target population and
the sample are poorly defined; The variables are ambiguous; The investigator does not describe with clarity the statistical procedures to asses the
validity of the data; The investigator does not give enough basic or
clinical information to support how he is going to interpret the data that
he is planning to obtain; The data forms and the instructions to the
research associates or the patients are insufficient or not appropriate.
The references. The investigator is unfamiliar with recent relevant
literature; The investigator does not differentiate the literature
pertinent to the research proposal from other articles. The time and the
budget. The time proposed for the research project is unrealistic; Exist
incompatibility among the time that is required for the project and the
time that the investigator and their collaborators have available; The
resources solicited are insufficient or in excess to complete the project.

FINAL COMMENTS

When the clinical research protocol is prepared by a professional member
of a clinical department from a teaching institution, various benefits can
be derived from this document. Among them are: It improves the patient
care since all patients included in the protocol will follow the same
guidelines, which have been thoroughly reviewed in advance. It improves
teaching, because the information for that particular disease as well as
the scientific process to be followed with the patients, are detailed in
the protocol. It improves research, since the data for the research
project is being captured by the investigator and his collaborators in a
systematic way, while clinical work continues being normal. Once the data
collection has concluded, he only has to follow the protocol guidelines to
present and analyze the data. When the time comes to write the report for

Table 6. Existing schemes used to classify the clinical research reports.

'Species' and 'phyla' of research architecture (21)	Clinical Research Designs (25)	Research Designs in Physiotherapy (5)	Research strategies in Clinical Medicine (18)
Cross-sectional studies	Cross-sectional	Experimental	I. Longitudinal studies
Descriptive	Longitudinal	Descriptive	A. Prospective studies
Spectrum	Case-control	Surveys	1. Studies of deliberate intervention
Diagnostic marker	Cohort	Case studies	a. Sequential
Spectral marker	Clinical trials	Normative studies	i) Cross-over
Cross-association	Uncontrolled	Developmental studies	ii) Self-control
Concordance	Nonrandom controlled trials	Cross-sectional techniques	b. Parallel
Analytic	Randomized controlled trials		c. External controls
Experiment			2. Observational studies
Pathodynamic			a. Causes and incidence of disease
Pseudo-cohort			b. Deliberate but uncontrolled interventions
Cross-dependency			c. Natural history: prognosis
Trohoc			3. Pseudo-prospective
Cohort studies			B. Retrospective studies
Descriptive			1. Deliberate intervention
Predictive			2. Observational
Course			3. Pseudo-retrospective
Analytic			II. Cross-sectional studies
Clinical trials			A. Disease description
Transition			B. Diagnosis and staging
Outcome			1. Normal ranges
Miscellaneous			2. Disease severity
			C. Disease processes
			1. Exploratory
			2. Observational
			3. Case reports

publication, a good part of it is contained in the protocol; therefore, writing an original article for publication is greatly facilitated. A good protocol can not improve a bad idea but, it can make the road a lot easier towards obtaining an answer for a good one.

APPENDIX 1

The existing classification schemes are summarized in Table 6. Their reproducibility is affected by: a) The use of methodological designs and research objectives in the same classification scheme. This fact creates confusion when one attempts to define the pertenence to a category of one article that have two or more criteria at the same time (5,18,21,25); b) Some categories exist only because one article was used to define it (18,21). In theory, other non analyzed articles could have been added another category to the classification, making them partial schemes; c) Articles that contained several study designs are included into a single category. Therefore, they are not useful as exemplary models to follow (18,21); d) The inclusion criteria used to define some of the categories are ill defined. For example: Cross-sectional and Longitudinal criteria are based on the time period covered by the study (5,25), and some criteria are arbitrary i.e.: the classification of case reports are based on the number of cases instead of the methodological design (25). Some of these deficiencies have been previously pointed out (18,21,250). Recently, an improved proposal for more informative abstracts of clinical articles has being published (15).

Feinstein (21) analyzed the medical literature retrospectively based on his previous taxonomic classification (22,23,26) adding new categories. The inclusion criteria used to define some of the category boundaries includes methodological designs and research objectives, i.e.: Cohort studies and Outcome. (See table 6).

Fletcher and Fletcher (25), used a theoretical classification scheme based on methodological criteria, to describe the frequency with which various research designs have appeared in the clinical literature. The methodological criteria used to define the main categories, i.e.: Cross-sectional and Longitudinal are based on time. The classification branches are clear, but they have a limited capacity of resolution to point out specific study design due to the fact that they are too general. (See table 6).

Makrides and Richman (5), proposed a theoretical classification for research planning in physiotherapy. The scheme proposed comprehend methodological designs and research objectives in the same classification level, i.e.: Cross-sectional and Developmental studies. (See table 6).

Bailar et al (18), developed a classification from a retrospective analysis of 332 research reports. These were classified depending on the methodological criteria and/or research objectives. Example: Cross-sectional and Disease severity. Papers that contained two or more designs, Longitudinal and Cross-sectional, were assigned to one category only, giving preference to the former, and within each of these major branches to the class or subclass that occur first.

The authors thanks: G. González, M. Valladares, H. Náñez, F. Bosques, F. Salazar, J. Guerra, E. Garza and D. Stark for their valuable comments and stimulating discussions; S. Borrego for the statistical advice; J. Silva and M.T. Riojas for correcting and typing the manuscript.

REFERENCES

1. V. Booth, Writing a scientific paper, Biochemical Society Transactions, 3:1-26 (1975).
2. M. Fishbein, Medical writing: "The technic and the art", 4th ed. New York, Charles C. Thomas, (1972).
3. H.J. Edward, "How to write and publish papers in the Medical Sciences", ISI Press, (1982).
4. American Medical Association, "Manual for authors and editors. Editorial Style and Manuscript preparation", AMA, (1981).
5. L. Makrides and J. Richman, Writing the research proposal, in: Research methodology and applied statistics: a seven-part series, Physiotherapy Can, 33:163-8 (1981).
6. P. H. Lehmann, B.A. Townsend and P. Pizzolato, Guidelines for the presentation of research in the written form, Am J Clin Pathol, 89:130-137 (1988).
7. S. S. Siegelman, Advice to authors, Radiology, 66:278-280 (1988).
8. H. V. Wyatt, When does information become knowledge? Nature, 235:86-9 (1972).
9. B. Prince, L. Makrides and J. Richman, The literature search, in: Research methodology and applied statistics: a seven-part series, Physiotherapy Can, 32:201-6 (1980).
10. R. B. Haynes, K. A. McKibbon, D. Fitzgerald, G. H. Guyatt, C. J. Walker and D. L. Sackett, How to keep up with the medical literature: I. Why try to keep up and how to get started, Ann Intern Med, 105:149-53 (1986).
11. R. B. Haynes, K. A. McKibbon, D. Fitzgerald, G. H. Guyatt, C. J. Walker and D. L. Sackett, How to keep up with the medical literature: II. Deciding which journals to read regularly, Ann Intern Med, 105:309-12 (1986).
12. R. B. Haynes, K. A. McKibbon, D. Fitzgerald, G. H. Guyatt, C. J. Walker and D. L. Sackett, How to keep up with the medical literature: III. Expanding the number of journals you read regularly, Ann Intern Med, 105:474-8 (1986).
13. R. B. Haynes, K. A. McKibbon, D. Fitzgerald, G. H. Guyatt, C. J. Walker and D. L. Sackett, How to keep up with the medical literature: IV. Using the literature to solve clinical problems, Ann Intern Med, 105:636-40 (1986).
14. R. B. Haynes, K. A. McKibbon, D. Fitzgerald, G. H. Guyatt, C. J. Walker and D. L. Sackett, How to keep up with the medical literature. V. Access by personal computer to the medical literature, Ann Intern Med, 105:810-24 (1986).
15. Ad Hoc working group for critical appraisal of the medical literature, A proposal for more informative abstracts of clinical articles, Ann Intern Med, 106:598-604 (1987).
16. J. R. Platt, Strong Inference: Certain systematic methods of scientific thinking may produce much more rapid progress than others, Science, 146:347-53 (1964).
17. J. A. Bronowsky, Twentieth century image of man, Leonardo, 7:117-21 (1974).
18. J. C. Bailar III, T. A. Louis, P. W. Lavori and M. Polansky, A classification for biomedical research reports, N Engl J Med, 311:-1482-7 (1984).
19. L. Makrides and J. Richman. General principles and basic concepts, in: Research methodology and applied statistics: a seven-part series, Physiotherapy Can. 32:135-9 (1980).
20. L. Makrides and J. Richman, Experimental design, in: Research methodology and applied statistics: a seven-part series, Physiotherapy Can, 33:6-14 (1981).
21. A. R. Feinstein, Clinical biostatistics, XLIV. A survey of the research architecture used for publications in general medical journals, Clin Pharmacol Ther, 24:117-25 (1978).

22. A. R. Feinstein, Clinical biostatistics, XLII. The architecture of cross-sectional research (Part I), Clin Pharmacol Ther, 23:81-96 (1978).

23. A. R. Feinstein, Clinical biostatistics, XLIII. The architecture of cross-sectional research (Conclusion), Clin Pharmacol Ther, 23:481-93 (1978).

24. L. W. Sontag, The history of longitudinal research: implications for the future, Child Development, 42:987-1002 (1971).

25. R. H. Fletcher and S. W. Fletcher, Clinical research in general medical journals: A 30-year perspective, N Engl J Med, 301:180-3 (1979).

26. A. R. Feinstein, "Clinical Biostatistics", Mosby Co, St. Louis, Chapters 3-8 (1977).

27. J. M. Hershman and J. A. Pittman Jr, Response to synthetic thyrotropin-releasing hormone in man, J Clin Endocrinol Metab. 31:457-60 (1970).

28. W. B. Kannel and R. D. Abbott, Incidence and prognosis of unrecognized myocardial infarction, An update on the Framingham Study, N Engl J Med, 311:1144-7 (1984).

29. A. R. Feinstein. "Clinical Biostatistics", Mosby Co, St. Louis, chapters 7,14,21 (1977).

30. C. White and J. C. Bailar III, Retrospective and prospective methods of studying association in medicine, A.J.P.H., 46:35-44 (1956).

31. R. Ferber, P. Sheatsley, A. Turner and J. Waksberg, What is a survey?, American Statistical Association, 1-25 (1980).

32. E. R. Babbie, "Survey research methods", Wadsworth Pub. Co. Inc., Belmont Calif., U.S.A. (1973).

33. M. Buset, B. DesMarez, M. Baize, M. Bourgeois and M. Cremer, Bleeding esophagogastric varices: an endoscopic study, Am J Gastroenterol, 82:241-4 (1987).

34. R. S. Chang, J. P. Lewis and Ch. F. Abildgaard, Prevalence of oropharyngeal excreters of leukocyte-transforming agents among a human population, N Engl J Med, 289:1325-9 (1973).

35. E. Vayda, A comparison of surgical rates in Canada and in England and Wales, N Engl J Med, 289:1224-9 (1973).

36. W. H. Lamb, C. M. Lamb, F. A. Foord and R. G. Whitehead, Changes in maternal and child mortality rates in three isolated gambian villages over ten years, The Lancet, ii:912-3 (1984).

37. C. E. Walden, R. H. Knopp, P. W. Wahl, K. W. Beach and E. Strandness Jr, Sex differences in the effect of diabetes mellitus on lipoprotein triglyceride and cholesterol concentrations, N Engl J Med, 311:953-9 (1984).

38. H. F. Dorn, Some problems arising in prospective and retrospective studies of the etiology of disease, N Engl J Med, 261:571-9 (1959).

39. A. R. Feinstein, "Clinical Biostatistics", Mosby Co., St. Louis., Chapters 13,14 (1977).

40. R. I. Horwitz and A. R. Feinstein, Methodologic standards and contradictory results in case-control research, Am J Med, 66:556-64 (1979).

41. B. F. Polk, R. Fox, R. Brookmeyer, S. Kanchanaraksa, R. Kaslow, B. Visscher, Ch. Rinaldo and J. Phair, Predictors of the acquired immunodeficiency syndrome developing in a cohort of seropositive homosexual men, N Engl J Med, 316:61-6 (1987).

42. A. R. Feinstein, "Clinical Biostatistics", Mosby Co., St. Louis., Chapter 6 (1977).

43. W. E. Rousseau, G. R. Noble, G. E. Tegtmeier, M. C. Jordan and T. D. Chin, Persistence of poliovirus neutralizing antibodies eight years after immunization with live, attenuated-virus vaccine, N Engl J Med, 289:1357-9 (1973).

44. R. O. Bonow, K. M. Kent, D. R. Rosing, K. K. G. Lan, E. Lakatos, J. S. Borer, S. L. Bacharach, M. V. Green, S. E. Epstein, Exercise-induced ischemia in mildly symptomatic patients with coronary-artery disease and preserved left ventricular function, Identification of subgroups at risk of death during medical therapy, N Engl J Med, 311:1339-45 (1984).

45. R. R. Monson, L. Rosenberg, S. C. Hartz, S. Shapiro, O. P. Heinonen, D. Slone, Diphenylhydantoin and selected congenital malformations, N Engl J Med 289:1049-52 (1973).

46. E. R. Greenberg, A. B. Barnes, L. Resseguie, J. A. Barret, S. Burnside, L. L. Lanza, R. K. Neff, M. Stevens, R. H. Young and T. Colton, Breast cancer in mothers given diethylstilbestrol in pregnancy, N Engl J Med, 311:1393-8 (1984).

47. I. J. Chasnoff, W. J. Burns, S. H. Schnoll and K. A. Burns, Cocaine use in pregnancy, N Engl J Med, 313:666-9 (1985).

48. S. Lakhanpal, T. W. Bunch, D. M. Ilstrup and L. J. Melton III, Polymyositis-dermatomyositis and malignant lesions: Does an association exist?, Mayo Clin Proc, 61:645-53 (1986).

49. E. R. Farmer and E. B. Helwig, Metastatic basal cell carcinoma: A clinicopathologic study of seventeen cases, Cancer, 46:748-57 (1980).

50. The cancer and steroid hormone study of the centers for disease control and The National Institute of Child Health and Human Development. The reduction in risk of ovarian cancer associated with oral-contraceptive use, N Engl J Med, 316:650-5 (1987).

51. R. M. Simon. Design and conduct of clinical trials, in: "Cancer: Principles and practice of oncology", Vol. 1, 2nd ed. V. T. DeVitta Jr, S. Hellman, S. A. Rosenberg, eds., J.B. Lippincott, Philadelphia (1985).

52. J. L. Lortat-Jacob, G. Mathe and J. Servier, International meeting on comparative therapeutic trials, Biomedicine, 28:2-63 (1978).

53. R. Peto, M. C. Pike, P. Armitage, N. E. Breslow, D. R. Cox, S. V. Howard, N. Mantel, K. McPherson, J. Peto and P. C. Smith, Design and analysis of randomized clinical trials requiring prolonged observation of each patient, I. Introduction and design, Br J Cancer. 34:585-612 (1976).

54. R. Peto, M.C. Pike, P. Armitage, N. E. Breslow, D. R. Cox, S. V. Howard, N. Mantel, K. McPherson, J. Peto and P. G. Smith, Design and analysis of randomized clinical trials requiring prolonged observation of each patient, II. Analysis and examples, Br J Cancer. 35:1-39 (1977).

55. C. L. Meinert and S. Tonascia, Clinical Trials: Design, conduct and analysis, Oxford University Press, (1986).

56. A. B. Hill, The clinical trial, Brit Med Bull, 7:278-82 (1951).

57. L. Lasagna, The controlled clinical trial: theory and practice, J Chron Dis, 1:353-67 (1955).

58. M. B. Emanuel, Basic concepts of clinical trial design, Clinical research reviews, 1:83-6 (1981).

59. A. V. Peterson Jr and L. D. Fisher, Teaching the principles of clinical trials design and management, Biometrics, 36:687-97 (1980).

60. Principles for the clinical testing of cariostatic agents. "Adapted from a conference held at The American Dental Association", Chicago, Ill. Oct. 1968;6-32.

61. T. A. Louis, J. C. Bailar III and P. Lavori, Experimental designs for clinical investigations, in: "Proceedings of the second world conference on clinical pharmacology and therapeutics", L. Lemberger, M. M. Reidenberg, eds., Washington, D.C. July 31 August 15, 1983. Bethesda, Md.: American Society for pharmacology and experimental therapeutics, 19-30 (1984).

62. E. R. Homan, Quantitative relationships between toxic doses of antitumor chemotherapeutic agents in animals and man, Cancer Chemother Rep, 3:13-9 (1972).

501

63. C. J. Williams and S. K. Carter, Management of trials in the development of cancer chemotherapy, Br J Cancer, 37:434-47 (1978).

64. L. I. Goldberg, G. H. Besselaar, J. D. Arnold, L. Lemberger, J. R. Mitchell and T. L. Whitsett, Panel 1: Phase I investigations, Clin Pharmacol Ther, 18:643-6 (1975).

65. J. R. Wittenborn, Guidelines for clinical trials of psychotropic drugs, Pharmakopsychiat, 10:207-31 (1977).

66. C. T. Dollery. The assessment of efficacy, toxicity and quality of care in long-term drug treatment, in: Ciba Foundation Symposium 44, Research and Medical Practice their interaction, Elsevier, 73-81 (1976).

67. L. E. Hollister, B. L. Martz, E. A. Carr, H. D. Cohn, R. Crout and J. Levine. Panel 2: Phase II investigations, Clin Pharmacol Ther, 18:647-9 (1975).

68. D. L. Azarnoff, W. B. Abrams, J. Cuttner, W. L. Hewitt and H. F. Hailman, Panel 3: Phase III investigations, Clin Pharmacol Ther, 18:650-2 (1975).

69. B. Blackwell, P. D. Stolley, R. Buncher, C. R. Klimt, R. Temple, D. Venn and W. M. Wardell, Panel 4: Phase IV investigations, Clin Pharmacol Ther, 18:653-6 (1975).

70. D. P. Byar, R. M. Simon, W. Friedewald, J. J. Schlesselman, D. L. DeMets, J. H. Ellenberg, M. H. Gail and J. H. Ware, Randomized clinical trials: prospectives on some recent ideas, N Engl J Med, 795:74-80 (1976).

71. T. C. Chalmers, J. B. Block and S. Lee, Controlled studies in clinical cancer research, N Engl J Med, 287:75-8 (1972).

72. F. J. Ingelfinger, The randomized clinical trial, N Engl J Med, 287:100-1 (1972).

73. M. A. Zelen. New design for randomized clinical trials, N Engl J Med, 300:1242-5 (1979).

74. H. Sacks, T. C. Chalmers and H. Smith Jr, Randomized versus historical controls for clinical trials, Am J Med, 72:233-40 (1982).

75. J. B. Block, M. Schneiderman, T. C. Chalmers and S. Lee, Nonevaluable patients in clinical cancer research, Cancer, 36:1169-73 (1975).

76. H. S. Sacks, T. C. Chalmers and H. Smith Jr, Sensitivity and specificity of clinical trials: Randomized vs historical controls, Arch Intern Med, 143:753-5 (1983).

77. M. C. Weinstein. Allocation of subjects in medical experiments, N Engl J Med, 291:1278-85 (1974).

78. H. Sacks, S. Kupfer and T. C. Chalmers, Are uncontrolled clinical studies ever justified?, N Engl J Med, 303:1067 (1979).

79. A. R. Feinstein and C. K. Wells, Randomized trials vs. historical controls: the scientific plagues of both houses, Trans Assoc Am Physicians, 90:239-47 (1977).

80. H. M. Kantarjian, E. H. Estey, W. Plunkett, M. J. Keating, R. S. Walters, S. Iacobini, K. B. McCredie, E. J. Freireich, Phase I-II clinical and pharmacologic studies of high-dose cytosine arabinoside in refractory leukemia, Am J Med, 81:387-94 (1986).

81. M. L. Zucker, C. Trowbridge, J. Woodroof, S. B. Chernoff, L. Reynoso and C. A. Dujovne, Low- vs High- dose aspirin. Effects on platelet function in hyperlipoproteinemic and normal subjects, Arch Intern Med, 146:921-5 (1986).

82. T. D. Bradley, I. G. Brown, R. F. Grossman, N. Zamel, D. Martínez, E. A. Phillipson and V. Hoffstein, Pharyngeal size in snorers, nonsnorers, and patients with obstructive sleep apnea, N Engl J Med, 315:1327-31 (1986).

83. M. Markman, V. Sheidler, D. S. Ettinger, S. A. Quaskey and D. Mellits, Antiemetic efficacy of dexamethasone. Randomized, double blind, crossover study with prochlorperazine in patients receiving cancer chemotherapy, N Engl J Med, 311:549-52 (1984).

84. J. Storr, E. Barrel, W. Barry and W. Lenney, Effect of a single oral dose of prednisolone in acute childhood asthma, The Lancet, i:879-81 (1987).

85. T. C. Chalmers, P. Celano, H. S. Sacks and H. Smith Jr, Bias in treatment assignment in controlled clinical trials, N Engl J Med, 309:1358-61 (1983).

86. A. Mindel, E. Allason-Jones and I. Barton, M. Jeavons, G. Kinghorn, P. Woolley, A. Faherty, P. Williams and G. Patou, Treatment of first-attack genital herpes-acyclovir versus inosine pranobex, The Lancet, i:1171-3 (1987).

87. L. E. Hill, A. J. Nunn and W. Fox, Matching quality of agents employed in "double-blind" controlled clinical trials, The Lancet, i:352-6 (1976).

88. T. A. Louis, P. W. Lavori, J. C. Bailar III and M. Polansky, Crossover and self-controlled designs in clinical research, N Engl J Med, 310:24-31 (1984).

89. J. Poloniecki, R. Hews and N. Barker, A review of cross-over trials, The Statistician, 31:71-80 (1982).

90. J. E. Grizzle, The two-period change-over design and its use in clinical trials, Biometrics, 21:467-80 (1965).

91. M. Hills and P. Armitage, The two-period cross-over clinical trial, Br J Clin Pharmac, 8:7-20 (1979).

92. B. W. Brown Jr, The crossover experiment for clinical trials, Biometrics, 136:69-79 (1980).

93. L. M. David and G. Sanders, CO_2 laser blepharoplasty: a comparison to cold steel and electrocautery, J Dermatol Surg Oncol, 13:110-4 (1987).

94. R. Peto, Clinical trial methodology, Biomedicine, 28:24 (1978).

95. P. Armitage, "Sequential medical trials", 2nd ed., John Wiley & Sons. New York (1975).

96. F. J. Anscombe, Sequential medical trials, American Statistical Association Journal, 58:365-83 (1963).

97. G. L. Elfring and J. R. Schultz, Group sequential designs for clinical trials, Biometrics, 29:471-7 (1973).

98. J. W. Tukey, Some thoughts on clinical trials, especially problems of multiplicity, Science, 198:679-84 (1977).

99. S. J. Pocock, Size of cancer clinical trials and stopping rules, Br J Cancer, 38:757-66 (1978).

100. R. L. Mowery and O. D. Williams, Aspects of clinic monitoring in large-scale multiclinic trials, Clin Pharmacol Ther, 25:717-9 (1979).

101. F. L. Ferris and F. Ederer, External monitoring in multiclinic trials: Applications from opthalmologic studies, Clin Pharmacol Ther, 25:720-3 (1979).

102. C. S. Good, (Ed.) "The principles and practice of clinical trials", Churchill Livingstone, (1976).

103. M. W. Pozen, R. B. D'Agostino, W. P. Selker, P. A. Sytkowski and W. B. Hood Jr. A predictive instrument to improve coronary-care-unit admission practices in acute ischemic heart disease: A prospective multicenter clinical trial, N Engl J Med, 310:1273-8 (1984).

104. W. Van der Linden, Pitfalls in randomized surgical trials, Surgery, 87:258-62 (1980).

105. L. Bonchek, Are randomized trials appropriate for evaluating new operations?, N Engl J Med, 301:44-5 (1979).

106. D. H. Spodick, Numerators without denominators: there is no FDA for the surgeon, JAMA, 232:35-6 (1975).

107. J. W. Love, Drugs and operations: some important differences, JAMA, 232:37-8 (1975).

108. D. H. Spodick, W. Aronow, B. Barber, H. Blackburn, D. Boyd, C. R. Conti, J. P. LoGerfo, B. Lown, V. S. Mathur, H. D. McIntosh, T. A.

Preston, A. Selzer, T. Takaro, Standards for surgical trials, <u>Ann Thorac Surg</u>, 27:284 (1979).

109. D. H. Spodick, Revascularization of the heart-numerators in search of denominators, <u>Amer Heart J.</u>, 81:149-57 (1971).

110. D. H. Spodick, The surgical mystique and the double standard: controlled trials of medical and surgical therapy for cardiac disease: analysis, hipothesis, proposal, <u>Am Heart J.</u> 85:579-83 (1973).

111. M. Gail, R. Williams, D. P. Byar and Ch. Brown, How many controls?, <u>J Chron Dis</u>, 29:723-31 (1976).

112. P. W. Lavori, T. A. Louis, J. C. Bailar III and M. Polansky, Designs for experiments-parallel comparison of treatment, <u>N Engl J Med.</u> 309:1291-8 (1983).

113. M. Lipkin and H. Newmark, Effect of added dietary calcium on colonic epithelial-cell proliferation in subjects at high risk for familial colonic cancer, <u>N Engl J Med</u>, 313:1381-4 (1985).

114. F. A. Oski, B. E. Marshall, P. J. Cohen, H. J. Sugerman and L. D. Miller, The role of the left-shifted or right-shifted oxygen-hemoglobin equilibrium curve, <u>Ann Int Med</u>, 74:44-6 (1971).

115. B. M. Greene, H. R. Taylor, E. W. Cupp, R. P. Murphy, A. T. White, M. A. Aziz, A. Schulz-Key, S. A. D'Anna, H. S. Newland, L. P. Goldschmidt, Ch. Aver, A. P. Hanson, S. V. Freeman, E. W. Reber, P. N. Williams, Comparison of ivermectin and diethylcarbamazine in the treatment of onchocerciasis, <u>N Engl J Med</u>, 313:133-8 (1985).

116. J. C. Bailar III, T. A. Louis, P. W. Lavori and M. Polansky, Studies without internal controls, <u>N Engl J Med</u>, 311:156-62 (1984).

117. E. A. Gehan and E. J. Freireich, Non-randomized controls in cancer clinical trials, <u>N Engl J Med</u>, 290:198-203 (1974).

118. L. Lasagna, Historical controls: the practitioner's clinical trials, <u>N Engl J Med</u>, 307:1339-40 (1982).

119. M. E. Gurney, A. C. Belton, N. Cashman and J. P. Antel, Inhibition of terminal axonal sprouting by serum from patients with amyotrophic lateral sclerosis, <u>N Engl J Med</u>, 311:933-9 (1984).

120. R. L. Ackoff and N. Sasieni, Fundamentals of operation research, <u>John Wiley and Sons, Inc.</u>, N.Y. p.60 (1968).

121. D. L. Caspar, "Design and assembly of organized biological structures", Molec Arch in Cell Physiol. Symp Soc Gen Physiol. N.Y. Prentice-Hall, p.202 (1966).

122. A. R. Feinstein, Clinical biostatistics. XLI. Hard science, soft data, and the challenges of choosing clinical variables in research, <u>Clin Pharmacol Ther</u>, 22:485-98 (1977).

123. J. Richman, L. Makrides and B. Prince, Measurement procedures in research, in: Research methodology and applied statistics: a seven part-series. <u>Physiotherapy Can.</u> 32:253-7 (1980).

124. S. S. Stevens, On the theory of scales of measurement, <u>Science</u>, 103:677-80 (1946).

125. L. Cañedo, "Investigación Clínica", 2nd. ed. México, <u>Interamericana</u> (In press), (1988).

126. R. DerSimonian, J. Charette, B. McPeek and F. Mosteller, Reporting on methods in clinical trials, <u>N Engl J Med</u>, 306:1332-7 (1982).

127. A. R. Feinstein, "Clinical biostatistics", Mosby Co. St. Louis, Chap. 2 (1977).

128. J. A. Amrhein, G. J. Klingensmith, P. C. Walsh, V. A. McKusick and C. J. Migeon, Partial androgen insensitivity: the Reifenstein Syndrome Revisited, <u>N Engl J Med</u> 297:350-6 (1977).

129. R. J. Sherins, C. L. Olweny and J. L. Ziegler, Gynecomastia and gonadal dysfunction in adolescent boys treated with combination chemotherapy for Hodgkin's disease, <u>New Engl J Med</u> 299:12-6 (1978).

130. P. F. Coccia, W. Krivit, J. Cervenka, C. Clawson, J. H. Kersey, T. H. Kim, M. E. Nesbit, N. K. C. Ramsay, P. I. Warkentin, S. L. Teitelbaum, A. J. Kahn, D. M. Brown, Successful bone-marrow

transplantation for infantile malignant osteopetrosis, <u>N Engl J Med</u>, 302:701-8 (1980).

131. R. Makuch and R. Simon, Sample size requirements for evaluating a conservative therapy, <u>Cancer Treat Rep</u>, 62:1037-40 (1978).

132. R. Simon, Confidence intervals for reporting results of clinical trials, <u>Ann Intern Med</u>, 105:429-35 (1986).

133. J. D. Gibbons and J. W. Pratt, P-values: Interpretation and methodology, <u>The American Statistician</u>, 29:20-5 (1975).

134. K. J. Rothman, A show of confidence, <u>N Engl J Med</u>, 299:1362-3 (1978).

135. J. A. Freiman, T. C. Chalmers, H. Smith Jr and R. R. Kuebler, The importance of beta, the type II error and sample size in the design and interpretation of the randomized control trial: Survey of 71 "Negative" trials, <u>N Engl J Med</u>, 299:690-4 (1978).

136. A. S. Detsky and D. L. Sackett, When is a "negative" clinical trial big enough? How many patients you need depends on what you found, <u>Arch Intern Med</u>, 145:709-12 (1985).

137. M. J. Young, E. A. Bresnitz and B. L. Strom, Sample size nomograms for interpreting negative clinical studies, <u>Ann   Intern Med</u> 99:248-51 (1983).

138. R. H. Fletcher, S. W. Fletcher and E. H. Wagner, "Clinical epidemiology: The essentials", 2nd. edition, <u>Williams and Wilkins</u> (1988).

139. M. G. Natrella, "Experimental Statistics", Handbook No. 91, National Bureau of Standards, Washington, U.S.A. (1963).

140. E. A. Gehan, The determination of the number of patients required in a preliminary and a follow-up trial of a new chemotherapeutic agent, <u>J Chron Dis</u>, 13:346-53 (1961).

141. R. W. Makuch and R. M. Simon, Sample size considerations for non-randomized comparative studies, <u>J Chron Dis</u>, 33:175-81 (1979).

142. M. Halperin, E. Rogot, J. Gurian and F. Ederer, Sample sizes for medical trials with special reference to long-term therapy, <u>J Chron Dis</u>, 21:13-24 (1968).

143. B. S. Pasternack, Sample sizes for clinical trials designed for patient accrual by cohorts, <u>J Chron Dis</u>, 25:673-81 (1972).

144. B. S. Pasternack and H. S. Gilbert, Planning the duration of long-term survival time studies designed for accrual by cohorts, <u>J Chron Dis</u>, 24:681-700 (1971).

145. S. L. George and M. M. Desu, Planning the size and duration of a clinical trial studying the time to some critical event, <u>J Chron Dis</u>, 27:15-24 (1974).

146. J. M. Lachin, Sample size determinations for r x c comparative trials, <u>Biometrics</u>, 33:315-24 (1977).

147. M. Gail and J. J. Gart, The determination of sample sizes for use with the exact conditional test in 2 x 2 comparative trials, <u>Biometrics</u>, 29:441-8 (1973).

148. M. Gail, The determination of sample sizes for trials involving several independent 2 x 2 tables, <u>J Chron Dis</u>, 26:669-73 (1973).

149. N. S. Weiss, Clinical Epidemiology, The study of the outcome of illness, New York, Oxford University Press, (1986).

150. S. J. Pocock, M. D. Hughes and R. J. Lee, Statistical problems in the reporting of clinical trials. A survey of three Medical Journals, <u>N Engl J Med</u>, 317:426-32 (1987).

151. F. Mosteller, J. P. Gilbert and B. McPeek, "Reporting standards and research strategies for controlled trials", Controlled Clinical Trials, 1:37-58 (1980).

152. P. Wright and J. Haybittle, Designs of forms for clinical trials, <u>Br Med J</u>, 2:529-30,590-2,650-1 (1979).

153. R. Stamler, Appendix 1: A proposed mechanism and set of criteria for the evaluation of the scientific contribution of individual investigators in collaborative studies, including large clinical trials, <u>Clin Pharmacol Ther</u>, 25:671-2 (1979).

154. B. J. Culliton, Psychosurgery: National commission issues surprisingly favorable report, _Science_, 194:299-301 (1976).

155. R. N. Smith, Safeguards for healthy volunteers in drug studies, _The Lancet_, ii:449-50 (1975).

156. T. Prout, Patient recruitment: problems and solutions, _Clin Pharmacol Ther_, 25:679-80 (1979).

157. J. A. Schoenberger, Recruitment in the coronary drug project and the aspirin myocardial infarction study, _Clin Pharmacol Ther_, 25:681-4 (1979).

158. G. W. Benedict, LRC coronary prevention trial: Baltimore, _Clin Pharmacol Ther_, 25:685-8 (1979).

159. W. S. Agras and G. Marshall, Recruitment for the coronary primary prevention trial, _Clin Pharmacol Ther_, 25:688-90 (1979).

160. G. Croke, Recruitment for the national cooperative gallstone study, _Clin Pharmacol Ther_, 25:691-4 (1979).

161. T. E. Prout, Other examples of recruitment problems and solutions, _Clin Pharmacol Ther_, 25:695-8 (1979).

162. K. McPherson, Statistics: the problem of examining accumulating data more than once, _N Engl J Med_, 290:501-2 (1974).

163. A. G. Campbell, Infants, children, and informed consent, _Br Med J_, 3:334-8 (1974).

164. F. Mosteller, Problems of omission in communications, _Clin Pharmacol Ther_, 25:761-6 (1979).

165. E. H. Lennette, A. Balows, W. J. Hausler Jr and H. J. Shadomy, Manual of Clinical Microbiology 4th Ed. American Society for Microbiology, Chapters 9-14, Washington, D.C. (1985).

166. Guidelines for research involving recombinant DNA molecules. Fed Regist. 49, Part IV, No. 227, 46266 (1984).

167. Committee on Hazardous Substances in the Laboratory. Prudent practices for handling hazardous chemicals in laboratories. National Academy Press. Washington, D.C. (1981).

168. Committee on Hazardous Substances in the Laboratory. Prudent practices for disposal of chemicals from laboratories. National Academy Press. Washington, D.C. (1983).

169. B. N. Ames, R. Magaw and L. S. Gold, Ranking possible carcinogenic hazards, _Science_, 236:271-9 (1987).

170. P. Slovic, Perception of risk, _Science_, 236:280-5 (1987).

171. L. B. Lave, Health and safety risk analysis: information for better decisions, _Science_, 236:291-5 (1987).

172. H. K. Beecher, Ethics and clinical research, _N Engl J Med_, 274:1354-60 (1966).

173. J. P. Gilbert, B. McPeek and F. Mosteller, Statistics and ethics in surgery and anesthesia, _Science_, 198:684-9 (1977).

174. Department of health, education, and welfare. Protection of human subjects, policies and procedures. Federal register. 38:31737-49 (1973).

175. J. Van Eys, "Research on children: Medical imperative, ethical quandaries, and legal constraints", University Park Press, Baltimore, London, Tokyo, (1978).

176. The National Commission for the protection of human subjects of biomedical and behavioral research: Research Involving Children. U.S. Department of Health, Education, and Welfare. Publication no. (OS) 77-0004, (1977).

177. J. Fletcher, Abortion, euthanasia, and care of defective newborns, _N Engl J Med_, 292:75-8 (1975).

178. R. Q. Marston, Research on minors, prisoners and the mentally ill, _N Engl J Med_, 288:158-9 (1973).

179. R. J. Malovany, O. Rosen, E. Messenger, S. Harrison, P. Wayne, L. Green, B. Wald, J. Cohen, Human rights and research, _N Engl J Med_, 288:1305 (1973).

180. F. Rosner, The ethics of randomized clinical trials, _Am J Med_, 82:283-90 (1987).

181. A. Schafer, The ethics of the randomized clinical trial, N Engl J Med, 307:719-24 (1982).
182. H. Brody, The lie that heals: the ethics of giving placebos, Ann Intern Med, 97:112-8 (1982).
183. H. K. Beecher, Surgery as placebo: a quantitative study of bias, JAMA, 176:88-93 (1961).
184. P. Meier, Terminating a trial-the ethical problem, Clin Pharmacol Ther, 25:633-40 (1979).
185. Ch. R. Klimt and P. L. Canner, Terminating a long-term clinical trial, Clin Pharmacol Ther, 25:641-6 (1979).
186. S. E. Luria, Biological aspects of ethical principles, The Journal of Medicine and Philosophy, 1:332-6 (1976).
187. J. Monod, Chance and Necessity New York. Alfred A. Knopf.(1971).
188. A. Cournand. The code of the scientist and its relationship to ethics, Science 198:699-705 (1977).
189. G. S. Stent, The dilemma of science and morals, Genetics, 78:41-51 (1974).
190. B. H. Gray, R. A. Cooke and A. S. Tannenbaum, Research involving human subjects: the performance of institutional review boards is assessed in this empirical study, Science, 201:1094-101 (1978).
191. J. Katz, "Experimentation with human beings: the authority of the investigator, subject, professions and state in the human experimentation process", Russell Sage Foundation, New York (1973).
192. B. H. Gray, "Human subjects in medical experimentation", John Wiley & Sons, New York (1975).
193. R. A. Greenwald, M. K. Ryan and J. E. Mulvihill, "Human subjects research", Plenum Press, New York (1982).
194. L. Mackrides and J. Richman, Ethics in human research, in: Research methodology and applied statistics: a seven-part series, Phisiotherapy Can, 33:89-94 (1981).
195. P. A. Freund, "Experimentation with Human Subjects", George Braziller, New York, (1970).
196. N. Howard-Jones and Z. Bankowski, "Medical experimentation and the protection of human rights", XIIth CIOMS round table conference. Council for international organizations of medical sciences and the Sandoz Institute for Health and Socio-economical studies, Geneva, (1979).
197. K. Vaux, "Biomedical ethics: Morality for the New Medicine", New York, Hagerstown, San Francisco, London. Harper & Row, (1976).
198. W. C. Weinstein, H. V. Fineberg, A. S. Elstein, H. S. Frazier, D. Neuhauser, R. R. Neutra, B. J. McNeil, Clinical Decision Analysis. W.B. Saunders Company. Philadelphia, London, Toronto, Chapter 6, (1980).
199. B. J. McNeil, E. Keeler and J. Adelstein, Primer on certain elements of medical decision making, N Engl J Med, 293:211-5 (1975).
200. D. F. Ransohoff and A. R. Feinstein, Is decision analysis useful in clinical medicine?, J Biol Med, 49:165-8 (1976).
201. S. G. Pauker and J. P. Kassirer, Decision analysis, N Engl J Med, 316:250-8 (1987).
202. J. P. Kassirer, A. J. Moskowitz, J. Lau and S. G. Pauker, Decision analysis: a progress report, Ann Intern Med, 106:275-91 (1987).
203. T. A. Louis, F. Mosteller and B. McPeek, Timely topics in statistical methods for clinical trials, Ann Rev Biophys Bioeng, 11:81-104 (1982).
204. D. G. Altman, S. M. Gore, M. J. Gardner and S. J. Pocock, Statistical guidelines for contributors to medical journals, Br Med J, 286:1489-93 (1983).
205. Sh. M. Gore, I. G. Jones and E. C. Rytter, Misuse of statistical methods: critical assessment of articles in BMJ from January to March 1976, Br Med J 1:85-7 (1976).

206. H. M. Schoolman, J. M. Becktel, W. R. Best and A. F. Johnson, Statistics in medical research: Principles versus practices, J Lab & Clin Med 71:357-67 (1968).

207. J. Richman and L. Makrides, Overcoming statistics anxiety: statistical considerations when planning research, in: Research methodology and applied statistics: a seven-part series. Physiotherapy Can. 32:321-9 (1980).

208. L. E. Moses, Statistical concepts fundamental to investigations, N Engl J Med, 312:890-7 (1985).

209. J. D. Emerson and G. A. Colditz, Use of statistical analysis in The New England Journal of Medicine, N Engl J Med, 309:709-13 (1983).

210. L. R. Elveback, How high is high? A proposed alternative to the normal range, Mayo Clin Proc, 47:93-7 (1972).

211. H. R. Black, H. Quallich and C. B. Gareleck. Racial differences in serum creatine kinase levels, Am J Med, 81:479-87 (1986).

212. A. W. Root, Endocrinology of puberty. I. Normal sexual maturation, J Pediatr, 83:1-19 (1973).

213. S. J. Cutler, S. W. Greenhouse, J. Cornfield and M. A. Schneiderman, The role of hypothesis testing in clinical trials, J Chron Dis, 857-82 (1966).

214. K. Godfrey, Simple linear regression in medical research, N Engl J Med, 313:1629-36 (1985).

215. M. Hollander and D. A. Wolfe, "Non-parametric statistical methods", New York. John Wiley & Sons, Inc. (1973).

216. N. Mantel and W. Haenszel, Statistical aspects of the analysis of data from retrospective studies of disease, J Nat Cancer Inst, 22:719-48 (1959).

217. O. S. Miettinen, Matching and design efficiency in retrospective studies, N Engl J Med, 91:111-8 (1970).

218. J. Cornfield and N. Haenszel, Some aspects of retrospective studies, J Chron Dis, 11:523-34 (1960).

219. A. R. Feinstein and J. M. Esdaile, Incidence, prevalence and evidence: Scientific problems in epidemiologic statistics for the occurrence of cancer, Am J Med, 82:113-23 (1987).

220. R. F. Morton and J. R. Hebel, A study guide to epidemiology and biostatistics. University Park Press. Baltimore, (1980).

221. R. M. Poses, R. D. Cebul, M. Collins and S. S. Fager, The importance of disease prevalence in transporting clinical prediction rules: The case of streptococcal pharyngitis, Ann Intern Med, 105:586-91 (1986).

222. A. R. Feinstein, Clinical biostatistics, XIV. The purposes of prognostic stratification, Clin Pharmacol Ther, 13:285-97 (1972).

223. A. R. Feinstein, Clinical biostatistics. XV. The process of prognostic stratification (Part 1), Clin Pharmacol Ther, 13:442-57 (1972).

224. E. A. Gehan, T. L. Smith and A. U. Buzdar, Use of prognostic factors in analysis of historical control studies, Cancer Treat Rep, 64:373-9 (1980).

225. A. R. Shapiro, The evaluation of clinical predictions: A method and initial application, N Engl J Med, 296:1509-14 (1977).

226. J. H. Wasson, H. C. Sox, R. K. Neff and L. Goldman, Clinical prediction rules: Applications and methodological standards, N Engl J Med, 313:793-9 (1985).

227. A. Kong, O. Barnett, F. Mosteller and C. Youtz. How medical professionals evaluate expressions of probability, N Engl J Med, 315:740-4 (1986).

228. R. I. Horwitz and A. R. Feinstein, Alternative analytic methods for case-control studies of estrogens and endometrial cancer, N Engl J Med, 299:1089-94 (1978).

229. R. S. Galen and S. R. Gambino, "Beyond Normality: The predictive value and efficiency of medical diagnosis", New York. John Wiley & Sons, Inc. (1975).

230. P. F. Griner, R. J. Mayewski, A. I. Mushlin and P. Greenland, Selection and interpretation of diagnostic tests and procedures, Principles and applications, Ann Int Med, 94:553-600 (1981).

231. G. N. Eaves, Who reads your project-grant application to the National Institutes of Health? Federation Proceedings. 31:2-9 (1972).

232. E. M. Allen, Why are research grant applications disapproved?, Science, 132:1532-4 (1960).

233. E. Braunwald, K. J. Isselbacher, R. G. Petersdorf, J. P. Wilson, J. B. Martin, Fauci As eds. Harrison's. Principles of internal medicine 11th ed. New York. McGraw-Hill Co. (1987).

234. J. B. Wyngaarden, L. H. Smith, eds. "Cecil text-book of medicine 17th ed. Philadelphia. W.P. Saunders Co. (1985).

235. J. H. Stein, W. J. Daly, J. D. Easton, M. J. Cline, J. J. Hutton, P. O. Kohler, R. A. O'Rourke, M. A. Sande, J. H. Stein, S. S. Trier, N. J. Zvaifler, Internal medicine 2nd. ed. Boston. Little, Brown and Co. (1987).

236. S. I. Schwartz, G. T. Shires, F. C. Spencer, F. H. Storer, eds. Principles of surgery. 4th ed. New York. McGraw-Hill. Book Co. (1984).

237. S. L. Robbins, "Pathologic basis of disease", Philadelphia. W.B. Saunders Co. (1984).

238. A. Goodman Gilman, L. S. Goodman, T. W. Rall, F. Murad F eds. Goodman and Gilman The pharmacological basis of therapeutics. 7th ed. New York. McMillan Pub. Co. (1985).

239. I. Davidson, J. B. Henry, eds. "Clinical diagnosis by laboratory methods", 14th ed. Philadelphia. W.B. Saunders Co. (1970).

240. A. C. Guyton. "Text book of Medical Physiology", 5th ed. Philadelphia. W.B. Saunders Co. (1976).

241. C. E. Welch, G. V. Rodkey and P. R. Gryska, A thousand operations for ulcer disease, Ann Surg, 204:454-67 (1986).

242. G. L. Colice, G. L. Chappel, S. M. Frenchman and D. A. Solomen, Comparison of computerized tomography with fiberoptic bronchoscopy in identifying endobronchial abnormalities in patients with known or suspected lung cancer, Am Rev Resp Dis, 131:397-400 (1985).

243. F. C. Notzon, P. J. Placek and S. M. Taffek, Comparisons of national cesarean-section rates, N Engl J Med, 316:386-9 (1987).

244. R. J. Knudson, J. W. Bloom, D. E. Knudson and W. T. Kaltenborn, Subclinical effects of smoking, Physiologic comparison of healthy middle-aged smokers and nonsmokers and interrelationships of lung function measurements, Chest, 86:20-9 (1984).

245. E. VanSonnenberg, C. C. Neff and R. C. Plister, Life-threatening hypotensive reactions to contrast media administration: comparison of pharmacologic and fluid therapy, Radiology, 162:15-9 (1987).

246. T. Aikawa, H. Sairenji, S. Furuta, K. Kiyasawa, T. Shikata, M. Imai, Y. Miyakawa, Y. Yanase, M. Mayumi, Seroconversion from hepatitis Be antigen to anti-HBe in acute hepatitis B virus infection, N Engl J Med, 298:439-41 (1978).

247. J. Dryjanski, J. W. Gold, M. T. Ritchie, R. C. Kurtz, S. L. Lim and D. Armstrong, Criptosporidiosis. Case report in a health team worker, Am J Med, 80:751-2 (1986).

248. J. W. Gurney, W. C. Harrison, K. Sears, R. A. Robbins, Ch. A. Dobry and S. I. Rennard, Bronchoaveolar lavage: radiographic manifestations, Radiology, 163:71-74 (1987).

249. N. L. Benowitz, P. Jacob III, L. T. Kozlowski and L. Yu, Influence of smoking fewer cigarettes on exposure to tar, nicotine, and carbon monoxide, N Engl J Med, 315:1310-13 (1986).

250. J. R. Boen, Clinical research in general medical journals, (letter), N Engl J Med, 301:1292 (1979).

PARTICIPANTS

STUART A. AARONSON
National Cancer Institute
Laboratory of Cellular and
Molecular Biology
Room 1E24, Bldg. 37
Bethestda, Maryland   20892

EFRAIN AZMITIA
New York University
1009 Main/Biology Department
100 Washington square E.
New York, New York 10003

ANGELO AZZI
Institute of Biochemistry
and Molecular Biology
University of Bern
CH-3012 Bern
Bühlstrasse 28
Switzerland

HUGO A. BARRERA-SALDAÑA
Unidad de Laboratorios de
Ingeniería y Expresión
Genéticas
Facultad de Medicina,
U.A.N.L.
Madero y Salvatierra
Monterrey, Nuevo León
Apartado Postal 1563

FEDERICO BERMUDEZ RATTONI
Instituto de Fisiología
Celular
Ciudad Universitaria - UNAM
México 04510, D.F.
Apartado Postal 70242

FRANCISCO BOLIVAR
Director Centro de
Investigación sobre
Ingeniería Genética y
Biotecnología, UNAM
Apartado Postal 510-3
Col. Miraval
62271 Cuernavaca, Mor.

BRUNO CALABRETTA
Temple University Health
Sciences Center
School of Medicine
Department of Pathology
3400 N. Broad Street
Philadelphia, Pa. 19140

LUIS E. CAÑEDO
Aida 63
Altavista San Angel
México D.F. 01060

ERNESTO CARAFOLI
Laboratorium fur Biochemie
Eidgenossische Technische
Hoschule
ETH - Zentrum
Universitatstrasse 16
CH - 8092
Zurich, Switzerland

SANDRA CARNEVALE CANTONI
Instituto Nacional de
Pediatría
Secretaría de Salud
Ave. Insurgentes Sur 3700
México 22, D.F.

THOMAS CASKEY, M.D.
Director
Institute for Molecular
Genetics
Baylor College of Medicine
One Baylor Plaza
Houston, Texas 77030

MARCELINO CEREJIDO
Centro de Investigaciones y
Estudios Avanzados
Instituto Politécnico
Nacional - CINVESTAV
Departamento de Bioquímica
Ave. IPN 2508 esq. Calzada
Ticomán
México 07000, D.F., Apdo.
Postal 14-440

JAMES CLEGG
Box 247
Bodega Marine Laboratory
University of California
Bodega Bay, Calif. 94923

ALEJANDRO CRAVIOTO
Director Científico
Instituto Nacional de
Ciencia y Tecnología - DIF
Ave. Liga del Imán No. 1 -
8o. piso
Col. Camisetal, Deleg.
Tlalpan
14410 México, D.F.

RAYMOND DAMADIAN
President
FONAR CORPORATION
110 Marcus Drive
Melville, New York 11747-
4212

RENE DRUCKER-COLIN
Instituto de Fisiología
Celular
Departamento de
Neurobiología
Ciudad Universitaria - UNAM
México 04510, D.F., Apdo.
Postal 70242

HENRYK EISENBERG
Department of Polymer
Research
The Weizmann Institute of
Science
Rehovot 76.100
POB 21, Israel

FREDERICK A. EISERLING
Molecular Biology Institute
University of California, LA
Los Angeles, Calif.   90024

GUILLERMO ELIZONDO RIOJAS
Massachusetts General
Hospital
Boston, Mass.
02114 U.S.A.

WILLIAM J. FREED
Chief Preclinical
Neuroscience Section
Department of Health and
Human Service
National Institute of Mental
Health
Saint Elizabeth Hospital
Washington, D.C.   20032

ROLAND GLASER
Section Biologie der
Humboldt-Universitat zu
Berlin
Invalidenstrasse 43
DDR - 104 Berlin
Democratic Republic of
Germany

MANUEL GONZALEZ GARAY
Unidad de Laboratorios de
Ingeniería y Expresión
Genéticas
Facultad de Medicina,
U.A.N.L.
Madero y Salvatierra
Monterrey, Nuevo León
Apartado Postal 1563

CARLTON F. HAZLEWOOD
Department of Physiology
Baylor College of Medicine
Texas Medical Center
Houston, Texas 77030 U.S.A.

ARMANDO ISIBASI
Laboratory of
Immunochemistry
Unidad de Investigación
Biomédica
Instituto Mexicano del
Seguro Social
P.O. Box 73-032
México, D.F.

JOSE JAZ
Director of International
Biomedical
Institute - I.B.M.I.
Via Putignani 12/A
70121 Barí, Italy

MARCO ANTONIO JOSE
VALENZUELA
Instituto Nacional de Salud
Pública
F. de P. Miranda No. 177
Col. Merced Gómez
Deleg. Alvaro Obregón
01600   México, D.F.

GREGORY KARCZMAR
University of California
Service
Veterans Administration
Medical Center
4150 Clement Street (11D)
San Francisco, California
94121

ARNOST KOTYK
Department of Cell
Physiology
Institute of Microbiology
Czechoslovakia Academy of
Sciences
142.20 Prague,
Czechoslovakia

JESUS KUMATE RODRIGUEZ
Secretario de Salud, S.S.A.
Lieja No. 7, 1er. piso
Col. Juárez, Deleg.
Cuauhtémoc
06696   México, D.F.

CARLOS LARRALDE RANGEL
Instituto de Investigaciones
Biomédicas
Depto. de Inmunología
Ciudad Universitaria - UNAM
México 04510, D.F., Apdo.
Postal 70228

JAVIER MARFIL RIVERA
Sub-Dirección de
Investigación y
Estudios de Post-Grado
Facultad de Medicina -
U.A.N.L.
Apartado Postal 4355-M
Monterrey, Nuevo León

LALITH MISRA, D.V.M., Ph. D.
Department of Medicine,
Hematology and Oncology
University of Texas Health
Science Center at Houston
6431 Fannin, Room 5016
Houston, Texas 77030

LINDA MUÑOZ ESPINOZA
Departamento de Unidad de
Hígado
Facultad de Medicina -
U.A.N.L.
Gonzalitos No. 235 Nte.,
Monterrey, Nuevo León

LESTER PACKER
Lawrence Berkeley Laboratory
Membrane Bioenergetics Group

Life Science Building
Berkeley, California 94720
U.S.A.

SERGIO PAPA
Institute of Medical
Biochemistry and Chemistry
University of Bari
Piazza G. Cesare - 70124
Bari, Italia

CHARLES A. PASTERNAK
St. George's Hospital
Medical School
Department of Biochemistry
University of London
London SW 17 ORE, United
Kingdom

DIANA RESENDEZ PEREZ
Unidad de Laboratorios de
Ingeniería y Expresión
Genéticas
Facultad de Medicina,
U.A.N.L.
Madero y Salvatierra
Monterrey, N. L.
Apartado Postal 1563

MARIO HENRY RODRIGUEZ
Director del Centro de
Investigación de Paludismo
(CIP)
6a. Avenida Nte. No. 19 Pte.
Anexo al Hospital Civil
Carman de Acebo
30700   Tapachula, Chiapas

FRANCISCO SANCHEZ-ANZALDO
Lab. Clínicos de Puebla
Blvd. Díaz Ordaz 808 Pte.
Col. Anzures
Puebla, Pue.   72530

GRADY F. SAUNDERS
Department of Biochemistry
and Molecular Biology - 117
M.D. Anderson Hospital and
Tumor Institute
University of Texas System
Cancer Center
1515 Holcombe Boulevard
Houston, Texas 77030

TREVOR SLATER
Department of Biophysics
Brunel University
Kingston Lane
Uxbridge, Middlesex
United Kingdom

DAVID STARK
G.I. Unit
Massachusetts General
Hospital
Boston, Mass.
02114 U.S.A.

JOSEPH M. TAGER
Laboratory of Biochemistry
University of Amsterdam
Academic Medical Centre
Meibergdreef 15
1105 Az Amsterdam
The Netherlands

LUIS E. TODD
Northgate Lane
Suite 717 Bldg. C
Laredo, Texas
78041   U.S.A.

OSCAR TORRES ALANIS
Departamento de Farmacología
y Toxicología
Facultad de Medicina -
U.A.N.L.
Gonzalitos No. 235 Nte.
Monterrey, Nuevo León

PAUL O.P. Ts'O
  Director
Division of Biophysics
The Johns Hopkins University
School of Hygiene and Public
Health
615, North Wolfe Street
Baltimore, Maryland, 21205
U.S.A.

SILVIO VARON.
Department of Biology
M-001, UCSD
School of Medicine
University of California,
San Diego
La Jolla, California 92093

DAN VASILESCU
Laboratoire de Biophysique
Universitè de Nice
Parc Valrose
06034 Nice, France

ANTONIO VELAZQUEZ
Instituto de Investigaciones
Biomédicas
U.N.A.M.
Apartado Postal 70228
México, D.F.

JESUS ZACARIAS VILLARREAL
Sub-Dirección de
Investigación y
Estudios de Post-Grado
Facultad de Medicina -
U.A.N.L.
Apartado Postal 4355-M
Monterrey, Nuevo León

CHARLES WEISSMAN
Institute of Molecular
Biology
University of Zurich
Hoenggerberg
8.093 Zurich, Switzerland

ULRICH ZIMMERMANN
Lehrstuhl fur Biotechnologie
der
Universitat Wurzburg
Rontgenring 11
D-8700 Wurszburg, Federal
Republic of Germany

toxic, 199
Paramixovirus, 112
Parasite-proteins, 297
Parkinson's disease, 91, 229
Pars compacta, 225
Parvalbumin, 133
Passive
  protection, 285
  transfer of protection, 287
Pathological tissues, 418
Pathophysiology of muscles, 421
Peripheral nerve, 243
Peritoneum, 462
Peroxidation, 193, 195, 220
Peroxyl products, 209, 211
pH, 99
Phagocytosis, 220
  cells, 313
Phenan-throline-cuprous
    complex, 357
Phenobarbital 3-
    methylcholanthrene, 191
Phenylketonuria, 87
Phorbol diesters, 115
Phosphatidylinositol-bis-
    phosphate (IP2), 115
Phosphoglycerate kinase, 79
Phosphorylation, 99, 165
Photosensitisation, 211
Physical exercise, 145
Pituitary gland
  necrosis, 445
  normal, 449
PKC, 115, 119, 120
  tumor promotion, 120
Placenta, 34, 400
  RNA synthesis inhibition, 401
Plasmid, 65
Plasmodium Malariae, 293
Plasmodium Vivax, 293, 295
Platelet aggregation, 119
Platelets, 397
Polycationic substrata, 237
Polylysine, 112
Polyunsaturated fatty acids, 211
Population dynamics of AIDS, 343
Porins, 281
Porphyrias, 211
Post surgical trauma,
    infection, 423
Potasium, 163
  channel, 97
Praziquantel, 325

Preeclampsia, 222
Probenecid, 227
Processes, 241
Promoter-operator region, 61
Prostaglandin cascade, 214
Prostaglandins, 115, 190, 196
Proteinaceous fluids, 420
Proteins
  histone, 1
  non-histone, 1
  heat-modifiable, 281
Proton
  imaging, 416
  relaxation times, 424, 425, 427
Pseudogene, 76
Pseudomonas Aeruginosa, 281
Psoralen, 366
Pufa-epoxides, 214
Pump
  Ca^{++}, 136
Quantum supermolecular
    computation, 178
Quasi-elastic neutron
    scattering, 405
Radiation, 210
Radicals
  alkoxil, 209, 220, 221
  hydroxil, 220
  peroxide, 220
  scavengers, 173
  superoxide, 220
  thiyl, 209
Radioprotector, 171
Radiotherapy, 172
Reactive oxygen species, 119
Receptor
  CD4
    Gp41, 356
    Gp120, 356
  site-ion channel, 99
  transferrin, 302
Recombinant DNA, 59
  techniques, 35
Reconnectivity, 265
Redox-reactions, 210
References, 495
Regeneration bridge, 243
Relaxation times, 415, 421, 461, 467
Repair of DNA, 173
Respiratory burst of
    neutrophils, 119
Respiratory electron
    transport system, 145, 199
Restriction fragment lenght
    polymorphisms, 79

522